Random Variable Theory

$F_X(\alpha)$ The cumulative distribution function of X, $F_X(\alpha) \triangleq P[X \le \alpha]$

$f_X(\alpha)$ The density function, $f_X(\alpha) \triangleq \dfrac{d}{d\alpha} F_X(\alpha)$; $P[\alpha < X < \alpha + \Delta\alpha]$ $= f_X(\alpha)\Delta\alpha$ for a continuous random variable

$p_X(\alpha)$ The mass function, $p_X(\alpha) \triangleq P(X = \alpha)$, used for a discrete random variable

$\overline{X}, \overline{X^2}, \sigma_X^2$ The mean, mean square, and variance, defined as $\int_{-\infty}^{\infty}\alpha f_X(\alpha)\,d\alpha$, $\int_{-\infty}^{\infty}\alpha^2 f_X(\alpha)\,d\alpha$ and $\int_{-\infty}^{\infty}(\alpha - \overline{X})^2 f_X(\alpha)\,d\alpha$

$\overline{g(X)}$ The expected value of $g(X)$, defined as $\int_{-\infty}^{\infty} g(\alpha)f_X(\alpha)\,d\alpha$ or, if $Y = K(X)$ as $\int_{-\infty}^{\infty} K(g(\beta))f_Y(\beta)\,d\beta$ using the Fundamental theorem

$F_{XY}(\alpha, \beta)$ The joint cumulative distribution function, $F_{XY}(\alpha, \beta) \triangleq P[(X \le \alpha)\cap(Y \le \beta)]$

$f_{XY}(\alpha, \beta)$ The joint density function, $f_{XY}(\alpha, \beta) \triangleq \dfrac{\delta^2}{\delta\alpha\,\delta\beta} F_{XY}(\alpha, \beta)$

$p_{XY}(\alpha_i, \beta_j)$ The joint mass function, $p_{XY}(\alpha, \beta) \triangleq P[(X = \alpha_i)\cap(Y = \beta_j)]$

$\overline{XY}$ The correlation of X and Y, defined as $\int_{-\infty}^{\infty}\int_{-\infty}^{\infty} \alpha\beta f_{XY}(\alpha, \beta)\,d\alpha\,d\beta$: also denoted $E(XY)$ and R_{XY}

$L_{XY}, \mathrm{cov}(X, Y)$ The covariance coefficient, defined as $\overline{(X - \overline{X})(Y - \overline{Y})}$

$\overline{g(X, Y)}$ The expected value of $g(X, Y)$, defined as $\iint g(\alpha, \beta) f_{XY}(\alpha, \beta)\,d\alpha\,d\beta$; or, if $X = F(U, V)$ and $Y = G(U, V)$ is given as $\iint g[F(\alpha, \beta), G(\alpha, \beta)]\,f_{UV}(\alpha, \beta)\,d\alpha\,d\beta$ by the Fundamental theorem

Probabilities,
Random Variables,
and Random Processes

Probabilities, Random Variables, and Random Processes

DIGITAL AND ANALOG

Michael O'Flynn
SAN JOSE STATE UNIVERSITY

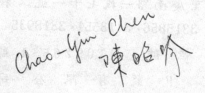

HARPER & ROW, PUBLISHERS, New York
Cambridge, Philadelphia, San Francisco,
London, Mexico City, São Paulo, Sydney

1817

Probabilities,
Random Variables,
and Random Processes

有著作權 • 不准翻印

台內著字第　　號

原著者： MICHAEL O'FLYNN

發行人：李　宗　興

發行所：新　智　出　版　社　有　限　公　司

行政院新聞局局版台業字第一五〇六號

台北市重慶南路一段七十一號二樓

電話： 3315856 • 3113551 • 3818935

郵　撥　帳　號： 0109265 - 1 號

印刷所：吉　豐　印　製　有　限　公　司

板橋市三民路二段居仁巷一弄五十三號

中華民國 72 年　　月　日初版

實價NT$

ISBN 0-07-050927-1

Contents

Part II   Random Variable Theory   95

Part III   An Introduction to
Random Processes   231

[1]Sections preceded by an asterisk may be omitted until encountered in practice.

Preface

The present requirement for an electrical engineer to understand noise and to carry out signal processing on waveforms necessitates a thorough comprehension of probability theory and random processes. This book was written to fill the needs of the senior electrical engineering student and those of the practicing engineer in communications and signal processing.

There are three major parts to the book:

Part I   Discrete Probability Theory
Part II   Random Variable Theory
Part III   An Introduction to Random Processes

Part I is written as an introduction to discrete probability, and the coverage of topics arises from the assumption that discrete probability is a prerequisite for signal processing and statistical communication theory. Set theory is reviewed, with an emphasis on its usage and that of closely related event theory. Counting, or permutations and combinations, is developed and utilized to enumerate points of event spaces. The definitions of probability and conditional probability of an event are given on a relative-frequency basis and compared with the axiomatic definitions. It is demonstrated how a thorough and clear grasp of these few definitions leads to the solution of any questions on assumed random phenomena. Throughout Part I the structure of solving a probabilistic problem is emphasized. Three stages in a solution are stressed: the choice of an appropriate event space, the assignment of probabilities to the points of this space, and the expression of a desired event in terms of the chosen space plus the assignment of a probability to it. The material is treated in general terms without emphasis on particular applications to electrical engineering. However, in the later stages, the reader is asked to focus on the sampling of waveforms, and the problems at the end of Chapter 3 include many of an applied nature.

Part II, "Random Variable Theory," assigns probabilities to numerically valued quantities through distribution, density, and mass functions of random variables, allowing the readers to use their calculus and function-theory background for solving probabilistic problems. At this stage, applications important to system theory and signal processing are discussed more frequently. Important statistics of a random variable such as its mean, mean square value, and variance are applied to the range and variation of values of a waveform; statistics of two random variables such as the correlation and covariance coefficients are applied to the rapidity of change of waveforms. A relationship between two random variables is associated with sampling the input and output of a system, while a relationship among groups of random variables involves multiple sampling of system input and output waveforms. This concentration of problems on sampling waveforms creates a bridge for the handling of random processes later in the text.

Part III, "An Introduction to Random Processes," provides the most structured and comprehensive introduction to the subject available. From the author's experience as a student and teacher, this material is probably the most challenging and fascinating of that encountered in an engineering curriculum. Chapter 6 defines, specifies, and classifies random processes and calls on the concepts from random variable theory to prove whether a process is nonstationary, first-order stationary, strictly stationary, or ergodic. Both continuous and discrete processes are covered on the assumption that many readers, despite the digital revolution, have difficulty with discrete waveforms. Chapter 7 is devoted to auto- and crosscorrelation functions of random processes. Many examples are solved for continuous processes; emphasis is placed on the development of an appreciation of what the terms randomness or lack of randomness mean when applied to a waveform. The properties of auto- and crosscorrelation functions for stationary random processes are covered in detail. Discrete autocorrelation functions are evaluated and analogous properties developed, as for the continuous case. The practical topic of estimating the autocorrelation function from a finite set of sampled values is discussed. At this stage, a tutorial treatment of time domain integrals in system analysis is given in the following order:

1. A review of convolution and correlation integrals for continuous, deterministic, finite-energy waveforms

2. Correlation integrals for periodic, finite-power waveforms
3. Correlation integrals for discrete or quantized waveforms
4. Correlation integrals for finite-power, noise waveforms, with consideration of the important parameters of sample length and sampling rate

Chapter 7 concludes by deriving input-output relations for linear time-invariant systems with either noise, deterministic signal-plus-noise or random signal-plus-noise inputs. Both continuous and discrete results are developed along parallel lines.

Chapter 8 is devoted to power spectral and cross-spectral densities, which define randomness through frequency. The chapter commences with a tutorial treatment of the transforms with which a modern student of system and communication theory is expected to be comfortable, fluent, and versatile. These are:

1. The one- and two-sided Laplace transforms
2. The one- and two-sided z transforms
3. The Fourier and discrete Fourier transforms and fast transform algorithms

As a minimum, it is expected that the reader is already familiar with the one-sided Laplace, Fourier, and if possible, one-sided z transforms. It is also desirable for readers to be already acquainted with residue theory and Laurent series, but for those who are not, Appendix B in this book enumerates important complex variable results.

The power spectral density is defined and interpreted physically for both continuous and discrete processes. Cross spectral densities are defined. The properties of spectral densities and the closely related power-transfer functions of system theory are listed and compared with the properties of correlation integrals from Chapter 7. Input-output spectral relations are derived for linear systems with either noise, or deterministic signal-plus-noise, or random signal-plus-noise inputs. Both continuous and discrete results are developed in detail.

Chapter 9 concludes the book with applications of the general system relations of Chapters 7 and 8. First, there is a description of a gaussian random process and an analysis is made of linear systems with gaussian inputs. The design of filters to achieve some optimum results is then discussed, and the important matched and Wiener filters are derived in both continuous and discrete forms. Finally, an autocorrelation function value at some time or a power spectral value at some frequency obtained from a section of data is discussed as a random variable.

There is a number of possible one-semester, three-unit courses for which the text may be used. At San Jose State University a three-unit, first-semester, senior-level course is given which covers Chapters 1 to 6 with a foray into Chapters 7 and 8. The first graduate course in the statistical communication sequence reviews Chapters 6 and 7, then thoroughly covers

Chapters 8 and 9 before making an in-depth study of filters and modulation theory. If a classic probability course from mathematics is a prerequisite, then a three unit course consisting of a rapid treatment of Chapters 4 and 5, and an in depth coverage of Chapters 6 through 9 is appropriate. Part I of the text requires approximately 11 one-hour lectures, while Parts II and III each require 14 to 16 one-hour lectures.

The book contains some sections marked by an asterisk in Chapters 4 and 5, which may be omitted but used as reference later in the text. For example, the Cauchy-Schwartz inequality, Cauchy inequality, or Schwartz inequality, derived in Section 4.5*, may be referred to during the discussion of the properties of crosscorrelation functions in Chapter 7 or the matched filter derivations in Chapter 9.

Drill sets are included throughout the book. Their numerical answers along with solutions to selected problems are given at the end of the text.

I would like to thank professors Aaron Collins of Tennessee Technological University, Marvin Siegel of Michigan State University, and Larry Schooley of the University of Arizona for very thorough and helpful reviews. I would also like to thank Evan Moustakes, chairman of the San Jose State University Electrical Engineering Department, for much encouragement and for use of his facilities and Tom Kailath of Stanford University for his comments and help. I am grateful to my students Rudy Maske and Robert R. Ortega, for correcting errors, and Pat Coles and Connie Mehalko, for their work on illustrations. For enduring my carelessness and poor handwriting and for their typing, I thank Cheryl Kidder, Kathleen Riggs, and Karen Stasko. For their helpfulness and skill, I am especially grateful to Carl McNair, electrical engineering editor, and Cynthia Indriso, project editor, both of Harper & Row, Publishers. Finally, I thank my sons, Michael and Brendan, for not being complete nuisances and for not complaining too loudly that they missed honing their probabilistic skills at the track.

MICHAEL O'FLYNN

Part I
DISCRETE
PROBABILITY THEORY

Chapter 1
Prerequisites of
Probability Theory

1.1 AN OVERVIEW OF DISCRETE PROBABILITY THEORY

Probability theory concerns itself with questions that do not have deterministic answers but are assumed to have answers whose statistical regularity may be estimated or predicted. Some questions of this nature are: "Will a 1 occur on the upturned face when a die is rolled?" and "Is the value of $\cos t$, for an arbitrary t, less than 0.717?" The answers are that a 1 may or may not occur on the die roll and based on fairness we say there is a 1-in-6 chance it will, and $\cos t$ may or may not be less than 0.717 and, based on fairness, there is a 3-in-4 chance it will be.

A random phenomenon is defined as a happening or experiment whose outcomes occur with statistical regularity. It is implied that the experiment may be repeated a large number of times and the statistical regularity of any possible outcome predicted. Often we make predictions based on visualization and the assumption of fairness. An experiment with different outcomes may be classified as a random phenomenon in two ways:

1. *As a result of statistical observation* (e.g., mortalities in life insurance, successes on an examination, the number of electrons emitted from a section of cathode area)

2. *As a result of assuming mathematical fairness* (e.g., dealt hands in poker, dice throwing)

Probability theory involves answering questions about a random phenomenon. Generally three stages are involved:

1. Listing an appropriate **event space** for the phenomenon. An event space is a set of outcomes, no two of which may occur on the same trial and one of which must occur. The description of an event space requires a knowledge of the "mathematics of counting" or "permutations and combinations," plus familiarity with the elements of set theory.
2. The assignment of a probability measure to the points of the event space using the definition of the **probability of an event**, which is defined if possible as the limit of the relative frequency of its occurrence as the number of trials approaches infinity.
3. Answering required questions concerning possible outcomes by describing them in terms of the chosen appropriate event space and utilizing the axioms of probability theory, which are soon to be encountered.

The concepts and ideas of discrete probability theory form the foundation for studying random variable theory and random or stochastic processes which constitute the necessary theoretical background for the vast field of communication theory in electrical engineering. The emphasis in Chapters 1 to 3 will be on the structure of an overall problem as opposed to obtaining a quick answer to a problem that is just a small part of a complete phenomenon. However, after much experience and problem solving the rapid solution to subproblems of a random phenomenon will come easily. As a preview example let us consider the question, "What is the probability of obtaining exactly two heads on three tosses of a coin?" Our question pertains to the random phenomenon or experiment of tossing a coin three times. The three stages for this problem are:

1. An event space describing the random phenomenon composed of eight compound points is

$$E = \{HHH, HHT, HTH, THH, HTT, THT, TTH, TTT\}$$

where (HTH) is the event the first toss is a head, the second a tail, and the third a head. The meaning of the other points should be self-evident.
2. The definition of the probability of an event, will allow us to assign a probability of $\frac{1}{8}$ to each of the eight points from symmetry.
3. Using set theory the event of exactly two heads on three tosses may be described as the union of three of the points of our event space, and the axioms of probability theory allows us to assign a probability of $\frac{3}{8}$ to this event.

For this problem, stage 1 was probably very clear and the chances of each outcome based on statistical regularity in stage 2 being one in eight, was more than likely acceptable. Perhaps stage 3 posed the most difficulty as we are indeed intuitively solving a problem in advance of having developed the necessary analytical tools. We will now proceed to systematically cover the field of discrete probability theory in the first three chapters.

The present chapter will be devoted to the prerequisite mathematical background for discrete probability theory—that is, set theory and its counterpart for dealing with outcomes of a random phenomenon, event theory, along with the mathematics of counting, which is also referred to as permutations and combinations. Counting will enable us to enumerate different outcomes of a phenomenon and set theory will facilitate relating different outcomes and expressing them in a way suitable for assigning probabilities.

Chapter 2 will be concerned with the definition of the probability of an outcome E of a random phenomenon, which is defined on a relative frequency basis as

$$P(E) = \lim_{N \to \infty} \frac{N_E}{N} \tag{1.1}$$

where N_E is the number of trials when E occurs based on the total number of trials N of the phenomenon, and "lim" denotes the limit of this ratio. The reader will be encouraged to dwell on and use this relative frequency definition, although it will be explained that there are instances where an axiomatic or alternative definition must be utilized. The axioms and laws of probability theory will be developed and used to assign a probability measure to the different outcomes of a phenomenon and to answer any probabilistic question concerning it. By the end of Chapter 2 the background to solve quite complex problems systematically will have been acquired. Chapter 3 is chiefly concerned with developing a problem-solving facility.

1.2 REVIEW OF SET THEORY AND EVENT THEORY FOR RANDOM PHENOMENA

In this section two almost interchangeable disciplines are discussed: "set theory" and "event theory" for random phenomena. Since it is assumed that the reader is already familiar with set theory, what is treated is done so in the nature of a review.

Set Theory Review

A set is defined as a collection of objects. A set that is complete so that only those objects in it need be considered is called a **space**. The objects of a set are called its **elements** and the elements may be simple or compound. For

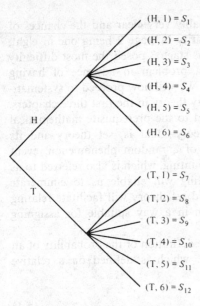

$(H, 1) = S_1$

$(H, 2) = S_2$

$(H, 3) = S_3$

$(H, 4) = S_4$

$(H, 5) = S_5$

$(H, 6) = S_6$

$(T, 1) = S_7$

$(T, 2) = S_8$

$(T, 3) = S_9$

$(T, 4) = S_{10}$

$(T, 5) = S_{11}$

$(T, 6) = S_{12}$

Figure 1.1   Tree diagram listing the 12 compound elements of a set.

example,

$$S = \{1, 2, 3, 4, 5, 6\}$$

is an explicit listing of the six simple elements of a set, whereas

$$S = \{x: 1 \leq x \leq 6, x \text{ an integer}\}$$

is a more compact description of this set. Moreover,

$$S = \{H1, H2, H3, H4, H5, H6, T1, T2, T3, T4, T5, T6\}$$

or

$$S = \{X, Y: X = H \text{ or } T; 1 \leq Y \leq 6, Y \text{ an integer}\}$$

is an example of a set with 12 compound elements each of size 2. If any confusion existed in the notation H1 or H2 we could also use H, 1 or H, 2—and to avoid confusion with commas we might write {(H, 1), (H, 2), and so forth.[1] Whenever the elements of a set are compound, the utilization of the concept of a tree or sequential diagram is very convenient for enumerating the points of S. The tree for this set is shown in Figure 1.1.

It is important to realize that the order of listing the elements of a set is of no consequence, because an element belongs to a set or it does not. However, a specific compound element of a set may involve order. For

---

[1] The elements of a set are not usually bracketed, inasmuch as H1 or H, 1 is usually an element and (H1) or (H, 1) is a subset.

example, the compound element $(1,2,3)$ could be different from the compound element $(2,1,3)$ depending on what we mean to imply. Sets are often further classified as being **discrete** (countable) or as being **continuous** (noncountable).

The set

$$S = \{x: 1 \leq x \leq 6, x \text{ an integer}\}$$

is discrete, whereas the set

$$S = \{x: 0 < x < 1\}$$

which would correspond to the points on that part of the real axis $0 < x < 1$, is continuous and the elements are said to be noncountable.

The elements of a set are also referred to as members or points and we will use the three terms "element," "point," and "member" interchangeably.

THE ALGEBRA OF SETS

The important operation definitions used in set theory will now be enumerated:

Two sets A and B are equal if they contain the same elements.
A is a **subset** of B or $A \subset B$ if every element of A is an element of B. If $A \subset B$ and $B \subset A$, then $A = B$.
The **complement** of a set A, denoted $\bar{A}$ or A^c, consists of all the points of the appropriate space S not in A.
The **union** of two sets A and B denoted $A \cup B$, consists of all the elements of the space S that belong to at least one of the sets A or B.
The **intersection** of two sets A and B denoted $A \cap B$, consists of all the points of the space S common to both A and B.

There are many interchangeable notations for these operations; $A + B$ for the union and AB for the intersection are widely used.

Two additional sets that are important in set theory are the complete set S itself and the impossible or null set $\varnothing$. These are defined by

$$A \cap \bar{A} \overset{\triangle}{=} \varnothing \tag{1.2}$$

and

$$A \cup \bar{A} \overset{\triangle}{=} S \tag{1.3}$$

Two sets A and B are said to be **mutually exclusive** if

$$A \cap B = \varnothing \tag{1.4}$$

In order to prove any relation in set theory, it is customary to establish a set of axioms and to prove the relation by use of the axioms. A convenient but

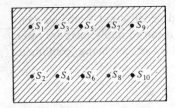

Figure 1.2   Venn diagram of S, consisting of 10 points.

not unique set of axioms with their names are as follows:

$$A \cup B = B \cup A \qquad\qquad \text{A commutative law}$$
$$A \cup (B \cup C) = (A \cup B) \cup C \qquad\qquad \text{An associative law}$$
$$A \cap (B \cup C) = (A \cap B) \cup (A \cap C) \qquad\qquad \text{A distributive law}$$
$$(\bar{\bar{A}}) = A \qquad\qquad \text{Double complementation}$$
$$\overline{A \cap B} = \bar{A} \cup \bar{B} \qquad\qquad \text{One of De Morgan's theorems}$$
$$A \cap \bar{A} = \varnothing \qquad\qquad \text{Theorem of the excluded middle}$$
$$A \cap S = A \qquad\qquad \text{Main absorption theorem}$$

$$(1.5)$$

A convenient pictorial representation of a space and its subsets is given by a Venn diagram. A plane figure is used to represent the space and its elements may be indicated by points in the plane figure or by subareas. For example, Figure 1.2 shows a space composed of 10 mutually exclusive points, indicated by $S_1, S_2, \ldots, S_{10}$. If the set A is defined as

$$A = S_1 \cup S_2 \cup S_3 \cup S_4$$

and the set B is defined as

$$B = S_2 \cup S_5 \cup S_6$$

then A, B, $A \cap B$, and $A \cup B$ are represented by Venn diagrams as shown in Figure 1.3.

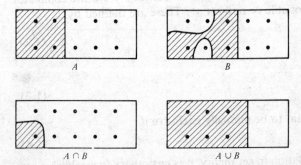

Figure 1.3   Venn diagram representation of A, B, $A \cap B$, and $A \cup B$ for the space of Figure 1.2.

Any relationship in set theory may be proved by means of our seven axioms or may loosely be appreciated by the use of Venn diagrams. Since the requirement of algebra is assuming more importance in engineering because of the field of digital systems, without digressing too much, some time will be devoted to algebraic proofs and demonstrations. For any sets A, B, and C defined on some space S, the theorems shown in Table 1.1 could be proved.

Because students of boolean algebra seem to have difficulty in recognizing and applying the negative absorption theorem,

$$A \cup (\bar{A} \cap B) \equiv A \cup B$$

we will prove it and demonstrate it with the aid of a Venn diagram.

Example 1.1

Prove the theorem

$$A \cup (\bar{A} \cap B) \equiv A \cup B$$

using the axioms of Eq. 1.5, and then demonstrate the proof on a Venn diagram.

SOLUTION

Let us consider the right-hand side $A \cup B$. Now

$$B = B \cap S$$

Table 1.1 SOME THEOREMS OF SET THEORY

THEOREM	TYPE OF OPERATION
$A \cup B \equiv B \cup A$ (axiom) $A \cap B \equiv B \cap A$	commutative theorems
$A \cup (B \cup C) \equiv (A \cup B) \cup C \equiv A \cup B \cup C$ (axiom) $(A \cap B) \cap C \equiv A \cap (B \cap C) \equiv A \cap B \cap C$	associative theorems
$A \cap (B \cup C) \equiv (A \cap B) \cup (A \cap C)$ (axiom)	distributive theorem
$A \cup (A \cap B) \equiv A$ $A \cap (A \cap B) \equiv A \cap B$ $A \cap S \equiv A$	absorption theorems
$A \cup \bar{A} \equiv S$ $A \cap \bar{A} \equiv \emptyset$ (axiom) $(\bar{\bar{A}}) \equiv A$ (axiom)	complementation theorems
$(\overline{A \cup B}) \equiv \bar{A} \cap \bar{B}$ (axiom) $(\overline{A \cap B}) \equiv \bar{A} \cup \bar{B}$	De Morgan's rules on complementation
$A \cup (\bar{A} \cap B) \equiv A \cup B$ $A \cap (\bar{A} \cup B) \equiv A \cap B$	theorems on negative absorption

and S may be written as

$$S = A \cup \bar{A}$$
$$\therefore B = B \cap (A \cup \bar{A})$$
$$= (B \cap A) \cup (B \cap \bar{A})$$

using the distributive axiom. Also,

$$A \cup B = A \cup [(B \cap A) \cup (B \cap \bar{A})]$$
$$= [A \cup (B \cap A)] \cup [A \cup (B \cap \bar{A})]$$

using the associative axiom.

Since $(B \cap A)$ is a subset of A, then

$$A \cup (B \cap A) = A$$

and

$$[A \cup (B \cap A)] \cup [A \cup (B \cap \bar{A})] = A \cup [A \cup (B \cap \bar{A}]$$
$$= A \cup (B \cap \bar{A})$$

which is identical to the left-hand side of the axiom. Next we will consider a Venn diagram illustration of this theorem. Figure 1.4a shows A, B, and $\bar{A} \cap B$, whereas Figure 1.4b shows $A \cup (\bar{A} \cap B)$ and $A \cup B$. It is clear that $A \cup (\bar{A} \cap B)$ and $A \cup B$ are equal.

The Algebra of Events

In probability theory we will use event theory, which will now be discussed, and set theory almost interchangeably. Moreover, we will assume it to be understood from the context whether we are referring to events or to sets. At this stage let us consider the conceptual difference between the two.

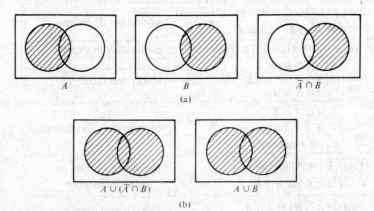

Figure 1.4 A Venn diagram demonstration of the negative absorption theorem. (a) A, B, and $\bar{A} \cap B$. (b) $A \cup (\bar{A} \cap B)$ and $A \cup B$.

An event is defined as an outcome of a random phenomenon. If the outcome is described by the set A, then we will use A to indicate both the set and the event it describes. However the descriptive language for sets and events is different, as we will demonstrate by defining some operations for events and comparing them to their set counterparts.

The *complement* of an event A, denoted $\bar{A}$, is the event that A does not occur and it is described by the set $\bar{A}$, which consists of all the points of S not in A.

The *union* of two events $A \cup B$ is the event that either A or B (which includes both) occur and it is described by the set $A \cup B$, which consists of all the points of S belonging to at least A or B.

The *intersection* of two events $A \cap B$ is the event that both A and B occur and it is described by the set $A \cap B$, which consists of all the points of S belonging to both A and B.

We denote the certain event S as $A \cup \bar{A}$ where A is any event and the impossible event $\varnothing$ as $A \cap \bar{A}$. In general an event is defined by a subset of a space, describing the different possible outcomes of a random phenomenon of interest.

For any random phenomenon an **event space** is defined as a listing of mutually exclusive, collectively exhaustive outcomes. This means, respectively, that no two outcomes may occur on the same trial of the phenomenon and that one outcome from the event space must occur.

As an example of an event space let us consider the transmission and reception of messages in a communication system, which is schematically indicated in Figure 1.5. For simplicity it is assumed that three messages— M_1, M_2, and M_3 —are transmitted with probabilities P_{T1}, P_{T2}, and P_{T3}, respectively. The reception of messages is a random phenomenon if the event space

$$E = \{(M_1, M_1), (M_1, M_2), (M_1, M_3), (M_2, M_1), (M_2, M_2),$$
$$(M_2, M_3), \ldots, (M_3, M_3)\}$$

has a probability measure $P_1, P_2, \ldots, P_9$ associated with it. The compound

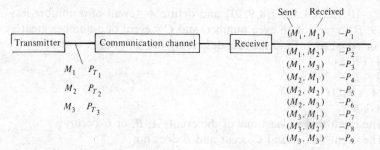

Figure 1.5   A schematic diagram of a communication channel in which three messages are transmitted.

point (M_1, M_1) denotes that the event M_1 is sent and M_1 is received and (M_1, M_2) the event that M_1 is sent and the message M_2 received. The meaning of the other seven outcomes should be obvious. If we define two new events of interest—A the event of an error at the receiver and B the event that the message M_2 is received—then in terms of our event space we may express them as

$$A = A \cap S$$

which for our space E becomes

$$A = (M_1, M_2) \cup (M_1, M_3) \cup (M_2, M_1) \cup (M_2, M_3)$$
$$\cup (M_3, M_1) \cup (M_3, M_2)$$

and

$$B = B \cap S$$

which for our space is

$$B = (M_1, M_2) \cup (M_2, M_2) \cup (M_3, M_2)$$

We may also require $C = A \cap B$ and ask for a verbal description of it.

$$A \cap B = (M_1, M_2) \cup (M_3, M_2)$$

and this is the event of falsely receiving M_2.

Set theory and event theory will be utilized for two of the three stages in the solution of probabilistic questions concerning a random phenomenon. In stage 1 set theory is utilized to represent the mutually exclusive, collectively exhaustive outcomes of a phenomenon, and in stage 3 set theory operations are used to express desired outcomes such as the event of an error in a communication system in terms of the outcomes of the event space. Before set theory is applied to these tasks the mathematics of counting will be discussed. This will enable us to enumerate, or at least indicate, large numbers of points in a space.

DRILL SET: SET THEORY

1. Let $S = \{0, 1, 2, 3, 4, 5, 6, 7, 8, 9, 10\}$ and define $A =$ event of a number less than 9, $B =$ event of an even number, and $C =$ event of a number greater than or equal to 3. Find the sets describing the following events:
   (a) $A \cap B \cap C$
   (b) $\overline{A \cap B}$
   (c) $\overline{(A \cap B)} \cup C$
   (d) $\overline{A} \cap \overline{B}$
   (e) The event that at least one of the events A, B, or C occurs
   (f) The event that A and C occur and B does not
2. Sketch each of the events of Problem 1 above on a Venn diagram.
3. Prove or disprove the following statements using the axioms of Eq. 1.5

and demonstrate with Venn diagrams.

(a) $\overline{A \cap B} = \overline{A} \cup \overline{B}$

(b) $\overline{A} \cup (\overline{A \cap B}) = \overline{A \cap B}$

(c) $\overline{A} \cap (A \cap B) = \overline{A}$

1.3 THE MATHEMATICS OF COUNTING, OR PERMUTATIONS AND COMBINATIONS

The actual enumeration or counting of different possible outcomes of an experiment is very important. From intelligent counting procedures, formulas of much versatility can be developed. This section will review some of the mathematics of permutations and combinations and the use of sequential trees to enumerate points in a space or complete set. The following notation will be used throughout:

$$n! \stackrel{\triangle}{=} n \times (n-1) \times (n-2) \times \cdots \times 3 \times 2 \times 1 \tag{1.6}$$

$n!$ is called "n factorial" and is defined for positive integers. Also $0! \stackrel{\triangle}{=} 1$.

$$(M)_n \stackrel{\triangle}{=} M \times (M-1) \times (M-2) \times \cdots \times (M-n+1)$$

$$= \frac{M!}{(M-n)!} \tag{1.7}$$

$(M)_n$ is called "M truncated n" and defined for $M \geq n$ and both positive integers.

$$\binom{n}{r} \stackrel{\triangle}{=} \frac{n!}{r!(n-r)!} = \frac{(n)_r}{r!} \tag{1.8}$$

and is called "n choose r," where $n \geq r$ and both are positive integers. For example,

$$6! = 6 \times 5 \times 4 \times 3 \times 2 \times 1 = 720$$

$$(6)_3 = 6 \times 5 \times 4 = 120$$

and

$$\binom{6}{2} = \binom{6}{4} = \frac{6!}{4!2!} = 15$$

The notations $^M P_n$ and $^n C_r$ are sometimes used in place of $(M)_n$ and $\binom{n}{r}$, respectively. This is done where $^M P_n$ results from a permutation problem and $^n C_r$ results from a combination problem. We will rarely use these notations, however.

Permutations

The question "How many permutations of size p can be formed from n objects?" asks, "In how many different ways can the n objects be arranged in rows where each row is to contain p objects?" Therefore a permutation of size p is an ordered array of p objects.

As a result of constructing sequential or tree diagrams and building from the case of permutations of small size, some important results will be developed.

PERMUTING DIFFERENT OBJECTS

Example 1.2

How many permutations of size 2 may be formed from five different objects A, B, C, D, and E, where it is assumed that each object may occur only once in any permutation?

SOLUTION

The enumeration or counting of the different permutations of size 2 might proceed as follows:

(a) There are four different permutations of size 2, with object A occupying position 1 of each row or permutation, as is shown in the thumbnail sketch. These permutations are AB, AC, AD, and AE.

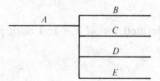

(b) Similarly there are four permutations with B in position 1, four with C in position 1, four with D in position 1, and finally four with object E in position 1.

In all, there are 5×4 permutations of size 2 that can be formed from five different objects. The complete sequential diagram of these permutations is shown in Figure 1.6.

Example 1.3

How many permutations of size 3 can be formed from five different objects?

SOLUTION

Utilizing the knowledge gained about permutations of size 2 this problem may be solved as follows:

(a) Consider the subquestion, "How many permutations of size 3 may be formed from five different objects where object A occupies position 1?" This is equivalent to asking, "How many permutations

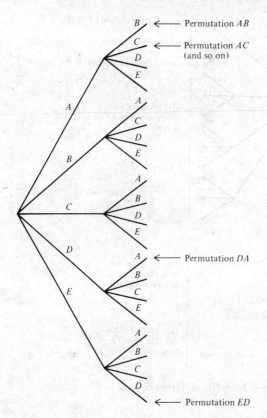

Figure 1.6   Enumeration of the different permutations of size 2 that may be formed from five different objects.

of size 2 may be formed from the four objects B, C, D, and E?" The answer is $(4)_2$.

(b) Similarly there are $(4)_2$ permutations of size 3 with B in position 1, $(4)_2$ with C in position 1, $(4)_2$ with D in position 1 and finally $(4)_2$ with E in position 1. In all, there are $5 \times (4)_2$ or $5 \times 4 \times 3 = (5)_3$ permutations of size 3 that can be formed from five different objects. A sequential diagram enumerating some of these permutations is shown in Figure 1.7.

Reflection on Example 1.3 should quickly lead to the conclusion that $(n)_3 = n(n-1)(n-2)$ permutations of size 3 can be formed from n different objects. *An extension of this thinking would lead to the conclusion that $(n)_r$ permutations of size r or $n!$ permutations of size n may be formed from n different objects.*

PERMUTING OBJECTS, SOME OF WHICH ARE IDENTICAL

Examples 1.2 and 1.3 were concerned with permuting different objects. However, the case where some of the objects are identical occurs often in

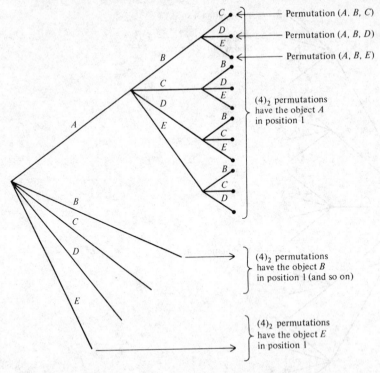

Figure 1.7 Permutations of size 3 from five different objects.

practice. One of the more important cases is that

$$\frac{n!}{a_1!a_2! \cdots a_m!} \tag{1.9}$$

permutations of size n may be formed from n objects, a_1 of which are identical of type 1, a_2 identical of type 2, and a_m identical of type m, where

$$\sum_{i=1}^{m} a_i = n$$

To develop a feeling for this formula a simple case involving small numbers will be treated.

Example 1.4

How many "words" of eight letters may be formed from three A's, two B's, C, D, and E? Assume that any ordered array of eight letters constitutes a word.

SOLUTION

For the moment let it be assumed that the three A's and the two B's are distinguishable, as shown:

$$A_1, A_2, A_3, B_1, B_2, C, D, E$$

The number of eight-letter words that may be formed is 8!.

Since in reality the A's are not distinguishable, the number of different words becomes 8!/3!. (Why?)

Since the B's are not distinguishable, the number of words is further reduced to 8!/3!2!, which is our result. This result is in conformity with Eq. 1.9, which says that

$$\frac{8!}{3!2!1!1!1!}$$

permutations of size 8 may be formed from eight objects, three of which are of type 1, two of type 2, and each of the others different.

It would be a beneficial exercise for the student to count or enumerate the 8!/3!2! words systematically. An intelligent procedure is:

1. Enumerate all the words with the A's in the first three positions.
2. Enumerate all the words with the A's in the first two positions and not in the third position and then continue logically to exhaust all the different permutations.

Another important formula is that p^m permutations of size m may be formed from a large number of objects, where we have p different types and at least m of each type. For example we can form 2^8 words of size eight using the bit 0 and the bit 1.

Combinations

When discussing permutations we were interested in ordered samples or arrays, but when we deal with combinations or sets, we are interested only in the composition of a sample. For example ABC, ACB, BAC, BCA, CAB, and CBA are six different permutations of size 3 that can be formed, but all six represent only one combination or set of size 3. We define a combination of size n as a group of n objects, where order is of no consequence.

The tree concept may again be used to count combinations intelligently. As with permutations a few examples will be solved to illustrate an inductive approach for their understanding.

Example 1.5

How many combinations or sets of size 2 may be formed from six different objects A, B, C, D, E, and F?

SOLUTION

The enumeration or counting of combinations of size 2 might proceed as follows. First, there are five different sets of size 2 that include the object A, as is shown in the accompanying sketch—that is, AB, AC, AD, AE, and AF.

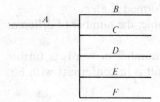

Next, there are four different sets that include B and exclude A as shown in the second sketch—that is, BC, BD, BE, and BF.

Similarly, there are three different sets including C and excluding A and B. Also there are two sets including D and excluding A, B, and C and one set including E and excluding A, B, C, and D. The student should draw the tree diagrams for these cases.

The total number of sets of size 2 that can be formed from six different objects is $5+4+3+2+1=15$, which also equals $\binom{6}{2}$.

It can be shown that the total number of sets of size 2 that can be formed from n different objects is $(n-1)+(n-2)+\cdots+2+1$ or $\binom{n}{2}$. This important result

$$1+2+3+\cdots+(n-1)=\binom{n}{2}$$

is clear from the fact the left-hand side is an arithmetic progression, with the sum S_n given by

$$S_n = \frac{1+(n-1)}{2} \times (n-1) = \frac{n(n-1)}{2} = \binom{n}{2}$$

Example 1.6

How many combinations or sets of size 3 can be formed from six different objects A, B, C, D, E, and F?

SOLUTION

Utilizing the knowledge gained about sets of size 2, the problem may be solved as follows:

(a) There are $\binom{5}{2}$ sets of size 3, each containing the object A, that may be formed from six different objects. (Be positive of this based on Example 1.5.)

(b) Similarly, there are $\binom{4}{2}$ sets of size 3, each containing object B and excluding object A, that may be formed from six different objects, and there are $\binom{3}{2}$ sets of size 3 containing C and excluding objects A and B and $\binom{2}{2}$ or one set of size 3 containing object D and excluding objects A, B, and C.

These combinations are shown counted in Figure 1.8. The total number of sets of size 3 is,

$$\binom{5}{2}+\binom{4}{2}+\binom{3}{2}+\binom{2}{2}=10+6+3+1$$
$$=20$$

Again, $\binom{2}{2}+\binom{3}{2}+\binom{4}{2}+\binom{5}{2}$ is the series $1+3+6+10$, and it will be left as a challenging exercise to derive the resultant sum as $\binom{6}{3}$.

An important result involving combinations or binomial coefficients, is that

$$\binom{n}{r}=\binom{n-1}{r-1}+\binom{n-2}{r-1}+\cdots+\binom{r-1}{r-1}\tag{1.11}$$

This physically states that if we were to enumerate the number of combinations of size r that could be formed from n different objects, there are $\binom{n-1}{r-1}$ sets containing object 1, $\binom{n-2}{r-1}$ containing object 2 and excluding object 1, and so on. It is left as an exercise to show that this series

$$\binom{r-1}{r-1}+\binom{r}{r-1}+\cdots+\binom{n-1}{r-1}$$

may be reduced to

$$1+r+\frac{1}{2}r(r+1)+\frac{1}{3!}r(r+1)(r+2)+\cdots+\binom{n-1}{r-1}$$

which has the sum $\binom{n}{r}$. (See problem 6b at the end of Chapter 1.)

The important result

$$\binom{n}{r}=\text{number of combinations of size } r \text{ that}$$
$$\text{can be formed from } n \text{ different objects}\tag{1.12}$$

is reiterated.

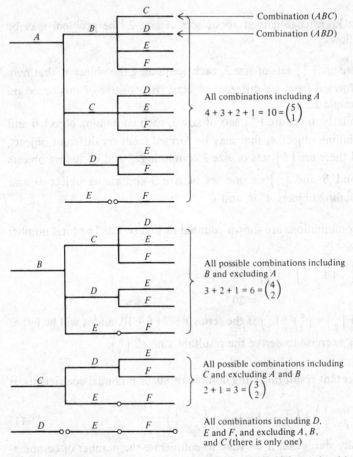

Figure 1.8 Combinations of size 3 that can be formed from six objects.

RELATIONSHIP BETWEEN PERMUTATIONS AND COMBINATIONS

An important and easily derived relationship exists between the number of permutations of size r that may be formed from n different objects and the number of combinations of size r that may be formed. Let ^nC_r denote the number of combinations and ^nP_r the number of permutations. From any combination of size r, $r!$ permutations may be formed.

$$\therefore {}^nP_r = r! {}^nC_r \tag{1.13}$$

using our results from this section we see that

$$\frac{(n)_r}{\binom{n}{r}} = \frac{n \times (n-1) \cdots (n-r+1)}{1} \times \frac{r!(n-r)!}{n!} = r!$$

as predicted by Eq. 1.13.

IMPORTANT COMMENT ABOUT THE USE OF $\binom{n}{r}$

In probability theory $\binom{n}{r}$ will occur repeatedly in two guises:

1. As the number of sets of size r that may be formed from n different objects
2. As the number of permutations of size n that can be formed from n objects, r of which are identical of type 1 and $(n-r)$ identical of type 2. We could use ^nC_r and $n!/r!(n-r)!$, which is a special case of Eq. 1.9, to distinguish these two applications, but we will not do so. It is left to the reader to realize from the context whether a combination or permutation result is being referred to.

Some Additional Counting Problems

A few general counting formulas were developed in the two preceding subsections. Some of these are: $(n)_r$ for the number of permutations of size r that may be formed from n different objects; $n!/n_1!n_2!\cdots n_m!$ for the number of permutations of size n that can be formed from n objects if n_1 are of type 1, n_2 of type 2, and so on; and $\binom{n}{r}$ for the number of combinations of size r that can be formed from n different objects or the number of permutations of size n that can be formed from n objects, r of which are of type 1 and $n-r$ of type 2. As a minimum requirement we should handle these formulas confidently and physically feel them by closing our eyes and quickly constructing the appropriate tree diagrams.

Many counting problems do not neatly fall under the category of belonging to a specific case for which we have a definite formula, and for these we should be able to generate the answer by subdividing the problem into stages. A number of problems of this nature will now be solved.

Example 1.7

How many permutations of size 4 can be formed from 12 different types of objects, subject to the condition that we have an unlimited supply of each and further subject to the condition each permutation must contain three of one type object and one of another type?

SOLUTION

Let us label our 12 types of objects A, B, C, D, E, F, G, H, I, J, K, and L and assume that we have available an unlimited supply of each letter. First we will solve for the number of different combinations. Two possible combinations are $AAAB$ and $BBBA$. With some thought we can see there are $2\binom{12}{2}$ different combinations possible, subject to our restriction. (Pause and be absolutely sure of this. If you do not see it, count.)

For each combination, such as $AAAB$, we can form $4!/3!1!$ or $\binom{4}{3}$ permutations. Therefore, the total number of permutations is

$$\binom{4}{3} \times 2 \times \binom{12}{2} = 4 \times 2 \times 66 = 528$$

In the solution of this problem we used the permutations formula $\binom{4}{3}$, the combination formula $\binom{12}{2}$, and the basic counting factor 2.

Example 1.8

In how many different ways may we achieve the result of exactly seven heads on 10 tosses of a coin?

SOLUTION

With thought we should see that this is equivalent to asking, "How many permutations of size 10 can we form from 10 objects, 7 of which are heads and 3 of which are tails?" Thus, the number of results is given by

$$\frac{10!}{7!3!} = \frac{10 \times 9 \times 8}{3 \times 2}$$
$$= 120$$

Example 1.9

(a) In how many ways can 12 boys be divided into three teams, each containing four players?

(b) After solving part (a), solve the general partitioning problem: How many ways can you divide n different objects into r groups, where one group contains n_1 objects, another n_2, and the last group n_r objects, where

$$\sum_{i=1}^{r} n_i = n$$

SOLUTION

(a) From 12 boys, we can form $\binom{12}{4}$ teams of size 4. Corresponding to any specific team, say $ABCD$, we could form $\binom{8}{4}$ other teams of size 4. Corresponding to any two specific teams, $ABCD$ and $EFGH$, we can form $\binom{4}{4}$ or one other team of size 4. Therefore we can form

$$\binom{12}{4} \times \binom{8}{4} \times \binom{4}{4}$$

groups of three teams with four to a team.

(b) The student should argue that the answer is

$$\binom{n}{n_1} \times \binom{n-n_1}{n_2} \times \binom{n-n_1-n_2}{n_3} \times \cdots \times \binom{n_r}{n_r}$$

It should be noted that this is identical to

$$\binom{n}{n_r} \times \binom{n-n_r}{n_{r-1}} \times \cdots \times \binom{n_1}{n_1}$$

since we might have formed a group of size n_r first. If there is difficulty in seeing this, just consider a special case, such as

$$\binom{8}{4} \times \binom{4}{3} \times \binom{1}{1}$$

and show that it is the same as

$$\binom{8}{1} \times \binom{7}{3} \times \binom{4}{4}$$

which would correspond to partitioning eight objects into three groups of size 4, 3, and 1, respectively.

Finally this section on miscellaneous counting will be concluded with a difficult problem.

Example 1.10

In how many different ways can a person be dealt a hand containing exactly 2 pairs in draw poker?

SOLUTION

In case the game of poker is not familiar, the problem will be restated in general terms: "Consider a deck consists of 52 different cards divided into four equal groups called suits, which have distinctive emblems called spades, hearts, diamonds, and clubs (denoted S, H, D, and C). Assume that in each suit the cards are numbered from 1 to 13, inclusive. If a person is dealt five cards face down and then picks them up and looks at them, in how many ways could he or she obtain a hand containing 2 pairs (a pair means two fours or two sixes, etc.) and a fifth card that has a different number?" Each different group of five cards is called a hand.

First it should be clear that in all there are $\binom{52}{5}$ different hands that can be dealt. We need to know how many of these satisfy the requirement of 2 pairs. Using our knowledge of counting we see that

One hand satisfies the requirement of containing (6H 6S 8H 8D 9C).

$\binom{4}{2} \times \binom{4}{2}$ hands contain two sixes, two eights, and 9C (be very sure of this).

$\binom{4}{2} \times \binom{4}{2} \times 44$ hands contain two sixes, two eights, and a fifth card that is not a 6 or an 8.

$\binom{13}{2} \times \binom{4}{2} \times \binom{4}{2} \times 44$ hands contain exactly 2 pairs.

Our solution is $\binom{13}{2} \times \binom{4}{2} \times \binom{4}{2} \times 44 = 123{,}552$ ways, out of the total possible $\binom{52}{5} = 2{,}598{,}960$ hands that can be dealt.

Examples 1.7 to 1.10 should have illustrated how just a few permutation and combination formulas can be combined to solve many complex counting problems. These examples were solved slowly, and in the future the solutions will be written down rapidly. If a student has done little counting previously, facility and confidence with counting problems requires from 30 to 60 hours of work. The drill problems and the problems at the end of the chapter should be solved carefully, with each step stated in words until such facility is achieved.

DRILL SET: THE MATHEMATICS OF COUNTING

1. The official result of a race consists of posting the first four finishers in order. Assume that a race has ten contestants.
   (a) How many different results can there be?
   (b) How many different results include competitor A in position 1?
   (c) How many results include A in one of the first four positions?
   (d) How many results include A and B in the first four positions?
   (e) How many results include A or B in the first four positions?
   Note: Do each part in as many different ways as you can.
2. Consider 10 executives, A to J.
   (a) How many different committees of size 4 may be formed?
   (b) How many of these committees include C and D?
   (c) How many include C or D but not both?
   (d) If designating a chairperson means that a committee of the same four people is different, modify your answers to (a), (b), and (c) to reflect this.
3. (a) How many different 8-bit computer words can be formed?
   (b) How many of these words will contain exactly three zeros.
   (c) How many will have a 1 in position 3 and end with two zeros.

1.4 THE ENUMERATION OF POINTS IN A SAMPLE SPACE

Section 1.2 reviewed set theory and event theory and Section 1.3 discussed counting using permutations and combinations. In this section we will return to random phenomena and use Section 1.3 to list points or sets that describe the outcomes of trials of a phenomenon and then utilize set theory to describe any event in terms of these points of S.

First, two vitally important terms associated with a random phenomenon will be defined. These are the **sample description space** and an **event space**. The sample description space S is defined as the finest grain, mutually exclusive, collectively exhaustive listing of all possible outcomes of the experiment. The term "finest grain" implies that each point or possible outcome represents a distinguishable outcome of the phenomenon within the limits of its definition. The term "mutually exclusive" implies that any two events of the space cannot occur at the same time on a trial of the experiment—or if s_i and s_j are points of S, then $s_i \cap s_j = \emptyset$. The term "collectively exhaustive" implies one of the events of the space must occur on a trial of the phenomenon or in set theory notation if $S = (s_1, s_2, \ldots, s_n)$ then we require that

$$s_1 \cup s_3 \cup s_3 \cup \cdots \cup s_n = \bigcup_{i=1}^{n} s_i = S$$

where S is the complete set. An event space E of a random phenomenon is defined as a mutually exclusive, collectively exhaustive listing of outcomes of the experiment. These very vital spaces, the sample description space S, and an event space E will be illustrated by some examples.

Example 1.11

Consider the random phenomenon of drawing a sample of 3 balls from an urn that contains 9 red, 3 blue, and 2 white balls. Set up the sample description space S. Assume that a ball is drawn, its color noted, and the ball retained; then a second ball is drawn, its color noted, and the ball retained; similarly, a third ball is drawn.

SOLUTION
A tree diagram of the different points or elements for the complete set S is shown in Figure 1.9. The sample description space consists of the 26 points,

$$S = \{(x, y, z): x = \text{R or B or W}, y = \text{R or B or W}, z = \text{R or B or W};$$

$$\text{exclude the point (W, W, W)}\}$$

For example, the event (R, W, W) is the event the first ball is red, the second is white, and the third is white. The meaning of the other events should be self-evident. Observing S we note that the number of points is equal to the

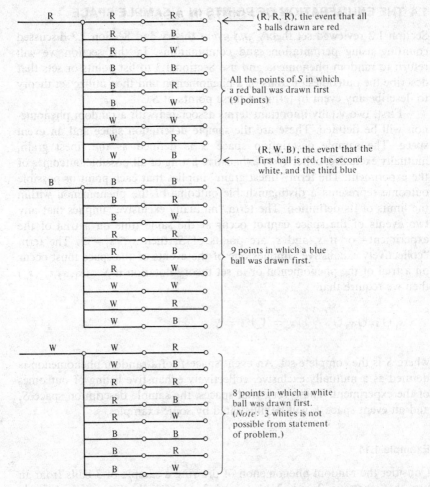

Figure 1.9 Different outcomes of drawing three balls consecutively from an urn in Example 1.11.

number of permutations of size 3 that can be formed from 14 objects, 9 of which are identical of type 1, 3 of type 2, and 2 of type 3.

It is now possible to express any outcome of the random phenomenon (consistent with the problem statement), as the union of points of S. For example, if we define the following events or outcomes:

A = the event the second ball drawn is red

B = the event exactly 2 balls drawn have the same color

C = the event the third ball drawn is not red

then
$$A = A \cap S$$
$$= \{(R,R,R) \cup (R,R,B) \cup (R,R,W) \cup (B,R,R) \cup (B,R,B)$$
$$\cup (B,R,W) \cup (W,R,R) \cup (W,R,B) \cup (W,R,W)\}$$

and the event A occurs on the experiment if any outcome described by one of the indicated elements of the set A occurs,

$$B = B \cap S$$
$$= \{(R,R,B) \cup (R,R,W) \cup (R,B,R) \cup (R,W,R) \cup (B,R,R)$$
$$\cup (W,R,R) \cup (B,B,R) \cup (B,B,W) \cup (B,R,B) \cup (B,W,B)$$
$$\cup (R,B,B) \cup (W,B,B) \cup (W,W,R) \cup (W,W,B) \cup (W,R,W)$$
$$\cup (W,B,W) \cup (R,W,W) \cup (B,W,W)\}$$

And C is the union of 17 points that the student should easily enumerate.

Example 1.12

Consider the random phenomenon of reaching in and simultaneously drawing a sample of 3 balls from an urn that contains 9 red, 3 blue, and 2 white balls. Set up the points in the sample description space that describes the color composition of the sample.

SOLUTION
In this case, S consists of nine points, each of which is a combination.

$$S = \{RRR, RRB, RRW, RBB, RWW, BBB, BBW, BWW, RBW\}$$

A tree diagram of the different points is shown in Figure 1.10. It is

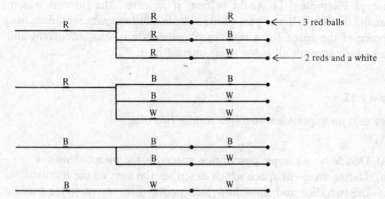

Figure 1.10   Points in a space, describing simultaneously drawing three balls from the urn of Example 1.12.

important to clearly realize RRB, RBR, and BRR are all the same point, which represents the event that the sample is composed of 2 red balls and 1 blue ball, unlike the situation in the previous problem, where each compound point involved order.

Comparing Examples 1.11 and 1.12 it can be seen that many events may be expressed in terms of the points of either space. Indeed, any event that can be expressed as the union of points of Figure 1.10 can also be expressed as the union of points of Figure 1.9. For example, (RBW), the event of 1 white, 1 red, and 1 blue ball in Figure 1.10, can be written in terms of the union of sequential events or points of Figure 1.9 as

$$(RBW) = (R, W, B) \cup (R, B, W) \cup (W, R, B) \cup (W, B, R)$$
$$\cup (B, R, W) \cup (B, W, R)$$

It is important to distinguish whether the elements or points of a space involve sequential order or not. It should be clear that the points of this space are mutually exclusive and collectively exhaustive and also finest grain within the context of the experiment. However, this particular space would be an event space for the experiment of Example 1-11, since it would not be finest grain.

Reconsidering Examples 1.11 and 1.12 it should be noted that if the random phenomenon is defined as the problem of drawing a sample of 3 balls without replacement from an urn containing 9 red, 3 blue, and 2 white balls, then the points of Example 1.11 shown in Figure 1.9 or the points of Example 1.12 shown in Figure 1.10 could be used as the space governing the problem. The choice of which space to use would depend on the interest in the experiment. If the interest was only in the color composition of the sample, either space would suffice; if the interest was in the color of the first or third ball drawn, then the space of Example 1.12 would be unsuitable, but that of Example 1.11 would suffice; if however, the interest was in whether the dirtiest ball was part of the sample, neither space would suffice. The choice of the space for a random phenomenon is very important and must take into account all subsequent interest in it.

Example 1.13

Assume that an experiment involves rolling two dice.

(a) Decide on a sample description space for the phenomenon.
(b) Define an event space which describes the sum of the numbers on the two dice and show how each point of the event space may be expressed, using set theory operations on the points from the sample description space.

SOLUTION

(a) The sample description space is shown in the tree diagram of Figure 1.11 and consists of 36 finest grain, mutually exclusive, collectively exhaustive points:

$$S = \{(x, y): 1 < x < 6, 1 < y < 6, \text{both an integer}\}$$

What we have really assumed here is that the two dice are distinguishable or that x refers to the first upturned number observed and y to the second.

(b) Obviously, the event space E will consist of 11 points:

$e_2 = $ event the sum is 2

$e_3 = $ event the sum is 3

$\vdots$

$e_{12} = $ event the sum is 12

(Note that for convenience we did not use e_1.) It is clear that

$$e_i \cap e_j = \emptyset$$

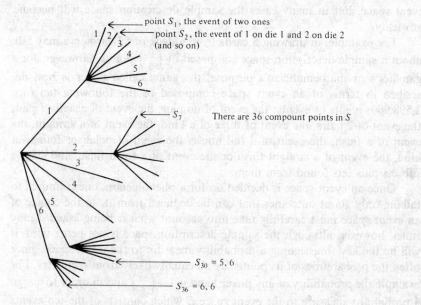

point S_1, the event of two ones
point S_2, the event of 1 on die 1 and 2 on die 2
(and so on)

S_7 There are 36 compound points in S

$S_{30} = 5, 6$

$S_{36} = 6, 6$

Figure 1.11   Tree diagram showing points of S for Example 1.13.

and

$$\bigcup_{i=2}^{12} e_i = S$$

is the certain event. Therefore, the points e_i are both mutually exclusive and collectively exhaustive. Since the space of part a was finest grain, we can express each point of E in terms of points of S as follows:

$$e_2 = s_1$$
$$e_3 = s_2 \cup s_7$$
$$e_4 = s_3 \cup s_8 \cup s_{13}$$
$$e_5 = s_4 \cup s_9 \cup s_{14} \cup s_{19}$$
$$\vdots$$
$$e_{12} = s_{36}$$

In the next chapter we will see that all probabilistic problems may be solved by means of a fundamental sample description space instead of an event space, but in many cases the sample description space will become unwieldy.

For example, in drawing 5 cards from a deck of 52 cards, we may talk about a sample description space composed of $\binom{52}{5}$ points. However, for a gambler's or mathematician's purpose, the game called poker is best described in terms of an event space composed of the following ten (not 2,598,960) points or events: the event of no pair, the event of exactly 1 pair, the event of 2 pairs, the event of three of a kind, the event of a straight, the event of a flush, the event of a full house, the event of poker or four of a kind, the event of a straight flush or the event of a royal flush, and simple subsets plus sets found from them.

Once an event space is decided on for a phenomenon, one is limited to talking only about outcomes that can be deduced from it, so the choice of an event space must carefully take into account what is being asked. Many times, however, although the sample description space is not being used, it will be the key to assigning a probability measure to the event space, since often the probabilities of its points are all equally likely from symmetry. For example the probability of any poker hand is $1 \div \binom{52}{5}$ and in order to assign a probability measure to the event space E which consists of the ten events we listed, each point of E is visualized as the union of equally likely points of S.

DRILL SET: SAMPLE AND EVENT SPACES FOR A RANDOM PHENOMENON

1. A play-off series consists of two teams playing the best of five games (i.e., the series concludes when either team wins its third game). If A and B meet in the series, draw a tree diagram of the different possible results.
2. Consider the compound points of Problem 1 as the sample description space of the random phenomenon.
   (a) Express the event that A wins in four games or less as the union of points of your space in Problem 1.
   (b) Express the event team B does not win the second game as the union of points in Problem 1.
   (c) Express the event the series lasts at least four games as the union of points in Problem 1.
3. Can you intuitively assign a probability measure to the sample space of Problem 1 assuming that the teams have equal ability? (*Hint*: With intuition we can conclude that a specific three-game result is twice as likely as any four-game result and that a specific four-game result is twice as likely as any five-game result.)

SUMMARY

Chapter 1 covered the basic prerequisite material for discrete probability theory. The language of set theory is used to describe events. The three basic operations for handling events are: the union of two events $A \cup B$, which is defined as the event that A or B occurs (this includes both); the intersection of two events $A \cap B$, which is defined as the event that A and B both occur; and the complement of an event $\overline{A}$ which is defined as the event that A does not occur. Set theory relations were described as being based on seven basic axioms;

$$A \cup B = B \cup A$$
$$A \cup (B \cup C) = (A \cup B) \cup C$$
$$A \cap (B \cup C) = (A \cap B) \cup (A \cap C)$$
$$\overline{(\overline{A})} = A$$
$$\overline{A \cap B} = \overline{A} \cup \overline{B}$$
$$A \cap \overline{A} = \varnothing$$
$$A \cap S = A$$

Using these axioms any theorem may be proved. In addition to proving relations using fundamental axioms, engineers often demonstrate results by means of Venn diagrams, which were treated in the chapter.

In order to describe the points of a space associated with the outcomes of a trial or an experiment that is assumed to be a random phenomenon, it is often required that we actually count or enumerate them in a logical manner. Section 1.3, on the mathematics of counting, developed by means of tree diagrams the important permutation and combination formulas: $(n)_r$, $n!/n_1!n_2!\cdots n_n!$, and $\binom{n}{r}$, whose meanings should by now be second nature. Some reasonably difficult counting problems were considered by subdividing each problem into parts and using the few basic formulas we are familiar with, over and over.

After the treatment of set theory and counting, attention was focused on random phenomena. The sample description space S was defined as the finest grain, mutually exclusive, and collectively exhaustive listing of outcomes and an event space was defined as a mutually exclusive and collectively exhaustive listing of outcomes for a phenomenon. For any probabilistic problem the choice of an event space will be determined by the interest in it or desired results required.

In summary we can say problems become probabilistic if we assume we are dealing with a random phenomenon. In order to answer any question concerning a phenomenon, we define its sample description space or an appropriate event space. If our space is properly chosen, then from set theory any event of interest, say A, may be expressed in terms of the space as $A = A \cap S$, which then becomes the union of some points of S. In the next chapter we will extend our treatment to include the assignment of a probability to any event defined on a phenomenon.

PROBLEMS

1. Discuss the following statements:
   (a) If a die is rolled, probability theory says the chance of a 1 is $\frac{1}{6}$.
   (b) If a die is rolled 6 million times, then each of the six numbers will occur 1 million times.
   It is hoped you will dispute statement (b). Modify it to convey how a layperson should phrase this statement.
2. Consider two points 1 and 2 joined by four possible communication paths A, B, C, and D as shown in Figure 1.12. Let A be the event that

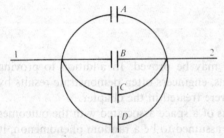

Figure 1.12   Two points connected by four different communication links.

link A is open and $\overline{A}$ be the event that it is closed and similarly define B, $\overline{B}$, C, $\overline{C}$, D and $\overline{D}$. Write out or indicate the event space E describing all possible conditions between points 1 and 2 at any time.

(a) Describe the event that links A and B are open and link C is closed in terms of points of E.

(b) Describe the event that link A is open and C is closed in terms of points of E.

(c) Can you solve this problem using a Karnaugh map for minterms from boolean algebra?

3. Prove or disprove the following, using the axioms of set theory:

$$(X \cup Y) \cap (\overline{X} \cup Y) \cap (X \cup Z) = \overline{X} \cup \left[X \cap (\overline{Y \cap Z}) \right]$$

$$(\overline{A \cup B \cup C}) = \overline{A} \cap \overline{B} \cap \overline{C}$$

Use two approaches: first, manipulate the axioms and see what results occur; second, refer both right- and left-hand sides to a fundamental space composed of mutually exclusive, collectively exhaustive points and see whether they contain the same points. (*Hint*: Think of Karnaugh maps and points such as $X \cap Y \cap Z$, $X \cap Y \cap \overline{Z}$, and so on.)

4. Two events are independent if the occurrence of one on a trial of an experiment of a phenomenon has no effect on whether the other occurs or not.

(a) Can two independent events be mutually exclusive?

(b) If A and B are mutually exclusive, are $\overline{A}$ and $\overline{B}$ mutually exclusive?

(c) If A and B are mutually exclusive and collectively exhaustive, are $\overline{A}$ and $\overline{B}$ mutually exclusive and collectively exhaustive?

5. The rank of a matrix is said to be p if it contains at least one $p \times p$ nonsingular submatrix (i.e., the determinant of the elements is nonzero) and no nonsingular submatrix of size $p + 1 \times p + 1$.

(a) How many submatrices of size $p \times p$ does an $m \times n$ matrix contain?

(b) If a student has to check the rank of a matrix by evaluating determinants, what is the least number of determinants that will have to be evaluated before he or she concludes that a 7×6 matrix has rank 3.

6. (a) Demonstrate the result

$$\binom{m+1}{n} = \binom{m}{n} + \binom{m}{n-1}$$

as the outcome of a physical problem involving combinations or permutations.

(b) Prove

$$\binom{r-1}{r-1} + \binom{r}{r-1} + \cdots + \binom{n-1}{r-1} \equiv \binom{n}{r}$$

7. A bridge hand is any set of 13 cards from a conventional deck of 52 cards. Therefore the event space E consists of $\binom{52}{13}$ points of size 13. State how many points of E must be unionized together to obtain
   (a) the event of 6 spades in a hand.
   (b) the event of two 6-card suits subject to the condition that the hand contains no hearts.

8. In how many ways can a person be dealt a hand containing three of a kind (such as three twos or threes) and no other pair in draw poker?

9. Consider a bin contains 50 resistors, 30 capacitors, and 20 inductors. An experiment consists in drawing a sample of size 5. Use the definition of probability and counting to
   (a) determine the number of samples containing three resistors and the probability of drawing a sample containing three resistors.
   (b) determine the number of samples containing three resistors, one capacitor, and one inductor, and the probability of drawing such a sample.
   (c) If you specifically know that one member of a sample is a resistor, what is the probability of having three resistors, one capacitor, and one inductor in your sample?

10. An experiment consists of drawing elements from a bin containing four resistors, two capacitors, and one inductor until two resistors are obtained. Carefully list the different possible outcomes of the phenomenon and construct a tree diagram of this space S.
    (a) Express the event that the first two elements drawn are different as the union of points of S.
    (b) Express the event that the first and last elements drawn are different in terms of points of S.
    (c) Thinking of the definition of the probability of an event, are all the outcomes of the points equally likely?
    (d) If the answer to (c) is no, intuitively decide which events are most and least likely.

Chapter 2
The Foundation of
Probability Theory

2.1 THE PROBABILITY OF AN EVENT AND AXIOMS

As previously stated, answering questions about assumed random phenomena involves three stages:

1. The listing of an appropriate *event space*
2. The assignment of a *probability measure* to the points of the event space
3. Obtaining probabilities for events of interest by expressing them in terms of the events of the event space and using the rules of probability theory

In Chapter 1 we considered stage 1 and the basic set theory to perform much of stage 3. In this chapter we will focus on stage 2 and the general solution of random phenomena problems using stage 3.

For any random phenomenon the **probability** of an event or outcome E, is defined on a relative-frequency basis as

$$P(E) \triangleq \lim_{N \to \infty} \frac{N(E)}{N}$$

$$= \lim_{N \to \infty} \frac{\text{number of trials that occur with } E}{\text{total number of trials } N} \qquad (2.1)$$

This definition of probability has philosophically led to great debate inasmuch as it implies either that the result of Eq. 2.1 may be visualized from fairness or predicted from a finite number of trials, since it is never possible to carry out an infinite number. There are many situations when we feel comfortable with this relative-frequency definition. For example, we say that Eq. 2.1 predicts the probability of a head on the toss of a coin as 0.5 and the probability of drawing a white ball from an urn with 6 white and 4 blue balls as 0.6. However we have made certain assumptions here, as in the coin tossing we assumed that the coin is structurally balanced and that knowing the toss starts with "tails up" has no effect on the outcome. Similar practical constraints are implied when we intuitively predict 0.6 for the probability of drawing a white ball from our urn. When we speak of "the probability that the height of a fifth grader is in a certain range," "the probability that a person chosen at random was born in June," or "the probability that 1000 electrons will be emitted in an extremely short time from a small cathode area," we feel less comfortable with Eq. 2.1 until swayed by statistical tables for fifth graders, and birth data for June (with restrictions), and with study we accept that cathode emissions obey a very famous Poisson probability law. When people talk about "the probability of landing a man on the moon" or "the probability of a nuclear accident," Eq. 2.1 cannot be used (why?), and indeed the term "probability" may be a misnomer.

If it is possible we will use the relative-frequency definition of probability, but an axiomatic definition is more widely accepted as a philosophical basis. The axiomatic definition of probability just assigns to each point s_i in a sample description space a number, $P(s_i)$, such that it obeys the following two conditions:

$$0 \leq P(s_i) \leq 1 \quad \text{and} \quad \sum_{i=1}^{n} P(s_i) = 1$$

In stage 2 of a probabilistic problem we will try to use a relative frequency approach for assigning a probability measure because it develops a strong intuitive feeling for the concept of probability. When answering questions pertaining to difficult events, however, we can appreciate the structure developed from the axiomatic approach. For a mature detailed discussion of the definition of probability the reader is referred to the work of A. Papoulis.[1] We will now try to cement this relative-frequency definition with a few problems.

[1] A. Papoulis, *Probability, Random Variables, and Stochastic Processes* (New York: McGraw-Hill, 1965), chap. 2.

Example 2.1

Using the definition

$$P(E) = \lim_{N \to \infty} \frac{N(E)}{N}$$

predict the probabilities of the following events:

(a) An even number on the roll of a die.
(b) Exactly three zeros in an 8-bit word.
(c) Two black balls, when 2 balls are drawn without replacement from an urn containing 7 black balls and 5 red balls.

SOLUTION

(a) P(even number) $= \frac{3}{6} = 0.5$.
(b) There are 256 different 8-bit words (or permutations of size 8) that can be formed from an unlimited supply of zeros and ones. Of these, $8!/3!5! = 56$ different words contain exactly three zeros. Assuming a trial of our random phenomenon implies to randomly generate one of the 256 words, we predict

$$P(\text{exactly three zeros and five ones}) = \tfrac{56}{256} = 0.22$$

(c) If 2 balls are drawn from an urn with 7 black balls and 5 red balls we may consider the problem in two ways. First, we imagine that the 7 black balls are numbered 1 through 7 and the 5 red balls numbered 8 through 12. If we reach in, draw out a ball, observe its number and without replacing it in the urn, draw out a second ball and observe its number, there are $(12)_2$ different samples possible and there are $(7)_2$ different samples of 2 black balls.

$$\therefore P(2 \text{ black balls}) = \frac{7 \times 6}{12 \times 11} = 0.32$$

Second, if we imagine we reach in and draw out 2 balls, then

$$P(2 \text{ black balls}) = \frac{\binom{7}{2}}{\binom{12}{2}} = 0.32$$

THE AXIOMS OF PROBABILITY THEORY

Directly following the definition of the probability of an event, a number of axioms are stated as being intuitively apparent and sufficient for obtaining any probabilistic rule or fact. For a random phenomenon with a sample space S these axioms are as shown.

Axiom 1

Where E_i is any event of S, $\boxed{P(E_i) \geq 0}$

Axiom 2

Where S is the certain event, $\boxed{P(S) = 1}$

Axiom 3

Where A and B are two events defined on S, $\boxed{\begin{aligned} &P(A \cup B) = P(A) + P(B), \\ &\text{if } A \cap B = \varnothing \end{aligned}}$

Axioms 1 and 2 should seem intuitively clear to the student. To visualize axiom 3, we might consider an experiment, that is carried out a large number of times N, and in which event A occurs N_1 times and event B occurs N_2 times. If A and B cannot occur at the same time or $A \cap B = \varnothing$, then A or B occurs $N_1 + N_2$ times, and

$$
\begin{aligned}
P(A \cup B) &= \lim_{N \to \infty} \frac{N_1 + N_2}{N} \\
&= \lim_{N \to \infty} \left(\frac{N_1}{N} + \frac{N_2}{N} \right) = \lim_{N \to \infty} \frac{N_1}{N} + \lim_{N \to \infty} \frac{N_2}{N} \\
&= P(A) + P(B)
\end{aligned}
$$

From these three simple axioms all the laws of probability may be derived. For example, let us find an expression for $P(A \cup B)$, where A and B are not mutually exclusive events on S, as is shown in the Venn diagram of Figure 2.1. From set theory,

$$ A \cup B = A \cup (B \cap \bar{A}) $$

where A and $(B \cap \bar{A})$ are mutually exclusive.

$$ \therefore P(A \cup B) = P(A) + P(B \cap \bar{A}) \tag{2.2} $$

by axiom 3.

If we use the set theory axiom, $B = B \cap S$, and write $S = A \cup \bar{A}$, then

$$ B = (B \cap A) \cup (B \cap \bar{A}) $$

which are mutually exclusive events.

$$ \therefore P(B \cap \bar{A}) = P(B) - P(A \cap B) $$

by axiom 3. Substitution of this result in Eq. 2.2 yields

$$ P(A \cup B) = P(A) + P(B) - P(A \cap B) \tag{2.3} $$

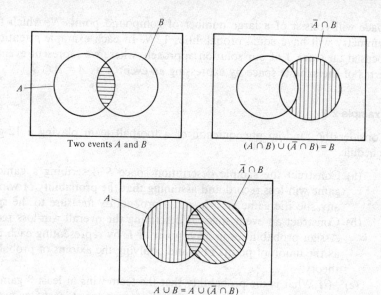

Two events A and B

$(A \cap B) \cup (\bar{A} \cap B) = B$

$A \cup B = A \cup (\bar{A} \cap B)$

Figure 2.1   Venn diagram for the probability of a union.

We note that this result is identical to axiom 3 when $A \cap B = \varnothing$. Figure 2.1 illustrates this proof with Venn diagrams.

DRILL SET: DEFINITION OF PROBABILITY AND THE AXIOMS

1. Use the definition of the probability of an event to assign probabilities to the following:
   (a) A computer word of 16 bits contains three zeros.
   (b) Three balls drawn from an urn containing 7 white and 3 red balls are white.
   (c) The sum of the numbers obtained when rolling two dice is 7.
   Explain carefully why your numerator-denominator ratios correspond to N_A and N for $N \rightarrow \infty$.
2. Derive a formula for $P(A \cup B \cup C)$ in terms of $P(A)$, $P(B)$, $P(C)$, $P(A \cap B)$, $P(A \cap C)$, $P(B \cap C)$, and $P(A \cap B \cap C)$, respectively.
3. Utilize your formula from Problem 2 to find the probability that the first or second or third toss of a coin is a head.

2.2   RANDOM PHENOMENA WITH EQUALLY LIKELY OUTCOMES

Before we discuss in the following sections the general assignment of a probability measure to the points of a space utilizing the concepts of independence and dependence, we will consider the solution of probabilistic questions where symmetry allows us to assume that all the points of S are equally likely by Eq. 2.1. In Examples 2.2 and 2.3 the sample description space will consist of a large number of compound points N, which from symmetry will have equal probabilities $1/N$. In each example attention is focused on stage 3 of our solution approach, where we represent events in terms of the sample space by expressing an event A as $A = A \cap S$.

Example 2.2

Consider the random phenomenon of a football team playing a 12-game schedule.

(a) Construct the sample description space S describing a game-by-game win-loss record and assuming that the probability of winning any specific game is $\frac{1}{2}$, assign a probability measure to the space.

(b) Construct an event space E describing the overall win-loss record. Assign probabilities to each point of E by representing each point as the union of points of S and applying the axioms of probability theory.

(c) (i) What is the probability that the team wins at least 8 games?

　　(ii) What is the probability that the team wins at least 5 games and at most 9 games?

　　(iii) What is the probability that the team wins its third and loses its ninth game? Solve this part in two ways, first by using the sample space of part (a) and second by creating a new random phenomenon on which the desired event is more easily described.

SOLUTION

(a) The sample description space for the possible game-by-game won-loss record based on a 12-game schedule is

$$S = \{a, b, c, d, e, f, g, h, i, j, k, l\}$$

where any of the letters may be W or N. If a is W, this is the event of winning the first game; if g is N, this is the event of not winning the seventh game. There are 2^{12} different points in S (start drawing a tree diagram), and by symmetry or fairness the probability of each point is

$$\frac{1}{2^{12}} = \frac{1}{4096}$$

(b) The event space describing the overall possible won-loss records is

$$E = \{(12W), (11W, 1N), (10W, 2N), \dots, (12N\text{'s})\}$$

or, more compactly,

$$E = \{k_1W, (12 - k_1)N : 0 < k_1 < 12, k_1 \text{ an integer}\}$$

For example, the point (10W, 2N) is the event of a "10 wins and 2 losses" record. Intuitively, it should be clear to the student that the 13 points of E are not equally likely with probabilities $\frac{1}{13}$. We will use our knowledge of permutations to assign probabilities to the points of E where P denotes probability.

$$P(12W) = \frac{1}{2^{12}} = \frac{1}{4096}$$

To find $P(11W, 1N)$ we first state that

$$P(\text{wins on first 11 games, loss on game 12}) = \frac{1}{2^{12}}$$

Now the event (11 wins and 1 loss) is the union of 12 points of S, since we can form $\binom{12}{1}$ or 12 permutations of size 12 from 11W's and 1L.

$$\therefore P(11W, 1N) = \binom{12}{1} \frac{1}{2^{12}} = \frac{12}{4096}$$

All the other probabilities should follow easily. (10W, 2N) is the union of $\binom{12}{2}$ points of S, and, by axiom 3,

$$P(10W, 2N) = \binom{12}{2} \frac{1}{2^{12}}$$

Similarly,

$$P(9W, 3N) = \binom{12}{3} \frac{1}{2^{12}}, \quad P(8W, 4N) = \binom{12}{4} \frac{1}{2^{12}},$$

$$P(7W, 5N) = \binom{12}{5} \frac{1}{2^{12}}, \quad P(6W, 6N) = \binom{12}{6} \frac{1}{2^{12}},$$

$$P(5W, 7N) = \binom{12}{7} \frac{1}{2^{12}}, \dots, P(12W) = \binom{12}{12} \frac{1}{2^{12}}$$

(c) (i) The event of winning at least 8 games may be expressed as the union of five mutually exclusive points of E as

(at least 8 wins)

$$= \{(8W, 4N) \cup (9W, 3N) \cup (10W, 2N) \cup (11W, 1N) \cup (12W)\}$$

$\therefore P(\text{at least 8 wins})$

$$= \left[\binom{12}{8} + \binom{12}{9} + \binom{12}{10} + \binom{12}{11} + \binom{12}{12} \right] \frac{1}{2^{12}}$$

$$= \frac{794}{4096}$$

(ii) This is very similar to the previous part, as the event

(at least 5 and at most 9 wins)

$$= \{(5W, 7N) \cup (6W, 6N) \cup (7W, 5N) \cup (8W, 4N) \cup (9W, 3N)\}$$

$\therefore P(\text{at least 5 wins and at most 9 wins})$

$$= \left[\binom{12}{5} + \binom{12}{6} + \binom{12}{7} + \binom{12}{8} + \binom{12}{9} \right] \frac{1}{2^{12}}$$

$$= \frac{3123}{4096}$$

(iii) We must express the event of winning game 3 and losing game 9 as the union of points of S. S contains 2^{12} points and 2^{10} of these have W in position 3 and N in position 9. Our desired event is the union of 2^{10} mutually exclusive points of S.

$$\therefore P(\text{winning game 3 and losing game 9}) = 2^{10} \left(\frac{1}{2^{12}} \right)$$

$$= 0.25$$

Alternatively, we could have created a new random phenomenon, which consists of the results of the third and ninth games on a schedule, where

$$S = \{(W, W), (W, N), (N, W), (N, N)\}$$

and for example (W, N) is the event of winning game 3 and losing game 9. All these points, by symmetry, are equally likely with probability 0.25.

$$\therefore P(W, N) = 0.25$$

as before.

Example 2.3

Consider the random phenomenon describing the different possible results of a race with 10 competitors, where the official result of a race consists in listing the first four finishers in order.

(a) Assuming that all the competitors have equal chances, set up the sample description space and assign a probability measure.
(b) What is the probability competitor A is first?
(c) What is the probability A finishes in the first four?
(d) What is the probability A or B finish in the first four positions?

SOLUTION

(a) Let the 10 competitors be named A, B, C, D, E, F, G, H, I, and J. The sample description space S describing the different possible results is

$$S = \{(a, b, c, d)\}$$

where a is any of the 10 competitors, b is any competitor except a, c is any competitor except a or b, d is any competitor except a, b, or c. S is composed of the $(10)_4$ permutations that can be formed. Each point of S is assigned a probability

$$\frac{1}{(10)_4} = \frac{1}{10 \times 9 \times 8 \times 7}$$

by symmetry.

(b) The event "A is first" is the union of $(9)_3$ points of S (those for which A is in position 1).

$$\therefore P(A \text{ is first}) = \frac{(9)_3}{(10)_4} = 0.1$$

The result 0.1 is also obvious by inspection, but we are trying to develop the facility of defining a space and expressing any desired event using the set theory axiom $A = A \cap S$, to yield it in terms of points of S.

(c) $P(A \text{ is in first four}) = 4(9)_3 / (10)_4 = 0.4$.

(d) Let us consider the event that A or B (and include both) finishes in the first four positions. With thought and maybe a review of permutations the student should show this is the union of

$$4(8)_3 + 4(8)_3 + \frac{4!}{1!1!2!}(8)_2$$

points of S.

$$\therefore P(A \text{ or } B \text{ is in first 4 positions}) = \frac{4(8)_3 + 4(8)_3 + 4!/2!(8)_2}{(10)_4}$$

$$= 0.67$$

Examples 2.2 and 2.3 demonstrated the general approach to answering probabilistic questions about a random phenomenon. The choice of an appropriate event space or the sample description space for the problem is important. The most common initial mistake made by a student is to decide incorrectly on the size of the points of S or E. In Example 2.2, since the phenomenon involved playing a 12-game schedule, then the points of S had to be of size 12. When a question is asked about an event A that does not involve 12 games, we must express A as $A \cap S$ to solve for it in terms of S. In summary, we can highlight two major concepts:

Carefully decide on the size of the points of S or E from the problem statement.

Always express any desired event in terms of S or E by $A = A \cap S$ or $A \cap E$.

At this state we must be flexible in visualizing the generality of problems. For example, Example 2.3 could be restated as:

Consider forming computer words of 12 bits.
(a) What is the probability a word contains at least eight ones?
(b) What is the probability a word contains at least five and at most nine ones?
(c) What is the probability the third bit is a 1 and the ninth bit is a 0?

DRILL SET: RANDOM PHENOMENA WITH EQUALLY LIKELY OUTCOMES

1. Consider the phenomenon of tossing a coin and rolling a die.
   (a) Construct the sample description space and assign probabilities to each point.
   (b) What is the probability the die number is less than 3.6?
   (c) What is the probability of a head on the coin toss and an even number on the die roll?
2. Consider the phenomenon of drawing three cards from a standard deck of 52 cards. Define an event space and assign a probability measure to it that allows us to evaluate the probabilities of the following events:
   (a) All 3 cards are black.
   (b) At least 1 card is a spade.

2.3 CONDITIONAL PROBABILITY: DEPENDENT AND INDEPENDENT EVENTS

The Conditional Probability of an Event

The **conditional probability** of an event A subject to the condition that event B has occurred is denoted $P(A/B)$ and is defined as

$$P(A/B) \triangleq \lim_{N_B \to \infty} \frac{N_{A \cap B}}{N_B} \tag{2.4}$$

where if N trials of the phenomenon occur, N_B is the number of times event B occurs and $N_{A \cap B}$ is the number of times events A and B occur. This concept, which is intuitive, represents the first vital stage where some students lose a physical grasp of understanding probability theory. Due to its importance we reiterate in words that the conditional probability of A subject to the condition that event B has occurred is defined on a relative-frequency basis, based on the trials N_B for which B occurs, as the ratio of the number of trials for which A occurs $N_{A \cap B}$ to the number of trials N_B. Philosophically this is no more difficult to visualize than an unconditional probability.

In order to think of two events A and B, we must relate both of them to a random phenomenon on whose space they are defined. Let us consider the rolling of a fair die where A is the event of obtaining a 6 and E_1 is the event of obtaining a number greater than 2. Consideration of Eq. 2.4 should intuitively yield

$$P(6/\text{number} > 2) = \tfrac{1}{4}$$

whereas $P(6) = \tfrac{1}{6}$ when no condition is imposed. It is clear that a condition tends to raise some probabilities and lower others. In this case the condition of the event E_1 that a number greater than two has occurred results in $P(1/E_1) = 0$, $P(2/E_1) = 0$, $P(3/E_1) = \tfrac{1}{4}$, $P(4/E_1) = \tfrac{1}{4}$, $P(5/E_1) = \tfrac{1}{4}$, and $P(6/E_1) = \tfrac{1}{4}$.

Example 2.4

Evaluate the probabilities of the following conditional events:

(a) Exactly three zeros in a computer word of 6 bits given the condition that the second and third bits are 0.

(b) $\sin\theta > 0.5$, given $\sin\theta > 0$.

SOLUTION

(a) There are 2^6 or 64 different computer words of 6 bits and 16 of them have the second and third bits equal to 0. Of these 16 words,

$4!/3!1!=4$ contain exactly three zeros.

$$\therefore P(\text{exactly three zeros/second and third bits are } 0)=\frac{4}{16}$$

$$=0.25$$

The unconditional probability is

$$P(\text{exactly three zeros})=\frac{6!/3!3!}{2^6}$$

$$=\frac{20}{64}$$

$$=0.312$$

The reader should not be surprised that in this case the condition made the event less likely. For our problem the ratio of $N:N_A:N_{A\cap B}:N_B$ is $64:20:16:4$.

(b) If we consider a thumbnail sketch of $\sin\theta$ versus θ, we can say

$$P\left(\frac{\sin\theta>0.5}{\sin\theta>0}\right)=\frac{150°-30°}{180°}$$

$$=0.67$$

whereas the unconditional probability was

$$P(\sin\theta>0.5)=\frac{150°-30°}{360°}$$

$$=0.33$$

In this case the condition made it twice as likely that $\sin\theta>0.5$.

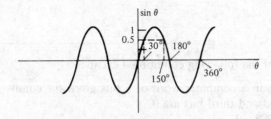

The Probability of a Compound Event

There is another important way of viewing Eq. 2.4; that is,

$$P(A/B)\triangleq\lim_{N_B\to\infty}\frac{N_{A\cap B}}{N_B}=\lim_{N\to\infty}\frac{N_{A\cap B}/N}{N_B/N}$$

Interpreting the numerator and denominator this yields,

$$P(A/B) = \frac{P(A \cap B)}{P(B)}$$

This formula may be arranged to yield the probability of the intersection of two events A and B as

$$P(A \cap B) = P(B)P(A/B) \tag{2.5}$$

In a purely symmetrical manner had we defined

$$P(B/A) = \lim_{N_A \to \infty} \frac{N_{A \cap B}}{N_A}$$

then $P(A \cap B)$ also may be expressed as

$$P(A \cap B) = P(A)P(B/A) \tag{2.6}$$

Generally the definitions of $P(A/B)$ and $P(B/A)$ given on an *axiomatic* basis are

$$P(A/B) \triangleq \frac{P(A \cap B)}{P(B)} \tag{2.7}$$

and

$$P(B/A) \triangleq \frac{P(A \cap B)}{P(A)} \tag{2.8}$$

However, if we are predicting their values from a relative-frequency point of view we think of

$$\frac{N_{A \cap B}}{N_B} \quad \text{and} \quad \frac{N_{A \cap B}}{N_A}$$

and statistically visualize them.

Dependent and Independent Events

We define two events A and B as being **dependent** if

$$P(A/B) \neq P(A) \quad \text{and} \quad P(B/A) \neq P(B) \tag{2.9}$$

provided that $P(A) \neq 0$ and $P(B) \neq 0$.

Similarly, two events are **independent** if the occurrence of one has no effect on the probability of the other or if

$$P(A/B) = P(A) \quad \text{and} \quad P(B/A) = P(B) \tag{2.10}$$

From Eqs. 2.5 and 2.9, where $P(A) \neq 0$ and $P(B) \neq 0$, we see that if two events are independent,

$$P(A \cap B) = P(A)P(B) \tag{2.11}$$

and this formula is also used as the definition of two independent events.

Using set theory, the concepts of conditional probabilities and dependent and independent events may be extended to three or more events. It should follow clearly that

$$P(A/B \cap C) = \lim_{N_{B \cap C} \to \infty} \frac{N_{A \cap B \cap C}}{N_{B \cap C}}$$

$$= \frac{P(A \cap B \cap C)}{P(B \cap C)} \tag{2.12}$$

and further we may write the definition for the probability of three events as

$$P(A \cap B \cap C) = P(B \cap C)P(A/B \cap C)$$

$$= P(B)P(C/B)P(A/B \cap C) \tag{2.13}$$

We could just as easily have derived $P(A \cap B \cap C)$ as

$$P(A \cap B \cap C) = P(A)P(B/A)P(C/A \cap B)$$

or

$$P(A \cap B \cap C) = P(C)P(A/C)P(B/A \cap C)$$

plus three other different versions. A general formula for the probability of the intersection of N events is

$$P(A_1 \cap A_2 \cap A_3 \cap \cdots \cap A_n)$$

$$= P(A_1)P(A_2/A_1)P(A_3/A_1 \cap A_2) \cdots P(A_n/A_1 \cap A_2 \cap \cdots \cap A_{n-1}) \tag{2.14}$$

and the theory of permutations should tell the reader how many different versions of this formula we can obtain.

Three events A, B, and C are defined as being independent if

$$P(A \cap B \cap C) = P(A)P(B)P(C) \tag{2.15}$$

and if, in addition, each pair of events is independent and $P(A) \neq 0$, $P(B) \neq 0$, and $P(C) \neq 0$. Similarly, n events $A_1, A_2, \ldots, A_n$ are independent if

$$P(A_1 \cap A_2 \cap \cdots \cap A_n) = P(A_1)P(A_2) \cdots P(A_n) = \prod_{i=1}^{n} P(A_i) \tag{2.16}$$

and each $n-1, n-2, \ldots, 2$ set of events are independent and $P(A_i) \neq 0$ for all i. The symbol Π indicates "take the product."

Example 2.5

Consider the experiment of uniformly sampling the waveform shown in Figure 2.2 between $t = 0$ and $t = 3$. Find the probabilities of the following

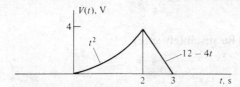

Figure 2.2   Diagram for Example 2.5.

events:

(a) A value less than 2.6 V is sampled from the t^2 portion.
(b) A sampled value of less than 2.6 V came from the t^2 portion; in other words, the probability that a value was from the t^2 portion given that it is less than 2.6 V.

SOLUTION

(a) For clarity let us provide the following definitions:

A = event of choosing a value less than 2.6 V

B = event of sampling a value from the t^2 portion of the waveform or corresponding to $0 < t < 2$ on the time axis

C = event a value less than 2.6 is sampled from the t^2 portion

By set theory,

$$C = B \cap A$$
$$\therefore P(C) = P(B)P(A/B)$$
$$= \frac{2}{3} \times \frac{\sqrt{2.6}}{2}.$$
$$= 0.54$$

We notice that this result could be obtained directly as

$$P(C) = \frac{\sqrt{2.6}}{3}$$
$$= 0.54$$

which corresponds to the formula $N_{A \cap B}/N$.

(b) We will carefully use our conditional probability formulas for this part. The required event, which we will denote by D, is

$$D = (\text{a value of less than 2.6 V came from } t^2 \text{ portion})$$

and its probability is

$$P(D) = P(B/A) \qquad \text{(Be absolutely sure)}$$

$$= \frac{P(B \cap A)}{P(A)}$$

$$= \frac{\frac{2}{3}\sqrt{2.6}/2}{P(A)}$$

$$= \frac{0.54}{P(A)}$$

Now, using the axiom $A = A \cap S$,

$$A = A \cap \left[(\text{sampling from } t^2 \text{ portion})\right.$$

$$\left. \cup (\text{sampling from linear portion})\right]$$

$$\therefore P(A) = \frac{2}{3}\frac{\sqrt{2.6}}{2} + \frac{1}{3}\frac{2.6}{4}$$

$$= 0.75$$

and our difficult question yields the solution

$$P(D) = \frac{0.54}{0.75} = 0.72$$

In conclusion, we notice that for this problem conditional questions involving sampling from a range of the time axis are simple, but conditional questions asking which range of time was chosen given information about the function value are complex and depend on the axiomatic interpretation of conditional probability. The formulas of this section (that is, Eq. 2.4 through 2.16) on conditional probability and the probability of compound points or the intersection of events will be heavily utilized in all problem solving, starting with Section 2.4.

DRILL SET: CONDITIONAL PROBABILITIES; DEPENDENT AND INDEPENDENT EVENTS

1. Use the formula $P(A/B) = N_{A \cap B}/N_B$ to find the following probabilities:
   (a) A computer word of 8 bits contains three zeros given that the first bit is a 1.
   (b) A computer word of 8 bits contains two consecutive zeros given that it contains exactly four zeros and four ones.

2. Redo Problem 1 using the formula

$$P(A/B) = \frac{P(A \cap B)}{P(B)}$$

3. Use the formula $P(A_1 \cap A_2 \cap \cdots \cap A_n) = P(A_1)P(A_2/A_1)\cdots$ $P(A_n/A_1 \cap A_2 \cap \cdots \cap A_n)$ to find the probability that 5 cards dealt from a standard deck of 52 cards are all black.

2.4 THE GENERAL SOLUTION OF A PROBLEM

Equations 2.4 through 2.16 are very important and should be intuitively acceptable. When we consider the general solution of random phenomena we will use them in different ways as follows:

1. The first stage in solving a problem always consists in listing the points of an event space or the sample description space. This requires just a knowledge of counting.
2. The second stage is the assignment of a probability measure to the points of a space. If the points are compound we will utilize the formula

$$P(A_1 \cap A_2 \cap A_3 \cap \cdots \cap A_n)$$

$$= P(A_1)P(A_2/A_1)\cdots P(A_n/A_1 \cap A_2 \cap \cdots \cap A_{n-1})$$

to find the probability of each point. This means that from symmetry we predict

$$\frac{N_{A_1}}{N}, \frac{N_{A_1 \cap A_2}}{N_{A_1}}, \frac{N_{A_1 \cap A_2 \cap A_3}}{N_{A_1 \cap A_2}}, \dots$$

if the events can be handled sequentially. Otherwise we must directly visualize

$$\frac{N_{A_1 \cap A_2 \cap \cdots \cap A_n}}{N}$$

3. The third stage of any problem involves solving for desired events or conditional events. To solve for the probability of an event A we express A in terms of points of S by the formula $A = A \cap S$, and to solve for a conditional probability, say $P(X/Y)$, we utilize the formula

$$P(X/Y) = \frac{P(X \cap Y)}{P(Y)} = \frac{P(E_1)}{P(E_2)}$$

where we express E_1, which denotes $X \cap Y$ in terms of points of S by $E_1 \cap S$, and E_2, which denotes Y in terms of S by $E_2 \cap S$ and hence find $P(E_1)$ and $P(E_2)$.

A few straightforward problems will now be solved in order to illustrate the use of this structure.

Example 2.6

If 2 balls are drawn without replacement from an urn that contains 10 balls, 7 of which are white and 3 of which are red, find:

(a) the probability that both balls drawn are white, and
(b) the probability the first ball is white, given the information that the second ball is white.

SOLUTION
The sample description space for the random phenomenon of drawing two balls from the urn is,

$$S = \{(W,W), (W,R), (R,W), (R,R)\}$$

Here (W,W) denotes the event of 2 white balls, (W,R) is the event the first ball drawn is white and the second is red, and the designation of the other two points or events should be obvious. A probability measure may now be assigned to these four mutually exclusive, collectively exhaustive points, using the formula $P(A \cap B) = P(A)P(B/A)$ to yield

$$P(W,W) = P(W)P(W/W) = \tfrac{7}{10} \times \tfrac{6}{9} = \tfrac{7}{15} \qquad \text{(Be sure)}$$

$$P(W,R) = P(W)P(R/W) = \tfrac{7}{10} \times \tfrac{3}{9} = \tfrac{7}{30}$$

$$P(R,W) = \tfrac{3}{10} \times \tfrac{7}{9} = \tfrac{7}{30}$$

$$P(R,R) = \tfrac{3}{10} \times \tfrac{2}{9} = \tfrac{1}{15}$$

Note that the sum of the probabilities is 1, which it should be for mutually exclusive events making up a complete space.

The solution to the two parts of the problem are as follows:

(a) $P(W,W) = \tfrac{7}{15}$.
(b) We require the probability of the conditional event that the first ball is white given that the second ball is white. By definition,

$$P(\text{first ball W}/\text{second ball W})$$

$$= \frac{P[\text{first ball W} \cap \text{second ball W}]}{P(\text{second ball W})}$$

$$= \frac{P(E_1)}{P(E_2)}$$

To illustrate the strict use of set theory, which is necessary for more difficult problems, we will express both E_1 and E_2 on S:

$$E_1 = E_1 \cap S = (W, W)$$

$$E_2 = E_2 \cap S = (W, W) \cup (R, W)$$

Hence

$$P(\text{first ball W}/\text{second ball W}) = \frac{P(W, W)}{P(W, W) + P(R, W)}$$

$$= \frac{\frac{7}{15}}{\frac{7}{15} + \frac{7}{30}}$$

$$= \frac{2}{3}$$

Historically, events such as we have just solved for have presented philosophical difficulties inasmuch as we are asking for the probability of a past event knowing the future.

Example 2.7

If 3 balls are drawn with replacement from an urn containing M balls of which M_W are white, find

(a) The probability that exactly 2 of the balls are white.
(b) The probability that exactly 2 are white given that at least 1 is white.

SOLUTION
Due to the condition of replacement between draws, the event of a white ball on any draw regardless of prior results is

$$P(W) = \frac{M_W}{M} = p$$

where p is used for convenience. Also the event of a nonwhite ball on any draw is

$$P(N) = \frac{M - M_W}{M} = q$$

where q is equal to $(1 - p)$. Often we say that p is the probability of a success and q is the probability of a failure.

The sample description space describing drawing a sample of size three from the urn is

$$S = \{(W, W, W), (W, W, N), (W, N, W), (W, N, N),$$

$$(N, W, W), (N, W, N), (N, N, W), (N, N, N)\}$$

For example, (N, W, N) is the event that the first ball is not white, the second is white, and the third is not white; the meaning of the other points should be self-evident.

The following probability measure may now be assigned.

$$P(W, W, W) = P(W)P(W/W)P(W/W \cap W)$$

Since $P(W/W) = P(W/W \cap W) = P(W)$ due to independence, then $P(W, W, W) = p^3$, and the probabilities of the other points are

$$P(W, W, N) = P(W, N, W) = P(N, W, W) = p^2 q$$

$$P(W, N, N) = P(N, W, N) = P(N, N, W) = pq^2$$

$$P(N, N, N) = q^3$$

As a check, we can show that the sum of all the probabilities on the sample space is

$$p^3 + 3p^2 q + 3pq^2 + q^3 = (p + q)^3 = 1^3 = 1$$

The answers to our problems are as follows:

(a) The event that exactly 2 balls are white is the union of 3 points on S, each with probability $p^2 q$:

$$\therefore P(\text{exactly } 2W) = 3p^2 q$$

(b) The conditional event that exactly 2 balls are white given at least 1 is white has the probability

$$P(\text{exactly } 2W/\text{at least } 1W) = \frac{P(\text{exactly } 2W \cap \text{at least } 1W)}{P(\text{at least } 1W)}$$

$$= \frac{3p^2 q}{1 - q^3}$$

The $1 - q^3$ factor comes about because the only point of S not included in the event of at least 1 white is the point (N, N, N), with probability q^3.

The solutions to Examples 2.5 and 2.6 are somewhat drawn out because the emphasis at this stage should be on the overall structure of questions concerning a random phenomenon. Developing an appreciation for—and a fluency in using—Eqs. 2.4 through 2.16 is essential. One final fundamental problem on the usage of these formulas will be solved.

Example 2.8

Consider the random phenomenon of rolling a tetrahedral die with faces numbered 1 through 4 and of drawing balls without replacement from an urn containing four balls numbered 1 through 4. If the number i is obtained on the die roll (the downward face) then i balls are drawn from the urn.

(a) Set up the sample description space and assign a probability measure.
(b) If one is only interested in the sum of the numbers of the balls drawn from the urn, set up an appropriate event space and, using part (a), assign probabilities to each point.
(c) If the sum of the numbers on the balls drawn was 6, what is the probability that a 2 had been obtained on the die roll?

SOLUTION

(a) For variety, the sample description space is shown by means of a tree diagram in Figure 2.3 and using the formula for a compound event, the probability of each point is written alongside. The sample space is unusual because there are points of many different sizes. There are in all 64 different points in S: 4 of size 2, 12 of size 3, and 24 each of size 4 and size 5. The student should verify the probabilities obtained using the formulas for dependent events. In Figure 2.3, for example, $(3,1,2,3)$ is the event of rolling a 3 on the die, and then drawing first the ball numbered 1 from the urn, then the ball numbered 2 and then the ball numbered 3 and its probability is

$$\tfrac{1}{4} \times \tfrac{1}{4} \times \tfrac{1}{3} \times \tfrac{1}{2} = \tfrac{1}{96}$$

(b) The event space describing the sum of the numbers on the balls drawn from the urn is

$$E = \{(1), (2), (3), (4), (5), (6), (7), (8), (9), (10)\}$$

For example, (5) is the event that the sum of all the numbers on the balls drawn from the urn is 5. Using set theory we can express each point of E such as e_i, as $e_i \cap S$ and obtain it as the union of points on S. Doing this we find,

$(1) = (1,1)$ and $P(1) = \tfrac{1}{16}$

$(2) = (1,2)$ and $P(2) = \tfrac{1}{16}$

$(3) = (1,3) \cup (2,1,2) \cup (2,2,1)$ and $P(3) = \tfrac{1}{16} + \tfrac{1}{24} = \tfrac{5}{48}$

$(4) = (1,4) \cup (2,1,3) \cup (2,3,1)$ and $P(4) = \tfrac{5}{48}$

$(5) = (2,1,4) \cup (2,2,3) \cup (2,3,2) \cup (2,4,1)$ and $P(5) = \tfrac{4}{48}$

$(6) = (2,2,4) \cup (2,4,2) \cup [(3,1,2,3) + (\text{five other such points})]$

and $P(6) = \tfrac{2}{48} + \tfrac{6}{96} = \tfrac{5}{48}$

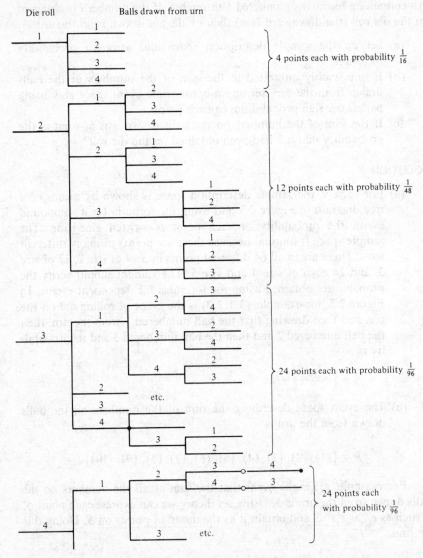

Die roll   Balls drawn from urn

4 points each with probability $\frac{1}{16}$

12 points each with probability $\frac{1}{48}$

24 points each with probability $\frac{1}{96}$

etc.

24 points each with probability $\frac{1}{96}$

etc.

24 points each with probability $\frac{1}{96}$

etc.

Figure 2.3   The sample description space for Example 2.8.

With similar discipline,

$$P(7) = \tfrac{5}{48}$$
$$P(8) = \tfrac{6}{96}$$
$$P(9) = \tfrac{6}{96}$$
$$P(10) = \tfrac{24}{96}$$

We can check the sum of the probabilities as

$$\tfrac{6}{96} + \tfrac{6}{96} + \tfrac{10}{96} + \tfrac{10}{96} + \tfrac{8}{96} + \tfrac{10}{96} + \tfrac{10}{96} + \tfrac{6}{96} + \tfrac{6}{96} + \tfrac{24}{96} = \tfrac{96}{96} = 1$$

(c) We now require the conditional probability that a 2 had been obtained on the die given the sum of the numbers on the balls was 6.

$$P(2 \text{ on die}/\text{sum is } 6) = \frac{P(2 \text{ on die} \cap \text{sum is } 6)}{P(\text{sum is } 6)} = \frac{P(X)}{P(Y)}$$

We must express X and Y as $X \cap S$ and $Y \cap S$, since the event space of part (b) is not appropriate. (Why?)

$$P(2 \text{ on die}/\text{sum is } 6)$$

$$= \frac{P[(2,2,4) \cup (2,4,2)]}{P[(2,2,4) \cup (2,4,2) \cup P(3,1,2,3) \cup \text{five similar points}]}$$

$$= \frac{\tfrac{2}{48}}{\tfrac{2}{48} + \tfrac{6}{96}}$$

$$= 0.4$$

Note: Our sample space would have been less cumbersome if we had used combinational type points, such as $(3,1,2,3)$, to mean 3 on the die and balls numbered 1, 2, and 3 being taken from the urn in any order. The probability of this point may be found as $\tfrac{1}{16}$. This notation would have reduced our sample space to 15 points and have been sufficient to handle parts (b) and (c) of the problem.

2.5 THE RELATIONSHIP BETWEEN DIFFERENT EVENT SPACES DESCRIBING THE SAME RANDOM PHENOMENON AND BAYES' THEOREM

This chapter will be concluded by material that utilizes strongly the ideas of independence and dependence. A general relationship will be found for expressing an event that is given on one event space E_1 on another event space E_2 defined for the same phenomenon. Practically, this will be a means to interpret the probability of what may be a directly conceptually difficult event in terms of its conditional probabilities on a simpler event space. The material will then culminate with the famous theorem of Bayes. The results derived here are all implied in Section 2.3, but the clarity of structure is very important.

Consider a random phenomenon for which two different event spaces, composed of mutually exclusive, collectively exhaustive points are defined. For simplicity we will consider the event space E_1 to be composed of four points

$$E_1 = \{A_1, A_2, A_3, A_4\}$$

where

$$\bigcup_{i=1}^{4} A_i = E_1 = S$$

and the event space E_2 to be composed of three points

$$E_2 = \{B_1, B_2, B_3\}$$

where

$$\bigcup_{i=1}^{3} B_i = E_2 = S$$

A Venn diagram of such a situation is shown in Figure 2.4. By inspection it is clear that for any specific event B_j,

$$B_j = B_j \cap S$$

and expressing S as

$$S = A_1 \cup A_2 \cup A_3 \cup A_4$$

we find that

$$B_j = B_j \cap (A_1 \cup A_2 \cup A_3 \cup A_4)$$
$$= (B_j \cap A_1) \cup (B_j \cap A_2) \cup (B_j \cap A_3) \cup (B_j \cap A_4)$$

More compactly, we can say that

$$B_j = \bigcup_{\text{all } i} (B_j \cap A_i)$$

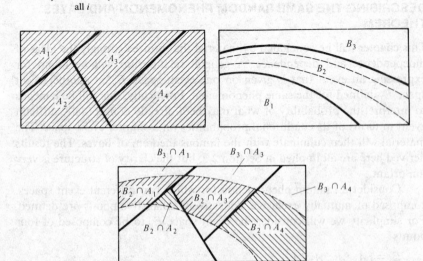

Figure 2.4   Two event spaces defined on the same S.

and, similarly,

$$A_j = \bigcup_{\text{all } i} (A_j \cap B_i) \qquad \text{for any } j$$

For example, if $j = 2$, then B_2 may be expressed as the union of $(B_2 \cap A_1)$, $(B_2 \cap A_2)$, $(B_2 \cap A_3)$, and $(B_2 \cap A_4)$, which are mutually exclusive events, as shown in Figure 2.4 and

$$P(B_2) = P(B_2 \cap A_1) + P(B_2 \cap A_2) + P(B_2 \cap A_3) + P(B_2 \cap A_4)$$

or

$$P(B_2) = \sum_{\text{all } i} P(B_2 \cap A_i)$$

If B_2 is a very difficult event to think about, but A_1, A_2, A_3, and A_4 and the conditional events (B_2/A_i) for all i are clearly defined, then

$$P(B_2 \cap A_i) = P(A_i)P(B_2/A_i)$$

and $P(B_2)$ is

$$P(B_2) = \sum_{\text{all } i} \left[P(A_i)P(B_2/A_i) \right]$$

In general, for any B_j,

$$P(B_j) = \sum_{\text{all } i} P(A_i)P(B_j/A_i) \qquad (2.17)$$

Equation 2.17 is used whenever its evaluation is easier than finding $P(B_j)$ directly. Also, conditional probability $P(A_k/B_j)$ may be very difficult to find directly as

$$P(A_k/B_j) = \frac{P(A_k \cap B_j)}{P(B_j)}$$

and so it may alternatively be expressed as

$$P(A_k/B_j) = \frac{P(A_k)P(B_j/A_k)}{\displaystyle\sum_{\text{all } i} P(A_i)P(B_j/A_i)} \qquad (2.18)$$

This philosophically famous and exciting theorem, called **Bayes' theorem**, has a stormy history including religious implications concerning predestination. The theorems of this section will now be made more meaningful by working some examples.

Example 2.9

Consider the problem of choosing a ball at random from one of three urns, where urn I has 6 red and 4 blue balls, urn II has 5 red and 3 blue balls, and urn III has 9 red and 1 blue ball.

Define the event space E_1 as the event of choosing a ball from urn I or the event of a ball from urn II or the event of a ball from urn III. Define the event space E_2 as the event of choosing a red ball or the event of choosing a blue ball.

    (a) Use the preceding theory to find the probability that a ball drawn at random is red.

    (b) Find the probability that a ball had been drawn from urn I, given the fact that a red ball had been selected.

SOLUTION

    (a) Let $E_1 = \{(I), (II), (III)\}$ and $E_2 = \{(R), (B)\}$, where the meaning of the points is self-evident. On consideration we see that E_1 is an easy event space, with each point having a probability of $\frac{1}{3}$, whereas E_2 is a difficult event space to visualize. Expressing (R) as $(R) \cap E_1$ yields

$$(R) = \{[(R) \cap (I)] \cup [(R) \cap (II)] \cup [(R) \cap (III)]\}$$

and since this is the union of three mutually exclusive points,

$$P(R) = P(I)P(R/I) + P(II)P(R/II) + P(III)P(R/III)$$

$$= \tfrac{1}{3}\left(\tfrac{6}{10} + \tfrac{5}{8} + \tfrac{9}{10}\right)$$

$$= \tfrac{17}{24} \qquad \text{(utilizing Eq. 2.17)}$$

    (b) For this problem,

$$P(I/R) = \frac{P(I \cap R)}{P(R)}$$

which, by Bayes' theorem is

$$P(I/R) = \frac{\tfrac{1}{3} \times \tfrac{6}{10}}{\left(\tfrac{1}{3} \times \tfrac{6}{10}\right) + \left(\tfrac{1}{3} \times \tfrac{5}{8}\right) + \left(\tfrac{1}{3} \times \tfrac{9}{10}\right)}$$

$$= \tfrac{24}{58}$$

On reflection we can see that the foregoing example could also have been solved by the techniques of the previous section, by defining

$$S = \{(I,R), (II,R), (III,R), (I,B), (II,B)\ (III,B)\}$$

and relating all events to S. In many cases the relating of different event spaces is much more rapid than solving in terms of the fundamental sample space, although such would not have been the case for this problem.

For our next example let us consider the application of 2.17 and 2.18 to the problem of sampling a time waveform, which has different analytical expressions over different ranges.

Example 2.10

Consider the waveform shown in Figure 2.5. Define the event space E_1 as consisting of the event A_1 of sampling the waveform in the range $0<t<2$, or the event A_2 of sampling the waveform in the range $2<t<3$. Also define the event space E_2 as consisting of the event B_1 of obtaining a waveform value less than 0.6 or the event B_2 of obtaining a waveform value greater than 0.6. Use the formulas of this section to evaluate the following:

(a) The probability that a waveform value randomly sampled between $t=0$ and $t=3$ is less than 0.6.
(b) If the waveform value is less than 0.6, the probability that it had been sampled in the range $0<t<2$.

SOLUTION
Consider the two event spaces

$$E_1 = \{A_1 = (0<t<2), A_2 = (2<t<3)\}$$
$$E_2 = \{B_1 = (x(t)<0.6), B_2 = (x(t)>0.6)\}$$

E_1 is a simple event space and

$$P(A_1) = \tfrac{2}{3} \quad \text{and} \quad P(A_2) = \tfrac{1}{3}$$

The event space E_2 is difficult to assign a probability measure to directly, but its conditional probabilities, knowing that an event from E_1 has

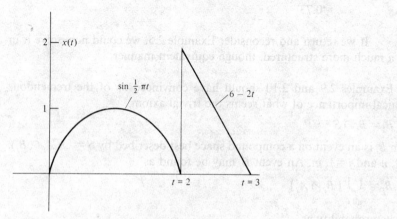

Figure 2.5   The waveform of Example 2.10.

occurred, are easy to evaluate.

(a) $P(x(t)<0.6)=P(B_1)$. Here B_1 may be written as

$$B_1 = B_1 \cap E_1$$

$$=(B_1 \cap A_1) \cup (B_1 \cap A_2)$$

Using Eq. (2.17),

$$P(B_1)=P(A_1)P(B_1/A_1)+P(A_2)P(B_1/A_2)$$

Observing Figure 2.5,

$$P(B_1)=\frac{2}{3}\times\frac{2\sin^{-1}0.6}{180°}+\frac{1}{3}\left(\frac{0.6}{2}\right)=0.37$$

(b) The difficult question—"Given that a value sampled is less than 0.6, what is the probability that it came from the region $0<t<2$?" —is answered by Bayes' theorem.

$$P[(0<t<2)/(x(t)<0.6)]$$

$$=\frac{P(A_1 \cap B_1)}{P(B_1)}$$

$$=\frac{P(A_1)P(B_1/A_1)}{P(A_1)P(B_1/A_1)+P(A_2)P(B_1/A_2)}$$

$$=\frac{0.67[2\times(36.9/180)]}{0.67[2\times(36.9/180)+0.33(0.6/2)]}$$

$$=0.73$$

If we return and reconsider Example 2.5, we could now solve it in a much more structured, though equivalent manner.

Examples 2.9 and 2.10 should have convinced us of the tremendous practical importance of what seems the trivial axiom,

$$B_i = B_i \cap S$$

when B_i is an event on a compound space best described by $S = \cup(A_i \cap B_j)$, $i=1, n$ and $j=1, m$. An event B_j may be found as

$$B_j = \bigcup_{\text{all } i} (B_j \cap A_i)$$

and its probability as

$$P(B_j)=\sum_{\text{all } i} P(A_i)P(B_j/A_i)$$

This is a powerful formula when B_j is difficult to envision directly but $P(A_i)$ and $P(B_j/A_i)$ are straightforward. In these cases we use Eq. 2.17. Similarly, we can visualize the usefulness of Bayes' theorem for finding $P(A_i/B_j)$. With such complicated events the axiomatic definitions are always used.

DRILL SET: RELATIONSHIP BETWEEN DIFFERENT EVENT SPACES

Consider that one of three levels denoted A_1, A_2, and A_3, is transmitted in a three-level digital communication system. Let the three corresponding signal levels at the receiver be B_1, B_2, and B_3. Assume that the a priori (before the fact) probabilities are $P(A_1)=0.5$, $P(A_2)=0.3$, and $P(A_3)=0.2$ and that the transition probabilities due to noise are as shown in the transition tree diagram. For example $P(B_1/A_1)$ denotes the probability that B_1 is received when A_1 is sent is 0.7.

(a) Sketch the following event spaces on a Venn diagram and assign probabilities to them:

$$A \triangleq \{A_1, A_2, A_3\}$$

$$B \triangleq \{B_1, B_2, B_3\}$$

$$C \triangleq \{A_i B_j: \quad i=1,3; j=1,3\}$$

where $A_i B_j$ is the event wherein A_i is sent and B_j received.

(b) Form the tree diagram for the a posteriori (after the fact) probabilities $P(A_i/B_j)$, for all i and j, as in Figure 2.6.

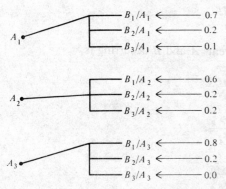

Figure 2.6   Tree diagram for transition probabilities.

SUMMARY

Answering questions about a random phenomenon involves the previously enumerated three stages:

1. Deciding on and using counting theory and set theory notation to list the points of a sample description or an appropriate event space
2. Assigning a probability measure to the events defined by the points of the space
3. Expressing desired unconditional and conditional events in terms of the space and assigning probabilities to them

Chapter 1 concluded with the acquired ability to carry out stage 1 and to understand the necessary set theory, to express different events in terms of a chosen space. In Chapter 2 we filled in the pieces required, in order to answer probabilistic questions. The relative-frequency definitions of probability are normally used to assign a probability measure to a space, and the axiomatic definitions are invoked to find desired probabilities of events that are related to the space of stage 1.

Using the relative-frequency interpretation three very important probabilities were defined as:

1. The probability of an event A is

$$P(A) = \lim_{N \to \infty} \frac{N_A}{N}$$

2. The conditional probability of an event A given that B has occurred is

$$P(A/B) = \lim_{N_B \to \infty} \frac{N_{A \cap B}}{N_B}$$

while from symmetry,

$$P(B/A) = \lim_{N \to \infty} \frac{N_{A \cap B}}{N_A}$$

3. The probability of two events A and B occurring is

$$P(A \cap B) = \lim_{N \to \infty} \frac{N_{A \cap B}}{N}$$

$$= \frac{N_A}{N} \cdot \frac{N_{A \cap B}}{N_A}$$

$$= \frac{N_B}{N} \cdot \frac{N_{A \cap B}}{N_B}$$

In the formulas for $P(A)$, $P(A/B)$, $P(B/A)$, and $P(A \cap B)$, N_A denotes the number of trials for which event A occurs, N_B the number of trials for which event B occurs, and $N_{A \cap B}$ the number of trials for which both events A and B occur, based on a large number of trials N of a phenomenon. The physical appreciation of these definitions and their use in assigning a probability measure to the outcomes of an event space is vital. If the outcomes of an event space are simple, we assign a probability to each point by visualizing N_A/N. If the points are sequential in nature, then for example $P(A_1 \cap A_2)$ may be found using $N_{A_1 \cap A_2}/N$, or by sequentially visualizing N_{A_1}/N and $N_{A_1 \cap A_2}/N_{A_1}$.

Stage 3 of any problem is to determine the probabilities of events and conditional events of interest. The structure of set theory is very important at this stage, as is the axiomatic version of the formula $P(A/B)$,

$$P(A/B) = \frac{P(A \cap B)}{P(B)} = \frac{P(C)}{P(B)}$$

This axiomatic definition of conditional probability states that if we have an event space

$$E = \bigcup_{i=1}^{n} E_i$$

where the E_i's are mutually exclusive, then the probability of any conditional event of interest (A/B) may be found. The numerator C is expressed by the axiom $C = C \cap S$ to yield some points of E, and hence it has a probability equal to the sum of the probabilities of these points. The denominator event B may similarly be interpreted. Of course, for any unconditional event its probability is found by using the axiom $A \cap S = A$ to express it as the union of points of E and hence using axiom 3 of probability theory. The use of $P(A/B)$ implies that an intelligent choice of E was made in stage 1, so $(A \cap B) \cap E$ and $B \cap E$ result in points of E. Basically a probabilistic question involves just the use of these relative-frequency and axiomatic formulas. All the philosophy of discrete probability theory are embodied in them.

Section 2.5 developed formulas for assigning a probability to a complex event and a complex conditional event defined on a compound event space. The relationship between the probabilities of two event spaces

$$E_1 = \bigcup_{\text{all } i} A_i \quad \text{and} \quad E_2 = \bigcup_{\text{all } i} B_i$$

where the events of E_2 are difficult to visualize, but those of E_1 are simple, is

$$P(B_j) = \sum_{i=1}^{n} P(A_i) P(B_j/A_i)$$

and Bayes' theorem yields

$$P(A_K/B_j) = \frac{P(A_K)P(B_j/A_K)}{\sum\limits_{i=1}^{n} P(A_i)P(B_j A_i)}$$

All the necessary material of discrete probability theory has now been concluded, but it would be foolhardy to feel at all expert. Confidence and comfort with the subject requires anywhere from 100 to 200 hours of problem solving. The next chapter will devote itself to problem solving and to the topics of fairness and information content, which are closely related to probability theory.

The problems in Chapter 2 are deliberately fairly easy and basic. We must be philosophically very secure and comfortable with the definitions. The framework has been established and, although we will proceed to more difficult problems and applications on and involving discrete probability theory, we must realize that there are only a few formulas governing the whole field.

PROBLEMS

Note: These problems are somewhat simple and fundamental. Since Chapter 3 is devoted to problem solving, it contains more challenging exercises.

1. Use the fundamental definition of

$$P(A) = \lim_{N \to \infty} \frac{N_A}{N}$$

to find the probabilities of the following events and in each case state clearly the infinite number of experiments N you visualize with the phenomenon and why your numerator and denominator correspond to N_A and N, respectively. (This is an important problem and we are more interested in philosophically understanding and verbalizing why our answer corresponds to N_A/N for a large N than in throwing out the correct answers without too much thought.)
   (a) Exactly 3 heads on 6 coin tosses.
   (b) More zeros than ones in a computer word of 9 bits.
   (c) Exactly 3 white balls in a sample of size 5 from an urn containing 4 white, 5 black, and 2 red balls, where a ball is replaced in the urn after it is drawn.
   (d) Exactly 3 white balls in a sample of size 5 from an urn containing 4 white, 5 black, and 2 red balls, where there is no replacement between withdrawals or the 5 balls are withdrawn at once.
2. Redo as many parts of Problem 1 as possible by defining random phenomena with equally likely outcomes. Decide on an event space E that involves the question asked and obtain the desired event as the

union of mutually exclusive points of E. Be structured and follow the format of writing out stage 1, stage 2, and stage 3 as followed in the text.

3. Use the fundamental definition of conditional probability

$$P(A/B) = \lim_{N_B \to \infty} \frac{N_{A \cap B}}{N_B}$$

to find the probabilities of the following events, and in each case state clearly the infinite number of experiments N_B you associate with the phenomenon and why your numerator and denominator correspond to $N_{A \cap B}$ and N_B, respectively.

(a) The event of 3 heads on 6 coin tosses given that the first 2 tosses are heads.

(b) The event of 3 heads on 6 coin tosses given that at least 2 tosses resulted in heads.

(c) The event of more zeros than ones in a computer word of 9 bits given that the first and last bits are zeros.

(d) The event of exactly 3 white balls on a sample of size 5 from an urn containing 4 white, 5 black, and 2 red balls, given that the sample contains at least 1 white, 1 black, and 1 red ball.

4. Redo Problem 3 by defining random phenomena that occur subject to conditions with equally likely outcomes. Decide on an event space E that involves the questions asked and obtain the desired event as the union of mutually exclusive points of the space.

5. Use the formulas for dependent and independent events

$$P(A_1 \cap A_2 \cap A_3 \cdots \cap A_n)$$
$$= P(A_1)P(A_2/A_1) \cdots P(A_n/A_1 \cap A_2 \cap \cdots \cap A_{n-1})$$

or

$$P(A_1 \cap A_2 \cdots \cap A_n) = P(A_1)P(A_2) \cdots P(A_n)$$

plus expressing an event as the union of mutually exclusive events to evaluate the probabilities of the following events:

(a) A team A wins a five-game play-off series in four games assuming that the probability of winning any specific game is 0.6.

(b) Two sampled values spaced $\pi/2$ S apart from the waveform $\sin t$ are both negative.

6. If $P(A) = p$, $P(A/B) = q$ and $P(B/A) = r$ what relations between the numbers p, q, and r will hold for the following cases? (For example, is $p \geq q$, or is $p + q$ or $p + r = 1$, and so on?)

(a) The events A and B are mutually exclusive.

(b) The events A and B are mutually exclusive and collectively exhaustive.

(c) The event A is a subevent of B.

(d) The event B is a subevent of A.

(e) The events $\bar{A}$ and $\bar{B}$ are mutually exclusive.

7. Consider A and B, defined in general on some space of a phenomenon. If possible, state relations between A and B such that
   (a) $P(A/B) > P(A)$.
   (b) $P(A/B) < P(A)$.
   (c) $P(A/B) = P(A)$.

8. Consider a best-of-five-games playoff series between teams A and B, where it is assumed that for any game the probability of A winning is $\frac{3}{5}$ and the probability of B winning is $\frac{2}{5}$. Let us consider the following event spaces E_1 and E_2. E_1 is defined as the different game by game won-loss conclusions. For example one point is, "A wins the first two games, loses the third and then wins the fourth game." We call this $AABA$. E_2 is defined as the event the series ends with A winning in three or four or five games, or by B winning in three or four or five games.
   (a) Find the probability measure for E_1 and hence find it also for E_2.
   (b) Find the probability A wins, given that the series lasts at least four games. Do this using both spaces E_1 and E_2.
   (c) Find the probability the series lasts exactly four games given that A will win.

9. Consider a general periodic waveform

$$g(t) = \sum_{-\infty}^{\infty} x(t - nT)$$

as is shown sketched in Figure 2.7. Let us refer to $x(t)$ as "period 1," $x(t - T)$ as "period 2," and so on. We now consider the phenomenon A of sampling the waveform value at some time $0 < t < T$, where all times are equally likely, and phenomenon B of sampling the waveform τ s later.
   (a) What is the probability that both values are sampled from the first period for $\tau = T/4, T/2, 3T/4$, and a general value τ, $0 < \tau < T$?
   (b) What is the probability that both values are sampled from different periods—that is, that the second value comes from the waveform $x(t - T)$, for $\tau = T/4, T/2, 3T/4$, and a general τ, $0 < \tau < T$?
   (c) Graph your results to parts (a) and (b) as a function of τ, $0 < \tau < T$. *Note*: This is a very important problem, which is 90% language, and for later applications we should be very fluent with these ideas.

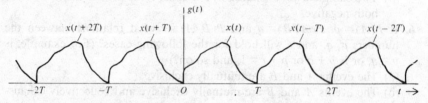

Figure 2.7   A general periodic waveform.

Chapter 3
General Formulation and Solution of Problems and Topics Related to Probability Theory

3.1 SOLVED PROBLEMS WITH COMMENTS

This section involves no new theory or concepts that were not developed in Chapter 2, but will consist of solving a smorgasbord of problems with the following purposes:

1. Developing problem-solving techniques and cementing the concepts and structure of the previous chapters.
2. Showing extensions of discrete probability theory to new situations, such as continuous-valued phenomena and phenomena with a countable infinity of points.
3. Meeting new nomenclature that will be used later when describing noise waveforms or random processes.

The author does not believe that one learns from reading someone else's solutions to problems; however, deep intelligent perusal can pay great dividends. Besides, there is nothing more irritating to an undergraduate than a text that solves only a few simple plug-in problems and then bombards the reader with hundreds of extremely difficult problems at the end of each chapter.

The summary of Chapter 2 should be used as a guide, as we proceed with the solution of Examples 3.1 to 3.5. Each example will be solved in our three structured stages:

1. Set up an appropriate event space.
2. Assign a probability measure to the event space.
3. Relate desired events to the points of the event space and find their probabilities.

Now let us proceed.

Example 3.1

Consider the experiment of generating random numbers (with an infinite number of decimal digits) between 0 and 1.

(a) Set up an event space describing the possible outcomes.
(b) What is the probability of generating a number between 0.412 and 0.634?
(c) Given that a number greater than 0.12 was generated, what is the probability that it was less than 0.84?

SOLUTION

(a) The sample description space in this case would consist of an uncountable infinite number of points,

$$S = \{x: \ 0 \leq x \leq 1\}$$

and the probability of any point is 0. There is only probabilistic meaning to taking on a value in a range of x. Therefore, an event space is defined that consists of outcomes, each defining the random number falling in a range. For example, we can consider an event space E with three generated ranges. Let $E = \{E_1, E_2, E_3\}$, where E_1 is the event of choosing a number x where $0 < x \leq a$, E_2 is the event of choosing a number x where $a < x \leq b$, and E_3 is the event of choosing a number x where $b < x \leq 1$.

Since all numbers between 0 and 1 are assumed equally likely, then it should be clear the probabilities of these ranges are $P(E_1) = a/1.0 = a$, where the trivial factor 1.0 is written in the denominator because if a random number was generated between 0 and L, then it would have had a probability $P(E_1) = a/L$.

$$P(E_2) = \frac{b-a}{1.0} = b - a$$

and

$$P(E_3) = \frac{1-b}{1.0} = 1-b$$

where $1 > b > a > 0$.

(b) If we let $a = 0.412$ and $b = 0.634$, then the event that the number lies between 0.412 and 0.634 is E_2. In this case,

$$P(E_2) = 0.634 - 0.412$$

$$= 0.222$$

(c) We require the probability of the conditional event that a number is less than 0.84 given that it is greater than 0.12. That is,

$$P[(x < 0.84)/(x > 0.12)] = \frac{P(0.12 < x < 0.84)}{P(x > 0.12)}$$

$$= \frac{0.72}{0.88}$$

$$= 0.82$$

In this problem we have succeeded in handling an infinite number of continuous points by attaching probabilities to ranges.

Example 3.2

Consider the experiment of sampling the waveform $x(t)$ shown in Figure 3.1, where a value of time is uniformly chosen as $0 < t < 3$ and then the

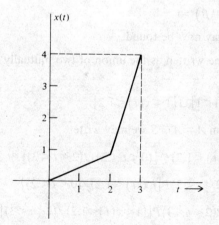

Figure 3.1   The waveform of Example 3.2.

waveform value $x(t)$ is noted. Evaluate the probabilities of the following events:

(a) The value of $x(t)$ is less than 1.2.
(b) The value of $x(t)$ is greater than 1.2 given that it is at least 0.4.

SOLUTION

Considering the questions asked, the phenomenon of sampling $x(t)$ will be described by the event space

$$E = \{(x_1, y_1), (x_1, y_2), (x_2, y_3), (x_2, y_4)\}$$

where

x_1 is the event that a point is chosen on the time axis $0 < t < 2$
x_2 is the event that a point is chosen on the time axis $2 < t < 3$
y_1 is the event that $0 < x(t) < \alpha$, for $0 < \alpha < 1$
y_2 is the event that $\alpha < x(t) < 1$
y_3 is the event that $1 < x(t) < \beta$
y_4 is the event that $\beta < x(t) < 4$

The compound point (x_1, y_1) describes the event $x_1 \cap y_1$, and similarly for all (x_i, y_j). The probabilities of the four compound points of our space may be found using conditional probability as:

$$P(x_1, y_1) = \frac{2\alpha}{3} \qquad P(x_1, y_2) = \frac{2}{3}(1 - \alpha)$$

$$P(x_2, y_3) = \frac{1}{3}\left(\frac{\beta - 1}{3}\right) \qquad P(x_2, y_4) = \frac{1}{3}\left(\frac{4 - \beta}{3}\right)$$

The sum of the probabilities of these four points is

$$\tfrac{2}{3}\alpha + (\tfrac{2}{3} - \tfrac{2}{3}\alpha) + (\tfrac{1}{9}\beta - \tfrac{1}{9}) + (\tfrac{4}{9} - \tfrac{1}{9}\beta) = 1$$

The solutions to parts (a) and (b) may now be found.

(a) The event $[x(t) < 1.2]$ may be written as the union of two mutually exclusive events:

$$[x(t) < 1.2] = [0 < x(t) < 1] \cup [1 < x(t) < 1.2]$$

Additionally using the axiom $A = A \cap S$ we may write

$$[x(t) < 1.2] = (x(t) < 1.2) \cap [(0 < t < 2) \cup (2 < t < 3)]$$

$$\therefore P[x(t) < 1.2)] = P(0 < t < 2)P[(x(t) < 1.2)/(0 < t < 2)]$$

$$+ P(2 < t < 3)P[(1 < x(t) < 1.2)/(2 < t < 3)]$$

$$= \frac{2}{3} \times 1 + \frac{1}{3}\left(\frac{1.2 - 1}{3}\right) = 0.67 + 0.02 = 0.69$$

(b) $P[(x(t)<1.2)/(x(t)>0.4)] = \dfrac{P[(x(t)<1.2) \cap (x(t)>0.4)]}{P[x(t)>0.4]}$

$= \dfrac{P[0.4<x(t)<1.2]}{P[x(t)>0.4]}$

Using the probability measure from our event space, this becomes

$$\frac{\frac{2}{3} \times 0.6 + \frac{1}{3} \times [(1.2-1)/3]}{\frac{2}{3} \times 0.6 + \frac{1}{3} \times 1} = \frac{0.42}{0.73} = 0.57$$

Example 3.3

Given a group of four people, construct an event space from which it is possible to find:

(a) The probability that exactly two have the same birth month.
(b) The probability that at least two have the same birth month.
(c) The probability that no two have the same birth month.

For simplicity, erroneously assume that the probability a person has a particular birth month is $\frac{1}{12}$.

SOLUTION
Our problem is a variation of the famous birthday problem posed in many classic probability texts, "How many persons must there be in a room so there is a 50% chance at least two have the same birthday?" The answer to this problem is 23.[1]

The event space for the problem is probably best described in words as

$$E = \begin{cases} \begin{array}{ccc} E_1 & E_2 & E_3 \\ \text{(all four} & \text{(exactly three} & \text{(exactly two have same} \\ \text{have same} & \text{have same} & \text{birth month and the} \\ \text{birth month),} & \text{birth month),} & \text{others have two different} \\ & & \text{birth months),} \end{array} \end{cases}$$

$$\begin{array}{cc} E_4 & E_5 \\ \text{(two have one birth month} & \text{(all have} \\ \text{and the other two} & \text{different} \\ \text{have another birth month),} & \text{birth months)} \end{array}$$

Looking at these events, an honest reader should concede how easy it might have been to omit event E_4. Consequently, checking the sum of the

[1] E. Parzen, *Modern Probability Theory and Its Applications* (New York: Wiley, 1960), p.46.

probabilities is very important when the description of the points of E is difficult.

All of these points are mapped from a sample description space S, where

$$S = \{(a, b, c, d); \text{ where any of the letters may be one of the 12 months}\}$$

and a refers to the birth month of one specific person, b to the birth month of another specific person, and similarly for c and d.

There are 12^4 points in S, each with a probability of $1/12^4$, and the probabilities of the points of E will be found by building on the points of S.

Point E_1

The probability of the event E_1 is found in stages as follows:

$$P\left(\begin{array}{llll} \text{1st person has} & \text{2d has} & \text{3d has} & \text{4th has} \\ \text{January as birth month,} & \text{January,} & \text{January,} & \text{January} \end{array}\right)$$

$$= \frac{1}{12^4}$$

$$\therefore P\left(\begin{array}{l} \text{all four have the} \\ \text{same birth month} \end{array}\right) = \frac{12}{12^4} = \frac{1}{12^3}$$

as there are 12 different months.

Point E_2

To find the probability that three people have the same birth month, we again start with a point from the sample space S:

$$P(\text{January, January, January, February}) = \frac{1}{12^4}$$

Next, the event that three people have January for a birth month and one February is the union of $\binom{4}{3}$ points of S.

$$\therefore P(\text{three have January and one February}) = \frac{\binom{4}{3}}{12^4}$$

We now consider the event three have one birth month and the other a different one and we can see that there are $\binom{12}{2}$ different sets of 2 months that can be formed from 12 and for each such set say (January, February), we could have (January, January, January, February) or (January, February, February, February).

$$\therefore P\left(\begin{array}{l} \text{three have one birth month and} \\ \text{the other a different one} \end{array}\right) = \frac{2\binom{12}{2}\binom{4}{3}}{12^4} = \frac{44}{12^3}$$

The reader should ponder this result and be sure that the event is the union of $12 \times 11 \times 4$ mutually exclusive points of S. With facility in counting this could be done in one step to yield

$$P(XXXY) = \binom{4}{3} \times 1 \times \frac{1}{12} \times \frac{1}{12} \times \frac{11}{12} = \frac{44}{12^3}$$

as before. Such rapid solutions should only be carried out where there is absolutely no doubt about their correctness.

Point E_3

To find the probability that exactly two have one birth month and the other two have two different birth months, we again start with a point of S.

$$P(\text{January, January, February, March}) = \frac{1}{12^4}$$

and proceeding more rapidly we find that

$$P\left(\begin{array}{l} \text{two have same birth} \\ \text{month and other two have} \\ \text{different ones} \end{array}\right) = \frac{3\binom{12}{3}\dfrac{4!}{2!1!1!}}{12^4}$$

$$= \frac{660}{12^4}$$

It is left as an exercise for the student to understand each factor in the numerator. Particularly the factor "3." If it is difficult, actually start counting for small cases.

Point E_4

To find the event that two have one birth month and the other two share a different birth month, we proceed with

$$P(\text{January, January, February, February}) = \frac{1}{12^4}$$

$$P(E_4) = \frac{\binom{12}{2}\dfrac{4!}{2!2!}}{12^4} = \frac{33}{12^3}$$

where it is hoped that the numerator factors are clear.

Point E_5

$$P\left(\begin{array}{l} \text{all have different} \\ \text{birth months} \end{array}\right) = \frac{\binom{12}{4}4!}{12^4} = \frac{990}{12^4}$$

As a check to our work we show

$$P(E_1) + P(E_2) + P(E_3) + P(E_4) + P(E_5) = \frac{1 + 44 + 660 + 33 + 990}{1728}$$

$$= 1$$

Considering our event space and its probability measure, we now answer all the required questions:

(a) $P\left(\begin{array}{l}\text{exactly two have}\\\text{same birth month}\end{array}\right) = P(E_3) = \frac{660}{1728} = 0.38$

(b) $P\left(\begin{array}{l}\text{at least two have}\\\text{same birth month}\end{array}\right) = \frac{1 + 44 + 660 + 33}{1728} = 0.43$

(c) $P\left(\begin{array}{l}\text{no two have same}\\\text{birth month}\end{array}\right) = P(E_5) = \frac{990}{1728} = 0.57$

This example was included because it is an excellent illustration of utilizing counting as developed in Section 1.3 of Chapter 1.

Example 3.4

Consider the problem of sampling parts from a large population until two good parts are obtained. Assume that 30% of the parts are defective.

(a) Set up the sample description space and assign a probability measure.
(b) Given that the second part sampled is defective, what is the probability of obtaining at least two defective parts before the second good one?

SOLUTION

(a) The sample description space is

$$S = \{(G,G), (G,D,G),\ldots,(G,D,\ldots,D,G),$$

$$(D,G,G), (D,G,D,G),\ldots,(D,G,D,\ldots,D,G),$$

$$(D,D,G,G), (D,D,G,D,G),\ldots,(D,D,G,D,\ldots,D,G),\ldots\}$$

This is a strange sample space because there are an infinite number of points for which the first part sampled is good, denoted by G; or when the first is defective, denoted by D, and the second part good, again we have an infinite number of points; or when the first two sampled are defective followed by a good, again we have an infinite number of points; and so on. If we proceed and start assigning probabilities assuming $P(G) = 0.7$, $P(D) = 0.3$, and also because of the large population stipulation we can assume independence or that $P(G/D)=P(G/G)=0.7$ and $P(D/D)=P(D/G)=0.3$, we obtain the following probabilities for the points of S:

$$P(G,G)=(0.7)^2$$

$$P(G,D,G)=(0.7)^2 0.3, \cdots, P\left(G,D \overset{n}{\cdots} D,G\right)=(0.7)^2 0.3^n$$

Therefore, the probability a G occurs first and last or the union of the points of the first row of our space is,

$$P(G \cdots G)=0.7^2+(0.7^2 \times 0.3)+ \cdots +(0.7^2 \times 0.3^n)+ \cdots$$

$$=0.7^2 \sum_{n=0}^{\infty} 0.3^n$$

$$=0.7^2 \left(\frac{1}{1-0.3}\right)$$

$$=0.7$$

using the result for a geometric progression. The probability that a defective part is sampled and the second part is good or the probability of the union of the points of the second row of our space is

$$P(DG \cdots G)=(0.3 \times 0.7^2)+(0.3^2 \times 0.7^2)+ \cdots +(0.3^n \times 0.7^2)$$
$$+ \cdots$$

$$=0.7^2 \sum_{n=1}^{\infty} 0.3^n$$

$$=0.3 \times 0.7$$

$$=0.21$$

Similarly, we can show the probability that m defective parts are sampled before a good part is sampled or the probability of the union of the points of the mth row of our space is

$$P(mD\text{'s},G \cdots G)=(0.3^m \times 0.7^2)+(0.3^{m+1} \times 0.7^2)+ \cdots$$

$$=0.7^2 \sum_{n=m}^{\infty} 0.3^n$$

$$=(0.3)^m \times 0.7 \qquad \text{for any } m=0,\infty$$

As a check we can see the sum of the probabilities for the rows of our space is

$$\sum_{m=0}^{\infty} (0.3)^m \times 0.7 = 0.7 \left(\frac{1}{1-0.3} \right)$$

$$= 1$$

(b) We require the probability of the conditional event of obtaining two defective parts before the second good one, given that the second part is defective. Using the notation A and B for the two events we have,

$$P(A/B) = P \frac{\left(\begin{array}{l} \text{two defective parts} \\ \text{occur before 2d good} \end{array} \cap \begin{array}{l} \text{2d part is} \\ \text{defective} \end{array} \right)}{P(\text{2d part is defective})}$$

$$P(A \cap B) = P[(G, D, D, G) \cup \cdots \cup (G, D, \cdots, D, G)$$

$$\cup (D, D, G, G) \cup \cdots \cup (D, D, G, D, \cdots, D, G)$$

$$\cup (D, D, D, G, G) \cdots \cup (D, D, D, G, \cdots, G), \text{etc.}]$$

$$= 0.7 \times 0.3^2 + 0.7 \times 0.3^2 (1 + 0.3 + 0.3^2 + \cdots)$$

$$= 0.7 \times 0.3^2 + 0.3^2$$

$$\therefore P(A/B) = \frac{0.7 \times 0.3^2 + 0.3^2}{0.3}$$

$$= 0.51$$

This example illustrates the case of a multicountable infinite number of points and still shows how the sum of all the probabilities converges to 1.

Finally we will present one more example, which is as much in preparation for notation prevalent in representing certain noise or random process waveforms as it is a challenge.

Example 3.5

Consider the idealized diode shown in Figure 3.2a, whose output i_o is defined as

$$i_o = i_{in} \qquad \text{if } i_{in} > 0$$

$$= 0 \qquad \text{if } i_{in} < 0$$

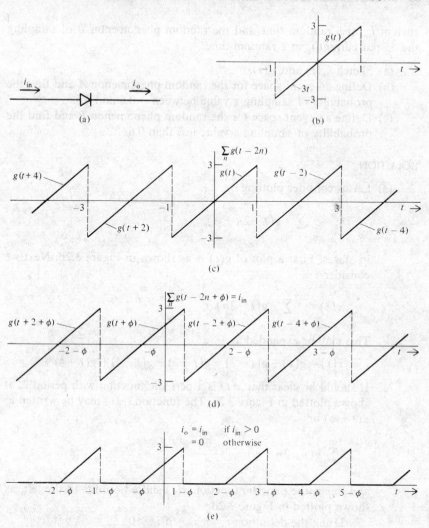

Figure 3.2   An ideal diode with its input and output waveforms for Example 3.5.

Given

$$i_{\text{in}}(t) = \sum_{n=-\infty}^{\infty} g(t - 2n + \phi)$$

where $g(t)$ is the deterministic function

$$g(t) = 3t \qquad \text{for } -1 < t < 1$$
$$= 0 \qquad \text{otherwise}$$

and ϕ is a uniform phase shift or any number chosen uniformly from $-1 < \phi < +1$, define the random phenomenon A of sampling the input current i_{in} at a random time and the random phenomenon B of sampling the output current i_o at a random time.

(a) Sketch $i_{in}(t)$ and $i_o(t)$.
(b) Define an event space for the random phenomenon A and find the probability of sampling a value between -0.6 and 0.5.
(c) Define an event space for the random phenomenon B and find the probability of sampling a value less than 0.6.

SOLUTION

(a) Let us consider plotting

$$i_{in} = \sum_{n=-\infty}^{\infty} g(t-2n+\phi)$$

in stages. First a plot of $g(t)$ is as shown in Figure 3.2b. Next we consider

$$x(t) = \sum_{n=-\infty}^{\infty} g(t-2n)$$

This may be expanded as

$$x(t) = g(t) + g(t-2) + g(t+2) + g(t-4) + g(t+4) + \cdots$$

It should be clear that $x(t)$ is a periodic function with period 2, as shown plotted in Figure 3.2c. The function $i_{in}(t)$ may be written as $x(t+\phi)$ or

$$\sum_{-\infty}^{\infty} g(t-2n+\phi)$$

and this is the periodic function $x(t)$ shifted ϕ units to the left, as shown plotted in Figure 3.2d.
   Using the definition $i_o = i_{in}$ if $i_{in} > 0$
   $\qquad\qquad\qquad\quad = 0$ if $i_{in} < 0$
the plot of $i_o(t)$ is as shown in Figure 3.2e.
(b) The random phenomenon A is defined as sampling the time waveform $i_{in}(t)$ at a random time. A trial of the phenomenon consists of uniformly choosing a time between $-\infty$ and $+\infty$ or, since we are dealing with a periodic waveform, choosing a time between $-1 < t < +1$, and the value for the trial is $i_{in}(t)$.

   From symmetry we can see from Figure 3.2 that all waveform values between -3 and $+3$ are equally likely, and we probably feel fortunate that the waveform was not sinusoidal or other than linear. The sample description space for the input values is $S = \{x, -3 < x < +3\}$, where all points are equally likely but have a probability of 0. To use discrete probability theory, we define the event space,

$$E = \{(-3 < x < a), (a < x < b), (b < x < 3)\}$$

where $-3 < a < b < 3$. The probability measure is

$$P(-3 < x < a) = \frac{a+3}{6} \qquad P(a < x < b) = \frac{b-a}{6}$$

$$P(b < x < 3) = \frac{3-b}{6}$$

We should check that the sum of the probabilities add to 1. Now

$$P(-0.6 < x < 0.5) = \frac{0.5 + 0.6}{6} = 0.18$$

On consideration of the waveform for i_o, we first note that the value $i_o = 0$ occurs half the time, while for the rest of the time any of the continuous values $0 < x < 3$ may occur. Therefore the sample description space for the random phenomenon B is

$$S = \{0, 0 < x < 3\}$$

where $P(0) = \frac{1}{2}$ and all the points $0 < x < 3$ are equally likely with the probability 0 since there is only probabilistic meaning to taking on a value in a range.

To use discrete probability theory, we could define the event space

$$E = \{(0), (0 < x < a), (a < x < b), (b < c < 3)\}$$

where $0 < a < b < 3$. The probability measure is $P(0) = \frac{1}{2}$, and the other probabilities may be found using the formula

$$P(A \cap B) = P(A)P(B/A)$$

to yield

$$P(0 < x < a) = \frac{1}{2}\left(\frac{a}{3}\right) \qquad P(a < x < b) = \frac{1}{2}\left(\frac{b-a}{3}\right)$$

$$P(b < x < 3) = \frac{1}{2}\left(\frac{3-b}{3}\right)$$

We can see that the probabilities of the four points of E add to 1:

$$\frac{1}{2} + \frac{a}{6} + \frac{b-a}{6} + \frac{3-b}{6} = \frac{3+a+b-a+3-b}{6} = 1$$

The derivation of these probabilities and the choice of our event space are worth detailed thought. Now $P(-0.6 < X < 0.5)$ is the union of mutually exclusive events,

$$(-0.6 < X < 0.5) = (X=0) \cup (0 < X < 0.5)$$

and

$$P(-0.6 < X < 0.5) = \frac{1}{2} + \frac{1}{2}\left(\frac{0.5}{3}\right) = 0.58$$

The purpose of this example was to introduce the system theory notation for a periodic waveform, to hint at the representation of a member of a random process, and to prepare for the type of discrete probability theory thinking with which we will be involved when analyzing noise waveforms in Section 3.3 of the text. However on removing all the camouflage, a fairly reasonable problem in discrete probability theory remained.

The preceding examples, along with those in Chapter 2, give the reader a demonstration of the techniques involved when solving questions involving random phenomena. We are emphasizing the structure of problem solving, which is what we must intrinsically understand later when we are involved with random variable theory and noise waveforms of random processes. Students in communication theory, for example, may rarely solve a direct problem in discrete probability theory, but its structure is the basis of everything they do. This is somewhat akin to an engineering student's reliance on calculus, or a calculus student's reliance on algebra, or an algebra student's reliance on handling numbers. If fundamentals are not understood we can only proceed on the results of constant exposure, regurgitation, and memorization. With dedicated problem solving, it is easily achievable to become quite expert by normal standards in discrete probability theory.

This chapter concludes with two topics that involve probabilities but use different terminologies. In everyday life, one deals constantly with odds and the value of a "game," whether it involves business decisions, casino gambling, the stock market or whatever. For this reason Section 3.2 will first treat the topic of fairness in gambling and the value of a game and then treat the concept of information content and the use of bits to measure it. This is very important in the field of information theory or in the computer world, where logical decisions are carried out by sequences of binary decisions.

3.2 TOPICS CLOSELY RELATED TO PROBABILITY THEORY

Fairness in Gambling

A game of chance is said to be fair if it is possible to wager a sum of money on a trial of the game in such a way that the original sum is returned no

Table 3.1   A "FAIR" GAME

OUTCOMES	ASSIGNED PROBABILITIES	ODDS	WAGER	RETURN
A	$\frac{1}{2}$	$\frac{1}{1}$	$50	100
B	$\frac{1}{3}$	$\frac{2}{1}$	$33.33	100
C	$\frac{1}{6}$	$\frac{5}{1}$	$16.66	100

matter what outcome occurs. We will see that the condition for fairness is that the sum of the bank-assigned probabilities add up to 1. In gambling the chances of an outcome on a trial is normally prescribed in terms of odds. The relationship between odds and probabilities is simple. If the probability of an event is X/Y, then the odds are said to be $(Y - X)/X$ or $(Y - X)/X$ to 1. For example, in rolling a die the probability of the event of a 1 is $\frac{1}{6}$, and the odds are $\frac{5}{1}$ or "five to one." The probability of obtaining a 4 or smaller number is $\frac{2}{3}$, and the odds are $\frac{1}{2}$. If the odds on an event are P/Q, then for a bet of Q units the bettor wins P units, or receives $P + Q$ units back.

Consider a hypothetical game or random phenomenon with three outcomes—A, B, and C—which have the assigned probabilities or odds shown in Table 3.1. If a sum of money (say $100) is wagered on a trial of the game with an amount bet on each outcome in proportion to the probability of the outcome, then no matter what outcome occurs the bettor receives his or her money back. This is a *fair* game.

The philosophy of bookies or casinos is simple; they work out the real probabilities or odds associated with a game, and then assign higher probabilities or lower odds to each event. Table 3.2 shows a different set of assigned probabilities and odds for the game of Table 3.1. Now if the bettor was to bet a sum of money (say $120), in proportion to the probabilities, he or she would receive less than the original wager back (in this case $100). The percentage loss on a game may be found by adding the probabilities. For our game the probabilities add to 1.2, which means that the game on an average returns $100 for every $120 wagered, which means that the game is about 17% unfair.

When deciding on the overall unfairness of a game we should also consider the unfairness of each individual outcome. For outcomes A, B, and C in our game of Table 3.2, each one is exactly 17% unfair so no optimum strategy exists for playing the game. In general, if the true probability is P_T

Table 3.2   A CASINO VERSION OF THE GAME OF TABLE 3.1

OUTCOMES	PROBABILITIES	ODDS	WAGER	RETURN
A	0.60	$\frac{2}{3}$	$ 60	100
B	0.40	$\frac{3}{2}$	$ 40	100
C	0.20	$\frac{4}{1}$	$ 20	100
	1.2		$120	

and the assigned probability is P_A, then the percentage unfairness for that outcome is

$$\frac{P_A - P_T}{P_A} \times 100$$

Until the seventeenth century most recorded wagering was between gentlemen, and the games were always fair. With the original emergence of casinos, it was agreed by bettors and mathematicians that about a 3% profit to the casino was sufficient for providing facilities and acting as a bank. The traditional games of dice, black jack, and roulette (one zero) give about a 3% edge to the bank. However in the twentieth century three groups became interested in gambling as a business: businessmen, the underworld, and the government, plus their unions and intersections, and the result has become disastrously unfair for bettors. Profits of 30% are not uncommon on the new so-called fun games such as keno, while slot machines take up to 50% in profits. The concept of gambling is now different, since one of the participants realistically cannot lose and the other cannot win—which from a moral point of view negates its purpose. In addition to games of chance, many types of insurance such as insurance on small personal loans or on furnishings are about 60 to 90% unfair to the insuree. This traditional change in the concept of fairness and fair play is probably a reflection of the morality of our times.

We will now consider a few modern real-life examples.

Example 3.6

Consider the game of football parlay, based on college and professional games, which is played legally in Nevada and illegally in most other places each football weekend. The casino or bookie odds from the front of a typical card are shown in Table 3.3. On the back of each card is always listed the point spread for more than 90 games. For each listed set of odds, evaluate the unfairness of the game.

SOLUTION
The term "3 teams, 5 to 1" means that for any specific set of three games the casino odds for picking the three winning teams is $\frac{5}{1}$. Of course if Notre Dame is playing a team like San Jose State, the point spread may be as much as 30 points. This means that Notre Dame wins from a betting point of view if the team wins the game by more than 30 points. San Jose wins if its score is less than 30 points behind Notre Dame's. The game does not count if Notre Dame wins by exactly 30 points.

Table 3.4 shows the assigned casino probabilities, the actual probabilities and the expected casino profit on each possible wager. For example, the probability of correctly forecasting four games is $\frac{1}{16}$. The assigned casino probability however is only $\frac{1}{11}$. Therefore on an average for every 16 people

Table 3.3   ODDS FROM A FOOTBALL PARLAY CARD*

3 teams	5 to 1
4 teams	10 to 1
5 teams	15 to 1
6 teams	25 to 1
7 teams	35 to 1
8 teams	60 to 1
9 teams	80 to 1
10 teams	100 to 1

*Minimum $2; tie considered on game; 3 teams with a tie, 7 to 5.

who bet, say $1, one wins and is paid $11. This means the casino makes $5 for every $16 wagered or a profit of 31.3%. Another way of arriving at this number is that for this line there are 16 different outcomes, each with an assigned probability of $\frac{1}{11}$ and the sum of the probabilities is $\frac{16}{11}$. Observing each line it is alarming how unfair the odds become as the number of teams in a parlay increases. It is hoped that engineering students will avoid this kind of gambling—not necessarily for religious reasons but as a result of calculating the unfairness of a game.

Example 3.7

One of the most widely played casino games is keno. Figure 3.3 shows a typical casino card and Figure 3.4 shows the payoffs associated with some of the different ways in which the game may be played. Keno represents a typical twentieth-century casino version of a betting game. Analyze the unfairness of the game for a few situations.

SOLUTION

First, the game of keno will be explained from a mathematical point of view. Consider an urn containing 80 balls numbered from 1 to 80. Draw 20 balls from the urn, while pausing and announcing the result after each draw, to create a furor of excitement. This constitutes one play or trial of the game. In Figure 3.4 is shown different possible ways in which to play the game. For example, "mark 3 spots" means that a bettor chooses 3 of the 80 numbers and buys a 70¢, $1, $1.40, or $5 ticket. For example, if a woman buys a 70¢ ticket, the payoffs after a trial of the phenomenon are $30.00 if all three numbers occur among the 20 drawn and 70¢ if 2 of her numbers are among the 20 drawn. Otherwise she loses her wager. The interpretation of the meanings and payoffs for "mark 1 spot" through "mark 15 spots" should now be obvious. We will evaluate the fairness of keno for some situations.

MARK 1 SPOT

In "mark 1 spot" the casino odds on any number occurring in the 20 chosen is "2 to 1," or the assigned probability is $\frac{1}{3}$. The fair odds are "3 to 1" or the

Table 3.4 CASINO PROFIT ON PARLAY FOOTBALL

PARLAYS	CASINO ASSIGNED PROBABILITY	ACTUAL PROBABILITY	EXPECTED CASINO PROFIT
3 teams	$\frac{1}{6}$	$\frac{1}{8}$	$\frac{2}{8} \times 100 = 25\%$
4 teams	$\frac{1}{11}$	$\frac{1}{16}$	$\frac{5}{16} \times 100 = 31.3\%$
5 teams	$\frac{1}{16}$	$\frac{1}{32}$	$\frac{16}{32} \times 100 = 50\%$
6 teams	$\frac{1}{26}$	$\frac{1}{64}$	$\frac{38}{64} \times 100 = 59.4\%$
7 teams	$\frac{1}{36}$	$\frac{1}{128}$	$\frac{92}{128} \times 100 = 71.9\%$
8 teams	$\frac{1}{61}$	$\frac{1}{256}$	$\frac{195}{256} \times 100 = 76.2\%$
9 teams	$\frac{1}{81}$	$\frac{1}{512}$	$\frac{431}{512} \times 100 = 84.2\%$
10 teams	$\frac{1}{101}$	$\frac{1}{1024}$	$\frac{923}{1024} \times 100 = 90.1\%$

probability is $\frac{1}{4}$. Therefore, for every \$80 bet the casino pays back \$60, and thus the casino profit is 25%. This result is also given by the formula

$$\left[\left(\tfrac{1}{3} - \tfrac{1}{4} \right) \div \tfrac{1}{3} \right] \times 100 = 25\%$$

MARK 2 SPOTS

The number of different 2-spot combinations that can occur is $\binom{80}{2} = 40 \times 79$.
The number of winning 2-spot combinations on any trial is $\binom{20}{2} = 190$. The casino pays \$12 for every dollar bet on a winning combination.

$$\therefore \% \text{ casino profit} = \frac{(40 \times 79) - (12 \times 190)}{40 \times 79} \times 100$$
$$= 27.8\%$$

MARK 3 SPOTS

Let us calculate the number of possible combinations and the number of winning combinations possible.

1	2	3	4	5	6	7	8	9	10
11	12	13	14	15	16	17	18	19	20
21	22	23	24	25	26	27	28	29	30
31	32	33	34	35	36	37	38	39	40
41	42	43	44	45	46	47	48	49	50
51	52	53	54	55	56	57	58	59	60
61	62	63	64	65	66	67	68	69	70
71	72	73	74	75	76	77	78	79	80

Figure 3.3 A typical keno card.

MARK 1 SPOT

Winning Spot	.70¢ Ticket Pays	$1.40 Ticket Pays	Regular $1.00 Ticket Pays
1	2.10	4.20	3.00

MARK 2 SPOTS

Winning Spot	.70¢ Ticket Pays	$1.40 Ticket Pays	Regular $1.00 Ticket Pays
2	8.50	17.00	12.00

MARK 3 SPOTS

Winning Spot	.70¢ Ticket Pays	$1.40 Ticket Pays	Regular $1.00 Ticket Pays
2	.70	1.40	1.00
3	30.00	60.00	42.00

MARK 4 SPOTS

Winning Spot	.70¢ Ticket Pays	$1.40 Ticket Pays	Regular $1.00 Ticket Pays
2	.70	1.40	1.00
3	3.00	6.00	4.00
4	75.00	150.00	113.00

MARK 5 SPOTS

Winning Spot	.70¢ Ticket Pays	$1.40 Ticket Pays	Regular $1.00 Ticket Pays
3	.70	1.40	1.00
4	6.50	13.00	9.00
5	580.00	1,160.00	820.00

MARK 6 SPOTS

Winning Spot	.70¢ Ticket Pays	$1.40 Ticket Pays	Regular $1.00 Ticket Pays
3	.40	.80	1.00
4	2.50	5.00	3.00
5	70.00	140.00	90.00
6	1,300.00	2,600.00	1,800.00

MARK 7 SPOTS

Winning Spot	.70¢ Ticket Pays	$1.40 Ticket Pays	Regular $1.00 Ticket Pays
4	.70	1.40	1.00
5	16.00	32.00	20.00
6	260.00	520.00	410.00
7	6,000.00	12,000.00	8,100.00

MARK 8 SPOTS

Winning Spot	.70¢ Ticket Pays	$1.40 Ticket Pays	Regular $1.00 Ticket Pays
5	6.30	12.60	9.00
6	63.00	126.00	90.00
7	1,155.00	2,310.00	1,650.00
8	12,600.00	25,000.00	18,000.00

MARK 9 SPOTS

Winning Spot	.70¢ Ticket Pays	$1.40 Ticket Pays	Regular $1.00 Ticket Pays
5	2.10	4.20	3.00
6	31.50	63.00	45.00
7	234.50	469.00	335.00
8	3,290.00	6,580.00	4,700.00
9	12,950.00	25,000.00	18,500.00

MARK 10 SPOTS

Winning Spot	.70¢ Ticket Pays	$1.40 Ticket Pays	Regular $1.00 Ticket Pays
5	1.40	2.80	2.00
6	14.00	28.00	20.00
7	99.40	198.80	142.00
8	700.00	1,400.00	1,000.00
9	3,150.00	6,300.00	4,500.00
10	13,300.00	25,000.00	19,000.00

MARK 11 SPOTS

Winning Spot	.70¢ Ticket Pays	$1.40 Ticket Pays	Regular $1.00 Ticket Pays
6	7.00	14.00	10.00
7	52.50	105.00	75.00
8	266.00	532.00	380.00
9	1,400.00	2,800.00	2,000.00
10	8,750.00	17,500.00	12,500.00
11	13,650.00	25,000.00	19,500.00

MARK 12 SPOTS

Winning Spot	.70¢ Ticket Pays	$1.40 Ticket Pays	Regular $1.00 Ticket Pays
6	4.00	8.00	6.00
7	20.00	40.00	28.00
8	160.00	320.00	200.00
9	700.00	1,400.00	850.00
10	1,800.00	3,600.00	2,400.00
11	12,500.00	25,000.00	13,000.00
12	25,000.00	25,000.00	25,000.00

Figure 3.4 Different plays in the keno game.

$$\text{Number of possible 3 spots} = \binom{80}{3}$$

$$= 82{,}160$$

$$\text{Number of winning 3 spots} = \binom{20}{3}$$

$$= 1140$$

Number of winning 2 spots on 3 spot tickets

$$= \binom{20}{2} \times 78 - 1140 \qquad \text{(Be sure)}$$

$$= 12{,}680$$

$$\% \text{ casino profit} = \frac{[82{,}160 - (1140 \times 42) + 12{,}680]}{82{,}160}$$

$$= 26.3\%$$

These calculations indicate that the casino profit on the game of keno is more than 25%. The working out of fairness for the higher mark spots is quite challenging and requires computer help for the manipulations. One practical, if costly, way to avoid the mathematics is to wager on each outcome and note the amount of money returned.

The Concept of Information Content (A Comment)

In communication theory, information is defined as a quantitative measure of what is conveyed to a receiver by a transmitter. If a transmitter sends one of a set of messages, $[s_1, s_2, \ldots, s_n]$, with transmission probabilities $P(s_1)$ $\cdots P(s_n)$, then the reception of a message that has a small probability conveys much information to the receiver, whereas the reception of a likely event or message conveys little information. Initially one might be inclined to define the information I of a message s_i as

$$I = \frac{1}{P(s_i)}$$

so that the reception of a sure event would convey zero information and the reception of a message with a probability of zero would convey infinite information. On reflection we would like to go further and require the reception of two mutually exclusive events to convey information equal to the sum of the separate events. Since

$$\frac{1}{P(s_i)} + \frac{1}{P(s_j)} \neq \frac{1}{P(s_i) + P(s_j)}$$

and $P(s_i) + P(s_j)$ is the probability of the occurrence of the union of s_i and s_j, we cannot use $1/P(s_i)$ as the information content of the occurrence of s_i.

If we formally decide to define information in such a way that (a) the information conveyed by a sure event is 0, (b) the information conveyed by an impossible event is infinite, and (c) the information conveyed by the occurrence of the union of two mutually exclusive events is the sum of the information of the separate events, then the definition

$$I = \log_K \frac{1}{P(S_i)}$$

for any base K will satisfy these criteria.

The commonly used definition in information theory and in digital logic is

$$I = +\log_2 \frac{1}{P(S_i)} \text{ bits}$$

For example, the information conveyed by knowing a decimal digit is 3.3 bits, and knowing that a 5-bit word consists of five zeros is 5 bits. Some more exotic statistics are that a page of text conveys thousands of bits of information and the information of a television picture might be as high as 10^6 bits—even though it may seem to be 0 during prime time.

SUMMARY

The first three chapters comprise the topic of discrete probability theory, which is the essential background for fields utilizing probabilistic thinking. A few of these fields are industrial sampling and testing, semiconductor statistics, statistical thermodynamics, and the topics for which this text is being specifically geared: communication theory and signal processing. Any philosophical weakness or lack of skill in problem solving will constantly plague us in later studies. It is heartening to realize and appreciate that probability theory is based on just a few axiomatic formulas that we have endeavored, by stressing a relative-frequency interpretation, to make intuitively acceptable.

The relative-frequency formulas defining the probability of an unconditional, conditional, and compound event are

$$P(A) = \lim_{N \to \infty} \frac{N_A}{N}$$

$$P(A/B) = \lim_{N \to \infty} \frac{N_{A \cap B}}{N_B}$$

and

$$P(A \cap B) = \lim_{N \to \infty} \frac{N_{A \cap B}}{N} \quad \text{or} \quad \lim_{N \to \infty} \frac{N_A}{N} \cdot \frac{N_{A \cap B}}{N_A}$$

The latter form for a compound event is used when the former is too cumbersome.

Using set theory structure, where a sample description space S and its probability measure are assumed known, we axiomatically "redefine" or interpret these formulas as follows:

$$A = A \cap S$$

$$= \bigcup_{i=1}^{n} s_i \quad \text{for some } n$$

where $s_1, \ldots, s_n$ are points of S whose union make up A. Then

$$P(A) = \sum_{i=1}^{n} P(s_i)$$

$$A \cap B = (A \cap B) \cap S$$

$$= \bigcup_{i=1}^{m} s_i \quad \text{for some } m$$

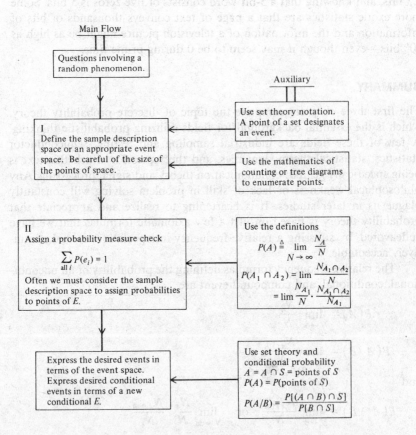

Figure 3.5   A flowchart summary for solving discrete probability problems.

where $s_1,\ldots,s_m$ are points of S and $m \leq n$. Then

$$P(A \cap B) = \sum_{i=1}^{m} P(s_i)$$

$$P(B/A) = \frac{P(A \cap B)}{P(A)}$$

$$= \frac{\sum_{i=1}^{m} P(s_i)}{\sum_{i=1}^{n} P(s_i)}$$

Figure 3.5 shows, in flowchart form, the different stages of answering questions concerning a random phenomenon. These stages are labeled under the heading "Main Flow." In addition, under the heading "Auxiliary" are shown the disciplines and formulas required to complete each stage. Careful perusal of this chart should philosophically bring together all the material of Chapters 1 through 3.

PROBLEMS

1. Philosophically, certain probabilistic problems cause concern to a layperson. In World War II straws often were drawn to see who went behind enemy lines; if the last man was handed the short straw, he felt cheated. Similarly, the draft system lottery was criticized during the Vietnam War. As a model let us consider the phenomenon of drawing a ball from an urn with 6 red and 5 blue balls. It is obvious that the probability of a red ball is $\frac{6}{11}$: Prove that if the phenomenon consists of drawing a ball and throwing it away and then drawing another ball the probability it is red is still $\frac{6}{11}$.

   Set up an urn model to explain why a G.I. being handed the short straw from 6 straws was not cheated. Assume that five soldiers had, previously drawn straws.

2. Assume that a pole-vaulter, Art, fails to leave the ground 20% of the time; the other 80% of the time he clears a height from 14 to 16 ft, where all subranges of equal width between 14 and 16 ft are equally likely.

   (a) What is the probability that Art clears between 14 ft 6 in. and 15 ft 2 in.?

   (b) Given that Art leaves the ground, what is the probability that he clears between 14 ft 6 in. and 15 ft 2 in.? Do this part in two ways:

      (i) Using the formula $P(A/B)$.

      (ii) Defining a new conditional space with its probability measure.

3. It is required to rewrite Problem 2 in a manner to stimulate a student with a budding interest in signal analysis and signal processing. We desire to consider the sampling of a periodic waveform $y(t)$, as shown in Figure 3.6, with period 1, where the phenomenon consists in sampling the waveform at some time $0 < t < 1$. Define the functional expression for the waveform $0 < t < 1$ so that the phenomenon is identical to that of Problem 2.

   Is your solution to this problem unique? List some operations that will not affect your solution, such as "reflecting $y(t)$," "time shifting," "time scaling," and so on.

4. In an Irish village, 70% of the men and 60% of the women speak Gaelic. If there are 400 men and 500 women in a village and the chances are 80% that a villager tells the truth, answer the following:

   (a) What is the probability that a woman who says she speaks Gaelic actually does?

   (b) What is the probability that a man who claims to speak Gaelic actually does?

   (c) Check your results to (a) and (b) by showing that the probabilities of points on an appropriate event space sum to 1.

5. Consider two urns with the following composition: urn I contains 7 red and 3 black balls, urn II contains 6 red and 8 black balls. Our random phenomenon consists of choosing one of the two urns and then drawing a sample (without replacement) of 3 balls. If 3 red balls are drawn, what is the probability they came from urn I?

6. Assume that a class contains 12 students of about equal ability. If the probability of a student obtaining an "A" is $\frac{1}{5}$, a "B" is $\frac{2}{5}$, and a "C" is $\frac{2}{5}$. what is the probability that 3 students will obtain A's, 4 will obtain B's, and 5 will obtain C's? What does this probability become, given at least two students receive C's?

7. Consider the problem of sampling five transistors from a large population, of which 20% are defective.

   (a) What is the probability that more good than defective transistors are sampled?

   (b) Given that at least one defective transistor is sampled, what is the probability of exactly three good transistors being sampled?

   (c) If three different samples of size 5 are chosen, what is the probabil-

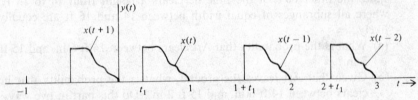

Figure 3.6   Periodic waveform for Problem 3.

ity that at least two of the samples contain more good than defective transistors?

8. Consider a department in which all of its graduate students range in age from 22 to 31 (inclusive). Given also that it is twice as likely a student's age is between 27 and 31 (inclusive) as between 22 and 26 (inclusive) and that in each group all specific annual ages are equally likely, find the probability that in a class of 9 a majority are over 25 years old.

9. Rapidly try to write down the answers for the probabilities of the following events (at most 2 minutes per part):
   (a) a 7 when rolling two dice
   (b) no 1 when rolling two dice
   (c) exactly 6 heads on 9 tosses
   (d) at least three zeros in a 5-bit computer word
   (e) exactly 5 black cards in a bridge hand
   (f) 4 of a kind on a draw poker hand
   (g) a waveform value less than 0.5 when sampling the waveform $\sin(t - \frac{1}{6}\pi)$

10. Consider the periodic waveform

$$y(t) = \sum_{n=-\infty}^{\infty} x(t - 3n) \qquad n \text{ an integer}$$

where

$$x(t) = 1 \qquad 0 < t < 1$$
$$= 0 \qquad 1 < t < 3$$

Define the random phenomenon of sampling $y(t)$ at $t = t$ and at $t = t + \tau$, where t is uniformly chosen as $0 < t < 3$.
   (a) Sketch $y(t)$.
   (b) Assign a probability measure to the event space E for the two sampled values when $\tau = 0.5$:

$$E = \{(0,0), (0,1), (1,0), (1,1)\}$$

   where, for example, $(0, 1)$ is the event that the first sampled value is 0 and the sampled value 0.5 s later is 1.
   (c) Assign a probability to E for all τ, $0 < \tau < 3$.
   (d) Given that the first sampled value is 1 and $\tau = \frac{1}{2}$, what is the probability that the second sampled value is 0?

11. Consider the periodic waveform

$$y(t) = \sum_{\substack{n=-\infty \\ \text{integer}}}^{\infty} x(t - 3n)$$

where

$$x(t) = 4t \qquad -1.5 < t < 1.5$$
$$= 0 \qquad \text{otherwise}$$

Define the random phenomenon of sampling $y(t)$ at $t = \tau$ and at $t = t + \tau$, where t is uniformly chosen as $0 < t < 3$. For this waveform repeat parts (b), (c), and (d) of Problem 10 where now

$$E = \{(x_1 < 0, x_2 < 0), (x_1 < 0, x_2 > 0), (x_1 > 0, x_2 < 0), (x_1 > 0, x_2 > 0)\}$$

where x_1 refers to $x(t)$ and x_2 to $x(t + \tau)$.

Part II
RANDOM VARIABLE
THEORY

Chapter 4
Single Random Variables, Associated Functions, and Their Usage

4.1 MASS, DENSITY, AND DISTRIBUTION FUNCTIONS

Definition of a Random Variable

Up to this point probability theory has consisted of a random phenomenon and a mathematical model defined in terms of a set of mutually exclusive, collectively exhaustive events with a probability measure. Such a model may be used to answer probabilistic questions about any events that may be described in terms of the event space. The sample description space is the most fundamental event space and provides a one-to-one mapping between the outcomes of the experiment and its points. The elements of the sample space may be numbers or colors or words or compound points that describe the possible outcomes of the phenomenon. In order to use function theory and calculus it will be desirable to have numbers associated with the points of S. For this reason *a random variable X associated with a sample space is defined by a rule that assigns a number to each point of S*. As an example we will consider many random variables that may be associated with a sample description space.

Example 4.1

For the experiment of throwing a fair die, show how the points of S map into points of the real axis for the following rules and state whether they define random variables.

(a) The variable X, defined by the rule that maps the points of S onto the real axis with a value equal to the number of the top face of the die.

(b) The variable Y, defined by the rule that maps the points of S onto the real axis with a value equal to 0 if the top face is even and equal to 1 if the top face is odd.

(c) The variable Z, defined by the rule that maps the points of S onto the real axis subject to the rule that a value 6 is assigned if the top face is less than or equal to 4 and a value of 8 is assigned if the top face is greater than 4.

(d) The variable P, defined by the rule that maps even points of S into the point 0 on the real axis and maps points of S for which the top face is greater than 2 into the point 1 on the real axis. Also map the point for which the top face is 1 into 1 on the real axis.

SOLUTION

The whole purpose of this very simple question is to cement in our minds the underlying definition given for a random variable. The sample description space we are dealing with is

$$S = \{(1), (2), (3), (4), (5), (6)\}$$
$$\quad\ \uparrow\ \ \ \ \uparrow\ \ \ \ \uparrow\ \ \ \ \uparrow\ \ \ \ \uparrow\ \ \ \ \uparrow$$
$$\quad\ S_1\ \ \ S_2\ \ \ S_3\ \ \ S_4\ \ \ S_5\ \ \ S_6$$

(a) The mapping defined by the rule is shown. The rule defines a random variable because given a point of S we get a unique number on the real axis for X.

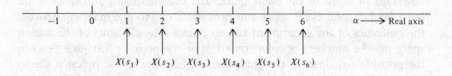

(b) The rule for Y maps three points of S into the point 0 on the real axis and three points into the point 1 on the real axis. This is in conformity with the definition of a random variable, so Y is a random variable defined on S.

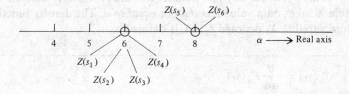

(c) The plot of the rule Z is shown and with consideration is seen to define Z as a random variable, as there is a unique answer for $Z(s_i)$ for all i.

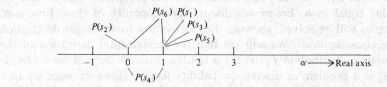

(d) The rule P maps the points S_2, S_4, and S_6 from S into the point 0 on the real axis and the points S_1, S_3, S_4, S_5, and S_6 into the point 1 on the real axis as shown plotted. There are two contradictions involved here, as we are given $P(s_4)=0$ and also $P(s_4)=1$ and, in addition, $P(s_6)=0$ and $P(s_6)=1$. Since a unique number is not assigned to s_4 or to s_6 by the rule, P is *not* a random variable.

The definition given for a random variable is abstract, but we will see that random variables normally are closely associated with reality. The most common choice is that if a random phenomenon yields numerical outcomes for a quantity, then we define the random variable associated with the quantity by the rule $X(s_i)=s_i$. For example, if it is decided that "the heights of people" is a random phenomenon, then we would let X be the random variable describing height and for the point 70 (in.) in our sample space we could let $X=70$. For this reason students are often confused between the physical quantity and the random variable assigned to it. Mathematically, any other function of height yielding a unique real value is a random variable, but in practice such a random variable would rarely be used.

The Cumulative Distribution, Density, and Mass Functions

A random variable X is characterized by three functions that allow for ready evaluation of any probabilistic question about the random variable. The most fundamental function is the **cumulative distribution function** (or distribution function) of the random variable, denoted $F_X(\alpha)$. $F_X(\alpha)$ is defined as

$$F_X(\alpha) \triangleq P(X \leq \alpha) \tag{4.1}$$

or, in words, $F_X(\alpha)$ for any real number α is the probability that the random variable X takes on a value *less than* or *equal to* α. The **density function** of a random variable X, denoted $f_X(\alpha)$, is defined as

$$f_X(\alpha) \triangleq \frac{d}{d\alpha} F_X(\alpha) \tag{4.2}$$

or in words, $f_X(\alpha)$ is the derivative of the distribution function. The *mass function* of a random variable (if appropriate) denoted $p_X(\alpha)$ is defined as

$$p_X(\alpha) = P(X = \alpha) \tag{4.3}$$

or in words $p_X(\alpha)$ for any real number α, is the probability that X takes on a value equal to α. Before we discuss the properties of these functions, examples will be solved, showing their derivation from sample description space considerations. We will see that the fundamental derivation of the cumulative distribution $F_X(\alpha)$ for a random variable defined on a sample space, is a problem in discrete probability theory. However, once we find $F_X(\alpha)$ and its derivative the density function $f_X(\alpha)$, it is then much easier to answer a multitude of probabilistic questions about the random variable using $F_X(\alpha)$ or $f_X(\alpha)$ instead of using discrete probability theory. Similarly, we will see that the mass function $p_X(\alpha)$, can also be obtained using discrete probability theory and that it will not exist for a sample space with a continuous range of points, where there is only meaning to the probability of assuming a value within a range. Examples will now be solved, showing the derivation of these functions.

Example 4.2

Consider the experiment of throwing two dice. Let X be the random variable which assigns to every point in S a value equal to the sum of the numbers on the upturned faces of the two dice.

(a) Show how the points of S are mapped onto the real axis for this random variable.

(b) Using discrete probability theory, evaluate $F_X(\alpha)$ for all α, where the real axis is called the α axis.

(c) Find $f_X(\alpha)$ and $p_X(\alpha)$ if appropriate.

SOLUTION

(a) The sample description space consists of the 36 points

$$S = \{(x, y): 1 \le x \le 6, 1 \le y \le 6, \text{both integers}\}$$

If the points are enumerated as shown in Figure 4.1a, then the mapping to the real axis subject to the definition of the random variable X is easily accomplished as in Figure 4.1b.

(b) $F_X(\alpha)$ can be determined over several ranges, as follows:

For the range $-\infty < \alpha \le 1.999$, by definition, $F_X(\alpha)$ asks the question, "What is the probability that the sum of the numbers on two dice takes on a value less than or equal to some number α between $-\infty$ and 1.99?" We see that for any such number the answer is 0, since the sum must be at least 2. Therefore, in this range

$$F_X(\alpha) = P(X \le \alpha)$$

$$= 0$$

Since, for the range $2 \le \alpha \le 2.99$, $F_X(\alpha)$ is defined as the probability of assuming a value less than or equal to a number in the given range, from set theory, the event $(X \le \alpha)$ may be written

$$(X \le \alpha) = (-\infty < X \le 1.99) \cup (1.99 < X \le \alpha)$$

Since a random variable is defined as assigning to points of S a unique point on the real axis, then the events $(-\infty < X \le 1.99)$ and $(1.99 < X \le \alpha)$ are mutually exclusive. Now using the axioms of probability theory,

$$F_X(\alpha) = F_X(1.99) + P(1.99 < X \le \alpha)$$

$$= 0 + \tfrac{1}{36}$$

Only the event of a two for the sum of the numbers on the dice is involved in the event $(1.99 < X \le \alpha)$. At this stage the student should be careful of a number of things. The use of equality signs is very vital and the appreciation that although we are talking about the range $(1.99 < \alpha \le 2.99)$ for α, in solving for $F_X(\alpha)$ we include $-\infty \le \alpha$ by definition of a distribution function. $F_X(\alpha)$ will now be

S_1	S_2	S_3	S_4	S_5	S_6
(1, 1)	(1, 2)	(1, 3)	(1, 4)	(1, 5)	(1, 6)
S_7	S_8	S_9	S_{10}	S_{11}	S_{12}
(2, 1)	(2, 2)	(2, 3)	(2, 4)	(2, 5)	(2, 6)
S_{13}	S_{14}	S_{15}	S_{16}	S_{17}	S_{18}
(3, 1)	(3, 2)	(3, 3)	(3, 4)	(3, 5)	(3, 6)
S_{19}	S_{20}	S_{21}	S_{22}	S_{23}	S_{24}
(4, 1)	(4, 2)	(4, 3)	(4, 4)	(4, 5)	(4, 6)
S_{25}	S_{26}	S_{27}	S_{28}	S_{29}	S_{30}
(5, 1)	(5, 2)	(5, 3)	(5, 4)	(5, 5)	(5, 6)
S_{31}	S_{32}	S_{33}	S_{34}	S_{35}	S_{36}
(6, 1)	(6, 2)	(6, 3)	(6, 4)	(6, 5)	(6, 6)

(a)

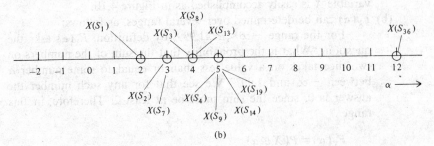

(b)

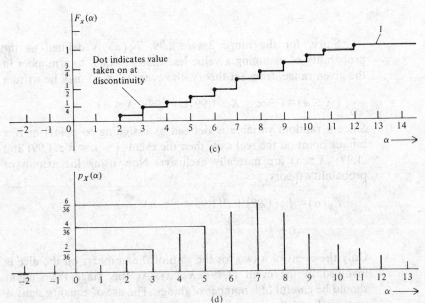

Figure 4.1 (a) The points of the sample space in Example 4.2. (b) The mapping of the points of S onto the real axis. (c) The cumulative distribution function $F_X(\alpha)$. (d) The mass function $p_X(\alpha)$.

determined much more rapidly for the remaining ranges of α, keeping in mind the care and discipline just indicated.

For the range $3 \leq \alpha \leq 3.99$,

$$F_X(\alpha) = P(X \leq \alpha)$$

$$= F_X(2.99) + P(3 \leq X \leq \alpha)$$

$$= \tfrac{1}{36} + P(\text{sum of dice is 3})$$

$$= \tfrac{1}{36} + P(s_2) + P(s_7)$$

$$= \tfrac{1}{12}$$

Similarly, for $4 \leq \alpha \leq 4.99$,

$$F_X(\alpha) = \tfrac{6}{36}$$

For $5 \leq \alpha < 5.99$,

$$F_X(\alpha) = \tfrac{10}{36}$$

For $6 \leq \alpha < 6.99$,

$$F_X(\alpha) = \tfrac{15}{36}$$

For $7 \leq \alpha < 7.99$,

$$F_X(\alpha) = \tfrac{21}{36}$$

For $8 \leq \alpha \leq 8.99$,

$$F_X(\alpha) = \tfrac{26}{36}$$

For $9 \leq \alpha \leq 9.99$,

$$F_X(\alpha) = \tfrac{30}{36}$$

For $10 \leq \alpha \leq 10.99$,

$$F_X(\alpha) = \tfrac{33}{36}$$

For $11 \leq \alpha < 11.99$,

$$F_X(\alpha) = \tfrac{35}{36}$$

For $x \geq 12$,

$$F_X(\alpha) = 1$$

A plot of $F_X(\alpha)$ is shown in Figure 4.1c, from which we see that it possesses special properties by virtue of being a cumulative distribution function. These will be developed in general later in the chapter, but it should be clear from discrete probability theory, that $F_X(\alpha)$ can never be negative, can never decrease in value when α is increased, and that its value can never exceed 1. It may be instructive for the student to formally prove these properties as an exercise.

(c) The question is to evaluate the density function

$$f_X(\alpha) \triangleq \frac{d}{d\alpha} F_X(\alpha)$$

and the probability mass function $p_X(\alpha) \triangleq P(X=\alpha)$. Since $F_X(\alpha)$ changes only by means of piecewise discontinuous jumps, at the moment we will pass on the derivation of $f_X(\alpha)$ until the concept of delta functions is developed later in the chapter. The use of delta functions will allow for indicating the derivative of a jump and is probably already familiar to the reader.

For this random phenomenon it is easily shown from the sample description space that the probability mass function $p_X(\alpha)$ defined as

$$p_X(\alpha) \triangleq P(X=\alpha)$$

is

$$p_X(2)=p_X(12)=\tfrac{1}{36} \qquad p_X(3)=p_X(11)=\tfrac{2}{36}$$

$$p_X(4)=p_X(10)=\tfrac{3}{36} \qquad p_X(5)=p_X(9)=\tfrac{4}{36}$$

$$p_X(6)=p_X(8)=\tfrac{5}{36} \qquad p_X(7)=\tfrac{6}{36}$$

otherwise

$$p_X(\alpha)=0$$

We note that finding $p_X(\alpha)$ is simpler than finding $F_X(\alpha)$. A plot of $p_X(\alpha)$ is shown in Figure 4.1d.

Observing our solution for the mass function $p_X(\alpha)$, there are certain general properties of mass functions that are apparent by definition. Obviously the value of $p_X(\alpha)$ is 0 or a positive number less than or equal to 1. If the random variable only assumes discrete values then the sum of all the probabilities $p_X(\alpha_i)$ must equal 1.

Before proceeding with deriving the general properties of cumulative distribution, density, and mass functions, a problem of a continuous nature plus a stranger problem of what will be called a "mixed" nature will be solved.

Example 4.3

Consider the experiment of spinning an infinitely finely calibrated wheel of fortune that takes on all values from 0 to 12. Assign values to the random variable X corresponding to the number obtained on the spin. Find $F_X(\alpha)$, $f_X(\alpha)$, and $p_X(\alpha)$.

SOLUTION

Since S consists of an infinite uncountable number of points, we can only define an event space with a probability measure for the experiment. For example, let $E = \{E_1, E_2, E_3\}$, where

E_1 = event of a number less than a
E_2 = event of a number between a and b
E_3 = event of a number between b and 12

Then

$$P(E_1) = \frac{a}{12} \qquad P(E_2) = \frac{b-a}{12} \qquad P(E_3) = \frac{12-b}{12}$$

assuming the wheel is fair and $0 < a < b < 12$.

Derivation of $F_X(\alpha)$, the Cumulative Distribution Function
    For the range $\alpha < 0$, by definition $F_X(\alpha) = P(X \le \alpha)$ and, since the smallest value the random variable can take on a trial of the phenomenon is 0, then

$$F_X(\alpha) = 0$$

For the range $0 < \alpha < 12$ we are posing the probabilistic problem: "Find $F_X(\alpha) \stackrel{\triangle}{=} P(X \le \alpha)$." From our event space the solution is

$$F_X(\alpha) = \frac{\alpha}{12}$$

Let us now examine the range $\alpha > 12$. Although it is never possible for X to assume a value greater than 12 we are asking to find $F_X(\alpha)$, where $\alpha > 12$.

$$\therefore F_X(\alpha) = P(X \le \alpha)$$
$$= F_X(12) + P(12 < X \le \alpha)$$
$$= 1 + 0 = 1$$

and we obtain this answer for any α up to $+\infty$.
    A plot of $F_X(\alpha)$ is shown in Figure 4.2a.

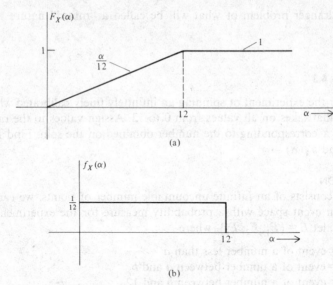

(a)

(b)

Figure 4.2   (a) The cumulative distribution function of Example 4.3. (b) The density function.

Summarizing, the mathematical formula for $F_X(\alpha)$ is

$$F_X(\alpha) = 0 \qquad \alpha < 0$$

$$= \frac{\alpha}{12} \qquad 0 < \alpha < 12$$

$$= 1 \qquad \alpha > 12$$

The student might notice that the necessity to watch equality signs carefully is no longer important, since there is only meaning to the probability of taking on a range of values.

Derivation of $f_X(\alpha)$, the Density Function
By definition, the density function is the derivative of $F_X(\alpha)$.

$$\therefore f_X(\alpha) \triangleq \frac{d}{d\alpha} F_X(\alpha)$$

$$= \tfrac{1}{12} \qquad 0 < \alpha < 12$$

$$= 0 \qquad \text{otherwise}$$

$f_X(\alpha)$ is shown plotted in Figure 4.2b.

Derivation of $p_X(\alpha)$, the Mass Function
For this continuous case we always have $p_X(\alpha) = 0$, where α is a value that can be assumed by X or not, since there is no probabilistic meaning to taking on one value out of an infinite possible set, all with equal probabilities. We will see that the smallest range we can talk about will be $\alpha < X < (\alpha + \Delta\alpha)$, and this has a probability of $f_X(\alpha)\Delta\alpha$.

Examples 4.2 and 4.3 introduced us to two types of random variables. In Example 4.2, we call X a **discrete random variable** as $F_X(\alpha)$ only changes in piecewise discontinuous jumps. The density function $f_X(\alpha)$ is not directly defined from calculus and the mass function $p_X(\alpha)$ has values at a finite number of discrete points. In Example 4.3 we encountered what we call a **continuous random variable** X. $F_X(\alpha)$ and $f_X(\alpha)$ are both continuous functions of α for all $-\infty < \alpha < \infty$ and $p_X(\alpha)$ is always 0. Next we will encounter a mixture of the two, called a "mixed" random variable.

Example 4.4

Consider the random phenomenon of the height cleared by a pole vaulter called A. Thirty percent of the time he misjudges his run and does not leave the ground; otherwise he clears from 13 to 15 ft, each point in the range having equal probability. Let X be the random variable that describes the height he clears and be defined by a one-to-one mapping from the phenomenon to the real axis. Find $F_X(\alpha)$, $f_X(\alpha)$, and $p_X(\alpha)$.

SOLUTION

Derivation of the Cumulative Distribution Function $F_X(\alpha)$

For the range $\alpha < 0$, $F_X(\alpha) = 0$, since obviously A never goes underground.

For the range $0 \leq \alpha < 13$,

$$F_X(0) = P(X \leq 0) = \tfrac{3}{10}$$

Now for any $\alpha > 0$ in this range,

$$(X \leq \alpha) = (X < 0) \cup (X = 0) \cup (0 < X \leq \alpha)$$

which are three mutually exclusive events.

$$\therefore F_X(\alpha) = 0 + \tfrac{3}{10} + 0$$
$$= \tfrac{3}{10}$$

For any α in the range $13 < \alpha < 15$,

$$F_X(\alpha) = F_X(13) + P(13 < X \leq \alpha)$$
$$= \frac{3}{10} + \left(\frac{7}{10}\right)\left(\frac{\alpha - 13}{2}\right)$$

The term 0.7 or $\tfrac{7}{10}[(\alpha - 13)/2]$ needs some explanation. Since the probability $X = 0$ is 0.3, then the probability of the event $(13 \leq \alpha \leq 15)$ is 0.7. Also, since all of the points of the range $13 < \alpha < 15$ are equally likely and this range is of width two units, the subrange $(\alpha - 13)$ will comprise $(\alpha - 13)/2$ of the total range and has a probability of $0.7[(\alpha - 13)/2]$ using conditional probability. Combining terms,

$$F_X(\alpha) = \frac{7\alpha}{20} - \frac{85}{20} \qquad 13 < \alpha < 15$$

For the range $\alpha > 15$, the event

$$(X \le \alpha) = (X \le 15) \cup (15 < X \le \alpha)$$

$$\therefore F_X(\alpha) = 1$$

A plot of $F_X(\alpha)$ is shown in Figure 4.3.

Derivation of the Density Function $f_X(\alpha)$

Because of the piecewise discontinuity at $\alpha = 0$, for the moment we cannot define

$$f_X(\alpha) \triangleq \frac{d}{d\alpha} F_X(\alpha)$$

at $\alpha = 0$. However, we can see that $f_X(\alpha)$ is defined at all other points.

$f_X(\alpha)$ is undefined   for $\alpha = 0$

$f_X(\alpha) = \frac{7}{20}$ $13 < \alpha < 15$

$\qquad\quad = 0$ otherwise

Derivation of the Mass Function $p_X(\alpha)$

In the case of the mass function there is again difficulty, since $p_X(0) = \frac{3}{10}$ but at all other points $p_X(\alpha) = 0$, although there are probabilities involving any event, including part or all of the range $13 < \alpha < 15$.

In order to handle distribution and density functions of all kinds of random variables by general mathematical formulas we will now develop the use of the delta function in probability theory. Then we will proceed to list important properties of $F_X(\alpha)$, $f_X(\alpha)$, and $p_X(\alpha)$. We should keep in mind, however, that this notation could be avoided and each type of random variable handled separately, using only the functions most useful for it, whether continuous, discrete, or mixed. However, the use of delta functions is widespread and now familiar to electrical engineering students in a basic circuits course.

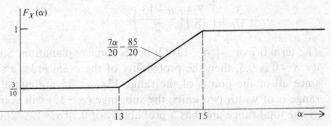

Figure 4.3   The mixed cumulative distribution function of Example 4.4.

THE DELTA FUNCTION IN PROBABILITY THEORY

The delta function denoted $\delta(x)$ is defined by the equation

$$\int_{-\infty}^{\infty} f(x)\delta(x-a)\,dx \overset{\triangle}{=} f(a) \tag{4.4}$$

where $f(x)$ is any function that is continuous at $x = a$ and $\delta(x-a)$ is called a delta function located at $x = a$. A practical engineering model $p_1(x)$ for $\delta(x)$ is a function of unit area centered about the origin, whose width W_1 is small compared to any range W_2 from $(a-0.5W_2)$ to $(a+0.5W_2)$, for which it is assumed that $f(x)$ varies little from $f(a)$. Then we can say that

$$\int_{-\infty}^{\infty} f(x)\delta(x-a)\,dx = \int_{a-W_1/2}^{a+W_1/2} f(x)p_1(x-a)\,dx$$

$$\approx f(a)\int_{a-W_1/2}^{a+W_1/2} p_1(x-a)\,dx = f(a)$$

A demanding mathematical model used for $\delta(x)$ is one that possesses the characteristics:

$$\delta(0) \to \infty$$

$$\delta(x) \to 0 \qquad x \neq 0$$

Area of $\delta(x) = 1$

All this says is that no matter how rapidly $f(x)$ is changing at $x = a$, we retain the result of Eq. 4.4 by the same procedure shown for our practical model. Appendix A gives a tutorial treatment of delta or singularity functions with many results that will be used throughout the text. If we define the unit step function $u(x)$ as

$$u(x) = 1 \qquad x > 0$$

$$= 0 \qquad x < 0$$

then, by definition,

$$\int_{-\infty}^{x} \delta(\alpha - a)\,d\alpha = 1 \qquad \text{if } x > a \qquad \text{(using Eq. 4.4)}$$

$$= 0 \qquad \text{if } x < a$$

We have shown that $u(x-a)$ is the integral of $\delta(x-a)$ and, conversely, $\delta(x-a)$ is the derivative of $u(x-a)$, as is shown in Figure 4.4, where $\delta(x-a)$ is indicated by its notation of a spike symbol at the point of its occurrence, with a number alongside indicating the weighting factor of the delta function. As a result of the integral-derivative pair relationship of Figure 4.4, it is easy to show that whenever we differentiate a piecewise discontinuity of size A at $x = a$, the delta function $A\delta(x-a)$ is obtained and whenever we integrate across a delta function with a weighting factor A, that is $A\delta(x-a)$ we obtain $Au(x-a)$ or a jump of A units at $x = a$.

As far as we are concerned there is nothing at all intimidating about delta functions. We will now reconsider parts of Examples 4.2, 4.3, and 4.4.

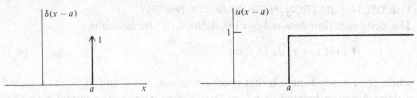

Figure 4.4 . The delta function and unit step function as a derivative-integral pair.

Example 4.5

(a) Reconsider the cumulative distribution function of Example 4.2 and find $f_X(\alpha)$ the density function.

(b) Repeat part (a) for $F_X(\alpha)$ of Example 4.4.

SOLUTION

(a) $F_X(\alpha)$ in Example 4.2 changes only by discrete jumps, and the density function $f_X(\alpha)$ defined by

$$f_X(\alpha) \triangleq \frac{d}{d\alpha} F_X(\alpha)$$

is a string of delta functions, each with a magnitude equal to the discontinuity in $F_X(\alpha)$.

$$\therefore f_X(\alpha) = \tfrac{1}{36}\delta(\alpha-2) + \tfrac{2}{36}\delta(\alpha-3) + \tfrac{3}{36}\delta(\alpha-4) + \tfrac{4}{36}\delta(\alpha-5)$$

$$+ \tfrac{5}{36}\delta(\alpha-6) + \tfrac{6}{36}\delta(\alpha-7) + \tfrac{5}{36}\delta(\alpha-8) + \tfrac{4}{36}\delta(\alpha-9)$$

$$+ \tfrac{3}{36}\delta(\alpha-10) + \tfrac{2}{36}\delta(\alpha-11) + \tfrac{1}{36}\delta(\alpha-12)$$

$f_X(\alpha)$ is shown plotted in Figure 4.5a. We notice the similarity of $f_X(\alpha)$ to $p_X(\alpha)$, the mass function. Instead of $f_X(\alpha)$ containing discrete values, it has delta functions with the probabilities of the mass function as weighting factors. So we associate discrete probabilities with delta functions in $f_X(\alpha)$.

(b) If we consider $F_X(\alpha)$ for Example 4.4 as shown plotted in Figure 4.3, we notice that it contains one discrete jump of 0.3 at $\alpha = 0$.

$$\therefore f_X(\alpha) = 0.3\delta(\alpha) + 0.35[u(\alpha-13) - u(\alpha-15)]$$

where $u(\alpha-13) - u(\alpha-15)$ is a short notation for indicating that the value 0.35 only exists from $13 < \alpha < 15$. $f_X(\alpha)$ is shown plotted in Figure 4.5b.

It is left as an exercise for the student to carefully demonstrate that

$$F_X(\alpha) \triangleq \int_{-\infty}^{\alpha^+} f_X(u)\, du$$

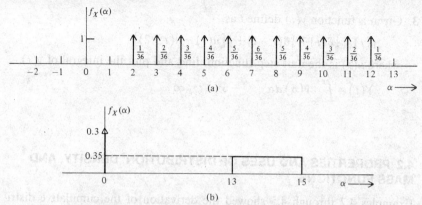

Figure 4.5   (a) The density function of Example 4.2. (b) The density function of Example 4.4.

does indeed yield the function of Figure 4.3, when it is evaluated for all α, $-\infty < \alpha < \infty$. It is worth noting that the fundamental definition of the integral of a function, which involves doing millions of problems as the upper limit of integration varies, is much mangled by students and even some graduate students at that. Also the use of α^+ implies that $F_X(\alpha)$ must include $P(X = \alpha)$.

DRILL SET: MASS, DENSITY, AND DISTRIBUTION FUNCTIONS

1. Consider the random phenomenon of uniformly sampling the waveform $x(t)$ between $t = 0$ and 3, where $x(t)$ is defined as

$$x(t) = 1 \qquad 0 < t < 1$$
$$= 2 \qquad 1 < t < 3$$

Define the random variable X as assuming the value of the waveform obtained on a trial. Find and sketch $p_X(\alpha)$, $F_X(\alpha)$, and $f_X(\alpha)$.

2. Consider the periodic waveform

$$x(t) = \sum_{n=-\infty}^{\infty} g(t - 4n)$$

where

$$g(t) = t \qquad -1 < t < 1$$
$$= 0 \qquad \text{otherwise}$$

is the input to a rectifier with output $y(t)$ defined by $y(t) = |x(t)|$. Define the experiments of sampling the input and output, where the random variables X and Y are associated (with a one-to-one mapping) with the input and output, respectively. Find $f_X(\alpha)$, $F_X(\alpha)$, $f_Y(\beta)$, and $F_Y(\beta)$.

3. Given a function $y(t)$ defined as

$$y(t) = \tfrac{1}{6}\delta(t) + \tfrac{1}{4}\delta(t-1) + \tfrac{1}{2}[u(t) - u(t-2)]$$

where $u(t)$ is the unit step function, find and plot the integral of $y(t)$,

$$Y(t) = \int_{-\infty}^{t} y(\alpha)\,d\alpha \qquad -\infty < t < \infty$$

4.2 PROPERTIES AND USES OF DISTRIBUTION, DENSITY, AND MASS FUNCTIONS

Examples 4.2 through 4.5 showed the derivation of the cumulative distribution function $F_X(\alpha)$, the density function $f_X(\alpha)$, and for the case of a discrete random variable the probability mass function $p_X(\alpha)$. A random variable was defined by a rule mapping the sample description space (or maybe an event space) onto the real axis. All the main properties of $F_X(\alpha)$, $f_X(\alpha)$, and $p_X(\alpha)$ may be derived basically from the axioms of probability theory:

$$P(E_i) \geq 0$$
$$P(S) = 1$$
$$P(E_i \cup E_j) = P(E_i) + P(E_j) \qquad \text{if } E_i \cap E_j = \varnothing$$

The Cumulative Distribution Function $F_X(\alpha)$

If a random variable X is defined on S, it must follow that the cumulative distribution function $F_X(\alpha)$, defined by $F_X(\alpha) = P(X \leq \alpha)$, possesses the following three properties:

1. $F_X(\beta) \geq F_X(\alpha)$ if $\beta > \alpha$. This is so because

$$(X \leq \beta) = (X \leq \alpha) \cup (\alpha < X \leq \beta)$$

$$\therefore F_X(\beta) = F_X(\alpha) + P(\alpha < X \leq \beta) \quad \text{and} \quad P(\alpha < X \leq \beta) \geq 0$$

We also note that nonintersecting ranges of α correspond to mutually exclusive events, since a point of S maps into only one point on the α axis.

2. $F_X(-\infty) = 0$. This property is obvious since all points of S map onto the real axis.

3. $F_X(+\infty) = 1$. This is true because $(X \leq \infty)$ contains the mapping of all points of S, and therefore $(X \leq \infty)$ is the sure event S.

These three properties are true whether the random variable is discrete, continuous, or mixed. They may be summarized by saying that $F_X(\alpha)$ is a nondecreasing function of α, starting at 0 for $\alpha = -\infty$ and culminating at 1 for $\alpha = \infty$.

The Density Function $f_X(\alpha)$

As a consequence of the three properties of $F_X(\alpha)$, it follows that the density function $f_X(\alpha)$, which is defined as $(d/d\alpha)F_X(\alpha)$, must possess the following two properties:

1. $f_X(\alpha) \geq 0$, since $F_X(\alpha)$ is nondecreasing.
2. $\int_{-\infty}^{\infty} f_X(\alpha)\, d\alpha = 1$, since

$$F_X(\alpha) = \int_{-\infty}^{\alpha^+} f_X(u)\, du$$

and therefore

$$F_X(\infty) = \int_{-\infty}^{\infty} f_X(\alpha)\, d\alpha = 1$$

These properties are true whether X is continuous, discrete, or mixed and whether we use singularity or delta functions.

When thinking of the density function of a *continuous* random variable

$$f_X(\alpha)\, d\alpha = P(\alpha < X < \alpha + d\alpha)$$

or we say, "The value of $f_X(\alpha)$ for any α is a measure of the probability of the range $\alpha < X < \alpha + d\alpha$." $f_X(\alpha_1)$ and $f_X(\alpha_2)$ would measure the relative probability of ranges $\Delta\alpha$ about α_1 and α_2, respectively.

The Mass Function $p_X(\alpha)$

The mass function $p_X(\alpha)$, defined as $p_X(\alpha) = P(X = \alpha)$, is useful only if the random variable is *discrete*, which means that $F_X(\alpha)$ changes only with discrete jumps. By definition, it must possess the following two properties:

1. $p_X(\alpha) \geq 0$.
2. $\sum_i p_X(\alpha_i) = 1$.

Its connection to $F_X(\alpha)$ is given by

$$F_X(\alpha) = \sum_i p_X(\alpha_i) \qquad \text{for } \alpha_i \leq \alpha$$

If $p_X(\alpha_i) > 0$, then the density function must contain $p_X(\alpha_i)\delta(\alpha - \alpha_i)$. The density function for a discrete random variable is

$$f_X(\alpha) = \sum_i p_X(\alpha_i)\delta(\alpha - \alpha_i)$$

So far this chapter has stressed the definition and understanding of the cumulative distribution, density, and mass functions for random variables.

The purpose of these functions is to allow a person to obtain answers rapidly to probabilistic questions pertaining to the random variable. An engineer may only want to use the previously developed results of Fermi-Dirac statistics or be given the mass functions for the number of electrons emitted from a cathode area or the density function for the velocities of emitted electrons. Their fundamental derivation from an assumed model and sample space, though fascinating, may be beyond the time requirements or even the abilities of the engineer. We will now consider a few problems where we are asked probabilistic questions about a random variable whose density function is given.

Example 4.6

The density function for a continuous random variable X is

$$f_X(\alpha) = Ae^{-\alpha} \qquad 1 < \alpha < 3$$
$$= 0 \qquad \text{otherwise}$$

(a) Find the value of A.
(b) What is the probability that the random variable takes on a value between 2 and 3?
(c) What is the probability that X takes on a value greater than 2 given that it has a value greater than 1.5?
(d) Find the conditional density function $f_X[(\alpha/X>2)]$.
(e) Find the cumulative distribution function $F_X(\alpha)$ and reanswer parts (b) and (c) using $F_X(\alpha)$.

SOLUTION

(a) In order for $f_X(\alpha)$ to be a density function we must have

$$\int_{-\infty}^{\infty} f_X(\alpha)\,d\alpha = 1 \quad \text{or} \quad \int_{1}^{3} Ae^{-\alpha}\,d\alpha = 1$$

$$\therefore A\left(-e^{-\alpha}\big|_1^3\right) = 1$$

or

$$A = \frac{1}{e^{-1} - e^{-3}} = \frac{1}{0.37 - 0.05} \approx 3.1$$

(b) $P(2 < X < 3) = \int_{2}^{3} 3.1 e^{-\alpha}\,d\alpha$

$$= 3.1(e^{-2} - e^{-3}) = 0.25$$

(c) $P[(2 < X < 3)/(X > 1.5)]$ is a question involving a conditional event, and using set theory as in discrete probability yields

$$P[(2<X<3)/(X>1.5)] = \frac{P[(2<X<3)\cap(X>1.5)]}{P(X>1.5)}$$

$$= \frac{P(2<X<3)}{P(X>1.5)}$$

$$= \frac{0.25}{\int_{1.5}^{3} 3.1e^{-\alpha}d\alpha} = \frac{0.25}{0.53} = 0.47$$

(d) To give probabilistic meaning to a density function we must consider integrating it over a range. Over a small range $\Delta\alpha$,

$$f_X(\alpha/X>2)\Delta\alpha = P[(\alpha<X<\alpha+\Delta\alpha)/(X>2)]$$

$$= \frac{P[(\alpha<X<\alpha+\Delta\alpha)\cap(X>2)]}{P(X>2)}$$

For $\alpha<2$,

$$f_X[\alpha/(X>2)] = 0$$

For $2<\alpha<3$,

$$f_X[(\alpha/(X>2)]\Delta\alpha = \frac{3.1e^{-\alpha}\Delta\alpha}{\int_{2}^{3} 3.1e^{-\alpha}d\alpha}$$

$$= 12.4e^{-\alpha}\Delta\alpha$$

$$\therefore f_X[(\alpha/(X>2)] = 12.4e^{-\alpha} \qquad 2<\alpha<3$$

For $\alpha>3$,

$$f_X[(\alpha/(X>2)] = 0$$

(e) With some work the student can find $F_X(\alpha)$ as follows:

$$F_X(\alpha)=0 \qquad \alpha<1$$

$$F_X(\alpha)=3.1(e^{-1}-e^{-\alpha}) \qquad 1<\alpha<3$$

and

$$F_X(\alpha)=1 \qquad \alpha>3$$

Now the previous questions will be reanswered.

(b) $P(2< X<3)= F_X(3)- F_X(2)$

$$=1-3.1(e^{-1}-e^{-2})$$

$$=0.25$$

(c) $P[(2< X<3)/(X>1.5)]=\dfrac{F_X(3)- F_X(2)}{1- F_X(1.5)}$

where we notice that $P(X>1.5)=1- P(X<1.5)$ since $(X>1.5)$ and $(X<1.5)$ are mutually exclusive, collectively exhaustive events.

$$\therefore P[(2< X<3)/(X>1.5)] = \dfrac{0.25}{1-0.47}$$

$$=0.47$$

It should be noticed that we have been careless in omitting equality signs. Although this is permissible for a continuous variable, it would be essential for us to keep track of these signs for a discrete or mixed random variable.

Example 4.7

Given that the density function for a mixed random variable X is

$$f_X(\alpha)=\tfrac{1}{3}\delta(\alpha)+\tfrac{1}{4}\delta(\alpha-3)+\tfrac{1}{12}[u(\alpha+3)-u(\alpha-2)]$$

find the probabilities of the following events;

(a) $(0< X\le2.6)$
(b) $[(0< X\le2.6)/(X\ge0)]$

SOLUTION
A thumbnail sketch of $f_X(\alpha)$ is shown.

(a) $P(0< X<2.6)= \displaystyle\int_{0^+}^{2.6}f_X(\alpha)\,d\alpha$

$$= \int_{0^+}^{2} \tfrac{1}{12}\, d\alpha$$

$$=0.167$$

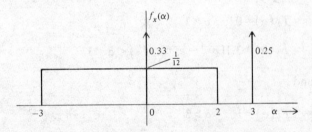

It is very important to realize that the delta function at the origin is not included in the integration.

(b) $P[(0<X<2.6)/(X\geq0)]=\dfrac{P[(0<X<2.6)\cap(X\geq0)]}{P(X\geq0)}$

$$=\dfrac{\displaystyle\int_{0^+}^2 \tfrac{1}{12}d\alpha}{\displaystyle\int_{0^-}^\infty f_X(\alpha)\,d\alpha}$$

$$=\dfrac{0.167}{0.33+0.25+0.167}$$

$$=0.22$$

In this case both delta functions were included in the denominator integral.

Examples 4.6 and 4.7 are illustrations of evaluating probabilistic questions about a random variable. All the required discipline was mastered in discrete probability theory. We notice that it is quicker to answer questions with $F_X(\alpha)$ than with $f_X(\alpha)$, since the density function requires integration. However, it is customary to represent random variables by their density functions because $f_X(\alpha)$ tends to be less cumbersome than $F_X(\alpha)$. If many questions are asked about a random variable, it becomes worthwhile to obtain and use the cumulative distribution function. Most simple books on statistics give tabular presentations of much used cumulative distribution functions (gaussian, binomial, and Poisson random variables) to facilitate problem solving or to enable education or social science students not versed in calculus to answer probabilistic questions. With the availability of calculators we will see that the only necessary table is one for the cumulative distribution function of the gaussian random variable.

In order to conclude this section, Table 4.1 in Section 4.5 shows a summary for the density functions of some much utilized random variables. In Section 4.5 the derivation of a number of these will be considered. From what we have done so far the student should understand the Bernoulli, binomial, and uniform functions and how they are involved in probability theory.

DRILL SET: PROPERTIES AND USES OF MASS, DENSITY, AND DISTRIBUTION FUNCTIONS

1. Given $F_X(\alpha)$ the cumulative distribution function of a random variable, find the probabilities of the following events in terms of $F_X(\alpha)$:
   (a) $(X<2)$ (b) $(X>3)$ (c) $(-1<X\leq2)$
   (d) $[(X>1)/(X\leq5)]$, $[(X<-1)\cup(X>+2)/(-3<X<1)]$

Note: Use $+$ and $-$ to account for lack or presence of equality signs, for example $P(X<2)=F(2^-)$, whereas $P(X\leq2)=F(2^+)$.

2. Given the density function

$$f_X(\alpha)=\tfrac{1}{6} \qquad 0<\alpha<2$$
$$=\tfrac{1}{3} \qquad 2<\alpha<4$$
$$=0 \qquad \text{otherwise}$$

(a) Find $P[(0.1<X<2)/(X<1)]$.

(b) Find and sketch $f_X(\alpha/X<1)$, $f_X(\alpha/X>3)$, and $f_X(\alpha/X=\tfrac{1}{4})$.

(c) Find and sketch $F_X(\alpha/X<1)$, $F_X(\alpha/X>3)$, and $f_X(\alpha/X=\tfrac{1}{4})$.

4.3 STATISTICS OF A RANDOM VARIABLE AND THE FUNDAMENTAL THEOREM

Statistics of a Random Variable

There are certain "statistics," or numbers, associated with random variables that give much information about them. For example, to a person who knows the density function they help to provide a physical picture and the ability to speak of important characteristics. To a statistician who is trying to decide if he or she is dealing with a random phenomenon and, if so, what the density function is for a random variable of interest, the experimental development of these statistics, or estimates of them, provides the basis for approximating the density function. Besides, many of these statistics—such as the mean, variance, and standard deviation—are part of our everyday life. We will now define some of the more important statistics of random variables.

The **expected** or **mean** or **average** value of a random variable X is defined as

$$\overline{X} \text{ or } E(X) \overset{\triangle}{=} \sum_{\text{all } i} \alpha_i p_X(\alpha_i) \tag{4.5}$$

for a discrete random variable and

$$\overline{X} \text{ or } E(X) = \int_{-\infty}^{\infty} \alpha f_X(\alpha)\, d\alpha \tag{4.6}$$

for any type random variable allowing for the use of delta functions. Formula 4.5 may easily be given a relative-frequency interpretation. If we consider that an experiment is performed N times on a random phenomenon and that α_i occurs N_1 times, α_2 occurs N_2 times, and so on, culminating with α_m occurring N_m times, then the average value obtained, which we will call $\overline{X}$, is

$$\overline{X} = \frac{\alpha_1 N_1 + \alpha_2 N_2 + \cdots + \alpha_m N_m}{N}$$

or

$$\bar{X} = \sum_{i=1} \alpha_i \frac{N_i}{N}$$

$$= \sum_{\text{all } i} \alpha_i p_X(\alpha_i) \qquad \text{as } \lim_{N \to \infty} \frac{N_i}{N} = p_X(\alpha_i)$$

For the continuous case the possible values are divided into ranges, say α_1 to $\alpha_1 + \Delta\alpha$, α_2 to $\alpha_2 + \Delta\alpha, \ldots, \alpha_m$ to $\alpha_m + \Delta\alpha$. If the different ranges occur $N_1, N_2, \ldots, N_m$ times, respectively, on N trials of the experiment, then if $\Delta\alpha$ is small over any range $\alpha_i < \alpha < \alpha_i + \Delta\alpha$, we can say that the value α_i is obtained N_i times. The average value $\bar{X}$ obtained is approximately

$$\frac{1}{N} \left(\sum_i \alpha_i N_i \right)$$

which becomes

$$\bar{X} = \sum_{\text{all } i} \alpha_i P(\text{range } \alpha_i < X < \alpha_i + \Delta\alpha)$$

$$= \sum_{\text{all } i} \alpha_i f_X(\alpha_i) \Delta\alpha$$

$$= \int_{-\infty}^{\infty} \alpha f_X(\alpha) \, d\alpha$$

as defined in Equation 4.6.

Before picturing what the mean value tells us, we can see that based on a relative-frequency interpretation, the expected value of any function of a random variable $g(X)$, denoted as $\overline{g(X)}$ or $E[g(X)]$, is defined as

$$\overline{g(X)} \text{ or } E[g(X)] \triangleq \sum_{\text{all } i} g(\alpha_i) p_X(\alpha_i) \tag{4.7}$$

for a discrete random variable and

$$\overline{g(X)} \text{ or } E[g(X)] \triangleq \int_{-\infty}^{\infty} g(\alpha) f_X(\alpha) \, d\alpha \tag{4.8}$$

for any random variable.

Some very useful special cases of Eqs. 4.7 and 4.8 are given in Eqs. 4.9 to 4.13. (We will only give the continuous formulas, using delta functions if necessary.) First,

$$E(X^2) \text{ or } \overline{X^2} \triangleq \int_{-\infty}^{\infty} \alpha^2 f_X(\alpha) \, d\alpha \tag{4.9}$$

where $\overline{X^2}$ is called the **mean square value**. Second,

$$E\left[(X - \bar{X})^2\right] \text{ or } \sigma_X^2 \triangleq \int_{-\infty}^{\infty} (\alpha - \bar{X})^2 f_X(\alpha) \, d\alpha \tag{4.10}$$

where σ_X^2 is called the **variance**. Third,

$$\sigma_X \overset{\Delta}{=} \sqrt{\int_{-\infty}^{\infty} (\alpha - \overline{X})^2 f_X(\alpha)\, d\alpha} \qquad (4.11)$$

where σ_X is called the **standard deviation** of X. Fourth,

$$E(X^n) \text{ or } \overline{X^n} \overset{\Delta}{=} \int_{-\infty}^{\infty} \alpha^n f_X(\alpha)\, d\alpha \qquad (4.12)$$

where $\overline{X^n}$ is called the **nth moment**. Finally,

$$E\left[(X - \overline{X})^n\right] \text{ or } \overline{(X - \overline{X})^n} \overset{\Delta}{=} \int_{-\infty}^{\infty} (\alpha - \overline{X})^n f_X(\alpha)\, d\alpha \qquad (4.13)$$

where $\overline{(X - \overline{X})^n}$ is called the **nth central moment**. If this were a textbook on the field of statistics, the present section would be expanded to probably the largest in the text. Since, however, we are primarily interested in probability theory, our treatment will be much briefer. An example will be solved in order to illustrate what some of these statistics should indicate to us.

Example 4.8

Given a random variable X and its density function

$$f_X(\alpha) = 1 \qquad 0 < \alpha < 1$$
$$\quad\; = 0 \qquad \text{otherwise}$$

and a second random variable Y and its density function

$$f_Y(\beta) = 0.25 \qquad -2 < \beta < 0$$
$$\qquad\;\; = 0.5 \qquad 0 < \beta < 1$$
$$\qquad\;\; = 0 \qquad \text{otherwise}$$

evaluate $\overline{X}$, $\overline{Y}$, $\overline{X^2}$, $\overline{Y^2}$, σ_X^2, and σ_Y^2, and comment on what they physically tell us about the random variables.

SOLUTION

A sketch of the density function $f_X(\alpha)$ is shown. Using the appropriate formulas we obtain

$$\overline{X} = \int_0^1 \alpha \cdot 1 \, d\alpha = 0.5$$

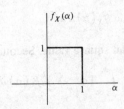

for the mean,

$$\overline{X^2} = \int_0^1 \alpha^2 \cdot 1 \, d\alpha = 0.33$$

for the mean square, and

$$\sigma_X^2 = \int_0^1 (\alpha - 0.5)^2 \, d\alpha = 0.083$$

for the variance. A sketch of the density function for $f_Y(\beta)$ is also shown.

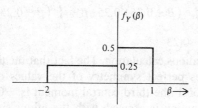

Our statistics now are as follows:

$$\overline{Y} = \int_{-2}^0 \beta(0.25) \, d\beta + \int_0^1 \beta(0.5) \, d\beta$$

$$= 0.125(-4) + 0.25$$

$$= -0.25$$

$$\overline{Y^2} = \int_{-2}^0 \beta^2(0.25) \, d\beta + \int_0^1 \beta^2(0.5) \, d\beta$$

$$= 0.66 + 0.17$$

$$= 0.83$$

and, with some work,

$$\sigma_Y^2 = \int_{-\infty}^\infty (\beta + 0.25)^2 f_Y(\beta) \, d\beta$$

becomes 0.76.

On consideration of our results we see that the average value for the symmetrical density function $f_X(\alpha)$ was midway between the endpoints, whereas for $f_Y(\alpha)$ it was at $Y = -0.25$. The average corresponds to the center of gravity if we consider that the density function assigns a mass density to the real axis. Statistically, $\overline{X}$ is the value of A that minimizes $(X - A)$ with value 0. On reflection of the meaning of the variance, we see that it indicates the central tendencies of values from the mean. A narrow density function has a smaller variance ($\sigma_X^2 = 0.08$) than a wide density function ($\sigma_Y^2 = 0.76$). The mean and the variance are the statistics that we visualize most readily, and in the accompanying drill set and homework a better grasp of them will be developed.

To a statistician higher-order central moments indicate properties such as symmetry or skewness of a density function about its mean. If we reconsider the density functions of Example 4.8, we find that the third central moments are

$$E(X - \overline{X})^3 = \int_0^1 (\alpha - 0.5)^3 \, d\alpha$$
$$= (0.25\alpha^4 - 1.5\alpha^3 + 0.375\alpha^2 - 0.125\alpha)|_0^1$$
$$= 0$$

and

$$E(Y - \overline{Y})^3 = \int_{-2}^0 (\beta + 0.25)^3 0.25 \, d\beta + \int_0^1 (\beta + 0.25)^3 0.5 \, d\beta$$
$$= -0.23$$

after a somewhat tedious calculation. The fact that the third central moment for X is 0 indicates perfect symmetry of the values of $f_X(\alpha)$ about their mean. In the case of Y the third central moment is -0.23, which indicates that the density function is skewed, with the minus sign resulting because most values are to the left of $\overline{Y}$.

Higher-Order Statistics

The problem of evaluating moments is so important that special functions are defined for calculating the general nth moment of a density function. The nth moment of a random variable is

$$\overline{X^n} = E(X^n) = \int_{-\infty}^{\infty} \alpha^n f_X(\alpha) \, d\alpha$$

It can be shown that if we define the function

$$X(p) \triangleq \int_{-\infty}^{\infty} f_X(\alpha) e^{-j2\pi p \alpha} \, d\alpha \tag{4.14}$$

that

$$E(X^n) = \int_{-\infty}^{\infty} \alpha^n f_X(\alpha) \, d\alpha = \frac{1}{(-2\pi j)^n} X^n(0) \tag{4.15}$$

where $X^n(0)$ is the nth derivative of $X(p)$ evaluated at $p = 0$. To an electrical engineer $X(p)$ is just the Fourier transform of $f_X(\alpha)$, where p is used to indicate frequency instead of f to avoid confusion with the f of $f_X(\alpha)$. The Fourier transform $X(p)$ is called the **moment-generating function** for the density function $f_X(\alpha)$. Students who have encountered Fourier transforms previously should check the solution to Example 4.8 for the means and variances using Eq. 4.15. Mathematicians tend to use a slightly different version of Eq. 4.15 for their moment-generating function. In addition, it can be shown that the version of a moment-generating function for discrete random variables corresponds to the z transform of electrical engineering. Since this is an introductory text, the enormous application of the Fourier and z transforms in evaluating moments will reluctantly not be pursued.

Rapid Calculation of Relations Between Moments

There is a number of important relations that can be developed involving the different statistics of a random variable and in addition there is a special shorthand algebra whose extension will be of paramount importance when dealing with second-order statistics for two random variables and in finding averages involving noise waveforms later in the text.

A very important law involving the variance is that, for any random variable,

$$\sigma_X^2 = \overline{X^2} - \overline{X}^2 \tag{4.16}$$

which, in words, says that the variance is the mean square value minus the mean value squared. An application of this in circuits is that if our waveform is sinusoidal, $x(t) = X_m \cos(\omega t + \phi)$, then the root-mean-square (rms) value turns out to be $X_m/\sqrt{2}$. Returning from our digression, we will prove Eq. 4.16 in a longhand manner and then discuss some shorthand notation.

$$\sigma_X^2 = \int_{-\infty}^{\infty} (\alpha - \overline{X})^2 f_X(\alpha)\, d\alpha$$

$$= \int_{-\infty}^{\infty} \alpha^2 f_X(\alpha)\, d\alpha - \int_{-\infty}^{\infty} 2\alpha \overline{X} f_X(\alpha)\, d\alpha + \int_{-\infty}^{\infty} \overline{X}^2 f_X(\alpha)\, d\alpha$$

$$= \overline{X^2} - 2\overline{X} \int_{-\infty}^{\infty} \alpha f_X(\alpha)\, d\alpha + \overline{X}^2 \int_{-\infty}^{\infty} f_X(\alpha)\, d\alpha$$

$$= \overline{X^2} - 2\overline{X} \cdot \overline{X} + \overline{X}^2 \cdot 1 \quad \text{(since the area of any density function is 1)}$$

$$= \overline{X^2} - \overline{X}^2$$

In our shorthand notation we would just write

$$E\left[(X - \overline{X})^2\right] = E(X^2 - 2\overline{X}X + \overline{X}^2)$$

$$= E(X^2) - 2\overline{X}E(X) + E(\overline{X})^2 = \overline{X^2} - \overline{X}^2$$

This implies the following rules, which are easily proved, for taking expected values:

$$E(C) = C \tag{4.17}$$

$$E[Cg(X)] = CE[g(X)] \tag{4.18}$$

and

$$E[g_1(X) + g_2(X)] = E[g_1(X)] + E[g_2(X)] \tag{4.19}$$

where C is a constant and X is a random variable. To illustrate the use of this notation further we will consider another example.

Example 4.9

Express the third central moment $E[(X - \overline{X})^3]$ in terms of the third, second, and first moments.

SOLUTION

$$E\left[(X-\overline{X})^3\right]=E\left[X^3-3X^2\overline{X}+3X(\overline{X})^2-\overline{X}^3\right]$$
$$=\overline{X^3}-3\overline{X}E(X^2)+3\overline{X}^3-\overline{X}^3$$
$$=\overline{X^3}-3\overline{X}\cdot(\overline{X^2})+2\overline{X}^3$$

This is the formula the author used when finding the third central moment for Y in Example 4.8.

DRILL SET: STATISTICS OF A RANDOM VARIABLE

1. Prove $\sigma_X^2=\overline{X^2}-\overline{X}^2$ for discrete random variables.
2. Consider $f_X(\alpha)$ as shown plotted. Find $\overline{X}$, $\overline{X^2}$, and σ_X^2.

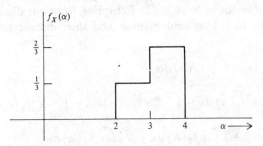

3. If $f_X(\alpha)$ is as shown, with minimum work predict the answers from the results of Problem 2. Check your answers.

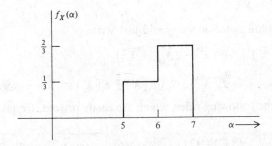

The Fundamental Theorem for One Random Variable

The fundamental theorem states that if a random variable Y is defined as a function of another random variable X as

$$Y=g(X)$$

then statistics involving Y may be evaluated, if so desired, in terms of the density function of X.

For example, if $Y=g(X)$,

$$\overline{Y}\overset{\Delta}{=}\int_{-\infty}^{\infty}\alpha f_Y(\alpha)\,d\alpha$$

using the density function $f_Y(\alpha)$, and

$$\overline{Y} = \overline{g(X)} = \int_{-\infty}^{\infty} g(\alpha) f_X(\alpha) \, d\alpha \quad .$$

using the fundamental theorem. Similarly,

$$\overline{Y^2} = \int_{-\infty}^{\infty} \alpha^2 f_Y(\alpha) \, d\alpha = \int_{-\infty}^{\infty} g^2(\alpha) f_X(\alpha) \, d\alpha$$

or

$$\overline{\sin Y} = \int_{-\infty}^{\infty} \sin \alpha \, f_Y(\alpha) \, d\alpha = \int_{-\infty}^{\infty} \sin[g(\alpha)] \, f_X(\alpha) \, d\alpha$$

The most general statement of the fundamental theorem is:
If $Y = g(X)$ then

$$\overline{K(Y)} \triangleq \int_{-\infty}^{\infty} K(\alpha) f_Y(\alpha) \, d\alpha = \int_{-\infty}^{\infty} K[g(\alpha)] f_X(\alpha) \, d\alpha \qquad (4.20)$$

The great simplification of work obtained by utilizing this theorem will be demonstrated by an example.

Example 4.10

Given that $Y = \sin X$ and that X is a random variable with density function

$$f_X(\alpha) = \frac{1}{2\pi} \qquad 0 < \alpha \le 2\pi$$

find $\overline{Y}$ and $\overline{Y^2}$.

SOLUTION
Using the fundamental theorem,

$$\overline{Y} = \int_{-\infty}^{\infty} \sin \alpha \, f_X(\alpha) \, d\alpha$$

$$= \int_{0}^{2\pi} \sin \alpha \, \frac{1}{2\pi} \, d\alpha = 0$$

$$\overline{Y^2} = \int_{-\infty}^{\infty} \sin^2 \alpha \, f_X(\alpha) \, d\alpha$$

$$= \int_{0}^{2\pi} \sin^2 \alpha \, \frac{1}{2\pi} \, d\alpha$$

$$= \frac{1}{2\pi} \int_{0}^{2\pi} \frac{1}{2} (1 - \cos 2\alpha) \, d\alpha$$

$$= \frac{1}{2}$$

In the next section we will study the problem of "given $Y = g(X)$ and $f_X(\alpha)$, find $f_Y(\beta)$," but at the moment we can appreciate the tremendous saving in not handling $\overline{Y}$ and $\overline{Y^2}$ directly.

Statistics for Sampling a Waveform Using the Fundamental Theorem

We will now use the fundamental theorem to develop a technique for finding statistics involved in sampling a periodic waveform or any waveform over a range of time. This problem, which is very important in communication theory, will be derived in general and then applied to a few specific cases. Let us consider a general periodic function,

$$y(t) = \sum_{-\infty}^{\infty} g(t - nT) \qquad n \text{ an integer}$$

where $g(t)$ is defined as being such that

$$g(t) \neq 0 \qquad 0 < t < T$$
$$= 0 \qquad \text{otherwise}$$

A typical sketch of $y(t)$ might be as shown in Figure 4.6.

If Y is the random variable that describes sampling the periodic waveform uniformly, in order to find statistics associated with Y we need only consider one period. Let X be the random variable associated with uniformly choosing a point between 0 and T on the time axis. The random variables Y and X are related by $Y = g(X)$. The density function for X is

$$f_X(\alpha) = \frac{1}{T} \qquad 0 < \alpha < T$$
$$= 0 \qquad \text{otherwise}$$

and, since $Y = g(X)$, any statistic associated with sampling the waveform may be found using $f_X(\alpha)$ and the fundamental theorem. For example,

$$\overline{Y} = \int_0^T g(\alpha) \frac{1}{T} d\alpha$$

and

$$\overline{Y^2} = \int_0^T g^2(\alpha) \frac{1}{T} d\alpha$$

and any statistic of Y may be found as

$$\overline{K(Y)} = \int_0^T K[g(\alpha)] f_X(\alpha) d\alpha$$

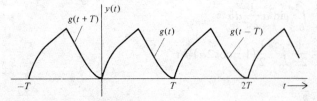

Figure 4.6   A typical periodic waveform.

Example 4.11

Consider the two periodic waveforms defined as follows:

$$\text{waveform 1:} \quad y_1(t) = \sin t$$

and

$$\text{waveform 2:} \quad y_2(t) = \sum_{n=-\infty}^{\infty} g_2(t - 2n)$$

where

$$g_2(t) = t \quad 0 < t < 1$$
$$= 0 \quad \text{otherwise}$$

Define the random variable X as the value obtained when $y_1(t)$ is uniformly sampled between 0 and 2π and the random variable Y as the value obtained when $y_2(t)$ is uniformly sampled between 0 and 2.

(a) Evaluate the following: $\overline{X}$, $\overline{X^2}$, σ_X, $\overline{Y}$, $\overline{Y^2}$, and σ_Y.
(b) Consider each waveform on a time basis and find its average, mean square, and rms values as you might in a basic circuits course.
(c) Are the statistics of the random variables and the time averages the same?

SOLUTION

(a) A plot of the two periodic waveforms is shown in Figure 4.7. From the problem statement the random variable X consists of sampling $y_1(t)$, where t is uniformly chosen and $0 < t < 2\pi$. If we define the random variable K as the value obtained by sampling the time axis uniformly between 0 and 2π, then the relationship between X and K is $X = \sin K$. Obviously, the density function for K is

$$f_K(\alpha) = \frac{1}{2\pi} \quad 0 < \alpha < 2\pi$$

and $\overline{X}$, $\overline{X^2}$, and σ_X may be obtained from the fundamental theorem using the density function of K and the fact that $X = \sin K$

$$\overline{X} = \int_0^{2\pi} \sin \alpha \frac{1}{2\pi} d\alpha = 0$$

$$\overline{X^2} = \int_0^{2\pi} \sin^2 \alpha \frac{1}{2\pi} d\alpha = \frac{1}{2}$$

and

$$\sigma_X = \sqrt{\tfrac{1}{2} - 0} = 0.72$$

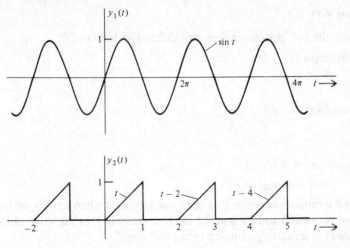

Figure 4.7   The periodic waveforms of Example 4.11.

The procedure for evaluating statistics on Y is very similar, but we must be careful. If we define W as the random variable describing the value obtained by uniformly choosing a point on the time axis $0 < t < 2$, then we have the relation between Y and W as follows:

$$Y = W \qquad \text{if } W \text{ assumes a value } 0 < W < 1$$

and

$$Y = 0 \qquad \text{if } W \text{ assumes a value } 1 < W < 2$$

The density function for W is

$$f_W(\beta) = \tfrac{1}{2} \qquad 0 < \beta < 2$$

Using the fundamental theorem the required statistics for Y are

$$\bar{Y} = \int_0^2 g(\beta) f_W(\beta) \, d\beta$$

$$= \int_0^1 \beta \left(\frac{1}{2}\right) d\beta + \int_1^2 0 \left(\frac{1}{2}\right) d\beta$$

$$= \frac{1}{4}$$

$$\overline{Y^2} = \int_0^1 \beta^2 \left(\frac{1}{2}\right) d\beta + \int_1^2 0^2 \left(\frac{1}{2}\right) d\beta$$

$$= \frac{1}{6}$$

and

$$\sigma_Y = \sqrt{\overline{Y^2} - \overline{Y}^2} = \sqrt{\tfrac{1}{6} - \tfrac{1}{16}} = 0.32$$

These statistics could have been found directly by obtaining $f_Y(\beta) = \tfrac{1}{2} + \tfrac{1}{2}\delta(\beta)$, $0 < \beta < 1$, and using this density function. The student should recalculate the values obtained.

(b) The problem now is the already familiar one from a basic circuits course of finding the average value of each waveform, the mean square value of each waveform, and the rms value based on taking time averages.

By definition, the time average of a waveform is

$$\widetilde{y_1(t)} = \lim_{T \to \infty} \frac{1}{2T} \int_{-T}^{T} y_1(t)\, dt$$

where the symbol $\sim$ is used to denote a *time* and not a statistical average. Since $y_1(t)$ is periodic, we may average over any one period instead of a very long time and

$$\widetilde{y_1(t)} = \frac{1}{2\pi} \int_{0}^{2\pi} \sin t\, dt = 0$$

Also,

$$\widetilde{y_2(t)} = \frac{1}{2}\left[\int_{0}^{1} t\, dt + \int_{1}^{2} 0\, dt \right] = \frac{1}{4}$$

The mean square values are

$$\widetilde{y_1^2(t)} = \frac{1}{2\pi} \int_{0}^{2\pi} \sin^2 t\, dt = \frac{1}{2}$$

and

$$\widetilde{y_2^2(t)} = \frac{1}{2}\left[\int_{0}^{1} t^2\, dt + \int_{1}^{2} 0\, dt \right] = \frac{1}{6}$$

and the rms values or standard deviations are

$$y_{1_{rms}} = \sqrt{ \frac{1}{2\pi} \int_{0}^{2\pi} \sin^2 t\, dt } = 0.72$$

$$= \sigma_x \quad (\text{since } \overline{X} = 0)$$

and

$$y_{2_{rms}} = \sqrt{\frac{1}{2}\left[\int_0^1 t^2\, dt + \int_1^2 0\, dt\right]} = \frac{\sqrt{6}}{6} = 0.4$$

and $\sigma_Y = \sqrt{0.4^2 - 0.06} = 0.32$, as before.

(c) The answer is yes. The statistical averages based on uniformly sampling a time waveform over a period, an exact multiple of a period, or from $-\infty$ to ∞, gives identical results to the time averages of the waveform already familiar to the reader. We note that the rms value equals the standard deviation only when the mean value is 0.

Example 4.11 was conceptually very important in that we tied together our previously understood concept of averages for time waveforms with a probabilistic and more fundamental definition for averages based on statistical definitions. This will serve as a basis for the general time domain, Monte Carlo, or computer analysis of what will be called "ergodic random processes."

DRILL SET: THE FUNDAMENTAL THEOREM

Consider the periodic waveform

$$x(t) = \sum_{-\infty}^{\infty} g(t - 4n)$$

where

$$g(t) = 2t \qquad -1 < t < +1$$
$$= 0 \qquad \text{otherwise}$$

(a) Sketch $x(t)$, its output $y(t)$ to a rectifier defined as $y(t) = |x(t)|$, and its output $z(t)$ to a square-law device defined by $z(t) = x^2(t)$.

(b) Define the following random variables with one-to-one mappings from the phenomenon of sampling the time axis and waveforms. Let P define sampling the time axis uniformly for all the waveforms, and let X, Y, and Z be associated with sampling $x(t)$, $y(t)$, and $z(t)$, respectively. Use the fundamental theorem to find $\overline{X}$, $\overline{X^2}$, σ_X^2, $\overline{Y}$, $\overline{Y^2}$, σ_Y^2, $\overline{Z}$, $\overline{Z^2}$, and σ_Z^2.

4.4 THE DENSITY FUNCTION FOR A FUNCTION OF A RANDOM VARIABLE

This section will consider the following problem:

> Given a random variable X and its associated density function $f_X(\alpha)$ and also a second random variable Y defined as $Y = g(X)$, where $g(X)$ is some function of X, find the density function $f_Y(\beta)$.

If we were interested only in a few statistics of Y, it would not be necessary to find $f_Y(\beta)$ inasmuch as the statistics could be found using the fundamental theorem. However, we can see the practical nature of this problem because if we have a system with

	and	
	a system	from which
input $x(t)$	rule	the output $y(t)$

may be found, then if $f_X(\alpha)$ is the density function for sampling $x(t)$, we might require to find $f_Y(\beta)$ the density function for sampling the output $y(t)$. Some important applications are linear systems and nonlinear systems such as rectifiers, square-law detectors, and so on.

One specific procedure for finding $f_Y(\beta)$ is to utilize the following steps:

Step 1 Express $F_Y(\beta) \triangleq P(Y \leq \beta)$ as a probabilistic statement about the random variable X.

$$F_Y(\beta) = P[g(X) \leq \beta] = \int_{\substack{\text{all points} \\ g(\alpha) \leq \beta}} f_X(\alpha)\, d\alpha$$

Step 2 Differentiate the answer for $F_Y(\beta)$ with respect to β to find $f_Y(\beta)$. To illustrate this procedure a few examples will be solved.

Example 4.12

Given

$$f_X(\alpha) = 1 \qquad 0 < \alpha < 1$$
$$ = 0 \qquad \text{otherwise}$$

and another random variable Y defined as $Y = 4X + 5$, find $f_Y(\beta)$ for all β.

SOLUTION

$$F_Y(\beta) = P(4X + 5 \leq \beta)$$
$$= P\left(X \leq \frac{\beta}{4} - \frac{5}{4}\right)$$

Now we must consider the general ranges of β of for which $F_Y(\beta)$ will exist. Since $f_X(\alpha)$ is 0, for $\alpha<0$, then Y takes on no value less than 5. Also, since X cannot take on a value greater than 1, Y takes on no greater value than 9. $F_Y(\beta)$ will now be found for all ranges of β.

For $\beta<5$,

$$F_Y(\beta)=0 \quad (\text{since } Y>5)$$

The range $5<\beta<9$ is arrived at by substituting $X=0$ and $X=1$ into $4X+5$.

$$F_Y(\beta)= P\left(X\le \frac{\beta}{4} - \frac{5}{4}\right)= \int_0^{\beta/4-5/4} 1\,d\alpha$$

$$= \frac{\beta}{4} - \frac{5}{4} \quad 5<\beta<9$$

For $\beta>9$,

$$F_Y(\beta)=1 \quad (\text{since } Y\le 9)$$

The density function $f_Y(\beta)$ may now be found by differentiation.

$$f_Y(\beta)=0 \quad \beta<5$$
$$= \tfrac{1}{4} \quad 5<\beta<9$$
$$=0 \quad \text{otherwise}$$

$f_X(\alpha)$, $F_Y(\beta)$, and $f_Y(\beta)$ are shown plotted in Figure 4.8a, b, and c, respectively.

With thought we can realize that a linear transformation of the type $Y= aX + b$ could be obtained by the operations of widening $f_X(\alpha)$ by a (if $a>1$ and positive) and then shifting the result b units to the right.

Example 4.13

Given $f_X(\alpha)=\tfrac{1}{2}e^{-|\alpha|}$ and $Y= X^2$, find $f_Y(\beta)$.

In engineering this problem might arise as follows: Given that the density function of a random variable sampling the input of a square-law detector is $f_X(\alpha)=\tfrac{1}{2}e^{-|\alpha|}$, find the density function for the random variable sampling the output waveform.

SOLUTION

The function $f_X(\alpha)$ is shown sketched in Figure 4.9a. Since $Y= X^2$, the smallest possible value Y can assume is 0. This leads to the conclusion that

$$F_Y(\beta)=0 \quad \beta<0$$

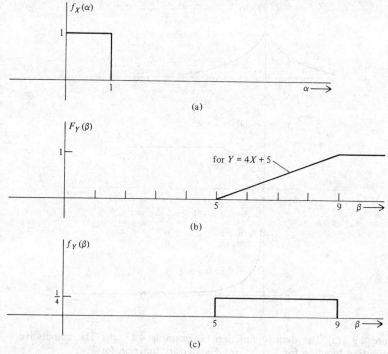

Figure 4.8 (a) The density function of Example 4.12. (b) The cumulative distribution function for $Y = 4X + 5$. (c) The density function for Y.

For the range $0 < \beta < \infty$,

$$F_Y(\beta) = P(X^2 < \beta) = P\left(-\sqrt{\beta} < X < +\sqrt{\beta}\right)$$

$$= \int_{-\sqrt{\beta}}^{\sqrt{\beta}} \frac{1}{2} e^{-|\alpha|} \, d\alpha$$

$$= 2 \int_0^{\sqrt{\beta}} \frac{1}{2} e^{-\alpha} \, d\alpha$$

$$= 1 - e^{-\sqrt{\beta}} \qquad \beta > 0$$

$f_Y(\beta)$ may now be found by differentiation as

$$f_Y(\beta) = \frac{e^{-\sqrt{\beta}}}{2\sqrt{\beta}} \qquad \beta > 0$$

Sketches of $F_Y(\beta)$ and $f_Y(\beta)$ are shown in Figure 4.9b and c, respectively. $f_Y(\beta)$ is interesting because $f_Y(0) = \infty$, but still on integration we find that the integral from 0 to any β is finite, and this is the discrete probability of the event $(0 < Y \leq \beta)$.

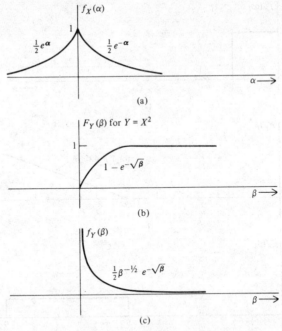

Figure 4.9 (a) The density function of Example 4.13. (b) The cumulative distribution function for $Y = X^2$. (c) The density function for Y.

General formulas may be developed for $f_Y(\beta)$ given $Y = g(X)$, but we must be careful to correctly interpret the mapping involved by the statement "$P[g(X) < \beta]$."

Case of One-to-One Mapping

If $Y = g(X)$ implies there is a one-to-one mapping between the random variables—that is, $X = g^{-1}(Y)$—then the statement

$$F_Y(\beta) = P[g(X) \leq \beta]$$

$$= \int_{-\infty}^{g^{-1}(\beta)} f_X(\alpha)\, d\alpha \quad \text{or} \quad \int_{g_2^{-1}(\beta)}^{\infty} f_X(\alpha)\, d\alpha$$

and on differentiating, using Leibnitz' rule, we obtain

$$f_Y(\beta) = f_X\big(g_1^{-1}(\beta)\big)\frac{d}{d\beta}\big[g_1^{-1}(\beta)\big]$$

or

$$f_Y(\beta) = -f_X\big(g_2^{-1}(\beta)\big)\frac{d}{d\beta}\big[g_2^{-1}(\beta)\big] \tag{4.21}$$

For example, $Y = 2X + 3$ yields $X = 0.5Y - 1.5$, or $g_2^{-1}(\beta) = 0.5\beta - 1.5$; and $Y = -4X + 3$ yields $X = -0.25Y + 0.75$ or $g_2^{-1}(\beta) = -0.25\beta + 0.75$.

Case of Multiple Mapping

If $Y = g(X)$ implies that there is not a one-to-one mapping between the random variables but that $Y = g(X)$ may lead to relations such as $X = g_1^{-1}(\beta)$ and $g_2^{-1}(\beta)$ as in Example 4.13, then the statement

$$F_Y(\beta) = P[g(x) \le \beta]$$

may result in a relation such as

$$F_Y(\beta) = \int_{g_2^{-1}(\beta)}^{g_1^{-1}(\beta)} f_X(\alpha) \, d\alpha$$

and using Leibnitz' rule twice we obtain

$$f_Y(\beta) = f_X\big(g_1^{-1}(\beta)\big) \frac{d}{d\beta}\big(g_1^{-1}(\beta)\big) - f_X\big(g_2^{-1}(\beta)\big) \frac{d}{d\beta}\big(g_2^{-1}(\beta)\big) \quad (4.22)$$

and we must carefully state for what range of β this density function exists.

Although these theoretical formulas are interesting, initially deriving a density function by systematically carrying out step 1 of this section to find $F_Y(\beta)$ and then differentiating the answer as step 2, is much less mechanical and more probabilistically enlightening than using Eqs. 4.21 and 4.22.

The Density Function for Sampling a Waveform

This section will be concluded by applying the preceding theory to the problem of sampling a waveform in some time interval and finding the density function for the random variable describing the sampled values. An example should illustrate the general procedure.

Example 4.14

Consider the waveform $x(t)$ shown in Figure 4.10a. Let X be the random variable that describes with a one-to-one mapping the experiment of uniformly choosing a point on the time axis $0 < t < 3$. Let Y be the random variable that describes with a one-to-one mapping the value of the waveform $x(t)$ corresponding to a value of X. Use the theory of this section to find $f_Y(\beta)$.

SOLUTION

The density function for uniformly sampling the time axis $0 < t < 3$ is

$$f_X(\alpha) = \tfrac{1}{3} \qquad 0 < \alpha < 3$$
$$= 0 \qquad \text{otherwise}$$

The random variable Y that describes sampling $x(t)$ is related to X by the functional relation

$$Y = X^2 \qquad 0 < X < 1$$
$$= 1 - X \qquad 1 < X < 2$$
$$= 0 \qquad 2 < X < 3$$

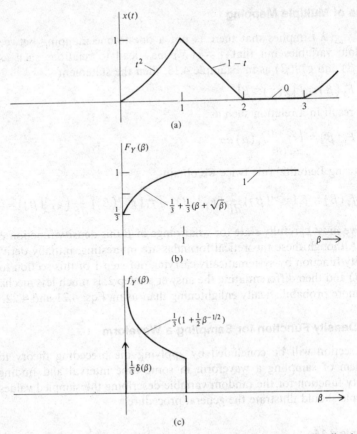

Figure 4.10   (a) The waveform of Example 4.14. (b) The cumulative distribution function for sampling it uniformly between 0 and 3. (c) The density function for sampling it.

The cumulative distribution function $F_Y(\beta)$ will now be found for all different possible ranges of β.

For range 1 ($\beta < 0$),

$$F_Y(\beta) = 0$$

For range 2 ($\beta = 0$),

$$F_Y(0) = P(Y \le 0) = P(2 < X < 3)$$

$$= \int_2^3 \frac{1}{3} d\alpha$$

$$= \frac{1}{3}$$

For range 3 $(0^+ < \beta < 1)$,

$$F_Y(\beta) = P(Y \le \beta)$$

$$= P(2 < X < 3) \cup [(0 < X < 1) \cap (X^2 < \beta)]$$
$$\cup [(1 < X < 2) \cap (1 - X) < \beta]$$
$$= \tfrac{1}{3} + P(0 < X < 1)P[(X^2 < \beta)/(0 < X < 1)]$$
$$+ P(1 < X < 2)P\{[(1 - X) < \beta]/(1 < X < 2)\}$$
$$= \tfrac{1}{3} + \tfrac{1}{3}\sqrt{\beta} + \tfrac{1}{3}[\beta]$$
$$= \tfrac{1}{3} + \tfrac{1}{3}\left(\beta + \sqrt{\beta}\right)$$

For range 4 $(\beta > 1)$,

$$F_Y(\beta) = 1$$

The derivation for range 3 is worth detailed thought because since the waveform has different analytic expressions for different ranges of time, the functional relation between Y and X is more complicated than in previous problems. The cumulative distribution function $F_Y(\beta)$ is shown plotted in Figure 4.10b. The density function $f_Y(\beta)$ may now be found by differentiation.

$$f_Y(\beta) = 0.33\delta(\beta) \qquad \beta = 0$$
$$f_Y(\beta) = 0.33(1 + 0.5\beta^{-1/2}) \qquad 0^+ < \beta < 1$$
$$= 0 \qquad \text{otherwise}$$

The density function is shown plotted in Figure 4.10c. This is a very important practical problem, whether we are talking about sampling a section of a deterministic waveform, a periodic waveform, or a section of a noise waveform of sufficient length to give information about the entire waveform for all time.

DRILL SET: THE DENSITY FUNCTION FOR A FUNCTION OF A RANDOM VARIABLE

1. Given

$$f_X(\alpha) = e^{-\alpha} \qquad 0 < \alpha < \infty$$
$$= 0 \qquad \text{otherwise}$$

with minimum work find the density functions of $Y = 3X + 4$ and $Z = -2X - 3$. Check your result for $f_Z(\gamma)$ by using the formula $F_Z(\gamma) = P[g(X) \le \gamma]$ carefully.

2. Given

$$f_X(\alpha) = \tfrac{1}{6} \qquad -2 < \alpha < 4$$

and $\qquad = 0 \qquad$ otherwise

$$Y = X^2$$

find $f_Y(\beta)$. Check that your answer is a density function.

*4.5 SOME NOTED MASS AND DENSITY FUNCTIONS[1]

This section will enumerate and discuss the derivations of mass and density functions for some of the most commonly occurring random variables.

Discrete Random Variables

THE BERNOULLI RANDOM VARIABLE
The Bernoulli random variable is defined as having a mass function

$$\left. \begin{aligned} p_X(0) &= q \\ p_X(1) &= p \end{aligned} \right\} \qquad (4.23)$$

where $p + q = 1$. Its applications are numerous and include:

1. The event of a success on a trial of a game.
2. The event a digital device is in the 0 or 1 state.
3. The event of a red ball being drawn from an urn (with replacement), and so on.

Its mean and variance are easily found as $\overline{X} = p$, $\overline{X^2} = p$, and $\sigma_X^2 = p - p^2 = pq$. The mass function and these statistics are given in Table 4.1.

THE BINOMIAL RANDOM VARIABLE
The binomial random variable is defined as having a mass function given by

$$p_X(\alpha) = \binom{n}{\alpha}(p)^{\alpha}(q)^{n-\alpha} \qquad \alpha = 0, 1, 2, \ldots, n$$

$$= 0 \quad (\alpha > n) \cup (\alpha < 0) \qquad (4.24)$$

where α is an integer, and $p > 0$, and $q = 1 - p$. From our previous experience we see that this random variable may arise as

1. The event of α successes on n trials of a game where the probability of success on a trial is p.
2. The event of drawing α red balls from an urn in a sample of size n (with replacement), and so on.

[1]Sections preceded by an asterisk may be omitted until encountered later in the book.

Table 4.1   SOME IMPORTANT MASS AND DENSITY FUNCTIONS

DENSITY OR MASS FUNCTION	COMMENTS	THUMBNAIL SKETCH
$p_X(\alpha_i) = p \quad \alpha_i = 1$ $(\quad) = q \quad \alpha_i = 0$ $(\quad) = 0 \quad$ otherwise	This is a Bernoulli trial or an event with two outcomes, where the probability of a success is p.	 $\bar{X} = p \quad \sigma_X^2 = pq$
$p_X(\alpha_i) = \binom{n}{\alpha_i} p^{\alpha_i}(q)^{n-\alpha_i}$ $\alpha_i = 0, 1, 2, \ldots, n$	This is the binomial law or the event of α_i successes on n Bernoulli trials, where the probability of a success on a trial is p.	 $\bar{X} = np, \ \sigma_X^2 = npq$
$p_X(\alpha_i) = e^{-\lambda} \dfrac{\lambda^{\alpha_i}}{\alpha_i!}$ $\alpha_i = 0, 1, 2, \ldots, \infty$	λ represents the average number of events in some fixed time and this Poisson law governs electron emission from cathodes plus applications in many fields for the number of events in a time T.	 $\bar{X} = \lambda$ $\sigma_X^2 = \lambda$ The Poisson plot looks like binomial for $\lambda \gg 0$.
$f_X(\alpha) = \dfrac{1}{b-a} \quad a < \alpha < b$ $= 0 \quad$ otherwise	This is the uniform choice of a point in an interval.	
$f_X(\alpha) = \lambda e^{-\lambda\alpha} \quad \alpha > 0$ $= 0 \quad$ otherwise	This is called an exponential random variable. One application is the time between emission of particles by a radioactive atom.	
$f_X(\alpha) = \dfrac{1}{\sigma\sqrt{2\pi}}$ $\exp\left[-\dfrac{1}{2}\left(\dfrac{\alpha-m}{\sigma}\right)^2\right]$	This is the famous gaussian random variable law. It occurs naturally in nature by virtue of the central limit theorem.	 *Note*: The binomial and Poisson approximate the gaussian law for np and λ large.

The mean and variance of X will now be found.

$$\bar{X} = \sum_0^n \alpha \binom{n}{\alpha} p^\alpha q^{n-\alpha}$$

$$= 0\left(\frac{n!}{0!n!}\right) p^0 q^n + 1\left[\frac{n!}{1!(n-1)!}\right] pq^{n-1} + \cdots$$

$$= np\left(0 + q^{n-1} + (n-1)pq^{n-2} + \frac{1}{2!}(n-1)(n-2)p^2q^{n-2} + p^{n-1}\right)$$

Recalling the binomial expansion,

$$(a+b)^n = a^n + na^{n-1}b + \left[\frac{n(n-1)}{2!}\right]a^{n-2}b^2 + \cdots + b^n$$

then

$$\bar{X} = np(q+p)^{n-1}$$
$$= np \quad \text{(since } q+p=1)$$

The mean square value is

$$\overline{X^2} = \sum_{\alpha>0}^{n} \alpha^2 \binom{n}{\alpha} p^\alpha q^{n-\alpha}$$

and, by careful expansion, it can be shown that

$$\overline{X^2} = npq + n^2 p^2$$

and the variance is

$$\sigma_X^2 = npq$$

The mass function for the case $n \gg 1$ and the mean and variance are given in Table 4.1.

THE GEOMETRIC RANDOM VARIABLE

The geometric random variable is defined as having a mass function

$$p_X(\alpha) = pq^{\alpha-1} \quad \alpha = 1, 2, 3, \ldots, \infty \tag{4.25}$$

where α is a positive integer, $p > 0$, and $q = 1 - p$. This random variable describes the number of trials until a success is achieved on a trial of an experiment, where the probability of a success is p.

It will be left as an exercise to show that

$$\bar{X} = \frac{1}{p} \quad \text{and} \quad \sigma_X^2 = \frac{1-p}{p^2}$$

THE POISSON RANDOM VARIABLE

A Poisson random variable has a mass function of the form

$$p_X(\alpha) = \frac{e^{-\lambda T}(\lambda T)^\alpha}{\alpha!} \quad \alpha = 1, 2, 3, \ldots, \infty \tag{4.26}$$

Next to a gaussian random variable, the Poisson random variable is probably the most important in electrical engineering. Among its applications are the mass function for (1) the number of electrons emitted from a small section of cathode surface in a time T and (2) waiting-line problems such as the number of customers served, the number of telephone calls received, or the number of cars passing a point in a time T.

The Poisson function may be derived as an extension of the binomial function to the continuous case.

DERIVATION OF THE POISSON MASS FUNCTION

Let us consider a fixed time interval T and consider it divided into n subintervals of time ΔT, such that n is very large. Let us assume that either one event (i.e., an electron being emitted or a phone call being received) can occur with probability $p = \lambda \Delta T$ or no event can occur with probability $q = 1 - \lambda \Delta T$ in an interval ΔT. Also assume that the occurrence of an event in a small interval has no effect on subsequent time intervals. Using the binomial mass function, the probability of α events occurring in T, where $n = T/\Delta T$ approaches infinity, is

$$
\begin{aligned}
p_X(\alpha:T) & \\
&= \lim_{n \to \infty} \binom{n}{\alpha} (\lambda \Delta T)^\alpha (1 - \lambda \Delta T)^{n-\alpha} \qquad \alpha = 0, 1, 2, \dots, n \\
&= \lim_{n \to \infty} \frac{n!}{\alpha!(n-\alpha)!} \frac{(\lambda T)^\alpha}{n^\alpha} \left(1 + \frac{\lambda T}{n}\right)^{n-\alpha} \\
&= \frac{(\lambda T)^\alpha}{\alpha!} \lim_{n \to \infty} \underbrace{\frac{n(n-1)\cdots(n-\alpha+1)}{n^\alpha}}_{\text{Term 1}} \underbrace{\left(1 - \frac{\lambda T}{n}\right)^n}_{\text{Term 2}} \underbrace{\left(1 - \frac{\lambda T}{n}\right)^{-\alpha}}_{\text{Term 3}}
\end{aligned}
$$

$$(4.27)$$

For $n \to \infty$, Term 1 obviously approaches 1. We will now show that Term 2 approaches $e^{-\lambda T}$ for $n \to \infty$.

$$
e^{-x} = 1 - x + \frac{1}{2}x^2 - \frac{1}{3!}x^3 + \frac{1}{4!}x^4 + \cdots
$$

$$
\left(1 - \frac{x}{n}\right)^n = 1 - n\left(\frac{x}{n}\right) + \frac{n(n-1)}{2}\left(\frac{x}{n}\right)^2 - \frac{n(n-1)(n-2)}{3!}\left(\frac{x}{n}\right)^3 + \cdots
$$

and as $n \to \infty$ this becomes approximately

$$
\left(1 - \frac{x}{n}\right)^n \approx 1 - x + \frac{1}{2}x^2 - \frac{1}{3!}x^3 + \cdots = e^{-x}
$$

$$
\therefore \lim \left(1 - \frac{\lambda T}{n}\right)^n = e^{-\lambda T}
$$

For a finite α, Term 3 approaches one since

$$
\left(1 - \frac{\lambda T}{n}\right)^{-\alpha} \approx 1
$$

which is the first term in its binomial expansion.

$$
\therefore p_X(\alpha, T) = \frac{(\lambda T)^\alpha e^{-\lambda T}}{\alpha!} \qquad \text{for } \alpha = 0, 1, 2, \dots, \infty
$$

Before the availability of calculators the use of the Poisson mass function for a quick approximation of the binomial mass function was much used.

For example,

$$\binom{20}{7}(0.3)^7(0.7)^{13} \simeq \frac{e^{-2.1}(2.1)^7}{7!} = 0.19$$

using the fact that $np = \lambda T$. The time saving in using the right-hand expression should be clear.

We will now evaluate the mean, mean square value, and variance for the Poisson random variable.

$$\bar{X} = \sum_0^\infty \alpha \frac{(\lambda T)^\alpha e^{-\lambda T}}{\alpha!}$$

$$= e^{-\lambda T}\left[0 + \frac{\lambda T}{1} + \frac{(\lambda T)^2}{1} + \frac{(\lambda T)^3}{2} + \cdots\right]$$

$$\therefore \bar{X} = \lambda T e^{-\lambda T}\left[1 + \lambda T + \frac{(\lambda T)^2}{2!} + \cdots\right]$$

$$= \lambda T$$

Similarly,

$$\overline{X^2} = \sum_0^\infty \alpha^2 \frac{(\lambda T)^\alpha e^{-\lambda T}}{\alpha!}$$

$$= e^{-\lambda T}\left[0 + \lambda T + (\lambda T)^2 + \frac{(\lambda T)^3}{1} + \frac{(\lambda T)^4}{2!} + \cdots\right]$$

$$= (\lambda T)^2 e^{-\lambda T}\left(\frac{1}{\lambda T} + e^{\lambda T}\right)$$

$$= \lambda T + (\lambda T)^2$$

and

$$\sigma_X^2 = (\lambda T)^2 + \lambda T - (\lambda T)^2$$

$$= \lambda T$$

The Poisson mass function and its statistics are shown sketched in Table 4.1.

Example 4.15

Consider a Poisson noise waveform where the number of transitions k from the value 1 to 0 obeys the Poisson mass function

$$p(k \text{ transitions: } 4 \text{ s}) = \frac{e^{-3}(3)^k}{k!} \qquad k = 0, 1, 2, \ldots, \infty$$

Sketch part of this "noise waveform" and find the probability that $x(t)$ and $x(t + \tau)$ are both of value 1 for a general τ.

SOLUTION

Every 4 s ($T=4$, $\lambda=0.75$), either zero, one, two, three, or more transitions occur, with the following probabilities:

$$P(0 \text{ transitions}) = \frac{e^{-3}3^0}{0!} = 0.05$$

$$P(1 \text{ transition}) = \frac{e^{-3}3^1}{1} = 0.15$$

$$P(2 \text{ transitions}) = \frac{e^{-3}9}{2} = 0.22$$

$$P(3 \text{ transitions}) = \frac{e^{-3}27}{6} = 0.22$$

$$P(4 \text{ transitions}) = 0.17$$

$$P(5 \text{ transitions}) = 0.10$$

and so on. A sketch of part of a section of this Poisson noise waveform is shown in Figure 4.11.

$$P[(x(t)=1) \cap (x(t+\tau)=1)] = P(1)P(1/1:\tau)$$

Obviously $P(x(t)=1)=0.5$ and the conditional probability of obtaining the value 1 at $t+\tau$ given the value was 1 at time t, is the probability an even number of transitions occur in τ s.

$$P(1/1:\tau) = P(\text{even number of transitions})$$

$$= \left(1 + \frac{e^{-0.75\tau}(0.75\tau)^2}{2} + \frac{e^{-0.75\tau}(0.75\tau)^4}{4!} + \cdots\right)$$

$$= e^{-0.75\tau}\left(1 + \frac{(0.75\tau)^2}{2} + \frac{(0.75\tau)^4}{4!} + \cdots\right)$$

To find the sum of this series of even terms we will use the following

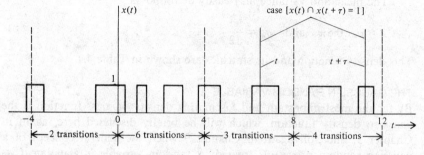

Figure 4.11   Section of Poisson noise waveform.

manipulation:

$$e^x = 1 + x + \frac{x^2}{2!} + \frac{x^3}{3!} + \cdots$$

$$e^{-x} = 1 - x + \frac{x^2}{2!} - \frac{x^3}{3!} + \cdots$$

$$\therefore 1 + \frac{x^2}{2!} + \frac{x^4}{4!} + \cdots = \frac{1}{2}(e^x - e^{-x})$$

and

$$P(1/1 : \tau) = e^{-0.75\tau}\left[\frac{1}{2}\left(e^{0.75\tau} - e^{-0.75\tau}\right)\right]$$

$$= \frac{1}{2}(1 + e^{-1.5\tau})$$

$$\therefore P[(x(t) = 1) \cap (x(t + \tau) = 1)] = \frac{1}{4}(1 + e^{-1.5\tau})$$

As a check we see that our answer is 0.5 if $\tau \to 0$ and 0.25 if $\tau \to \infty$. (Why is this a check?)

Continuous Random Variables

THE UNIFORM RANDOM VARIABLE

The uniform random variable is defined as having a density function

$$f_X(\alpha) = \frac{1}{w} \qquad a < \alpha < a + w$$

$$= 0 \qquad \text{otherwise} \tag{4.28}$$

A uniform density function is used when it is assumed that all continuous values in some range seem equally likely by symmetry. Some possible examples are:

1. The unknown phase angle of a sinusoid.
2. The error due to quantizing a signal.
3. A point randomly chosen on a segment of the time axis.

The mean and variance may easily be found as

$$\bar{X} = a + 0.5w \quad \text{and} \quad \sigma_X^2 = \frac{w^2}{12}$$

This density function and its statistics are shown in Table 4.1.

THE GAUSSIAN RANDOM VARIABLE

By far the most important and fascinating density or mass function is the gaussian density function, which will be briefly discussed here, again in Chapter 5, and still again in Chapter 9 when the general concept of a gaussian random process is treated. A random variable is gaussian if its density function is of the form,

$$f_X(\alpha) = \frac{1}{\sqrt{2\pi}\,\sigma} e^{-(\alpha-m)^2/2\sigma^2} \qquad -\infty < \alpha < \infty \tag{4.29}$$

where m and σ are real positive constants. Some notable characteristics of a gaussian random variable are

1. It occurs as a model for many random phenomena occurring in nature, heights of people, scores on tests, thermal noise, plasma current and voltage fluctuations in diodes, and so on.
2. It is completely specified by its mean and variance.
3. The density function for sampling the output of a linear system is gaussian if the density function for sampling the input is gaussian.
4. Its extension to a joint gaussian density function in Chapter 5 and a gaussian random process in Chapter 9 is of the utmost practical importance and analytical simplicity.

In order to answer any probabilistic questions about a gaussian random variable we need to relate the question to the standardized gaussian random variable,

$$f_s(\alpha) = \frac{1}{\sqrt{2\pi}} e^{-\alpha^2/2} \qquad -\infty < \alpha < \infty \tag{4.30}$$

The cumulative distribution function for the standardized gaussian random variable is

$$F_s(\alpha) = \int_{-\infty}^{\alpha} f_s(\alpha)\,d\alpha$$

and extensive tables of $f_s(\alpha)$ and $F_s(\alpha)$ are available in all classical texts and handbooks.[2] The statistics for the standardized gaussian random variables are

$$\overline{X_s} = 0 \quad \text{and} \quad \overline{\left(X_s - \overline{X}\right)^2} = \sigma_s^2 = 1$$

and the higher-order central moments yield the elegantly simple result,

$$\overline{\left(X - \overline{X}\right)^n} = 1 \times 3 \times 5 \times \cdots \times (n-1)$$

These statistics are not easily derived, however. It is left as an exercise to show that by simple transformation of variables the cumulative distribution function for any gaussian random variable as in Eq. 4.29 is

$$F_X(\alpha) = F_s\left(\frac{\alpha - m}{\sigma}\right) \tag{4.31}$$

---

[2]*CRC Handbook of Tables for Probability and Statistics*, 2d ed. (Cleveland: CRC Press, 1968), pp. 125–134.

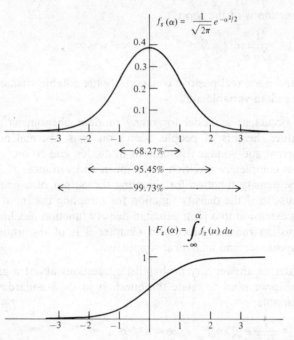

Figure 4.12   Normalized gaussian function and cumulative distribution function.

A simple sketch of $f_s(\alpha)$ and $F_s(\alpha)$ is shown in Figure 4.12 and a student calculated table of $F_s(\alpha)$ is shown in Table 4.2.

RELATIONSHIP BETWEEN THE GAUSSIAN, POISSON, AND BINOMIAL RANDOM VARIABLES

The envelopes of the mass functions for the binomial and Poisson random variables and a gaussian density function are approximately equal, as are

Table 4.2   TYPICAL $F_s(\alpha)$ VALUES

α	$F_s(\alpha)$	α	$F_s(\alpha)$
0	0.50	0.9	0.82
0.1	0.54	1.0	0.84
0.2	0.58	1.2	0.88
0.3	0.62	1.4	0.92
0.4	0.66	1.6	0.94
0.5	0.69	1.8	0.96
0.6	0.73	2.0	0.98
0.7	0.76	2.5	0.99
0.8	0.79	3.0	1.00

Note: If X is gaussian with mean $\overline{X}$ and variance σ^2,

$$F_X(\alpha) = F_s\left(\frac{\alpha - \overline{X}}{\sigma}\right)$$

their cumulative distribution functions, for the case $n \gg 1$ and p close to 0.5 for the binomial and $\lambda \gg 1$ for the Poisson. If $p_X(\alpha)$ is binomial with $\overline{X} = np$ and $\sigma^2 = n^2 p^2 q^2$, then for $n \gg 1$ it closely approximates the Poisson mass function with $\lambda = np$, whose envelope in turn approximates the gaussian random variable with $\overline{X} = \lambda$ or np and $\overline{(X - \overline{X})^2} = \lambda$. With the aid of a good calculator the reader could obtain intuitive insight by showing this for a few specific cases. The proof of these results we have quoted would involve showing that the characteristic functions of the binomial and Poisson random variables defined as

$$\phi_X(v) \triangleq \overline{e^{jv\alpha}}$$

$$= \sum_{\text{all } i} \alpha_i e^{jv\alpha_i} \tag{4.32}$$

approach the characteristic function for the gaussian random variable

$$\phi_X(v) = \int_{-\infty}^{\alpha} e^{jv\alpha} \frac{1}{\sqrt{2\pi}\,\sigma} e^{-(\alpha-m)^2/2\sigma^2} d\alpha \tag{4.33}$$

The evaluation of characteristics or moment-generating functions is very important but is not being pursued in this introductory treatment.

SUMMARY

A random variable X was defined by a *rule* that maps points of a sample description space onto the real axis. Three important functions were defined to facilitate answering probabilistic questions about a random variable. The cumulative distribution function $F_X(\alpha) \triangleq P(X \le \alpha)$ may always be obtained by solving a problem involving discrete probability theory; that is,

Given S the sample description space $$S = \bigcup_{\text{all } i} s_i$$ with a probability measure	and X defined by $X(s_i)$	we can solve for $$F_X(\alpha) \triangleq P(X \le \alpha)$$ for all α, where $-\infty < \alpha < +\infty$, by discrete probability theory

The density function $f_X(\alpha)$ is mechanically defined by $f_X(\alpha) \triangleq (d/d\alpha) F_X(\alpha)$. When we think about $f_X(\alpha)$ we visualize $f_X(\alpha) \Delta \alpha$ as the probability of the random variable assuming a value in the range $\alpha < X < \alpha + \Delta\alpha$ for the continuous case. Knowing $f_X(\alpha)$ we can find $F_X(\alpha)$ by integration as

$$\int_{-\infty}^{\alpha^+} f_X(u)\, du$$

where α^+ indicates the limit at α approached from the right. The $+$ sign is only necessary for discrete or mixed random variables. The mass function $p_X(\alpha) \triangleq P(X = \alpha)$ is used only for discrete random variables. $p_X(\alpha)$ may be found easily by discrete probability and

$$F_X(\alpha) = \sum_{\alpha_i \le \alpha} p_X(\alpha_i)$$

by definition. Although $F_X(\alpha)$ is the most fundamental function characterizing a random variable, in practice if possible we find $p_X(\alpha)$ or $f_X(\alpha)$ first and then $F_X(\alpha)$ since it is the most unwieldy of the three.

It was shown that given $f_X(\alpha)$ or $p_X(\alpha)$ for a random variable, it is a straightforward exercise using set theory and conditional probability to evaluate any probabilistic question concerning an event or a conditional event defined in terms of X.

For example,

$$P(\alpha_1 < X \le \alpha_2) = \int_{\alpha_1}^{\alpha_2^+} f_X(\alpha)\, d\alpha \quad \text{or} \quad \sum_{\alpha_1 < \alpha_j \le \alpha_2} p_X(\alpha_j)$$

$$P[(X > \alpha_1)/(X < \alpha_2)] = \frac{\int_{\alpha_1^+}^{\alpha_2^-} f_X(\alpha)\, d\alpha}{\int_{-\infty}^{\alpha_2^-} f_X(\alpha)\, d\alpha} \quad \text{if } \alpha_2 > \alpha_1$$

$$f_X(\alpha/(X > \alpha_1)) = 0 \qquad \alpha < \alpha_1$$

$$= \frac{f_X(\alpha)}{\int_{\alpha_1^+}^{\infty} f_X(\alpha)\, d\alpha} \qquad \alpha > \alpha_1$$

Section 4.3 discussed important "statistics" used to convey information about a random variable. The expected or mean value $\overline{X}$ was defined as

$$\int_{-\infty}^{\infty} \alpha f_X(\alpha)\, d\alpha$$

and is the average value of all the possible values for X that occur on a large number of trials of the experiment. Physically, if $f_X(\alpha)$ is a mass density, then $\overline{X}$ is its center of gravity. The mean square value $\overline{X^2}$ and σ_X^2 the variance were also defined. σ_X^2 indicates the central tendency of the values about the mean. The larger σ_X^2, the wider $f_X(\alpha)$ is. Higher-order moments indicate skewness of the density function about the mean value.

The fundamental theorem states that for a random variable Y whose density function is not known its statistics may be evaluated as

$$\overline{Y} = \int_{-\infty}^{\infty} K(\alpha) f_X(\alpha)\, d\alpha$$

or

$$\overline{K(Y)} = \int_{-\infty}^{\infty} K[g(\alpha)] f_X(\alpha)\, d\alpha$$

where $Y = g(X)$ and the density function for X is known. This theorem often allows for enormous labor saving by not having to find $f_Y(\alpha)$.

An important engineering application of the fundamental theorem involves finding statistics for the sampled value of a waveform $g(t)$ over a range T. In addition, it was noted that if $g(t)$ is considered periodic with period T, time averages yielded the same results as the statistical averages.

Section 4.4 discussed the problem of finding the density function of a random variable Y, which is defined as a function of a known random variable X. An important engineering application of this problem is to find the density function for sampling a waveform $g(t)$ defined for $a < t < a + T$. From probability theory the cumulative distribution function $F_Y(\beta)$ is found as

$$F_Y(\beta) = P[g(t) < \beta]$$
$$= P[t < g^{-1}(\beta)]$$
$$= \int \frac{1}{T} d\alpha \qquad \text{for all } \alpha < g^{-1}(\beta)$$

If $g^{-1}(t)$ does not define t with a one-to-one mapping between a and $a + T$, then great care must be exercised. If the more complex problem of sampling a waveform between a and $a + T$, where

$$g(t) = f_1(t) \qquad a < t < b$$
$$= f_2(t) \qquad b < t < T + a$$

is considered, then from probability theory,

$$F_Y(\beta) = P(a < t < b) P[f_1(t) \le \beta] + P(b < t < a + T) P[f_2(t) \le \beta]$$
$$= \frac{b - a}{T} \int_{\alpha \le f_1^{-1}(\beta)} \frac{1}{T} d\alpha + \frac{a + T - b}{T} \int_{\alpha \le f_2^{-1}(\beta)} \frac{1}{T} d\alpha$$

and

$$f_Y(\beta) = \frac{d}{d\beta} F_Y(\beta)$$

Chapter 4 concluded with a listing of some of the more commonly occurring random variables. Familiarity with the binomial, Poisson and gaussian random variables is an absolute minimum.

PROBLEMS

1. Consider the random phenomenon of rolling a tetrahedral die and if the number obtained is i, then drawing i balls from an urn containing 4 balls numbered from 1 to 4. Let X be the random variable describing the result of the die roll with a one-to-one mapping. Let Y be the random variable describing the sum of the numbers on the balls drawn from the urn with a one-to-one mapping.

   (a) Construct the sample description space of the phenomenon, assign a probability measure to it, and indicate the mapping of each point subject to $X(s)$ and $Y(s)$.

   (b) Find and plot, in whatever order is fastest, $F_X(\alpha)$, $f_X(\alpha)$, $p_X(\alpha)$, $F_Y(\beta)$, $f_Y(\beta)$, and $p_Y(\beta)$.

2. Consider the random phenomenon of tossing a coin three times. Define the random variable X such that X takes on a value equal to the number of heads on a trial. Find and plot $p_X(\alpha)$, $f_X(\alpha)$, and $F_X(\alpha)$.

3. Consider the deterministic waveform

$$x(t) = 2t \qquad 0 < t < 1$$
$$= 1 \qquad 1 < t < 2$$
$$= 0 \qquad \text{otherwise}$$

   Define the random phenomenon of sampling $x(t)$ uniformly from $t = 0$ to $t = 3$ and the random variable X as taking on a value equal to the waveform value sampled. Find $f_X(\alpha)$ and $F_X(\alpha)$.

4. Consider the periodic waveform

$$x(t) = \sum_{n = -\infty}^{\infty} g(t - 3n)$$

   where

$$g(t) = t^2 \qquad 0 < t < 1$$
$$= 1 \qquad 1 < t < 3$$
$$= 0 \qquad \text{otherwise}$$

   (a) Sketch $x(t)$.

   (b) Define X as the random variable sampling $x(t)$ uniformly with a one-to-one mapping. Find $F_X(\alpha)$ and $f_X(\alpha)$.

5. Given

$$f_X(\alpha) = C\alpha \qquad 0 < \alpha < 10$$
$$= 0 \qquad \text{otherwise}$$

   (a) Find the value C must assume.

   (b) Find $P(2 < X < 5)$.

   (c) $P[(X > 1)/(0 < X < 3)]$.

   (d) If A is the event $(0 < X < 2)$ and B is the event $(0.5 < X < 4)$ find the probabilities of the following: $A \cup B, A \cap B, (A/B), (B/A), (A^c/B),$ (A^c/B^c).

(e) $f_X(\alpha/X<2)$

6. Given the mass function for the random variable X describing the number of electrons emitted from a section of cathode area in some time T is

$$p_X(\alpha_i) = \frac{e^{-2}(2)^{\alpha_i}}{\alpha_i!} \qquad \alpha_i \geq 0 \text{ and an integer}$$

(a) Sketch the mass function up to $\alpha_i = 5$.
(b) Show that $\sum\limits_{\text{all } i} p_X(\alpha_i) = 1$.
(c) Given the condition A, that at least two electrons are emitted, derive the conditional mass function $p_X(\alpha_i/A)$ and plot it for $\alpha_i = 3, 4,$ and 5.

7. (a) Evaluate $\overline{X}, \overline{X^2}$, and σ_X^2 for the random variable of Problem 3.
   (b) Evaluate these averages on a time average basis and state whether or not you are surprised at the results. For example on a time average basis,

$$\overline{\overline{X^2}} = \frac{1}{3} \int_0^3 x^2(t)\, dt$$

8. (a) Evaluate $\overline{X}, \overline{X^2}$, and σ_X^2 directly using the density function $f_X(\alpha)$ of Problem 4.
   (b) Reevaluate $\overline{X}, \overline{X^2}$, and σ_X^2 by the fundamental theorem using the density function for uniformly sampling the time axis over its first period.

9. (a) Consider the waveform

$$x_1(t) = \sum_{n=-\infty}^{\infty} g(t-n)$$

where

$$g(t) = 10t \qquad 0 < t < \tfrac{1}{2}$$
$$= 0 \qquad \text{otherwise}$$

Evaluate $f_{X_1}(\alpha), \overline{X}_1, \overline{X_1^2}$, and $\sigma_{X_1}^2$ where X_1 is the random variable that samples $x_1(t)$ uniformly. Sketch $x_1(t)$.

(b) Repeat the questions of part (a) and sketch the function

$$x_2(t) = \sum_{n=-\infty}^{\infty} g(t-2n)$$

where

$$g(t) = 5t \qquad 0 < t < 1$$
$$= 0 \qquad \text{otherwise}$$

and X_2 is the random variable that samples $x_2(t)$ uniformly.

(c) Repeat the questions of part (a) and sketch $x_p(t)$ where

$$x_p(t) = \sum_{n=-\infty}^{\infty} g(t - pn)$$

where

$$g(t) = (10/p)t \qquad 0 < t < p/2$$
$$= 0 \qquad p/2 < t < p$$

and p is any fixed number $0 < p < \infty$, and X_p is the random variable that samples $x_p(t)$ uniformly.

(d) From what you have learned from (a), (b), and (c) explain what first-order statistics such as $\bar{X}$, $\overline{X^2}$, and σ_X^2 tell us about how rapidly a waveform is changing.

10. Consider the eight density functions shown plotted in Figure 4.13. For the statistics $\bar{X}$, $\overline{X^2}$, σ_X^2, $\overline{X^3}$, and $\overline{(X - \bar{X})^3}$ list for each the density function yielding the smallest value through the one yielding the largest value. If the value is zero indicate so. Do no calculations except as a last resort. For example, for $\bar{X}$ we will have $\bar{X}_2 = \bar{X}_3 = 0$, $\bar{X}_5 < 0$, and $\bar{X}_6 > 0$.

11. The time function $x(t) = \cos \omega_c t$ is the input to the rectifier shown. Find and sketch $y(t)$ and use the fundamental theorem to evaluate $\bar{Y}$, $\overline{Y^2}$, and σ_Y^2, where Y is the random variable defined as sampling the output.

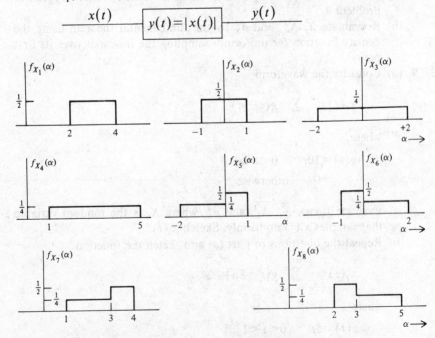

Figure 4.13 Density functions considered in Problem 10.

12. (a) Find the density function of the random variable X, which is defined as sampling the input waveform $x(t) = \cos \omega_c t$ of Problem 11.

    (b) Find the density function $f_Y(\beta)$ for the random variable Y of Problem 11.

13. Given that

$$x(t) = \sum_{n=-\infty}^{\infty} g(t - 3n)$$

where

$$g(t) = t \qquad -1 < t < 1$$
$$ = 0 \qquad \text{otherwise}$$

is the input to a rectifier defined as follows:

$$\underline{\quad x(t) \quad} \boxed{y(t) = |x(t)|} \underline{\quad y(t) \quad}$$

    (a) Plot $x(t)$ and $y(t)$.

    (b) Find $\bar{X}$, $\overline{X^2}$, σ_X^2, $\bar{Y}$, $\overline{Y^2}$, and σ_Y^2 using the fundamental theorem, where X and Y are the random variables that sample $x(t)$ and $y(t)$, respectively, uniformly over a period.

    (c) Find $f_Y(\beta)$ as a function of a known random variable in two different ways by relating it to X and to a random variable that samples the time axis of the output.

14. Consider that a channel transmits words of 8 bits. Assume that the probability a bit is incorrectly transmitted is 0.02, and this is independent for each bit. An error-correcting code is used at the receiver that can interpret the word correctly if at most 2 bits are in error.

    (a) Find the probability that a word is decoded correctly.

    (b) Find the probability that a message of four words is decoded correctly.

15. Assume that the arrival of telephone calls at a switchboard is Poisson with a rate of one per minute.

    (a) Find the probability that more than 60 calls arrive in an hour.

    (b) Find the probability that more than 60 calls arrive assuming that 20 will arrive.

16. Given that a random variable X is standard gaussian with 0 mean and unit variance, use the abbreviated table of the text to find the following:

    (a) $P(0.5 < X < 2)$

    (b) $P[(X < 2)/(X > 0.5)]$

    (c) $P[(X > 2)/(X < 0.5)]$

17. If Y is a gaussian random variable with mean μ and variance σ, express the following in terms of $F_s(\alpha)$ for the standard gaussian random

variable
(a) $P(Y<\beta)$
(b) $P(\alpha<Y<\beta)$
(c) $P[(\alpha<Y<\beta)/Y<0.5(\alpha+\beta)]$

18. If Y is gaussian with a mean of 50 and a variance of 5 find
(a) $P(Y>30)$
(b) $P[(Y>30)/(Y<54)]$

Chapter 5
Two Random Variables, Associated Functions and Usage

5.1 JOINT DISTRIBUTION, DENSITY, AND MASS FUNCTIONS

Two random variables defined on a sample description or event space are characterized by their joint distribution function $F_{XY}(\alpha, \beta)$ which is defined as

$$F_{XY}(\alpha, \beta) \triangleq P[(X \le \alpha) \cap (Y \le \beta)] \tag{5.1}$$

or, in words, $F_{XY}(\alpha, \beta)$ is the probability of the intersection of the two events $(X \le \alpha)$ and $(Y \le \beta)$.

The joint density function of two random variables X and Y is defined as

$$f_{XY}(\alpha, \beta) = \frac{\partial^2}{\partial \alpha \partial \beta} F_{XY}(\alpha, \beta) \tag{5.2}$$

This is the second partial derivative with respect to α and β of the joint distribution function. The joint mass function (which will exist for discrete random variables) is defined as

$$p_{XY}(\alpha_i, \beta_j) \triangleq P[(X = \alpha_i) \cap (Y = \beta_j)] \tag{5.3}$$

or, in words, the joint mass function is the probability that X equals α_i and Y equals β_j.

The conceptual understanding of the probabilistic information conveyed by joint distribution, density, and mass functions is vital for future studies. First, we must appreciate that given a random phenomenon and the probability measure for its sample space S, the evaluation of the cumulative distribution function for two random variables defined on S resolves itself to a problem in discrete probability theory. Second, it will be demonstrated that knowing the cumulative distribution or joint density or joint mass function of two random variables, it is possible to answer any probabilistic question about the random variables subject to any condition involving them.

Two fundamental problems will now be solved showing the derivation of $F_{XY}(\alpha, \beta)$ [and $f_{XY}(\alpha, \beta)$ or $p_{XY}(\alpha_i, \beta_j)$] using discrete probability. Really digging in and thoroughly understanding them will cement the definitions (given by Eqs. 5.1 through 5.3) in our minds and facilitate their usage in answering probabilistic questions.

Example 5.1

Consider the random phenomenon of tossing a coin and of drawing 1 or 2 balls from an urn that contains 3 balls numbered from 1 to 3. Let the random variable X be associated with the coin toss with mapping $X(T)=0$ and $X(H)=1$, where T and H indicate a tail and head, respectively. In the event the coin toss is a tail, 1 ball is drawn from the urn and in the event of a head 2 balls (without replacement) are drawn.

Let Y be the random variable associated with the urn draw and assume a numerical value equal to the sum of the numbers on the balls drawn, whether 1 or 2 balls are drawn.

(a) Find the joint mass function $p_{XY}(\alpha_i, \beta_j)$.
(b) Find the joint cumulative distribution function $F_{XY}(\alpha, \beta)$.

SOLUTION
The sample space for the experiment and its probability measure are as follows:

$$S = \{T1, T2, T3, H3, H4, H5\} \quad \text{and} \quad P(s) = \tfrac{1}{6}, \tfrac{1}{6}, \tfrac{1}{6}, \tfrac{1}{6}, \tfrac{1}{6}, \tfrac{1}{6}$$

For each point the initial letter indicates the coin toss and the number the sum of the balls drawn.

(a) For $p_{XY}(\alpha_i, \beta_j)$, the probabilities for all the possible points $X = \alpha_i$ and $Y = \beta_j$ are easily found and are shown in matrix form in Figure 5.1a. For example, $p_{XY}(1,4)$ is equal to $P(H4)$ from the sample space and has a value $\tfrac{1}{6}$.

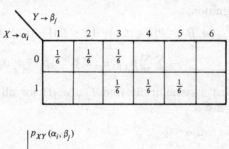

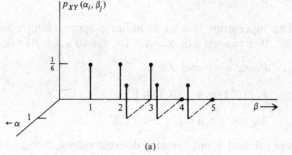

(a)

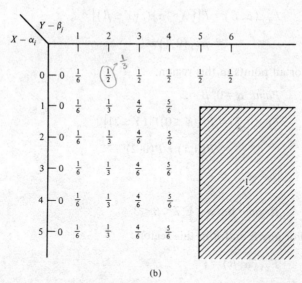

(b)

Figure 5.1 (a) The joint mass function of Example 5.1. (b) The joint cumulative distribution function.

(b) By definition,

$$F_{XY}(\alpha, \beta) = P[(X \le \alpha) \cap (Y \le \beta)]$$

$$= \sum \sum p_{XY}(\alpha_i, \beta_j) \qquad \beta_j \le \beta, \alpha_i \le \alpha$$

We must systematically find $F_{XY}(\alpha, \beta)$ for all $-\infty < \alpha < \infty$ and $-\infty < \beta < \alpha$.

Region: $\alpha < 0^- \cup \beta < 1^-$

The superscript is used to indicate approaching a value from the left. If $X = \alpha \le 0^-$ or $Y = \beta \le 1^-$, then $F_{XY}(\alpha, \beta) = 0$.

Point: $\alpha = 0$ *and* $\beta = 1$

$$F_{XY}(0, 1) = p_{XY}(0, 1) = \tfrac{1}{6}$$

Region: $0 < \alpha < 1^-, 1 < \beta < 2^-$

Since X and Y only assume discrete values, then

$$F_{XY}(\alpha, \beta) = P[(X \le \alpha) \cap (Y \le \beta)]$$

$$= p_{XY}(0, 1) = \tfrac{1}{6}$$

for all points in this region.

Point: $\alpha = 0$, $\beta = 2$

$$F_{XY}(0, 2) = P[(X \le 0) \cap (Y \le 2)]$$

$$= P(0, 1) + P(0, 2)$$

$$= \tfrac{1}{3}$$

Region: $0 < \alpha < 1, 2 < \beta < 3$

For any α and β in this region,

$$F_{XY}(\alpha, \beta) = \tfrac{1}{3}$$

Point: $\alpha = 0$, $\beta = 3$

$$F_{XY}(0, 3) = P[(X \le 0) \cap (Y \le 3)]$$

$$= P(0, 1) + P(0, 2) + P(0, 3)$$

$$= \tfrac{1}{2}$$

Region: $0 < \alpha < 1, \beta > 3$

$$F_{XY}(\alpha, \beta) = \tfrac{1}{2}$$

For example, $F_{XY}(0.6, 27)$ asks the question, "What is the probability X takes on a value less than or equal to 0.6 and Y takes on a value less than or equal to 27?" and this is $P(0,1) + P(0,2) + P(0,3)$.

Point: $\alpha = 1, \beta = 3$

$$F_{XY}(1,3) = P[(X \le 1) \cap (Y \le 3)]$$

$$= \sum \sum p_{XY}(\alpha_i, \beta_j) \qquad \alpha_i \le 1, \beta_j \le 3$$

$$= P(0,1) + P(0,2) + P(0,3) + P(1,3)$$

$$= \tfrac{2}{3}$$

Region: $1 \le \alpha < 2, \; 3 < \beta < 4$

$$F_{XY}(\alpha, \beta) = \tfrac{2}{3}$$

It is left as an exercise for the reader to derive $F_{XY}(\alpha, \beta)$ for all other α and β.

The joint mass and joint cumulative distribution functions are shown in matrix form in Figures 5.1a and b. $p_{XY}(\alpha_i, \beta_j)$ assigns real positive values to a specific number of points in the $\alpha\beta$ plane. $F_{XY}(\alpha, \beta)$ assigns a nonnegative value to every point and generates a nondecreasing two-dimensional function of α and β. Since the random variables assume only discrete values, in this problem, $F_{XY}(\alpha, \beta)$ changes only by discrete jumps and is a two-dimensional stairway in space. A plateau of height 1 exists for all $\alpha \ge 1$ and $\beta \ge 5$. Exploring the $\alpha\beta$ plane for any fixed α, such as $0 < \alpha < 1$, and varying β, where $-\infty < \beta < \infty$, will only lead to a plateau of height 0.5, whereas, for example, exploring for a fixed β, $4 < \beta < 5$ will involve a step of 0.5 at $\alpha = 0$; a step of $(0.83 - 0.5) = 0.33$ at $\alpha = 1$, and then a flat terrain of height 0.83 indefinitely.

Example 5.2

Consider the experiment of choosing two real decimal numbers in the following way. First a number is chosen uniformly between 0 and 10. If this number is less than 4, then the second number is chosen uniformly between 0 and 3. If in the first place a number greater than 4 is chosen, then the second number is chosen uniformly between 0 and 4. Let X be the random variable describing the first random number and let Y be the random variable describing the second random number selected with a one-to-one mapping from the experiment. Find

(a) $F_{XY}(\alpha, \beta)$ for all α and β.
(b) $f_{XY}(\alpha, \beta)$ for all α and β.

SOLUTION

(a) The event space for this problem and its probability measure will not be given because doing so is equivalent to solving the problem. A diagram is shown indicating this somewhat confusing-to-read experiment.

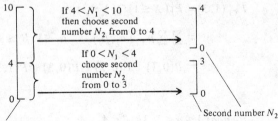

If we wanted to construct a practical model for carrying out this experiment we could use three infinitely finely calibrated wheels of fortune with scales from 0 to 3, 0 to 4, and 0 to 10, respectively. To find the first number we would spin the 0–10 wheel. If the result was less than 4 then we would obtain the second number from the 0–3 wheel, whereas if the first number was greater than 4, we would obtain the second number from the 0–4 wheel. Now we will carry out the formidable task of finding $F_{XY}(\alpha, \beta)$.

By definition,

$$F_{XY}(\alpha, \beta) \triangleq P[(X < \alpha) \cap (Y < \beta)]$$

$$= P(X \le \alpha)P[(Y \le \beta)/(X \le \alpha)]$$

using the conditional probability theory of events.

We will have to consider all the points $-\infty < \alpha < \infty$ and $-\infty < \beta < \infty$.

Range: $\alpha < 0$ or $\beta < 0$

If $\alpha < 0$ or $\beta < 0$, then

$$P(X \le \alpha) = 0 \quad \text{or} \quad P[(Y \le \beta)/(X \le \alpha)] = 0 \quad \text{and}$$

$$F_{XY}(\alpha, \beta) = 0$$

Range: $0 < \alpha < 4, \, 0 < \beta < 3$

$$F_{XY}(\alpha, \beta) = P(X \le \alpha)P[(Y \le \beta)/(X \le \alpha)]$$

$$= \frac{\alpha}{10} \cdot \frac{\beta}{3}$$

$$= \frac{\alpha\beta}{30}$$

since in this case the 0–3 wheel was used for the second number.

Range: $0 < \alpha < 4$, $\beta > 3$

$$F_{XY}(\alpha, \beta) = \frac{\alpha}{10} P[(Y \leq \beta)/(X \leq \alpha)]$$

$P[(Y \leq \beta)/(X \leq 4)] = 1$ if $\beta > 3$ since Y must take on a value between 0 and 3, which is always less than β.

$$\therefore F_{XY}(\alpha, \beta) = \frac{\alpha}{10}$$

This is reflected in the plot of $F_{XY}(\alpha, \beta)$ in Figure 5.2a.

Range: $4 < \alpha < 10$, $0 < \beta < 3$

Finding $F_{XY}(\alpha, \beta)$ in this range is challenging. We can say that

$$F_{XY}(\alpha, \beta) = P(X \leq \alpha) P[(Y \leq \beta)/(X \leq \alpha)]$$

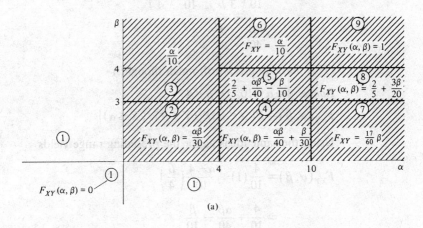

(a)

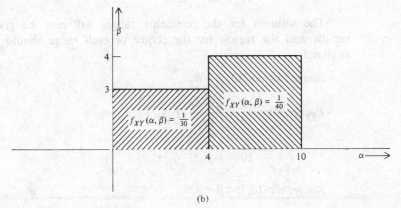

(b)

Figure 5.2 (a) The cumulative distribution function of Example 5.2. (b) The joint density function.

However, it is not directly possible to find $P[(Y \leq \beta)/(X \leq \alpha)]$, because knowing that X is less than some number α, which is greater than 4, does not indicate whether the 0–3 or 0–4 wheel was used for the second number. Using set theory we write

$$(X \leq \alpha) = (X \leq 4) \cup (4 < X \leq \alpha)$$

and, using the distributive property of set theory,

$$P[(Y \leq \beta)/(X \leq \alpha)] = \frac{P[(Y \leq \beta) \cap [(X \leq 4) \cup (4 \leq X < \alpha)]]}{P(X \leq \alpha)}$$

$$F_{XY}(\alpha, \beta) = P(X \leq 4)P[(Y \leq \beta)/(X \leq 4)]$$

$$+ P(4 < X \leq \alpha)P[(Y \leq \beta)/(4 < X \leq \alpha)]$$

$$= \frac{4}{10}\left(\frac{\beta}{3}\right) + \frac{\alpha - 4}{10}\left(\frac{\beta}{4}\right)$$

$$= \frac{\alpha\beta}{40} + \frac{\beta}{30}$$

(This derivation is worth much thought.)

Range: $4 < \alpha < 10$, $3 < \beta < 4$

$$F_{XY}(\alpha, \beta) = P(X \leq \alpha)P[(Y \leq \beta)/(X \leq \alpha)]$$

Using the same ideas as for the preceding range yields

$$F_{XY}(\alpha, \beta) = \frac{4}{10}(1) + \frac{\alpha - 4}{10}\left(\frac{\beta}{4}\right)$$

$$= \frac{4}{10} + \frac{\alpha\beta}{40} - \frac{\beta}{10}$$

The solution for the remaining ranges will now be given rapidly and the reason for the choice of each range should be analyzed.

Range: $4 < \alpha < 10$, $\beta > 4$

$$F_{XY}(\alpha, \beta) = \frac{4}{10} + \frac{\alpha - 4}{10}(1)$$

$$= \frac{\alpha}{10}$$

Range: $\alpha > 10$, $0 < \beta < 3$

$$F_{XY}(\alpha, \beta) = \frac{4}{10}\left(\frac{\beta}{3}\right) + \frac{6}{10}\left(\frac{\beta}{4}\right) = \frac{17}{60}\beta$$

Range: $\alpha > 10,\ 3 < \beta < 4$

$$F_{XY}(\alpha, \beta) = P(X \le \alpha)P[(Y \le \beta)/(X \le 10)]$$

since

$$(X \le \alpha) \cap (Y \le \beta) = [(X \le 4) \cap (Y \le \beta)] \cup [(4 < X \le \alpha) \cap (Y \le \beta)]$$

$$\therefore F_{XY}(\alpha, \beta) = \frac{4}{10}(1) + \frac{6}{10}\left(\frac{\beta}{4}\right)$$

$$= \frac{4}{10} + \frac{3\beta}{20}$$

Range: $\alpha > 10,\ \beta > 4$

$$F_{XY}(\alpha, \beta) = 1$$

The plot of $F_{XY}(\alpha, \beta)$ for all α and β in Figure 5.2a should be carefully studied.

This problem so far has been rather exhausting, and in general finding a joint cumulative distribution function involves a great deal of work. However, it conceptually resolves itself to repeatedly finding the probability of the intersection of two events using, if necessary, conditional probability and set theory to express the desired events in terms of the event space.

Cumulative distribution functions possess very interesting properties. If we consider $F_{XY}(\alpha, \beta)$ as the height of a point with coordinates α and β, then $F_{XY}(\alpha, \beta)$ never decreases as α or β increases. For the function of this example we might consider the consequences of an observer taking a few strolls in the $\alpha\beta$ plane.

(1) If we go from $\beta = -\infty$ to $\beta = +\infty$ for $\alpha = 2$, then for $-\infty < \beta < 0$ the terrain is flat at sea level. From $0 < \beta < 3$ the terrain is uphill with a slope

$$\frac{d}{d\beta}\left(\frac{2\beta}{30}\right) = \frac{1}{15}$$

At $\beta = 3$ the height is 0.2 and this is maintained from $3 < \beta < \infty$.

(2) If we go from $\alpha = -\infty$ to $\alpha = +\infty$, for $\beta = 3.5$, then for $-\infty < \alpha < 0$ the terrain is flat at sea level; from $0 < \alpha < 4$ the terrain is uphill with a slope of 0.1; from $4 < \alpha < 10$ the terrain is uphill with a slope of $7 \div 80$, and at $\alpha = 10$ we are at a height of $37 \div 40$, and we then proceed on flat terrain at this height from $10 < \alpha < \infty$.

(b) Now the density function is to be found as

$$f_{XY}(\alpha, \beta) = \frac{\partial^2}{\partial \alpha \, \partial \beta} F_{XY}(\alpha, \beta)$$

$f_{XY}(\alpha, \beta)$ will be nonzero only in the ranges ②, ④, and ⑤ of Figure 5.2a.

$$f_{XY}(\alpha, \beta) = \tfrac{1}{30} \qquad 0 < \alpha < 4, 0 < \beta < 3$$

$$f_{XY}(\alpha, \beta) = \tfrac{1}{40} \qquad 4 < \alpha < 10, 0 < \beta < 4 \text{ (for both ④ and ⑤)}$$

and

$$f_{XY}(\alpha, \beta) = 0 \qquad \text{otherwise}$$

This joint density function is shown plotted in Figure 5.2b. It can be seen that it is much more economical to characterize X and Y by $f_{XY}(\alpha, \beta)$ than by $F_{XY}(\alpha, \beta)$. If $f_{XY}(\alpha, \beta)$ gives the same information as $F_{XY}(\alpha, \beta)$, we will now see whether we can derive it directly.

Example 5.3

For the experiment of Example 5.2, derive $f_{XY}(\alpha, \beta)$ directly.

SOLUTION
Since for a continuous pair of random variables the joint density function is the second partial derivative of the cumulative distribution function, it follows that

$$P[(\alpha < X < \alpha + \Delta\alpha) \cap (\beta < Y < \beta + \Delta\beta)] = f_{XY}(\alpha, \beta) \Delta\alpha \Delta\beta \qquad (1)$$

Also, using conditional probability,

$$\begin{aligned} P[(\alpha &< X < \alpha + \Delta\alpha) \cap (\beta < Y < \beta + \Delta\beta)] \\ &= f_X(\alpha) \Delta\alpha \, P[(\beta < Y < \beta + \Delta\beta)/(\alpha < X < \alpha + \Delta\alpha)] \\ &= f_X(\alpha) \Delta\alpha \, f_Y[\beta/(\alpha < X < \alpha + \Delta\alpha)] \Delta\beta \\ &= f_X(\alpha) f_Y[\beta/(X = \alpha)] \Delta\alpha \Delta\beta \end{aligned} \qquad (2)$$

for $\Delta\alpha$ very small.

Equating (1) and (2) yields

$$f_{XY}(\alpha, \beta) = f_X(\alpha) f_Y[\beta/(X = \alpha)] \qquad (3)$$

If we consider the experiment of Example 5.2, we can find

$$\begin{aligned} f_X(\alpha) &= \tfrac{1}{10} \qquad 0 < \alpha < 10 \\ &= 0 \qquad \text{otherwise} \end{aligned}$$

and

$$\begin{aligned} f_Y[(\beta/(X = \alpha)] &= \tfrac{1}{3} \qquad 0 < \beta < 3, \text{ for any } \alpha, 0 < \alpha < 4 \\ &= 0 \qquad \beta < 0 \text{ or } \beta > 3, \text{ for any } \alpha, 0 < \alpha < 4 \end{aligned}$$

Similarly,

$$\begin{aligned} f_Y[(\beta/(X = \alpha)] &= \tfrac{1}{4} \qquad 0 < \beta < 4, \text{ for any } \alpha, 4 < \alpha < 10 \\ &= 0 \qquad \beta < 0 \text{ or } \beta > 4, \text{ for any } \alpha, 4 < \alpha < 10 \end{aligned}$$

Using the above, the joint density function may be found for all α and β as

$$f_{XY}(\alpha, \beta) = f_X(\alpha) f_Y(\beta/\alpha)$$

$$= \tfrac{1}{30} \qquad 0 < \alpha < 4, \, 0 < \beta < 3$$

$$= \tfrac{1}{40} \qquad 4 < \alpha < 10, \, 0 < \beta < 4$$

$$= 0 \qquad \text{otherwise}$$

which corresponds to the results in Figure 5.2b.

Example 5.3 was solved much more rapidly than Example 5.2 despite the fact we carried out the derivation $f_{XY}(\alpha, \beta) = f_X(\alpha) f_Y[\beta/(X = \alpha)]$ while doing the problem. However we should notice that it is much more mechanical and we are not as conscious of our basic discrete probability theory as we were in Example 5.2.

How Joint Density Functions Completely Characterize Random Variables

$F_{XY}(\alpha, \beta)$ or $f_{XY}(\alpha, \beta)$ completely characterize two random variables X and Y. By this we mean that they enable us to answer any probabilistic question about both of them or about either one, or any conditional question involving them.

Deriving $f_X(\alpha)$ and $f_Y(\beta)$ from $f_{XY}(\alpha, \beta)$

Consider the event $(\alpha_0 < X < \alpha_0 + \Delta\alpha)$. Based on the density function for X, the probability of this event is

$$P(\alpha_0 < X < \alpha_0 + \Delta\alpha) = f_X(\alpha_0) \Delta\alpha \tag{5.4}$$

Alternatively, based on the space for X and Y and using the set theory axiom $A \cap S = A$, we write

$$(\alpha_0 < X < \alpha_0 + \Delta\alpha) = (\alpha_0 < X < \alpha_0 + \Delta\alpha) \cap (-\infty < Y < \infty)$$

This allows us to express the probability of the event as

$$P(\alpha_0 < X < \alpha_0 + \Delta\alpha) = \int_{-\infty}^{\infty} \left[\int_{\alpha_0}^{\alpha_0 + \Delta\alpha} f_{XY}(\alpha, \beta) \, d\alpha \right] d\beta$$

$$= \int_{-\infty}^{\infty} f_{XY}(\alpha_0, \beta) \Delta\alpha \, d\beta$$

$$= \left[\int_{-\infty}^{\infty} f_{XY}(\alpha_0, \beta) \, d\beta \right] \Delta\alpha \tag{5.5}$$

Comparing Eqs. 5.4 and 5.5 we obtain the density function for X in terms of $f_{XY}(\alpha, \beta)$ as

$$f_X(\alpha) = \int_{-\infty}^{\infty} f_{XY}(\alpha, \beta) \, d\beta \tag{5.6}$$

Such a formula is almost intuitively obvious from set theory.

In a completely analogous way we can show that

$$f_Y(\beta) = \int_{-\infty}^{\infty} f_{XY}(\alpha, \beta) \, d\alpha \qquad (5.7)$$

Since we have already demonstrated that

$$f_{XY}(\alpha, \beta) = f_X(\alpha) f_Y[\beta/(X = \alpha)]$$
$$= f_Y(\beta) f_X[\alpha/(Y = \beta)]$$

we can obtain the conditional density functions as

$$f_X[\alpha/(Y = \beta)] = \frac{f_{XY}(\alpha, \beta)}{f_Y(\beta)} \qquad (5.8)$$

and

$$f_Y[\beta/(X = \alpha)] = \frac{f_{XY}(\alpha, \beta)}{f_X(\alpha)} \qquad (5.9)$$

These formulas, which look very simple, often turn out to be deceptive in practice. We will carefully solve a problem to illustrate their use and interpretation.

Example 5.4

As an illustration use the joint density function

$$f_{XY}(\alpha, \beta) = \tfrac{1}{30} \qquad 0 < \alpha < 4, 0 < \beta < 3$$
$$= \tfrac{1}{40} \qquad 4 < \alpha < 10, 0 < \beta < 4$$

from Example 5.3 and, using Eqs. 5.6 through 5.9,

(a) Find $f_X(\alpha)$ and $f_Y(\beta)$.
(b) Find $f_Y[\beta/(X = 1)]$, $f_Y[\beta/(X = 6)]$, $f_X[\alpha/Y = 2)]$, and $f_Y(\beta/3 < X < 6)$.
(c) Are the random variables X and Y independent?

In each case discuss the reasonableness of your results by referring to the physical experiment of Example 5.2.

SOLUTION

The joint density function is shown plotted on the $\alpha\beta$ plane in Figure 5.3a.

(a) To find $f_X(\alpha)$ we will evaluate

$$f_X(\alpha) = \int_{-\infty}^{\infty} f_{XY}(\alpha, \beta) \, d\beta \qquad \text{for all } \alpha, \ -\infty < \alpha < \infty$$

For $\alpha < 0$,

$$f_X(\alpha) = \int_{-\infty}^{\infty} f_{XY}(\alpha, \beta) \, d\beta$$

$$= \int_{-\infty}^{\infty} 0 \, d\beta \qquad \text{(See the line of integration in Figure 5.3a)}$$

$$= 0 \qquad \alpha < 0$$

For $0 < \alpha < 4$,

$$f_X(\alpha) = \int_{-\infty}^{\infty} f_{XY}(\alpha, \beta) \, d\beta$$

$$= \int_{0}^{3} \frac{1}{30} \, d\beta \qquad \text{(See the line of integration in Figure 5.3a)}$$

$$= \frac{1}{10} \qquad 0 < \alpha < 4$$

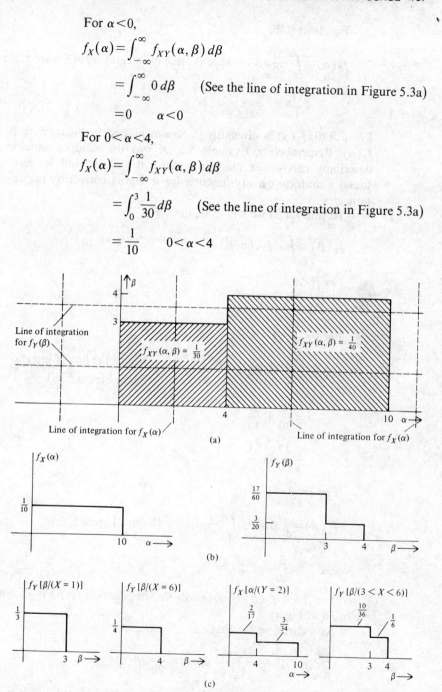

Figure 5.3   (a) The joint density function of Example 5.4. (b) The required density functions. (c) The required conditional density functions.

For $4 < \alpha < 10$,

$$f_X(\alpha) = \int_0^4 \frac{1}{40} \, d\beta \qquad \text{(See the line of integration in Figure 5.3a)}$$

$$= \frac{1}{10}$$

For $\alpha > 10$, $f_X(\alpha)$ is obviously 0. Shown plotted in Figure 5.3b is $f_X(\alpha)$. Remembering Example 5.2, X was the random variable describing choosing a number between 0 and 10, and we have found a uniform density function for it that is correct by inspection.

To find $f_Y(\beta)$ we will evaluate

$$f_Y(\beta) = \int_{-\infty}^{\infty} f_{XY}(\alpha, \beta) \, d\alpha$$

For $\beta < 0$,

$$f_Y(\beta) = 0$$

For $0 < \beta < 3$,

$$f_Y(\beta) = \int_0^4 \frac{1}{30} \, d\alpha + \int_4^{10} \frac{1}{40} \, d\alpha \qquad \begin{array}{l}\text{(Notice the line of integra-} \\ \text{tion in Figure 5.3a)}\end{array}$$

$$= \frac{4}{30} + \frac{6}{40}$$

$$= \frac{17}{60}$$

For $3 < \beta < 4$,

$$f_Y(\beta) = \int_0^4 0 \, d\alpha + \int_4^{10} \frac{1}{40} \, d\alpha \qquad \text{(Notice Figure 5.3a)}$$

$$= \frac{3}{20}$$

For $\beta > 4$, $f_Y(\beta)$ is obviously 0. The function $f_Y(\beta)$ is shown plotted in Figure 5.3c.

As a check we see that

$$\int_{-\infty}^{\infty} f_Y(\beta) \, d\beta = \int_0^3 \frac{17}{60} \, d\beta + \int_3^4 \frac{3}{20} \, d\beta$$

$$= \frac{17}{60} (3) + \frac{9}{60}$$

$$= 1$$

as it should be. It is left as a challenging exercise for the student to obtain this density function for Y directly from the experiment of Example 5.2 by using symmetry.

(b) Let us now find the conditional density functions $f_Y[\beta/(X=1)]$, $f_Y[\beta/(X=6)]$, $f_X[\alpha/(Y=2)]$, and $f_Y[\beta/(3<X<6)]$.

First,

$$f_Y[\beta/(X=1)] = \frac{f_{XY}(1,\beta)}{f_X(1)} \qquad \text{for all } \beta, \ -\infty<\beta<\infty$$

$$=0 \qquad \beta<0, \text{ as } f_{XY}(1,\beta)=0 \ \text{(Note Figure 5.3a)}$$

$$=\frac{1/30}{1/10}=\frac{1}{3} \qquad 0<\beta<3$$

$$=0 \qquad \beta>3$$

where we carefully considered the line $X=1$. Figure 5.3c shows a plot of $f_Y[\beta/(X=1)]$, which can be interpreted as saying, "When a 1 is obtained for the first number we choose the second number uniformly between 0 and 3." This is in conformity with the experiment.

Second,

$$f_Y[\beta/(X=6)] = \frac{f_{XY}(6,\beta)}{f_X(6)} \qquad -\infty<\beta<+\infty$$

$$=0 \qquad \beta<0$$

$$=\frac{1/40}{1/10}=\frac{1}{4} \qquad 0<\beta<4 \quad \text{(Note Figure 5.3a)}$$

$$=0 \qquad \beta>4$$

The function $f_Y[\beta/(X=6)]$ is shown plotted; it says, "When the number 6 is obtained for the first number, then the second number is chosen uniformly between 0 and 4."

Third,

$$f_X[\alpha/(Y=2)] \triangleq \frac{f_{XY}(\alpha,2)}{f_Y(2)} \qquad \text{for all } \alpha \qquad -\infty<\alpha<\infty$$

Observing the line $\beta=2$ we obtain

$$f_X[\alpha/(Y=2)]=0 \qquad \alpha<0$$

$$=\frac{1/30}{17/60}=\frac{2}{17} \qquad 0<\alpha<4$$

$$=\frac{1/40}{17/60}=\frac{3}{34} \qquad 4<\alpha<10$$

$$=0 \qquad \text{otherwise}$$

The function $f_X[\alpha/(Y=2)]$ is shown plotted in Figure 5.3c; it says, "Given that the second number was 2, it was 33% more likely that the first number was less than 4." The student should handwave this conclusion.

Fourth, to interpret the function $f_Y(\beta/3< X<6)$ from discrete probability we write

$$f_Y(\beta/3< X<6)\Delta\beta = P[(\beta<Y<\beta+\Delta\beta)/(3< X<6)]$$

$$f_Y(\beta/3< X<6)=\frac{\left[\int_3^6 f_{XY}(\alpha,\beta)\,d\alpha\right]}{\int_3^6\left[\int_{-\infty}^{\infty} f_{XY}(\alpha,\beta)\,d\beta\right]d\alpha}$$

$$=\frac{\int_3^6 f_{XY}(\alpha,\beta)\,d\alpha}{\int_3^6 \frac{1}{10}d\alpha}$$

We must now *carefully* consider this formula for all β, $-\infty<\beta<+\infty$. Obviously the answer is 0 for $\beta<0$ and $\beta>4$.

For $0<\beta<3$,

$$f_Y(\beta/3< X<6)=\frac{\int_3^4 \frac{1}{30}d\alpha+\int_4^6 \frac{1}{40}d\alpha}{0.3}$$

$$=\frac{10}{36}$$

For $3<\beta<4$,

$$f_Y(\beta/3< X<6)=\int_3^4 0\,d\alpha+\int_4^6 \frac{1}{40}d\alpha$$

$$=\frac{1}{6}$$

This conditional density function is shown plotted in Figure 5.3c and we note as a check that its area is 1.

(c) Two events A and B are independent if

$$P(A\cap B)= P(A)P(B)$$

It follows that any two statements about random variables are independent if

$$f_{XY}(\alpha,\beta)= f_X(\alpha)f_Y(\beta)$$

or

$$f_Y[\beta/(X=\alpha)]= f_Y(\beta) \quad\text{and}\quad f_X[\alpha/(Y=\beta)]= f_X(\alpha)$$

We can see immediately that X and Y are not independent for this problem as for example,

$$f_{XY}(1,1) = \tfrac{1}{30}$$

and

$$f_X(1)f_Y(1) = \tfrac{1}{10} \times \tfrac{17}{60} \neq \tfrac{1}{30}$$

A Comment on Examples 5.1 to 5.4

A thorough and overall understanding of the solutions to Examples 5.1 to 5.4 and a realization of their purpose is essential. Examples 5.1 and 5.2 illustrated how finding the joint cumulative distribution function is a problem in discrete probability theory, utilizing the concept of independent and dependent events by the formula $F_{XY}(\alpha, \beta) = F_X(\alpha)F_Y[\beta/(X \le \alpha)]$. The choice of different ranges on the $\alpha\beta$ plane working from $F_{XY}(-\infty, -\infty)$ to $F_{XY}(\infty, \infty)$ may be quite extensive due to the difficulty of finding $F_Y[\beta/(X \le \alpha)]$.

Example 5.3 illustrated the fact that a knowledge of density functions of one random variable may be utilized to find the joint density function $f_{XY}(\alpha, \beta)$ as $f_X(\alpha)f_Y[\beta/(X = \alpha)]$. In general this is much quicker to find than the joint cumulative distribution function as the condition $X = \alpha$ is more straightforward than $X \le \alpha$. The function $f_{XY}(\alpha, \beta)$ may occupy a much smaller region of the $\alpha\beta$ plane and is usually a simpler analytic function than is $F_{XY}(\alpha, \beta)$.

Example 5.4 illustrated that a joint density function (or joint cumulative distribution function) completely characterized two random variables. The individual density functions $f_X(\alpha)$ and $f_Y(\beta)$ may be found, as may the conditional density functions $f_X[\alpha/(Y = \beta)]$ and $f_Y[\beta/(X = \alpha)]$. Soon we will demonstrate that it is possible, given $f_{XY}(\alpha, \beta)$, to answer any probabilistic question concerning any event or conditional event involving the random variables.

Before we proceed with listing general properties of joint distribution, density, and mass functions, the case of discrete random variables will be discussed and an example solved.

Some Comments on Discrete Random Variables

If two random variables that can only assume discrete values are defined on the sample space of a random phenomenon, then we define $F_{XY}(\alpha, \beta)$ by

$$F_{XY}(\alpha, \beta) \triangleq P[(X \le \alpha) \cap (Y \le \beta)]$$

and from the set theory of events it will only be possible for $F_{XY}(\alpha, \beta)$ to change in jumps as α or β changes. The joint mass function is defined as

$$p_{XY}(\alpha, \beta) \triangleq P[(X = \alpha) \cap (Y = \beta)]$$

and will assume positive values less than or equal to 1 at different points in the $\alpha\beta$ plane. In a completely analogous and simpler manner because of the discreteness, corresponding to Eqs. 5.4 through 5.9, we can derive using the set theory of events and the mutually exclusiveness of points in the $\alpha\beta$ plane, the following:

$$p_X(\alpha_i) = \sum_{\text{all } j} p_{XY}(\alpha_i, \beta_j) \qquad \text{for any } i \qquad (5.10)$$

$$p_Y(\beta_j) = \sum_{\text{all } i} p_{XY}(\alpha_i, \beta_j) \qquad \text{for any } j \qquad (5.11)$$

$$p_X[\alpha_i/(Y=\beta_j)] = \frac{p_{XY}(\alpha_i, \beta_j)}{p_Y(\beta_j)} \qquad \text{for any } i \qquad (5.12)$$

and

$$p_Y[\beta_j/(X=\alpha_i)] = \frac{p_{XY}(\alpha_i, \beta_j)}{p_X(\alpha_i)} \qquad \text{for any } j \qquad (5.13)$$

Two discrete random variables are independent if

$$p_{XY}(\alpha_i, \beta_j) = p_X(\alpha_i)p_Y(\beta_j) \qquad \text{for all } i \text{ and } j$$

The actual derivation of a joint mass function or joint cumulative distribution function was carried out in Example 5.1, and now practice will be obtained utilizing Eqs. 5.10 through 5.13.

Example 5.5

Given a joint mass function as shown in the matrix of Figure 5.4a, find

(a) $p_X(\alpha)$ and $p_Y(\beta)$.
(b) $p_X[\alpha/(Y=3)]$, $p_Y[\beta/(X>0)\cap(2<Y\leq 4)]$.
(c) $P[(X<2)\cap(Y\leq 2)]$.
(d) $P[(X<2)/(Y\leq 2)]$.

SOLUTION
The meaning of the joint mass matrix should be self-evident. For example, $p_{XY}(2,1)=0.2$ and $p_{XY}(2,4)=0$.

(a) To solve for $p_X(0)$ we have from set theory that

$$(X=0) = (X=0) \cap (-\infty < Y + \infty)$$

$$\therefore (X=0) = (0,1) \cup (0,2) \cup (0,3) \cup (0,4)$$

and this yields

$$p_X(0) = 0.3$$

$$(X=1) = (1,1) \cup (1,2) \cup (1,3) \cup (1,4)$$

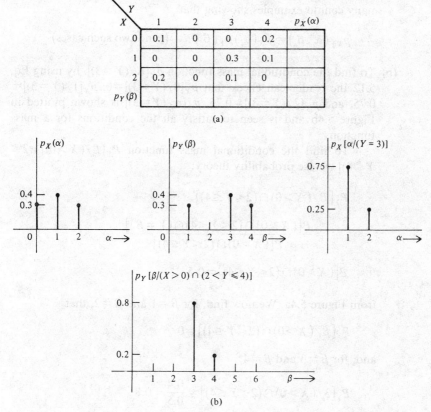

Figure 5.4   (a) The joint mass function for Example 5.5. (b) The required mass functions.

and

$$p_X(1)=0.4$$

Similarly,

$$p_X(2)=0.3$$

Had we constructed an extra column in Figure 5.4a and labeled it $p_X(\alpha)$, then the sum of the elements of the row for which $X=0$ yields $p_X(0)$, and similarly for the $X=1$ and $X=2$ rows. For this reason, $p_X(\alpha)$ is often referred to as a "marginal" mass function. This mass function is shown plotted in Figure 5.4b. Using reasoning identical to that used in finding $p_X(\alpha)$ we could evaluate $p_Y(1)=0.3$, $p_Y(2)=0$, $p_Y(3)=0.4$, and $p_Y(4)=0.3$. Figure 5.4b is a plot of $p_Y(\beta)$.

The two random variables are not independent, since there are many counterexamples showing that

$$p_{XY}(\alpha_i, \beta_j) \neq p_X(\alpha_i) p_Y(\beta_j) \qquad \text{(Find two such cases)}$$

(b) To find the conditional mass function $p_X[\alpha/(Y=3)]$, by using Eq. 5.12 the reader can check that $p_X[0/(Y=3)]=0$, $p_X[1/(Y=3)]=0.75$, and $p_X[2/(Y=3)]=0.25$. $p_X[\alpha/(Y=3)]$ is shown plotted in Figure 5.4b and is seen to satisfy all the conditions for a mass function.

To find the conditional mass function $P_Y[\beta/(X>0)\cap(2<Y<4)]$ we use probability theory:

$$P_Y[\beta/(X>0)\cap(2<Y\leq4)]$$

$$= \frac{P[(X>0)\cap(2<Y<4)\cap(Y=\beta)]}{P[(X>0)\cap(2<Y\leq4)]}$$

$$P[(X>0)\cap(2<Y\leq4)]=0.5$$

from Figure 5.4a. We now find, for $\beta=1$ and $\beta=2$, that

$$P_Y[\beta/(X>0)\cap(2<Y\leq4)]=0$$

and, for $\beta=3$ and $\beta=4$,

$$P_Y[3/(X>0)\cap(2<Y\leq4)]=\frac{0.4}{0.5}=0.8$$

$$P_Y[4/(X>0)\cap(2<Y\leq4)]=\frac{0.1}{0.5}=0.2$$

This conditional mass function is shown plotted in Figure 5.4b.
(c) The probability of the event $[(X<2)\cap(Y\leq2)]$ is found by intersecting it with all the points of the $\alpha\beta$ plane to yield

$$[(X<2)\cap(Y\leq2)]=(0,1)\cup(0,2)\cup(1,1)\cup(1,2)$$

$$\therefore P[(X<2)\cap(Y\leq2)]=0.1$$

(d) The probability of the conditional event $[(X<2)/(Y\leq2)]$ is, by conditional probability,

$$P[(X<2)/(Y\leq2)]=\frac{0.1}{0.3}=0.33$$

The overall simplicity of this problem compared to the continuous case should be evident.

DRILL SET: JOINT DISTRIBUTION, DENSITY, AND MASS FUNCTIONS

Consider the periodic waveform

$$x(t) = \sum_{-\infty}^{\infty} g(t - nb)$$

where

$$g(t) = 2 \quad 0 < t < \frac{b}{3}$$
$$= 0 \quad \text{otherwise}$$

as the input to a rectifier device defined by $y(t) = |x(t)|$. Define the experiment of sampling $x(t)$ at some time t_1 chosen uniformly and also the experiment of sampling $y(t)$ at some time t_2 chosen uniformly independent of time t_1. Let X and Y be the random variables describing the values $x(t_1)$ and $y(t_2)$, respectively, obtained.

(a) Find $p_{XY}(\alpha_i, \beta_j)$ and hence $F_{XY}(\alpha, \beta)$, and sketch both these functions on the $\alpha\beta$ plane.

(b) Assume that the random variables are such that X describes $x(t)$ and Y describes $y(t)$ at the same time t_1. Find $p_{XY}(\alpha_i, \beta_j)$ and $F_{XY}(\alpha_i, \beta_j)$.

5.2 PROPERTIES AND USAGE OF JOINT DISTRIBUTION, DENSITY, AND MASS FUNCTIONS

This section will enumerate the properties of $F_{XY}(\alpha, \beta)$, $f_{XY}(\alpha, \beta)$, and $p_{XY}(\alpha, \beta)$ and then culminate in solving some probabilistic questions while pinpointing the care and discipline required. Joint random variables are defined by rules mapping points of a sample description space onto the two-dimensional $\alpha\beta$ plane. All the properties of $F_{XY}(\alpha, \beta)$, $f_{XY}(\alpha, \beta)$, and $p_{XY}(\alpha, \beta)$ may be derived from the axioms of probability theory using the theory of sets and the concept of conditional probability.

The Joint Cumulative Distribution Function $F_{XY}(\alpha, \beta)$

If two random variables $X(s)$ and $Y(s)$ are defined on a sample space, it must follow that $F_{XY}(\alpha, \beta)$ possesses the following properties:

$$F_{XY}(-\infty, -\infty) = 0$$
$$F_{XY}(\alpha_1, \beta) \geq F_{XY}(\alpha_2, \beta) \quad \text{for } \alpha_1 > \alpha_2, \text{ any } \beta$$
$$F_{XY}(\alpha, \beta_1) \geq F_{XY}(\alpha, \beta_2) \quad \text{for } \beta_1 > \beta_2, \text{ any } \alpha$$
$$F_{XY}(\alpha, \infty) = F_X(\alpha)$$
$$F_{XY}(\infty, \beta) = F_Y(\beta)$$
$$F_{XY}(\infty, \infty) = 1$$
$$F_{XY}(\alpha, \beta) = \int_{-\infty}^{\beta^+} \int_{-\infty}^{\alpha^+} f_{XY}(\alpha, \beta) \, d\alpha \, d\beta \quad \text{for all random variables}$$

and

$$F_{XY}(\alpha, \beta) = \sum_{\substack{\text{all } j \\ \beta_j \leq \beta}} \sum_{\substack{\text{all } i \\ \alpha_i \leq \alpha}} p_{XY}(\alpha_i, \beta_j) \qquad \text{for discrete random variables}$$

As an example we will prove the property $F_{XY}(\alpha_1, \beta) > F_{XY}(\alpha_2, \beta)$ if $\alpha_1 > \alpha_2$.

PROOF

From set theory,

$$[(X \leq \alpha_1) \cap (Y \leq \beta)] = [(X \leq \alpha_2) \cup (\alpha_2 < X \leq \alpha_1)] \cap (Y \leq \beta)$$

$$\therefore F_{XY}(\alpha_1, \beta) = F_{XY}(\alpha_2, \beta) + P[(\alpha_2 \leq X \leq \alpha_1) \cap (Y \leq \beta)]$$

$$\geq F_{XY}(\alpha_2, \beta)$$

The Joint Density Function $f_{XY}(\alpha, \beta)$

The following properties follow directly from the definition that

$$f_{XY}(\alpha, \beta) \triangleq \frac{\partial^2}{\partial \alpha \partial \beta} F_{XY}(\alpha, \beta)$$

$$f_{XY}(\alpha, \beta) \geq 0 \qquad \text{for all } \alpha \text{ and } \beta$$

$$\int_{-\infty}^{\infty} \int_{-\infty}^{\infty} f_{XY}(\alpha, \beta) \, d\alpha \, d\beta = 1$$

and

$$f_{XY}(\alpha, \beta) \Delta \alpha \Delta \beta = P[(\alpha < X < \alpha + d\alpha) \cap (\beta < Y < \beta + d\beta)]$$

The use of joint density functions is uncommon for discrete random variables, but delta functions may occur in an expression for $f_{XY}(\alpha, \beta)$, such as

$$f_{XY}(\alpha, \beta) = f_X(\alpha) f_Y[\beta/(X = \alpha)]$$

$$= 1\delta(\beta - \alpha) \qquad 0 < \alpha < 1, \beta = \alpha$$

which says that X is uniform between 0 and 1 and $Y = X$ or its value on an experiment is the same as the value of X. In noise theory such functions will occur when periodic type waveforms are sampled at two points. (Concentrate on this concept, but if frustrated, skip it until we face it later on.)

The Joint Mass Function $p_{XY}(\alpha, \beta)$ for Discrete Random Variables

The joint mass function is defined as

$$p_{XY}(\alpha, \beta) \triangleq P[(X = \alpha) \cap (Y = \beta)]$$

and is only useful when X and Y map into discrete points of the $\alpha\beta$ plane. It follows from the definition that

$$p_{XY}(\alpha_i, \beta_j) \geq 0 \qquad \text{for any } \alpha_i \text{ or } \beta_j$$

$$\sum_{\text{all } j} \sum_{\text{all } i} p_{XY}(\alpha_i, \beta_j) = 1$$

and

$$F_{XY}(\alpha, \beta) = \sum \sum p_{XY}(\alpha_i, \beta_j) \qquad \beta_j \leq \beta, \alpha_i \leq \alpha$$

So far this chapter has stressed the definitions of $F_{XY}(\alpha, \beta)$, $f_{XY}(\alpha, \beta)$, and $p_{XY}(\alpha, \beta)$ and their derivation as a discrete probability problem from a sample description space. In practice, however, an engineer will often be given the appropriate joint function and then will use it to answer probabilistic questions. We touched on some of this already, particularly in Examples 5.3, 5.4, and 5.5, but we now concentrate on general questions about events and pinpoint the care that must be exercised in integration.

Finding Different Probabilities Directly from the Joint Density Function

Since $f_{XY}(\alpha, \beta)$ completely characterizes two random variables X and Y, any probabilistic question involving them may be solved by integrating *appropriately* over the $\alpha\beta$ plane. The only difficulties are the same as those encountered when evaluating areas or weighted areas in calculus. A few examples will illustrate the procedure.

Example 5.6

Given

$$f_{XY}(\alpha, \beta) = \tfrac{1}{30} \qquad 0 < \alpha < 4, 0 < \beta < 3$$
$$= \tfrac{1}{40} \qquad 4 < \alpha < 10, 0 < \beta < 4$$

evaluate, by integrating over the $\alpha\beta$ plane,

  (a) $P[(3 < X < 5) \cap (2 < Y < 4)]$
  (b) $P[(2X + Y) < 3]$
  (c) $P\{(X > 2)/[(2X + Y) < 3]\}$

SOLUTION

  (a) To find the probability of the event $(3 < X < 5) \cap (2 < Y < 4)$ we sketch it on an $\alpha\beta$ plane, weight it by $f_{XY}(\alpha, \beta)$, and integrate as in calculus. Figure 5.5a shows a shaded area that represents the portion of $(3 < X < 5) \cap (2 < Y < 4)$ over which the density function is nonzero. In general,

$$P[(3 < X < 5) \cap (2 < Y < 4)] = \int_2^4 \left[\int_3^5 f_{XY}(\alpha, \beta) \, d\alpha \right] d\beta$$

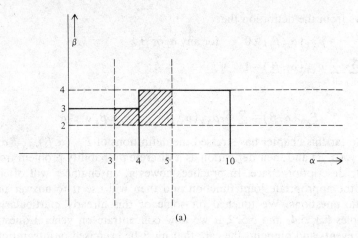

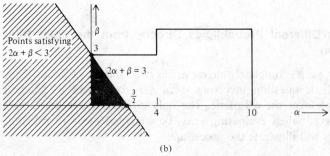

Figure 5.5   (a) The area of integration for which $f_{XY}(\alpha, \beta)$ is nonzero in Example 5.6a. (b) The area of integration for Example 5.6b.

and observing where $f_{XY}(\alpha, \beta) = 0$ we are left only with the integral:

$$P[(3 < X < 5) \cap (2 < Y < 4)] = \int_3^4 \left[\int_2^3 \frac{1}{30} d\beta \right] d\alpha + \int_4^5 \left[\int_2^4 \frac{1}{40} d\beta \right] d\alpha$$

$$= \frac{1}{12}$$

(b) To find $P[(2X + Y) < 3]$ the points of the α, β plane satisfying the statement $[(2\alpha + \beta) < 3]$ must be located. This involves plotting all lines $2\alpha + \beta = M$ up to $M = 3$, as is shown in Figure 5.5b.

$$\therefore P[(2X + Y) < 3]$$

$$= \int \int f_{XY}(\alpha, \beta) \, d\alpha \, d\beta \qquad \text{for all points } (2\alpha + \beta) < 3$$

Carefully observing the limits of integration, this becomes

$$P[(2X+Y)<3]=\int_0^{3/2}\left[\int_0^{3-2\alpha}\frac{1}{30}d\beta\right]d\alpha$$

$$=\int_0^{3/2}\frac{1}{30}(3-2\alpha)\,d\alpha$$

$$=\frac{1}{10}$$

(c) To find $P[(X>2)/(2X+Y<3)]$ we use conditional probability

$$P[(X>2)/(2X+Y<3)]=\frac{P\{(X>2)\cap[(2X+Y)<3]\}}{P[(2X+Y)<3]}$$

It is left as an exercise to show that $f_{XY}(\alpha,\beta)=0$ for the set of points defined by $\{(X>2)\cap[(2X+Y)<3]\}$.

$$\therefore P\{(X>2)/[(2X+Y)<3]\}=0$$

Example 5.7

Given

$$f_{XY}(\alpha,\beta)=\frac{1}{100}\qquad 0<\alpha<0.5\beta,\,0<\beta<20$$

evaluate

(a) $p[(X<4.2)\cap(Y>1)]$
(b) $p[(X<4.2)\cap(Y>1)/(X>2)]$
(c) $f_Y[\beta/(2<X<4)]$

SOLUTION

(a) The joint density function is shown plotted in Figure 5.6a, and the region $[(X<4.2)\cap(Y>1)]$ is also indicated.

$$\therefore p[(X<4.2)\cap(Y>1)]=\int_1^\infty\int_{-\infty}^{4.2}f_{XY}(\alpha,\beta)\,d\alpha\,d\beta$$

observing carefully where the density function is nonzero,

$$p[(X<4.2)\cap(Y>1)]=\int_1^{8.4}\left[\int_0^{0.5\beta}\frac{1}{100}d\alpha\right]d\beta$$

$$=0.17$$

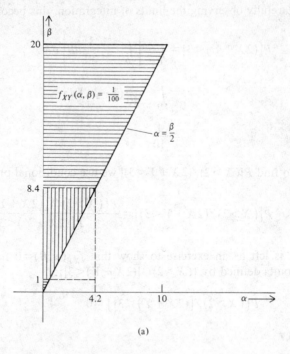

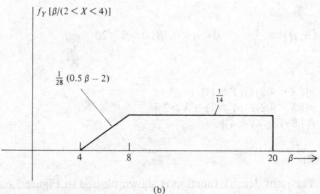

Figure 5.6 (a) The given joint density function and (b) required density function for Example 5.7.

(b) $p[(X<4.2)\cap(Y>1)/(X>2)]=\dfrac{p[(2<X<4.2)\cap(Y>1)]}{p[X>2]}$

$$=\dfrac{\displaystyle\int_{1}^{\infty}\left[\int_{2}^{4.2}f_{XY}(\alpha,\beta)\,d\alpha\right]d\beta}{\displaystyle\int_{2}^{\infty}\left[\int_{-\infty}^{\infty}f_{XY}(\alpha,\beta)\,d\beta\right]d\alpha}$$

observing very carefully where in Figure 5.6b the density function is nonzero, this becomes

$$p[(X<4.2)\cap(Y>1)/(X>2)]$$

$$= \frac{\int_4^{8.4}\left[\int_2^{0.5\beta}0.01\,d\alpha\right]d\beta + \int_{8.4}^{20}\left[\int_2^{4.2}0.01\,d\alpha\right]d\beta}{\int_4^{20}\left[\int_2^{0.5\beta}0.01\,d\alpha\right]d\beta}$$

$$= \frac{4.8+25.5}{64}$$

$$=0.47$$

(c) By definition,

$$f_Y[(\beta/(2<X<4)]\,\Delta\beta = \frac{p[(\beta<Y<\beta+\Delta\beta)\cap(2<X<4)]}{p(2<X<4)}$$

$$= \frac{\left[\int_2^4 f_{XY}(\alpha,\beta)\,d\alpha\right]\Delta\beta}{\int_{-\infty}^{\infty}\left[\int_2^4 f_{XY}(\alpha,\beta)\,d\alpha\right]d\beta}$$

On careful observation of Figure 5.6c the following results are obtained:
For $(\beta<4)\cup(\beta>20)$,

$$f_Y[\beta/(2<X<4)]=0$$

For $4<\beta<8$,

$$f_Y[(\beta/(2<X<4)] = \frac{\int_2^{0.5\beta}0.01\,d\alpha}{0.28} = \frac{1}{28}(0.5\beta-2)$$

For $8<\beta<20$,

$$f_Y[(\beta/(2<X<4)] = \frac{\int_2^4 0.01}{0.28} = \frac{1}{14}$$

$f_Y(\beta/2<X<4)$ is shown plotted in Figure 5.6b.

Consideration of Examples 5.6 and 5.7 should convince us that finding the probability of any event or conditional event resolves itself to evaluating a weighted planar integral or the ratio of two weighted planar integrals from calculus. We must firmly keep our eye on the definition of an event as a region of the $\alpha\beta$ plane and on subregions caused by different functional values of $f_{XY}(\alpha, \beta)$. We should never become too mechanical about the calculus, less we lose sight of the probabilistic aspects of a problem.

The Case of Three or More Random Variables

All the concepts developed in this chapter can readily be extended to the case of three or more random variables. For example, for three random variables X, Y, and Z defined on a sample space, the cumulative distribution function is defined as

$$F_{XYZ}(\alpha, \beta, \gamma) \triangleq P[(X \le \alpha) \cap (Y \le \beta) \cap (Z \le \gamma)] \tag{5.14}$$

for all α, β, γ in three-dimensional space, $-\infty < \alpha < \infty$, $-\infty < \beta < \infty$, and $-\infty < \gamma < \infty$.

The joint density function $f_{XYZ}(\alpha, \beta, \gamma)$ is defined as

$$f_{XYZ}(\alpha, \beta, \gamma) \triangleq \frac{\partial^3}{\partial \alpha \partial \beta \delta \gamma} F_{XYZ}(\alpha, \beta, \gamma) \tag{5.15}$$

For a set of discrete random variables the joint mass function is

$$p_{XYZ}(\alpha, \beta, \gamma) \triangleq P[(X = \alpha) \cap (Y = \beta) \cap (Z = \gamma)] \tag{5.16}$$

Philosophically these three functions may be found, given their definition on a sample space, although such a problem could very easily be cumbersome. Normally we find $f_{XYZ}(\alpha, \beta, \gamma)$ as

$$f_{XYZ}(\alpha, \beta, \gamma) = f_X(\alpha) f_Y[\beta/(X = \alpha)] f_Z\{\gamma/[(X = \alpha) \cap (Y = \beta)]\} \tag{5.17}$$

As an extension of the case of two random variables, we could show that the joint density function would enable us to answer any probabilistic question about the three random variables, or about any two of them, or about any one of them (subject to any conditions). The following relations are easily proved from event theory:

$$f_{XY}(\alpha, \beta) = \int_{-\infty}^{\infty} f_{XYZ}(\alpha, \beta, \gamma) \, d\gamma \tag{5.18}$$

and

$$f_X(\alpha) = \int_{-\infty}^{\infty} \left[\int_{-\infty}^{\infty} f_{XYZ}(\alpha, \beta, \gamma) \, d\gamma \right] d\beta \tag{5.19}$$

with similar obvious symmetric formulas for $f_{YZ}(\beta, \gamma)$, $f_{XZ}(\alpha, \gamma)$, $f_Y(\beta)$, and $f_Z(\gamma)$. Some conditional density functions are

$$f_{XY}[(\alpha, \beta)/(Z = \gamma_0)] = \frac{f_{XYZ}(\alpha, \beta, \gamma_0)}{f_Z(\gamma_0)} \tag{5.20}$$

and

$$f_X\{\alpha/[(Y=\beta_0)\cap(Z=\gamma_0)]\} = \frac{f_{XYZ}(\alpha,\beta_0,\gamma_0)}{f_{YZ}(\beta_0,\gamma_0)} \tag{5.21}$$

Example 5.8

Write down the solution in terms of a general $f_{XYZ}(\alpha,\beta,\gamma)$ for the probabilities of the following events:

(a) $P[(2<X<3)/(Y=1)\cap(1<Z<4)]$
(b) $P[(X<5)/(Z>6)]$

SOLUTION

(a) $P\{(2<X<3)/[(Y=1)\cap(1<Z<4)]\}$

$$= \frac{P[(2<X<3)\cap(Y=1)\cap(1<Z<4)]}{P[(Y=1)\cap(1<Z<4)]}$$

$$= \frac{\int_1^4\left[\int_2^3 f_{XYZ}(\alpha,1,\gamma)\,d\alpha\right]d\gamma}{\int_2^3\left[\int_{-\infty}^{\infty} f_{XYZ}(\alpha,1,\gamma)\,d\alpha\right]d\gamma}$$

The fact that the specific point $Y=1$ was included in the problem statement causes no difficulty since each event leads to a finite area.

(b) By inspection,

$$P[(X<5)/(Z>6)] = \frac{P[(X<5)\cap(Z>6)]}{P[(Z>6)]}$$

$$= \frac{\int_6^{\infty}\left\{\int_{-\infty}^{\infty}\left[\int_{-\infty}^5 f_{XYZ}(\alpha,\beta,\gamma)\,d\alpha\right]d\beta\right\}d\gamma}{\int_6^{\infty}\left\{\int_{-\infty}^{\infty}\left[\int_{-\infty}^{\infty} f_{XYZ}(\alpha,\beta,\gamma)\,d\alpha\right]d\beta\right\}d\gamma}$$

In both (a) and (b) it is obvious from the integral limits that the numerator event in each case is a subset of the denominator event.

To complete this section, the general conceptual case of n random variables for any n will be discussed. Since there are only 26 letters in the alphabet, the random variable notation uses one specific capital letter, with a subscript for each random variable. Therefore for n random variables defined on a sample space we use the notation $X_1, X_2,\ldots, X_n$ for the random variables and $\alpha_1, \alpha_2, \ldots, \alpha_n$ for their respective axes. Examples of density functions are $f_{X_1}(\alpha_1)$, $f_{X_1 X_2}(\alpha_1, \alpha_2)$, and $f_{X_1 X_2 X_3}(\alpha_1, \alpha_2, \alpha_3)$. For n random variables we denote the density function by vector notation,

$$f_{\mathbf{X}}(\boldsymbol{\alpha})$$

where

$$\mathbf{X} = (X_1, X_2, \ldots, X_n) \quad \text{and} \quad \boldsymbol{\alpha} = (\alpha_1, \alpha_2, \ldots, \alpha_n)$$

The vectors $\mathbf{X}$ and $\boldsymbol{\alpha}$ may be of any size. When we will deal with random processes, it will be important to visualize $\mathbf{X}$ as $n \to \infty$, but it will be really a conceptual problem. As a rather trivial example let us consider the joint density function, for sampling a dc waveform of value 1 V at five random points. With thought we arrive at the weird density function,

$$f_{X_1 X_2 X_3 X_4 X_5}(\alpha_1, \alpha_2, \alpha_3, \alpha_4, \alpha_5)$$
$$= f_{X_1}(\alpha_1) f_{X_2}[\alpha_2 / (X = \alpha_1)] \cdots$$
$$f_{X_5}\{\alpha_5 / [(X_1 = \alpha_1) \cap (X_2 = \alpha_2) \cap (X_3 = \alpha_3) \cap (X_4 = \alpha_4)]\}$$
$$= \delta(\alpha_1 - 1) \delta(\alpha_2 - \alpha_1) \delta(\alpha_3 - \alpha_2) \delta(\alpha_4 - \alpha_3) \delta(\alpha_5 - \alpha_4)$$

where we indicate integration must first be performed with respect to α_5, then α_4, then α_3, then α_2, and finally α_1.

DRILL SET: PROPERTIES OF JOINT DISTRIBUTION, DENSITY, AND MASS FUNCTIONS

1. Assume that $F_{XY}(\alpha, \beta)$ is the joint distribution function of X and Y and is a continuous function of both random variables. Express the probabilities of the following events in terms of $F_{XY}(\alpha, \beta)$
    (a) $[(X < 4) \cap (Y < 5)]$
    (b) $[(X < 3) \cap (Y > 2)]$
    (c) $\{[(2 < X < 4) \cup (X > 7)] \cap (3 < Y < 5)\}$
    (d) $[(2 < X < 4) \cap (Y < 1) / (X < 3)]$
    (e) $(\{[(1 < X < 5) \cup (X > 6)] \cap (1 < Y < 4)\} / [(-1 < X < 4) \cap (2 < Y < 7)])$
    Note: If, for example, we were asked to find $P[(1 < X < 3) \cap (Y < 4)]$, we could do it more easily graphically than by using set theory. If we view the sketch shown

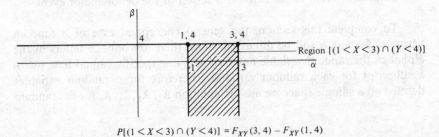

$$P[(1 < X < 3) \cap (Y < 4)] = F_{XY}(3, 4) - F_{XY}(1, 4)$$

2. Given the joint mass function $p_{XYZ}(\alpha_i, \beta_j, \gamma_k)$, give expressions for the following:

(a) $p_X(\alpha_i)$

(b) $P[(1 \leq X < 2.6) \cap (Y=4) \cap (3 < Z < 4.1)]$

(c) $p_X\{\alpha_i / [(1 < Y \leq 3) \cap (Z=2)]\}$

(d) If we know p_{XYZ} is 0 for $Z=2$, is the answer to (c) 0 or infinity?

3. (a) Evaluate $\displaystyle\int_{\alpha=0}^{2} \left[\int_{\beta=1}^{\beta=5} \frac{1}{2} \alpha \delta(\beta-3) \, d\beta \right] d\alpha.$

(b) Evaluate $\displaystyle\int_{\alpha=0}^{2} \left[\int_{\beta=1}^{2} \frac{1}{2} \alpha \delta(\beta-3) \, d\beta \right] d\alpha.$

5.3 STATISTICS OF TWO RANDOM VARIABLES

There is a number of "statistics" or "numbers" associated with two random variables, which give much information about the relationship between them or about the location, symmetry, or width of the joint density function. Some of the more important of these "statistics" will now be defined. The expected value of any function of two random variables, say $g(X, Y)$, is defined as

$$E[g(X,Y)] \triangleq \int_{-\infty}^{\infty} \int_{-\infty}^{\infty} g(\alpha, \beta) f_{XY}(\alpha, \beta) \, d\alpha \, d\beta \qquad (5.22)$$

for a continuous density function, or

$$\triangleq \sum_{\text{all } j} \sum_{\text{all } i} g(\alpha_i, \beta_j) p_{XY}(\alpha_i, \beta_j) \qquad (5.23)$$

for discrete random variables. The correlation of X and Y, denoted R_{XY}, is defined as

$$R_{XY} = E(XY) = \overline{XY} \triangleq \int_{-\infty}^{\infty} \int_{-\infty}^{\infty} \alpha \beta f_{XY}(\alpha, \beta) \, d\alpha \, d\beta \qquad (5.24)$$

and the covariance of X and Y, denoted L_{XY} or $\text{cov}(X, Y)$, is defined as

$$L_{XY} = \overline{(X - \overline{X})(Y - \overline{Y})} \triangleq \int_{-\infty}^{\infty} \int_{-\infty}^{\infty} (\alpha - \overline{X})(\beta - \overline{Y}) f_{XY}(\alpha, \beta) \, d\alpha \, d\beta \qquad (5.25)$$

For engineering applications, R_{XY} and L_{XY}, plus extensions of them called the autocorrelation function and covariance function $R_{XY}(\tau)$ and $L_{XY}(\tau)$, which we will use when studying noise waveforms, are the most widely used second-order statistics. In addition, statisticians define a large number of additional statistics, including moments and central moments, to which they give much physical interpretation. These are the joint moments,

$$m_{N,P} \triangleq E(X^N Y^P) \triangleq \int_{-\infty}^{\infty} \int_{-\infty}^{\infty} \alpha^N \beta^P f_{XY}(\alpha, \beta) \, d\alpha \, d\beta \qquad (5.26)$$

and the joint-centered moments,

$$\mu_{N,P} \triangleq E\Big[(X-\overline{X})^N(Y-\overline{Y})^P\Big]$$

$$\triangleq \int_{-\infty}^{\infty}\int_{-\infty}^{\infty}(\alpha-\overline{X})^N(\beta-\overline{Y})^P f_{XY}(\alpha,\beta)\,d\alpha\,d\beta \qquad (5.27)$$

where N and P are nonnegative integer numbers. For example, $R_{XY}=m_{11}$ and $L_{XY}=\mu_{11}$; and $\overline{X}=M_{10}$, $\overline{Y}=M_{01}$, $\overline{X^2}=M_{20}$, $\overline{Y^2}=M_{02}$, $\sigma_X^2=\mu_{20}$ and $\sigma_Y^2=\mu_{02}$. Two-dimensional moment-generating functions corresponding to two-dimensional Fourier transforms (which are somewhat awesome) are defined, from which all second-order and first-order statistics may be obtained.

Relative-Frequency Interpretation

Since the second-order statistics R_{XY} and L_{XY} are very important, we should dwell on their relative-frequency interpretation in order to make the meanings a part of us. Assume that we are dealing with a random phenomenon, with a sample space S, on which the random variables X and Y are defined. Assume also that a very large number of trials N ($N\to\infty$) of the phenomenon occur (or are visualized to occur), with the following observed results for events involving X, Y, and XY:

$N_{X1}, N_{X2},\ldots,N_{XN}$ are the number of occurrences of the possible numerical values $\alpha_1,\alpha_2,\ldots,\alpha_N$ that X takes on.

$N_{Y1}, N_{Y2},\ldots,N_{YN}$ are the number of occurrences of the possible numerical values $\beta_1,\beta_2,\ldots,\beta_M$ that Y takes on.

$N_{11}, N_{12},\ldots,N_{1M}; N_{21},\ldots,N_{2M};\ldots; N_{N1},\ldots,N_{NM}$ are the number of occurrences of the events $[(X=\alpha_i)\cap(Y=\beta_j)]$ for all i and j.

For example, N_{35} is the number of times $[(X=\alpha_3)\cap(Y=\beta_5)]$ occurs. The numbers given here allow us to evaluate any first-order or second-order statistic. *Very carefully* the following relative-frequency results may be found

$$R_{XY}=E(XY)$$

$$=\frac{\alpha_1\beta_1 N_{11}+\alpha_1\beta_2 N_{12}+\cdots+\alpha_1\beta_M N_{1M}+\cdots+\alpha_N\beta_M N_{NM}}{N}$$

$$=\sum_{\text{all }j}\sum_{\text{all }i}\alpha_i\beta_j p(\alpha_i,\beta_j)$$

in the discrete form of Eq. 5.24. Also

$$L_{XY}=E\Big[(X-\overline{X})(Y-\overline{Y})\Big]$$

$$=\frac{(\alpha_1-\overline{X})(\beta_1-\overline{Y})N_{11}+\cdots+(\alpha_N-\overline{X})(\beta_M-\overline{Y})N_{NM}}{N}$$

$$=\sum_{\text{all }j}\sum_{\text{all }i}(\alpha_i-\overline{X})(\beta_j-\overline{Y})p_{XY}(\alpha_i,\beta_j)$$

in accordance with the discrete form of Eq. 5.25, where, it is assumed, $\overline{X}$ and $\overline{Y}$ were first statistically evaluated. In order to show deep understanding we will now find $\overline{X}$ in two ways and see whether the results agree. First, finding $\overline{X}$, using the numbers obtained for the different values of $X, \alpha_1, \alpha_2, \ldots, \alpha_n$, yields

$$\overline{X} = \frac{\alpha_1 N_{X1} + \cdots + \alpha_N N_{XN}}{N}$$

$$= \sum_{\text{all } i} \alpha_i p_X(\alpha_i)$$

as we found in the last chapter. Second, finding $\overline{X}$ or M_{10} based on the number of occurrences of compound events such as $[(X = \alpha_i) \cap (Y = \beta_j)]$ yields

$$\overline{X} = \alpha_1 \frac{N_{11} + N_{12} + \cdots + N_{1M}}{N} + \alpha_2 \frac{N_{21} + N_{22} + \cdots + N_{2M}}{N}$$

$$+ \cdots + \alpha_N \frac{N_N + \cdots + N_{NM}}{N} \tag{5.28}$$

Showing some mathematical elegance this becomes

$$\overline{X} = \alpha_1 \frac{\displaystyle\sum_{K=1}^{M} N_{1K}}{N} + \alpha_2 \frac{\displaystyle\sum_{K=1}^{M} N_{2K}}{N} + \cdots + \alpha_N \frac{\displaystyle\sum N_{2K}}{N} \tag{5.29}$$

$$= \sum_i \sum_K \alpha_i \left(\frac{N_{iK}}{N} \right) = \sum_{\text{all } i} \sum_{\text{all } K} \left[\alpha_i p_{XY}(\alpha_i, \beta_K) \right] \tag{5.30}$$

This is a special case of the general formula given by Eq. 5.22. The result of Eq. 5.29 may also be interpreted term by term. Since

$$\sum_{K=1}^{M} N_{1K} = N_{X1} \qquad \text{(Be sure of this)}$$

or, in general,

$$\alpha_i \frac{\displaystyle\sum_{K=1}^{M} N_{iK}}{N} = \alpha_i \frac{N_{Xi}}{N} = \alpha_i p_X(\alpha_i) \qquad \text{for } i = 1, N$$

then Eq. 5.29 states

$$\overline{X} = \sum_{\text{all } i} \alpha_i p_X(\alpha_i)$$

in another form.

Results Involving Second-Order Statistics

Two random variables are **independent** if

$$f_{XY}(\alpha, \beta) = f_X(\alpha) f_Y(\beta)$$

and two random variables are **uncorrelated** if

$$E(XY) = E(X)E(Y)$$

It should be obvious that independence implies that random variables are uncorrelated, but the reverse is not necessarily true. Important results can be developed for the mean and variance of the sum of a number of random variables and—more important to us—for the product, using the condition of independence and uncorrelatedness.

The Mean and Variance of the Sum and Product of Random Variables

By definition,

$$
\begin{aligned}
E(X+Y) &\triangleq \int_{-\infty}^{\infty}\int_{-\infty}^{\infty}(\alpha+\beta)f_{XY}(\alpha,\beta)\,d\alpha\,d\beta \\
&= \int_{-\infty}^{\infty}\int_{-\infty}^{\infty}\alpha f_{XY}(\alpha,\beta)\,d\alpha\,d\beta + \int_{-\infty}^{\infty}\int_{-\infty}^{\infty}\beta f_{XY}(\alpha,\beta)\,d\alpha\,d\beta \\
&= \int_{-\infty}^{\infty}\alpha f_X(\alpha)\,d\alpha + \int_{-\infty}^{\infty}\beta f_Y(\beta)\,d\beta \\
&= \bar{X} + \bar{Y}
\end{aligned}
$$

$$\therefore \overline{X+Y} = \bar{X} + \bar{Y} \tag{5.31}$$

for any two random variables.

The derivation of a formula for $\mathrm{var}(X+Y)$ will be found using a shorthand notation, similar to that used in Chapter 4 for one random variable. If there is difficulty in following the proof, then it should be demonstrated longhand as for $E(X+Y)$.

$$
\begin{aligned}
\sigma_{X+Y}^2 &\triangleq \overline{\left[(X+Y)-(\bar{X}+\bar{Y})\right]^2} \\
&= \overline{(X+Y)^2 + (\bar{X}+\bar{Y})^2 - 2(X+Y)(\bar{X}+\bar{Y})} \\
&= \overline{X^2} + \overline{Y^2} + 2\overline{XY} - \bar{X}^2 - \bar{Y}^2 + 2\bar{X}\bar{Y} - 4\bar{X}\bar{Y} \\
&= \sigma_X^2 + \sigma_Y^2 + 2\overline{XY} - 2\bar{X}\bar{Y} \\
&= \sigma_X^2 + \sigma_Y^2 + 2L_{XY} \tag{5.32}
\end{aligned}
$$

If the two random variables are *uncorrelated*, then $2\overline{XY}=2\bar{X}\bar{Y}$, and

$$\sigma_{X+Y}^2 = \sigma_X^2 + \sigma_Y^2 \tag{5.33}$$

The expected value of the product of two random variables R_{XY} or $\overline{XY}$ from Eq. 5.24 is

$$R_{XY} = \int_{-\infty}^{\infty}\int_{-\infty}^{\infty}\alpha\beta f_{XY}(\alpha,\beta)\,d\alpha\,d\beta$$

If X and Y are independent, this becomes

$$R_{XY} = \int_{-\infty}^{\infty} \int_{-\infty}^{\infty} \alpha\beta f_X(\alpha) f_Y(\beta) \, d\alpha \, d\beta$$

$$= \int_{-\infty}^{\infty} \alpha f_X(\alpha) \, d\alpha \int_{-\infty}^{\infty} \beta f_Y(\beta) \, d\beta$$

$$\therefore R_{XY} = \overline{X}\,\overline{Y} \qquad \text{(when X and Y are independent)} \qquad (5.34)$$

Also let us consider the case of complete dependence $Y = X$. From the first-order statistics of Chapter 4, it is obvious that

$$R_{XY} = E(X \cdot X) = E(X^2) = \overline{X^2}$$

To develop facility with joint density functions we will derive this result from Eq. 5.24 as follows:

$$R_{XY} = E(XY) = \int_{-\infty}^{\infty} \int_{-\infty}^{\infty} \alpha\beta f_{XY}(\alpha, \beta) \, d\alpha \, d\beta$$

If $Y = X$, then

$$f_{XY}(\alpha, \beta) = f_X(\alpha) f_Y[\beta/(X=\alpha)]$$

Logically, $f_Y[\beta/(X=\alpha)] = \delta(\beta - \alpha)$. Hence

$$f_{XY}(\alpha, \beta) = f_X(\alpha)\delta(\beta - \alpha)$$

and $E(XY)$ becomes

$$R_{XY} = \int_{-\infty}^{\infty} \int_{-\infty}^{\infty} \alpha\beta f_X(\alpha)\delta(\beta - \alpha) \, d\alpha \, d\beta$$

$$= \int_{-\infty}^{\infty} \alpha f_X(\alpha) \left[\int_{-\infty}^{\infty} \beta\delta(\beta - \alpha) \, d\beta \right] d\alpha$$

$$= \int_{-\infty}^{\infty} \alpha f_X(\alpha)[\alpha] \, d\alpha$$

$$= \overline{X^2} \qquad (5.35)$$

The two results we have found state that for independent (or uncorrelated) random variables the correlation is the product of the mean values, and for completely dependent random variables $(Y = X)$ the correlation is the mean square value. The physical impact of this information will be of paramount importance when we study "randomness" of waveforms.

Similar results to those we have derived for the correlation may be found for the covariance L_{XY} [in mathematics this is denoted $\text{cov}(X, Y)$]. The student may easily show

$$L_{XY} \triangleq \overline{(X - \overline{X})(Y - \overline{Y})} = R_{XY} - \overline{X}\,\overline{Y}$$

and for *independent* or uncorrelated random variables

$$L_{XY} = 0 \qquad (5.36)$$

whereas for *completely dependent* random variables ($Y = X$), L_{XY} yields the definition of the variance of X

$$L_{XY} = \overline{X^2} - \overline{X}^2 = \sigma_X^2 \qquad (5.37)$$

A few challenging examples will be solved, with an inclination toward future applications involving noise theory.

Second-Order Statistics for Sampling a Periodic Waveform

Example 5.9

Consider the periodic waveform

$$x(t) = \sum_{n=-\infty}^{\infty} g(t - 2n) \qquad n \text{ an integer}$$

where

$$g(t) = 1 \qquad 0 < t < 1$$
$$= 0 \qquad \text{otherwise}$$

(a) Define the experiment of sampling $x(t)$ at two times differing by $\frac{1}{4}$ s and let K and θ be the random variables describing the respective values obtained. Find the joint mass function $p_{K\theta}(\alpha_i, \beta_j)$ and evaluate $R_{K\theta}$ and $L_{K\theta}$ or cov(P, θ). Are K and θ uncorrelated?

(b) Repeat part (a) for sampling two values spaced 1.2 s apart.

(c) Consider sampling two values spaced τ s apart and plot R_{XY} and L_{XY} as a function of τ. Discuss the reasonableness of your results.

SOLUTION

A sketch of $x(t)$ is shown in Figure 5.7a.

(a) The correlation $R_{K\theta} = \overline{K\theta}$ is defined as

$$\overline{K\theta} = \sum_{\text{all } j} \sum_{\text{all } i} \alpha_i \beta_j p_{K\theta}(\alpha_i, \beta_j)$$

The first task is to find $p_{K\theta}(\alpha_i, \beta_j)$. Now

$$p_{K\theta}(\alpha_i, \beta_j) = p_K(\alpha_i) p_\theta[\beta_j / (X = \alpha_i)]$$

and there are four possible values, $p_{K\theta}(0,0)$, $p_{K\theta}(0,1)$, $p_{K\theta}(1,0)$, and $p_{K\theta}(1,1)$. If θ samples the waveform a $\frac{1}{4}$ s later than K on a trial of the experiment, then if K is chosen at random we will show that

$$p_\theta[0/(K=0)] = \tfrac{3}{4} \qquad p[1/(K=0)] = \tfrac{1}{4} \qquad p[1/(K=1)] = \tfrac{3}{4}$$
$$p[0/(K=1)] = \tfrac{1}{4}$$

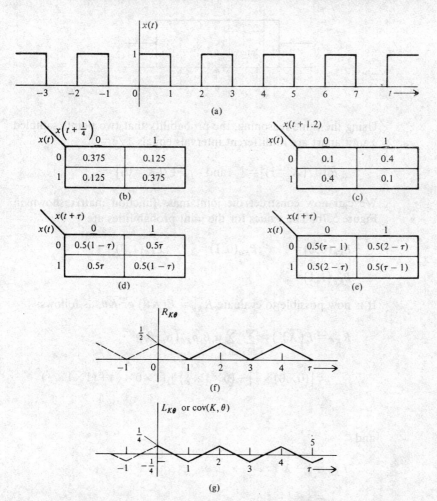

Figure 5.7   (a) The waveform of Example 5.9. (b) through (g) The joint mass function for sampling at two instants: (b) 0.25 s apart; (c) 1.2 s apart; (d) τ s apart, $0 < \tau < 1$; (e) τ s apart, $1 < \tau < 2$; (f) the correlation R_{XY}; (g) the covariance L_{XY}.

The derivation of these values is worth detailed thought. If $K = 1$, then it may be at any point in an interval where $x(t)$ is 1, as is shown in the thumbnail sketch. Obviously, if it is at the start of an interval $p_\theta[0/(K=1)] = 0$ and $p_\theta[1/(K=1)] = 1$. Since the interval is 1 unit long, then the probability that two values sampled $\frac{1}{4}$ unit apart are in the same interval equals $(1 - \frac{1}{4})/1 = \frac{3}{4}$.

$$\therefore p_\theta[0/(K=0)] = p_\theta[1/(K=1)]$$

$$= P(\text{both points in same interval})$$

$$= \tfrac{3}{4}$$

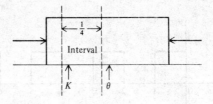

Using the same reasoning, the probability that two points sampled $\frac{1}{4}$ unit apart are in different intervals equals $\frac{1}{4}/1 = \frac{1}{4}$.

$$\therefore p_\theta[0/(K=1)] = \frac{1}{4} \quad \text{and} \quad p_\theta[1/(K=0)] = \frac{3}{4}$$

We can now construct the joint mass function matrix shown in Figure 5.7b. The values for the joint probabilities are

$$P_{K\theta}(0,0) = \frac{3}{8} \qquad P_{K\theta}(0,1) = \frac{1}{8} \qquad P_{K\theta}(1,0) = \frac{1}{8}$$

$$P_{K\theta}(1,1) = \frac{3}{8}$$

It is now possible to evaluate $R_{K\theta} = E(K\theta)$ or $\overline{K\theta}$, as follows:

$$R_{K\theta} = E(XY) = \sum_{\text{all } j} \sum_{\text{all } i} \alpha_i \beta_j p_{K\theta}(\alpha_i, \beta_j)$$

$$= \left[(0 \times 0) \times \frac{3}{8} \right] + \left(0 \times 1 \times \frac{1}{8} \right) + \left(1 \times 0 \times \frac{1}{8} \right) + \left(1 \times 1 \times \frac{3}{8} \right)$$

$$= \frac{3}{8}$$

and

$$L_{K\theta} = \frac{3}{8} - \overline{X}\,\overline{Y}$$

$$= \frac{3}{8} - \frac{1}{4} = \frac{1}{8}$$

We further see that $\overline{K}\overline{\theta} = \frac{1}{2} \times \frac{1}{2} = \frac{1}{4}$ does not equal $\overline{K\theta}$, and therefore the random variables are not uncorrelated or independent, which is not surprising.

(b) K and θ are now defined as sampling values spaced 1.2 s apart. It is required to find as in part (a) $p_K[0/(\theta=0)]$, $p_K[1/(\theta=0)]$, $p_K[(0/(\theta=1)]$, and $p_K[(1/(\theta=1)]$, and hence the joint mass function.

Since the time spacing between the sampled values is 1.2 s, then the sampled values are either from adjacent interval lengths (of size 1) called case A, or the second sampled value is from an interval which is 1 length further away again, called case B, as

Case A   Case B

shown in the thumbnail sketches.

$$P\left(\begin{array}{c}\text{samples are from}\\\text{adjacent interval lengths}\end{array}\right)=\frac{2-1.2}{1}=0.8\qquad\text{(Note case A)}$$

$$P\left(\begin{array}{c}\text{samples are not}\\\text{from adjacent intervals}\end{array}\right)=\frac{1.2-1}{1}=0.2\qquad\text{(Note case B)}$$

The student should really dwell on these results and is expected to encounter difficulties before achieving understanding. The required conditional probabilities—taking into account that samples from adjacent intervals have different values and samples from nonadjacent intervals have the same values—are

$$P[(0/(\theta=0)]=P[1/(\theta=1)]=0.2$$
$$P[1/(\theta=0)]=P[0/\theta=1)]=0.8$$

and the joint mass function matrix is shown in Figure 5.7c. We can now find the correlation as

$$R_{K\theta}=E(K\theta)$$
$$=(0\times0\times0.1)+(0\times1\times0.4)+(1\times0\times0.4)+(1\times1\times0.1)$$
$$=0.1$$

and the covariance is

$$L_{K\theta}=0.1-0.25=-0.15$$

The random variables are obviously uncorrelated, and with time we will develop the ability and a means of measuring by how much they are uncorrelated. We now proceed to part (c) of the problem, which represents a vital gap to be bridged in our analysis of random or noise waveforms.

(c) We are attempting to find $R_{K\theta}$, where the sampling interval between the two random variables is τ, for any general τ, $0<\tau<\infty$. We must divide τ into ranges $0<\tau<\tau_1$, $\tau_1<\tau<\tau_2$, and so on in such a way that we may solve for R_{XY} in terms of τ for each range.

For the range: $0 < \tau < 1$,

Here the upper limit of 1 on our range must be arrived at by inspecting the waveform. For a general τ in this range, based on our reasoning in part (a) we can say

$$P\left(\begin{array}{c}\text{both sampled values}\\ \text{are from the same interval}\end{array}\right) = P\left(\begin{array}{c}\text{both samples have}\\ \text{same value}\end{array}\right) = 1 - \tau$$

and

$$P\left(\begin{array}{c}\text{the sampled values are}\\ \text{from different intervals}\end{array}\right) = P\left(\begin{array}{c}\text{samples have}\\ \text{different values}\end{array}\right) = \tau$$

The joint mass function utilizing conditional probability may be found and is shown in Figure 5.7d. The correlation using Figure 5.7d is

$$\therefore R_{K\theta} = \overline{K\theta} = 1 \times 1 \times \tfrac{1}{2}(1 - \tau)$$

$$= \tfrac{1}{2}(1 - \tau)$$

For the range: $1 < \tau < 2$,

Using the same thought process as involved in part (a), we can show

$$P\left(\begin{array}{c}\text{samples are from}\\ \text{adjacent time intervals}\end{array}\right) = P\left(\begin{array}{c}\text{samples have}\\ \text{different values}\end{array}\right) = 2 - \tau$$

$$P\left(\begin{array}{c}\text{samples are not}\\ \text{from adjacent intervals}\end{array}\right) = P\left(\begin{array}{c}\text{samples have}\\ \text{the same value}\end{array}\right) = \tau - 1$$

The joint mass function is as shown in the accompanying matrix of Figure 5.7e. The correlation using Figure 5.7e is

$$R_{K\theta} = \overline{K\theta} = \tfrac{1}{2}(\tau - 1)$$

To finish our problem, we can use symmetry taking into account the periodicity of $x(t)$ to say: "$R_{K\theta}$ has the same value for any τ, $\tau + 2$, $\tau + 4$, or $\tau + 2n$, where n is any nonnegative integer."

It is now possible to plot $R_{K\theta}$ as a function of τ, $0 < \tau < \infty$, and this is shown in Figure 5.7f. It is seen that since $x(t)$ is periodic with period 2, then $R_{K\theta}$ for sampling at t and $t + \tau$ for a random t is a periodic function of τ with period 2. Indeed, by definition we can show that $R_{K\theta}$ for negative τ yields the same values as for positive τ, and therefore $R_{K\theta}$ is an even periodic function of τ.

The covariance $L_{K\theta}$ or $\text{cov}(K, \theta)$ for $0 < \tau < 1$ is

$$L_{K\theta} = R_{K\theta} - \overline{K}\overline{\theta}$$

$$= \tfrac{1}{2}(1-\tau) - \tfrac{1}{2} \times \tfrac{1}{2}$$

$$= \tfrac{1}{4} - \tfrac{1}{2}\tau$$

For $1 < \tau < 2$,

$$L_{K\theta} = \tfrac{1}{2}(\tau - 1) - \tfrac{1}{2} \times \tfrac{1}{2}$$

$$= \tfrac{1}{2}\tau - \tfrac{3}{4}$$

$L_{K\theta}$ is also a periodic function of τ for any periodic waveform $x(t)$ and is shown plotted in Figure 5.7g.

Example 5.9 is of paramount importance for two reasons. We are now merging strictly mathematical random variable theory with physical problems involving time waveforms, and in addition we are adjusting to somewhat tricky notation, which will soon have to be a standard part of our vocabulary. The amount of time necessary to deeply absorb this problem could be anywhere from two to five hours, but let us not dare to proceed without thorough conceptual as well as mechanical understanding. It is interesting to note that $R_{K\theta}$ for $\tau = 0, 2, 4, 6, \ldots$ is, by definition, K^2 and that $R_{K\theta}$ for $\tau = \pm 1, \pm 3, \pm 5, \ldots$ is 0. Later on, an engineer with physical understanding of the linear nature of the result might just sketch $R_{K\theta}$ versus τ using these two results.

One more example involving R_{XY} for sampling a periodic time waveform leading to a continuous or mixed density function will be solved.

Example 5.10

Given

$$x(t) = \sum_{-\infty}^{\infty} g(t-n) \qquad \text{for all } n$$

where

$$g(t) = t \qquad 0 < t < 1$$
$$= 0 \qquad \text{otherwise}$$

Consider the random phenomenon of sampling $x(t)$ uniformly over a period (or multiple of periods or a very long interval) at a time t and a time $t + \tau$. Let X and Y be the two random variables describing the sampled values. Evaluate R_{XY} and plot R_{XY} versus τ for all τ. Comment on your result.

SOLUTION

A plot of $x(t)$ is shown in Figure 5.8a, and it is seen to be periodic with period 1. We will now consider the evaluation of R_{XY} for a general τ, $0 < \tau < 1$.

Let X be the sampled value obtained when $x(t)$ is sampled uniformly over the range $0 < t < 1$ and Y be the sampled value $x(t + \tau)$. Obviously, since $x(t) = t$, $0 < t < 1$,

$$f_X(\alpha) = 1 \qquad 0 < \alpha < 1$$

$$= 0 \qquad \text{otherwise}$$

$$f_{XY}(\alpha, \beta: \tau) = f_X(\alpha) f_Y[\beta / (X = \alpha)]$$

In finding $f_Y[\beta / (X = \alpha)]$, two situations may arise: first, $x(t + \tau)$ may be from the same interval as $x(t)$, in which case $\beta = (\alpha + \tau)$; or second, $x(t + \tau)$ may be from the next interval and $\beta = (\alpha + \tau) - 1$, since from

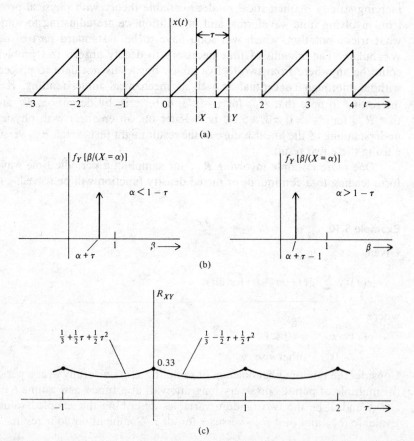

Figure 5.8 (a) The waveform of Example 5.10. (b) The conditional density function at $t + \tau$. (c) The correlation R_{XY}.

$1 < t < 2$, $x(t)$ is defined as $t - 1$.

$$\therefore f_Y[\beta/(x = \alpha)] = \delta[\beta - (\alpha + \tau)] \qquad \text{if } \alpha + \tau < 1$$

$$\text{or } \alpha < 1 - \tau$$

$$= \delta[\beta - (\alpha + \tau - 1)] \qquad \text{if } \alpha + \tau > 1$$

$$\text{or } \alpha > 1 - \tau$$

This conditional density function is shown sketched in Figure 5.8b.
The required joint density function is

$$f_{XY}(\alpha, \beta : \tau) = 1\delta[\beta - (\alpha + \tau)] \qquad 0 < \alpha < 1 - \tau$$

$$\beta = \alpha + \tau$$

and

$$f_{XY}(\alpha, \beta) = 1\delta[\beta - (\alpha + \tau - 1)] \qquad (1 - \tau) < \alpha < 1$$

$$\beta = \alpha + \tau - 1$$

This is probably the most challenging derivation so far in the text and is worth much thought, physical interpretation, and loss of sleep. If it is too difficult, consider a special case for τ—say, $\tau = \frac{1}{4}$. Now it should be clear that the probability X and Y sample the same interval is $\frac{3}{4}$ as if $x(t)$ is between 0 and $\frac{3}{4}$, then $x(t + \tau)$ is less than 1 and from the same interval. This implies that $P(Y < 1/X < \frac{3}{4}) = 1$ and, if $X = \alpha$, then $Y = \alpha + \tau$ for $\alpha < \frac{3}{4}$.
The correlation R_{XY} is

$$R_{XY} = \overline{XY} = \int_{-\infty}^{\infty} \int_{-\infty}^{\infty} \alpha \beta f_{XY}(\alpha, \beta) \, d\alpha \, d\beta$$

$$= \int_0^{1 - \tau} \left[\int_{-\infty}^{\infty} \alpha \beta \delta[\beta - (\alpha + \tau)] \, d\beta \right] d\alpha$$

$$+ \int_{1 - \tau}^1 \left\{ \int_{-\infty}^{\infty} \alpha \beta \delta[\beta - (\alpha + \tau - 1)] \, d\beta \right\} d\alpha$$

Integrating with respect to β and using the definition of a delta function, this becomes

$$R_{xy} = \overline{XY} = \int_0^{1 - \tau} \alpha(\alpha + \tau) \, d\alpha + \int_{1 - \tau}^1 \alpha(\alpha + \tau - 1) \, d\alpha$$

$$= \frac{1}{3} \alpha^3 \Big|_0^1 + \frac{1}{2} \alpha^2 \tau \Big|_0^1 - \frac{1}{2} \alpha^2 \Big|_{1 - \tau}^1$$

$$= \frac{1}{3} + \frac{1}{2} \tau - \frac{1}{2} + \frac{1}{2}(1 - \tau)^2$$

$$= \frac{1}{3} - \frac{1}{2} \tau + \frac{1}{2} \tau^2$$

The plot of R_{XY} versus τ for all τ is shown in Figure 5.8c. We can see that due to the periodicity of $x(t)$, R_{XY} must yield the same result when the sampled values are spaced τ, $\tau \pm 1$, or $\tau \pm n$ for n integer units apart. R_{XY} is periodic with the same period as $x(t)$.

Note: It would be easy to omit problems such as 5.9 and 5.10 from the text and to sacrifice very difficult problem solving for precise and clear elegance and theory, with minimum algebraic discomfort. However, our goal is to develop as much confidence as possible and to demonstrate this with problem-solving skill.

5.4 THE FUNDAMENTAL THEOREM FOR TWO OR MORE RANDOM VARIABLES

For the case of one random variable the fundamental theorem stated that

$$E[K(Y)] = \overline{K(Y)} = \int_{-\infty}^{\infty} K(\alpha) f_Y(\alpha)\, d\alpha$$

could be evaluated using the density function for X, as

$$\overline{K(Y)} = \int_{-\infty}^{\infty} K[g(\alpha)] f_X(\alpha)\, d\alpha$$

where $Y = g(X)$.

This theorem allowed for the evaluation of any "statistic" involving the random variable Y on the density function of a related random variable X and is very useful if $f_X(\alpha)$ is known and $f_Y(\alpha)$ has not yet been found.

The fundamental theorem for two random variables K and θ may be stated in one of many ways depending on the circumstances. In general,

$$E[g(K,\theta)] = \overline{g(K,\theta)} = \int_{-\infty}^{\infty} \int_{-\infty}^{\infty} g(\alpha,\beta) f_{K\theta}(\alpha,\beta)\, d\alpha\, d\beta$$

and the direct evaluation of $\overline{g(K,\theta)}$ by finding the density function of K and θ may be avoided by evaluating

$$\overline{g(K,\theta)} = \int_{-\infty}^{\infty} g(h_1(\alpha), h_2(\alpha)) f_X(\alpha)\, d\alpha \tag{5.38a}$$

if both random variables K and θ are related to a random variable X by the functional relationships $K = h_1(X)$ and $Q = h_2(X)$. Also,

$$\overline{g(K,\theta)} = \int_{-\infty}^{\infty} \int_{-\infty}^{\infty} g[h_1(\alpha,\beta), h_2(\alpha,\beta)] f_{XY}(\alpha,\beta)\, d\alpha\, d\beta \tag{5.38b}$$

if $K = h_1(X,Y)$ and $Q = h_2(X,Y)$ and the joint density function of X and Y is known.

Formulas 5.38a and 5.38b are supposedly intuitively obvious from a relative-frequency point of view, and we will try to demonstrate this by a very simple example.

Example 5.11

Consider the random phenomenon of uniformly sampling the waveform $x(t) = \sin t$ at two points spaced τ s apart. Let the random variables describing the sampled values (using a one-to-one mapping from S) be K and θ, respectively. Find $R_{K\theta} = E(K\theta)$.

SOLUTION

The direct formula for $K\theta$ is

$$R_{K\theta} = \overline{K\theta} = \int_{-\infty}^{\infty} \int_{-\infty}^{\infty} \alpha\beta f_{K\theta}(\alpha, \beta)\, d\alpha\, d\beta$$

and this involves the difficult problem of finding $f_{K\theta}(\alpha, \beta)$ for two sampled values of a sine wave as shown in Figure 5.9a.

Let us focus on the random variables X and Y, where X is the random variable describing choosing a point on the time axis $0 < t < 2\pi$ and Y is the random variable describing choosing a point on the time axis τ s later. The relationship between K and θ and X and Y is

$$K = \sin X$$
$$\theta = \sin Y$$

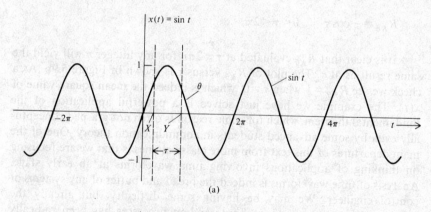

(a)

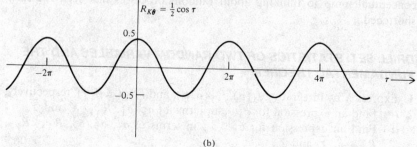

(b)

Figure 5.9   (a) The waveform of Example 5.11. (b) The correlation $R_{K\theta}$.

where we also have $Y = X + \tau$. For any τ, $0 < \tau < 2\pi$, we can now use the fundamental theorem to evaluate $\overline{K\theta}$; that is,

$$R_{K\theta} = \overline{K\theta} = \int_{-\infty}^{\infty} \int_{-\infty}^{\infty} \alpha \beta f_{K\theta}(\alpha, \beta) \, d\alpha \, d\beta \qquad \text{directly}$$

or, by the fundamental theorem,

$$\overline{K\theta} = \int_{-\infty}^{\infty} \int_{-\infty}^{\infty} \sin \alpha \sin \beta f_{XY}(\alpha, \beta) \, d\alpha \, d\beta$$

Using the relation $Y = X + \tau$ this becomes

$$\overline{K\theta} = \int_{0}^{2\pi} \sin \alpha \frac{1}{2\pi} \left\{ \int_{-\infty}^{\infty} \sin \beta \, \delta [\beta - (\alpha + \tau)] \, d\beta \right\} d\alpha$$

as $f_X(\alpha) = 1/2\pi$, $0 < \alpha < 2\pi$, and the value of $\beta = \alpha + \tau$.

$$\therefore \overline{K\theta} = \int_{0}^{2\pi} \frac{1}{2\pi} \sin \alpha \sin(\alpha + \tau) \, d\alpha$$

Using trigonometry this becomes

$$\overline{K\theta} = \frac{1}{2\pi} \int_{0}^{2\pi} \frac{1}{2} \cos(2\alpha + \tau) \, d\alpha + \frac{1}{2\pi} \int_{0}^{2\pi} \frac{1}{2} \cos \tau \, d\alpha$$

$$= 0 + \frac{1}{2} \cos \tau$$

$$R_{K\theta} = \frac{1}{2} \cos \tau \qquad 0 < \tau < 2\pi$$

It is clear that $R_{K\theta}$ evaluated at $\tau \pm 2\pi n$ for any integer n will yield the same result as at τ. The plot of $R_{K\theta}$ versus τ is shown in Figure 5.9b. As a check we see $R_{K\theta} = \frac{1}{2}$ when $\tau = 0$, which is indeed the mean square value of $x(t)$. The example we have just solved is a powerful application of the fundamental theorem, which for some reason is often not grasped conceptually even by some advanced students in communication theory. One of the main departures of this text from more classical ones is that we are focusing on thinking of applications involving time waveforms at an early stage. Analysis of time waveforms is indeed the bread and butter of any systems or control engineer. We may be having some difficulty, but already the conceptual jump to thinking about random processes has been radically shortened.

DRILL SET: STATISTICS OF TWO RANDOM VARIABLES AND THE FUNDAMENTAL THEOREM

1. Express $\overline{X}$ by integrating $f_X(\alpha)$, $f_{XY}(\alpha, \beta)$, and $f_{XYZ}(\alpha, \beta, \gamma)$, respectively.
2. (a) Find an expression for σ_{X-Y}^2 in terms of σ_X^2, σ_Y^2, R_{XY}, $\overline{X}$, and $\overline{Y}$.
   (b) Find an expression for σ_{X+Y+Z}^2 in terms of σ_X^2, σ_Y^2, σ_Z^2, R_{XY}, R_{YZ}, R_{XY}, $\overline{X}$, $\overline{Y}$, and $\overline{Z}$.

3. Consider the periodic waveform

$$x(t) = \sum_{-\infty}^{\infty} g(t-2n)$$

where

$$g(t) = 2t \qquad 0 < t < 2$$

If $x(t)$ is sampled at a general value of t uniformly chosen between 0 and 2 and simultaneously at $t + \tau$ and we define X and Y as random variables assuming the sampled values obtained:

(a) Find $f_{XY}(\alpha, \beta: \tau)$ and R_{XY} for all τ.

(b) Use the fundamental theorem and the random variables P and θ associated with sampling the time axis, where $f_{P\theta}(\alpha, \beta) = \frac{1}{2}\delta[\beta - (\alpha + \tau)]$, $0 < \alpha < 2$, $\beta = \alpha + \tau$ to find R_{XY}.

4. Consider passing the waveform $x(t) = \cos t$ through a rectifier with output defined by $y(t) = |\cos t|$. Use the fundamental theorem to find $R_{Y_1 Y_2}$ where Y_1 and Y_2 are random variables sampling $y(t)$ at $t = t$ and $t = t + \tau$.

5.5 DENSITY FUNCTIONS FOR FUNCTIONS OF RANDOM VARIABLES

In Chapter 4 the problem of finding the density function of a function $Y = g(X)$ of a random variable was discussed. In general it can be shown that given n random variables $X_1, X_2, \ldots, X_n$, their joint density function $f_{X_1, X_2, \ldots, X_n}(\alpha_1, \alpha_2, \ldots, \alpha_n)$, and m other random variables defined by

$$Y_1 = g_1(X_1, X_2, \ldots, X_n)$$

$$\vdots$$

$$Y_m = g_m(X_1, X_2, \ldots, X_n)$$

it is possible to find the joint density function $f_{Y_1, Y_2, \ldots, Y_m}(\beta_1, \beta_2, \ldots, \beta_m)$. This problem will be of paramount importance in the discussion of linear systems with random or signal plus noise inputs where the X random variables will be samples of the input waveform, the Y random variables will be samples of the output, and the actual system will allow for finding the functional relationship between the random variables. We are still some way from handling a general problem of such scope. We will, however, consider simpler cases of this problem based on knowing the density function of two random variables. We should keep in mind that if we only require a few statistics of a function or functions of two random variables, then using the fundamental theorem may be the quickest approach if we know the density function of related random variables.

The Density Function

Given two random variables X and Y and their joint density function $f_{XY}(\alpha, \beta)$, it is possible to find the density function $f_Z(\gamma)$ of any random variable Z that is a function of X and Y. The procedure for finding $f_Z(\gamma)$ is straightforward, and we may use the following steps:

1. Express $F_Z(\gamma)$ as a probabilistic statement about X and Y.
2. Find $f_Z(\gamma) = (d/d\gamma)F_Z(\gamma)$ by differentiation.

This procedure will be illustrated by a few examples. The only difficulties that will arise will be those that crop up when evaluating areas or weighted double integrals in calculus; for example,

$$\iint_A g(x, y)\, dx\, dy$$

where A is a portion of the xy plane.

Example 5.12

Given

$$f_{XY}(\alpha, \beta) = 0.25 \qquad 0 < \alpha \le 2, 0 < \beta \le 2$$

$$= 0 \qquad \text{otherwise}$$

find $f_Z(\gamma)$ if

(a) $Z = X + Y$.
(b) $Z = X - Y$.

SOLUTION

(a) Figure 5.10a shows a plot of $f_{XY}(\alpha, \beta)$ with an indication of the points obeying the events $Z = X + Y < \gamma$ for the different possible ranges of γ. $F_Z(\gamma)$ will now be found for each different appropriate range of γ.
For $\gamma < 0$,

$$F_Z(\gamma) = 0$$

For $0 < \gamma < 2$,

$$F_Z(\gamma) = p[X + Y < \gamma]$$

$$= \int_0^\gamma \int_0^{\gamma - \beta} 0.25\, d\alpha\, d\beta \qquad \text{(See Figure 5.10a)}$$

$$= 0.125\gamma^2 \qquad \text{(with some work)}$$

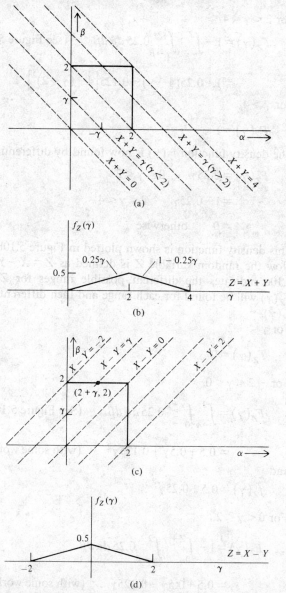

Figure 5.10 (a) The given joint density function of Example 5.12. (b) The density function for $Z = X + Y$. (c) The joint density function again. (d) The density function for $Z = X - Y$.

For $2 < \gamma < 4$,

$$F_Z(\gamma) = 1 - \int_{\gamma-2}^{2} \int_{\gamma-\beta}^{2} 0.25 \, d\alpha \, d\beta \qquad \text{(see Figure 5.10a)}$$

$$= 1 + 0.25(4 - \gamma) - 0.125\left[4 - (\gamma - 2)^2\right]$$

For $\gamma > 4$,

$$F_Z(\gamma) = 1$$

The density function $f_Z(\gamma)$ is now found by differentiation as

$$F_Z(\gamma) = 0.25\gamma \qquad 0 < \gamma < 2$$

$$= 1 - 0.25\gamma \qquad 2 < \gamma < 4$$

$$= 0 \qquad \text{otherwise}$$

This density function is shown plotted in Figure 5.10b.

(b) Now the random variable Z is defined as $Z = X - Y$, and Figure 5.10c indicates the different possible ranges for $Z = X - Y < \gamma$. $F_Z(\gamma)$ will be found for each range and then differentiated to yield $f_Z(\gamma)$.

For $\gamma < -2$,

$$F_Z(\gamma) = 0$$

For $-2 < \gamma < 0$,

$$F_Z(\gamma) = \int_{-\gamma}^{2} \int_{0}^{\gamma+\beta} 0.25 \, d\alpha \, d\beta \qquad \text{(See Figure 5.10c)}$$

$$= 0.5 + 0.5\gamma + 0.125\gamma^2 \qquad \text{(with some work)}$$

and

$$f_Z(\gamma) = 0.5 + 0.25\gamma$$

For $0 < \gamma < 2$,

$$F_Z(\gamma) = 1 - \int_{0}^{Z+\gamma} \int_{\gamma+\beta}^{2} 0.25 \, d\alpha \, d\beta$$

$$= 0.5 + 0.5\gamma - 0.125\gamma \qquad \text{(with some work)}$$

and

$$f_Z(\gamma) = 0.5 - 0.25\gamma$$

For $\gamma > 2$,

$$F_Z(\gamma) = 1 \quad \text{and} \quad f_Z(\gamma) = 0$$

The density function $f_Z(\gamma)$ is shown plotted in Figure 5.10d.

The Joint Density Function

The next general most complex case to be considered is: "Given two random variables

$$K = g(X, Y) \quad \text{and} \quad Q = h(X, Y)$$

where $f_{XY}(\alpha, \beta)$ is known, find $f_{K\theta}(r, s)$." If we consider two real numbers r and s, then

$$F_{K\theta}(r, s) \overset{\triangle}{=} P[(K \le r) \cap (Q \le s)]$$

$$= P[(g(X, Y) \le r) \cap (h(X, Y) \le s)] \tag{5.39}$$

The two requirements $g(\alpha, \beta) \le r \cap h(\alpha, \beta) \le s$ will define a range or a multiple of ranges which we will call D of the $\alpha\beta$ plane.

$$\therefore F_{K\theta}(r, s) = \int\int_D f_{XY}(\alpha, \beta) \, d\alpha \, d\beta$$

and the answer will be a function of r and s. Now, $f_{K\theta}(r, s)$ may be found by partial differentiation as

$$f_{K\theta}(r, s) \overset{\triangle}{=} \frac{\partial^2}{\partial r \, \partial s} F_{K\theta}(r, s)$$

In mathematics there exists a general procedure for quickly evaluating such relationships. The tool is the Jacobian, which we used when transforming coordinates in vector analysis.

RAPID DETERMINATION OF $f_{K\theta}(r, s)$ FROM $f_{XY}(\alpha, \beta)$

Given $r = g(\alpha, \beta)$ and $s = h(\alpha, \beta)$, the Jacobian is defined as

$$J(\alpha, \beta) \overset{\triangle}{=} \begin{vmatrix} \dfrac{\partial g}{\partial \alpha} & \dfrac{\partial g}{\partial \beta} \\[2ex] \dfrac{\partial h}{\partial \alpha} & \dfrac{\partial h}{\partial \beta} \end{vmatrix}$$

To find $f_{K\theta}(r, s)$ we first solve the equations

$$g(\alpha, \beta) = r \quad \text{and} \quad h(\alpha, \beta) = s$$

which will in general yield a set of solutions denoted by

$$(\alpha_1, \beta_1) \cdots (\alpha_n, \beta_n)$$

The density function for $f_{K\theta}(r, s)$ is given by

$$f_{K\theta}(r, s) = \frac{f_{XY}(\alpha_1, \beta_1)}{|J(\alpha_1, \beta_1)|} + \frac{f_{XY}(\alpha_n, \beta_n)}{|J(\alpha_1, \beta_1)|} \tag{5.40}$$

One simple example will be solved to illustrate the mechanical use of the theory just given.

Example 5.13

Given

$$f_{XY}(\alpha, \beta) = e^{-\alpha} e^{-\beta} \qquad \alpha > 0, \beta > 0$$

evaluate $f_{K\theta}(r, s)$, where the random variables K and θ are defined by $K = 2X + Y$ and $\theta = X - Y$.

SOLUTION

From the problem statement we have $g(\alpha, \beta) = 2\alpha + \beta$ and $h(\alpha, \beta) = \alpha - \beta$. Hence

$$J(\alpha, \beta) = \begin{vmatrix} 2 & 1 \\ 1 & -1 \end{vmatrix} = -3$$

Solving $2\alpha + \beta = r$ and $\alpha - \beta = s$ yields

$$\alpha_1 = \frac{\begin{vmatrix} r & 1 \\ s & -1 \end{vmatrix}}{-3} = \frac{1}{3}(r + s) \qquad \beta_1 = \frac{\begin{vmatrix} 2 & r \\ 1 & s \end{vmatrix}}{-3} = +\frac{1}{3}r - \frac{2}{3}s$$

which gives just one solution

$$f_{K\theta}(r, s) = \frac{f_{XY}(\alpha_1, \beta_1)}{|J(\alpha_1, \beta_1)|}$$

$$= \tfrac{1}{3}\exp[-\tfrac{1}{3}(r + s)]\exp(\tfrac{1}{3}r - \tfrac{2}{3}s) \qquad r > 0, \; -\infty < s < \infty$$

$$= \tfrac{1}{3}\exp(-\tfrac{2}{3}r + \tfrac{1}{3}s)$$

DRILL SET: DENSITY FUNCTIONS FOR FUNCTIONS OF RANDOM VARIABLES

1. Given

$$f_{XY}(\alpha, \beta) = \tfrac{1}{2} \qquad 0 < \alpha < \beta, 0 < \beta < 2$$

and

$$Z = X - Y$$

(a) Show on a sketch of the $\alpha\beta$ plane all the points satisfying $Z \leq -2$, $Z \leq 0$, and $Z \leq +2$.

(b) Evaluate $F_Z(\gamma)$ and $f_Z(\gamma)$ for $\gamma = -1$.

2. Given

$$f_{XY}(\alpha, \beta) = 1 \qquad 0 < \alpha < 1, 0 < \beta < 1$$

and two random variables Z and W defined as

$$Z = 4X + 1 \quad \text{and} \quad W = 5Y - 2$$

find $f_{ZW}(r, s)$.

*5.6 JOINTLY GAUSSIAN RANDOM VARIABLES

Gaussian random variables are of the utmost importance because they serve as models for sampling many practical noise waveforms, such as current fluctuations in cathode emission or shot noise, thermal voltage or current resistor fluctuations, laser noise, and so on. In addition we will see that if sampling the input waveform to a linear system yields gaussian random variables, so too will sampling the output.

JOINT GAUSSIAN DENSITY FUNCTION OF TWO RANDOM VARIABLES

Two random variables X_1 and X_2 are said to be jointly gaussian if their joint density function is

$$f_{X_1 X_2}(\alpha_1, \alpha_2) = \frac{1}{2\pi\sigma_1\sigma_2\sqrt{1-\rho^2}}$$

$$\times \exp\left[-\frac{\sigma_2^2(\alpha_1 - M_1)^2 - 2\rho\sigma_1\sigma_2(\alpha_1 - M_1)(\alpha_2 - M_2) + \sigma_1^2(\alpha_2 - M_2)^2}{2\sigma_1\sigma_2(1-\rho^2)} \right]$$

$$(5.41)$$

where M_1, M_2, σ_1, σ_2, and ρ are numbers with the restrictions $\sigma_1 > 0$, $\sigma_2 > 0$ and $0 < |\rho| < 1$. Since $f_{X_1 X_2}(\alpha_1, \alpha_2)$ is a joint density function it is possible (with labored effort), using the theory of this chapter, to find:

$$f_{X_1}(\alpha_1) = \frac{1}{\sqrt{2\pi}\,\sigma_1} \exp\left[-\frac{(\alpha_1 - M_1)^2}{2\sigma_1^2} \right] \qquad (5.42)$$

$$f_{X_2}(\alpha_2) = \frac{1}{\sqrt{2\pi}\,\sigma_2} \exp\left[-\frac{(\alpha_2 - M_2)^2}{2\sigma_2^2} \right] \qquad (5.43)$$

$$f_{X_1}[\alpha_1 / (X_2 = \alpha_2)] = \frac{1}{\sigma_1\sqrt{2\pi(1-\rho^2)}}$$

$$\times \exp\left\{ -\frac{[\alpha_1 - M_1 - \rho(\sigma_1/\sigma_2)(\alpha_2 - M_2)]^2}{2\sigma_1^2(1-\rho^2)} \right\}$$

$$(5.44)$$

$$f_{X_2}[\alpha_2 / (X_1 = \alpha_1)] = \frac{1}{\sigma_2\sqrt{2\pi(1-\rho^2)}}$$

$$\times \exp\left\{ -\frac{[\alpha_2 - M_2 - \rho(\sigma_2/\sigma_1)(\alpha_1 - M_1)]^2}{2\sigma_2^2(1-\rho^2)} \right\}$$

$$(5.45)$$

The mathematical manipulations involved in proceeding by integration from Eq. 5.41 to 5.45 are beyond the scope of this text. Observing Eqs. 5.41 through 5.45 we can make the following statements:

A jointly gaussian density function is characterized by the five numbers M_1, M_2, σ_1, σ_2, and ρ.

Both X_1 and X_2 are separately gaussian, where X_1 is characterized by its mean M_1 and variance σ_1^2, and X_2 is characterized by its mean M_2 and variance σ_2^2.

The conditional density functions $f_{X_1}[\alpha_1 /(X_2 = \alpha_2)]$ and $f_{X_2}[\alpha_2 /(X_1 = \alpha_1)]$ are characterized by their means and variances. For X_1 given X_2 is α_2 we find its mean as

$$\overline{[X_1/(X_2 = \alpha_2)]} = M_1 + \rho\left(\frac{\sigma_1}{\sigma_2}\right)(\alpha_2 - M_2) \tag{5.46}$$

and its standard deviation as

$$\sigma_{X_1/(X_2 = \alpha_2)} = \sigma_1\sqrt{1 - \rho^2} \tag{5.47}$$

The conditional density function for X_2 yields

$$\overline{[X_2/(X_1 = \alpha_1)]} = M_2 + \rho\left(\frac{\sigma_2}{\sigma_1}\right)(\alpha_1 - M_1) \tag{5.48}$$

and

$$\sigma_{X_2/(X_1 = \alpha_1)} = \sigma_2\sqrt{1 - \rho^2} \tag{5.49}$$

The second-order statistic $\text{cov}(X_1, X_2)$ or $L_{X_1 X_2}$ defined as

$$\overline{(X_1 - M_1)(X_2 - M_2)}$$

may be evaluated to yield

$$\text{cov}(X_1, X_2) = \rho\sigma_1\sigma_2 \tag{5.50}$$

or the number ρ may be written as

$$\rho = \frac{\text{cov}(X_1, X_2)}{\sigma_1\sigma_2} \tag{5.51}$$

where ρ is called the correlation coefficient and $|\rho| \leq 1$. An interesting conclusion from our results is: "When $\rho = 0$, $f_{X_1 X_2}(\alpha_1, \alpha_2) = f_{X_1}(\alpha_1)f_{X_2}(\alpha_2)$ and the random variables X_1 and X_2 are statistically independent."

There is a commonly used vector notation for gaussian random variables that will now be given and applied to the case of two random variables as in Eq. 5.41 and later to three or more random variables.

Denote the vectors or row matrices $\mathbf{X}$, $M_x(\boldsymbol{\alpha})$, and $\boldsymbol{\alpha}$ as follows:

$$\mathbf{X} = (X_1, X_2)$$
$$\boldsymbol{\alpha} = (\alpha_1, \alpha_2)$$
$$M_x(\boldsymbol{\alpha}) = (M_1, M_2)$$

and the **covariance matrix** (Λ_x) as

$$(\Lambda_x) \triangleq \left(\begin{array}{cc} \sigma_1^2 & \overline{(X_1 - M_1)(X_2 - M_2)} \\ \overline{(X_1 - M_1)(X_2 - M_2)} & \sigma_2^2 \end{array} \right) \tag{5.52}$$

where a general member is seen to be

$$\Lambda_{ij} = \overline{(X_i - M_i)(X_j - M_j)}$$

Using this matrix notation it is now possible to write Eq. 5.41, the joint gaussian density function, as

$$f_{\mathbf{X}}(\boldsymbol{\alpha}) = \frac{1}{2\pi |\Lambda_x|^{1/2}} \exp\left\{ -\tfrac{1}{2}\left[(\boldsymbol{\alpha} - M_x(\boldsymbol{\alpha}))(\Lambda_x)^{-1}(\boldsymbol{\alpha} - M_x(\boldsymbol{\alpha}))^T \right] \right\} \tag{5.53}$$

where $|\Lambda_x|$ denotes the value of the $n \times n$ determinant associated with (Λ_x), $(\Lambda_x)^{-1}$ is the inverse matrix of (Λ_x), and the superscript T implies taking the transpose of a matrix. The reader should very carefully expand Eq. 5.53 to obtain Eq. 5.41.

Since this material and the quoting of so many results may seem like a severe jolt, some intuitive comments will be made about jointly gaussian density functions. Figure 5.11 shows a rough sketch of jointly gaussian random variables for $\rho = \pm 1$ and $\rho = 0$. It can be seen that for $\rho = 0$ the density function is a perfectly symmetrical bell when the variances σ_X and σ_Y are equal and that this bell degenerates to a two-dimensional bell-type disk $f_Y(\beta)$ when $\rho = +1$ or $f_X(\alpha)$ when $\rho = -1$. The three-dimensional picture for $\rho = +0.9$ would correspond to the bell for $\rho = 0$ "squashed" about the central α axis and for $\rho = -0.9$ the bell would appear to be squashed about the central β axis.

Example 5.15

Given that two random variables are jointly gaussian with means $\overline{X}_1 = 3$ and $\overline{X}_2 = 2$ and with a covariance matrix

$$(\Lambda_X) = \begin{pmatrix} 2 & 1 \\ 1 & 4 \end{pmatrix}$$

find

(a) ρ, the correlation coefficient.
(b) $f_{X_1}(\alpha_1)$ and $f_{X_2}(\alpha_2)$.
(c) $f_{X_1}[\alpha_1 / (X_2 = 1)]$ and $f_{X_2}[\alpha_2 / (X_1 = 3)]$.

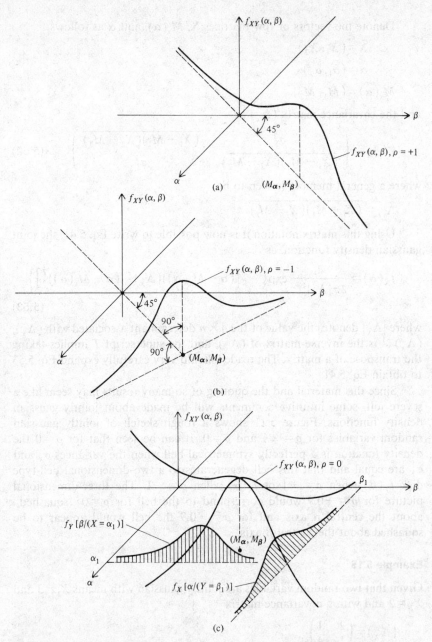

Figure 5.11   The gaussian joint density function for different values of ρ: (a) $\rho = +1$; (b) $\rho = -1$; (c) $\rho = 0$.

SOLUTION

By definition, a general element of (Λ) is

$$\Lambda_{ij} = \overline{(X_i - \overline{X}_i)(X_j - \overline{X}_j)}$$
$$= \overline{X_i X_j} - \overline{X}_i\,\overline{X}_j \qquad i \neq j$$
$$\Lambda_{ii} = \overline{X_i^2} - \overline{X}_i^{\,2}$$

and

$$\rho = \frac{\Lambda_{12}}{\sigma_1 \sigma_2}$$

In this case, $\Lambda_{11}=2$, $\sigma_1^2=2$, $\overline{X_1^2}=2+\overline{X}_1^{\,2}=11$, $\Lambda_{22}=4$, $\sigma_2^2=4$, $\overline{X_2^2}=4+\overline{X}_2^{\,2}$ $=8$, $\Lambda_{12}=1$, $\overline{X_1 X_2}-6=1$, and $\overline{X_1 X_2}=7$. We can now answer the three required questions.

(a) $\rho = \dfrac{\Lambda_{12}}{\sigma_1 \sigma_2} = \dfrac{1}{\sqrt{2} \times \sqrt{4}}$

$$= 0.35$$

(b) X_1 is a gaussian random variable with a mean of 3 and a variance of 2. X_2 is a gaussian random variable with a mean of 2 and a variance of 4.

(c) When X_2 is 1, the conditional density function for X_1 is gaussian with a mean

$$E[X_1/(X_2=1)] = \overline{X}_1 + \rho\left(\frac{\sigma_1}{\sigma_2}\right)(1 - \overline{X}_2)$$

$$= 2 + \frac{\sqrt{2}}{4}\left(\frac{\sqrt{2}}{\sqrt{4}}\right)(1-2)$$

$$= 1.75$$

and a standard deviation

$$\sigma_{(X_1/X_2)} = \sigma_1\sqrt{1-\rho^2}$$

$$= \sqrt{2}\,\sqrt{\tfrac{7}{4}}$$

$$= 1.32$$

When X_1 is 3, the conditional density function for X_2 is gaussian, with a mean

$$E\left[X_2/(X_1=3)\right]=\bar{X}_2+\frac{\sigma_1}{\sigma_2}\left(3-\bar{X}_1\right)$$

$$=2+\frac{\sqrt{2}}{4}\left(\frac{\sqrt{2}}{\sqrt{4}}\right)(3-2)$$

$$=2.25$$

and a standard deviation

$$\sigma_{X_2/X_1=3}=\sigma^2\sqrt{1-\rho^2}$$

$$=2\sqrt{\tfrac{7}{8}}$$

$$=1.87$$

JOINT GAUSSIAN DENSITY FUNCTION OF n RANDOM VARIABLES

Using the vector notation

$$\mathbf{X}=(X_1, X_2,\ldots,X_n)$$
$$\boldsymbol{\alpha}=(\alpha_1,\alpha_2,\ldots,\alpha_n)$$
$$M_x(\boldsymbol{\alpha})=(M_1, M_2,\ldots,M_n)$$

and the $n\times n$ covariance matrix (Λ_x) defined such that

$$\Lambda_{ij}=\overline{(X_i-M_i)(X_j-M_j)}$$

for all i and j, we say n random variables are jointly gaussian if their density function is

$$f_{\mathbf{X}}(\boldsymbol{\alpha})=\frac{1}{(2\pi)^{n/2}|\Lambda_x|^{1/2}}$$

$$\times\exp\left\{-\tfrac{1}{2}\left[(\boldsymbol{\alpha}-M_x(\boldsymbol{\alpha}))(\Lambda_x)^{-1}(\boldsymbol{\alpha}-M_x(\boldsymbol{\alpha}))^T\right]\right\}\qquad(5.54)$$

Considering this general nth-order joint density function it could be shown by repeated integration that each marginal density function is

$$f_{X_i}(\alpha_i)=\frac{1}{\sqrt{2\pi}\,\sigma_i}\exp\left[-\frac{(\alpha_i-M_i)^2}{2\sigma_i^2}\right]\qquad i=1,2,\ldots,n$$

and that the joint density for any k of these n random variables can be obtained by finding the appropriate $k\times k$ submatrix $(\Lambda_x)_K$ and writing Eq. 5.54 for this matrix.

The n random variables in Eq. 5.54 are *independent* if $\Lambda_{ij}=0$ for $i\neq j$.

Example 5.16

Given three random variables X_1, X_2, and X_3 are jointly gaussian with means all equal to 2 and a covariance matrix

$$(\Lambda_X) = \begin{pmatrix} 4 & 1 & 0.5 \\ 1 & 3 & 0.5 \\ 0.5 & -1 & 2 \end{pmatrix}$$

find

(a) $f_{X_1 X_2}(\alpha_1, \alpha_2)$ and $f_{X_1 X_3}(\alpha_1, \alpha_3)$.

(b) $f_{X_1}(\alpha_1)$, $f_{X_2}(\alpha_2)$, and $f_{X_3}(\alpha_3)$.

SOLUTION

(a) The appropriate Λ matrix for X_1 and X_2 is

$$(\Lambda_{1,2}) = \begin{pmatrix} 4 & 1 \\ 1 & 3 \end{pmatrix}$$

The inverse of this matrix is

$$(\Lambda_{1,2})^{-1} = \frac{1}{11}\begin{pmatrix} 3 & -1 \\ -1 & 4 \end{pmatrix}$$

and the joint density function for X_1 and X_2 is given by substituting this matrix and $|\Lambda_{1,2}|^{1/2} = \sqrt{11}$ into Eq 5.54. The appropriate (Λ) matrix for X_1 and X_3 is

$$\Lambda_{1,3} = \begin{pmatrix} 4 & 0.5 \\ 0.5 & 2 \end{pmatrix}$$

and the inverse of this matrix is

$$(\Lambda_{1,3})^{-1} = \frac{1}{7.75}\begin{pmatrix} 2 & -0.5 \\ -0.5 & 4 \end{pmatrix}$$

The joint density function for X_1 and X_3 is given by substitution of this matrix and $|\Lambda_{1,3}|^{1/2} = \sqrt{7.75}$ in Eq. 5.54.

(b) X_1 is a gaussian random variable with mean 2 and variance $\sigma_1^2 = 4$. X_2 is also gaussian with mean 2 and variance $\sigma_2^2 = 3$ and X_3 is also gaussian with mean 2 and variance $\sigma_3^2 = 2$.

We will conclude this section on jointly gaussian random variables by quoting without proof a theorem for finding the joint density function of a set of random variables that are defined as a linear combination of a set of gaussian random variables.

THEOREM

Any m random variables $\mathbf{Y}$

$$\mathbf{Y} = (Y_1, Y_2, \ldots, Y_m)$$

that are defined as a linear combination of n jointly gaussian random variables are themselves jointly gaussian, and their joint density function is

$$f_{\mathbf{Y}}(\boldsymbol{\beta}) = \frac{1}{(2\pi)^{m/2} |\Lambda_y|^{1/2}}$$

$$\times \exp\left\{-\frac{1}{2}\left[(\boldsymbol{\beta} - M_Y(\boldsymbol{\beta}))(\Lambda_y)^{-1}(\boldsymbol{\beta} - M_Y(\boldsymbol{\beta}))^T\right]\right\} \quad (5.55)$$

If $\mathbf{Y}$ is defined as

$$\begin{array}{cccc} \mathbf{Y}^T & = & (a) & \mathbf{X}^T \\ (m \times 1) & & (m \times 1)(n \times 1) \end{array}$$

the covariance Y matrix is

$$(\Lambda_Y) = (a)(\Lambda_X)(a)^T \quad (5.56)$$

and the means of the m random variables are

$$M_Y(\boldsymbol{\beta}) = (a)M_X(\boldsymbol{\alpha}) \quad (5.57)$$

This theorem will now be illustrated by an example.

Example 5.17

Given

$$\begin{pmatrix} Y_1 \\ Y_2 \end{pmatrix} = \begin{pmatrix} 1 & 2 & -1 \\ 1 & -1 & 2 \end{pmatrix} \begin{pmatrix} X_1 \\ X_2 \\ X_3 \end{pmatrix}$$

where

$$(\Lambda_x) = \begin{pmatrix} 2 & 0 & 0 \\ 0 & 1 & 0 \\ 0 & 0 & 2 \end{pmatrix}$$

and $\bar{X}_1 = \bar{X}_2 = \bar{X}_3 = 0$, find $f_{\mathbf{Y}}(\boldsymbol{\beta})$

(a) by evaluating the statistics of $\mathbf{Y}$ using the gaussian theorem.
(b) by mechanically using Eq. 5.56.

SOLUTION

(a) $\mathbf{Y}=0$, since $\mathbf{X}=0$. We require the elements of (Λ_{XY}):

$$\overline{Y_1^2} = \overline{(X_1+2X_2-X_3)^2}$$
$$= \overline{X_1^2}+4\overline{X_2^2}+\overline{X_3^2}$$
$$=2+4+2$$
$$=8=\Lambda_{11}$$

$$\overline{Y_2^2} = \overline{(X_1-X_2+2X_3)^2}$$
$$= \overline{X_1^2}+\overline{X_2^2}+4\overline{X_3^2}$$
$$=11=\Lambda_{22}$$

Now, to find Λ_{12} and Λ_{21},

$$\overline{(Y_1-\overline{Y_1})(Y_2-\overline{Y_2})} = \overline{Y_1 Y_2}+0$$
$$= \overline{(X_1+2X_2-X_3)(X_1-X_2+2X_3)}$$
$$= \overline{X_1^2}-2\overline{X_2^2}-2\overline{X_3^2}+0$$
$$=2-2-4$$
$$=-4=\Lambda_{12}=\Lambda_{21}$$

$$\therefore (\Lambda_y)=\begin{pmatrix} 8 & -4 \\ -4 & 11 \end{pmatrix}$$

and

$$f_{Y_1,Y_2}(\beta_1,\beta_2)=\frac{1}{2\pi\begin{vmatrix} 8 & -4 \\ -4 & 11 \end{vmatrix}^{1/2}}$$

$$\times\exp\left\{-\frac{1}{2}\left[(\beta_1,\beta_2)\begin{pmatrix} 8 & -4 \\ -4 & 11 \end{pmatrix}^{-1}\begin{pmatrix} \beta_1 \\ \beta_2 \end{pmatrix}\right]\right\}$$

$$\begin{vmatrix} 8 & -4 \\ -4 & 11 \end{vmatrix}=88+16=72$$

$$\therefore |\Lambda_y|^{1/2}=\sqrt{72}=8.48$$

$$(\Lambda_y)^{-1}=\frac{1}{72}\begin{pmatrix} 11 & 4 \\ 4 & 8 \end{pmatrix}^T$$

$$=\frac{1}{72}\begin{pmatrix} 11 & 4 \\ 4 & 8 \end{pmatrix}=\begin{pmatrix} 0.1 & 0.04 \\ 0.04 & 0.08 \end{pmatrix}$$

and

$$(\beta_1, \beta_2) \begin{pmatrix} 0.153 & 0.056 \\ 0.056 & 0.111 \end{pmatrix} \begin{pmatrix} \beta_1 \\ \beta_2 \end{pmatrix}$$
$$\underset{1\times2}{} \qquad \underset{2\times2}{} \qquad \underset{2\times1}{}$$

$$= (\beta_1, \beta_2) \begin{pmatrix} 0.153\beta_1 + 0.056\beta_2 \\ 0.056\beta_1 + 0.111\beta_2 \end{pmatrix}$$
$$\underset{1\times2}{} \qquad\qquad \underset{2\times1}{}$$

$$= (0.153\beta_1^2 + 0.056\beta_1\beta_2 + 0.056\beta_1\beta_2 + 0.111\beta_2^2)$$

$$= (0.153\beta_1^2 + 0.111\beta_1\beta_2 + 0.111\beta_2^2)$$

$$\therefore \boxed{ f_{Y_1Y_2}(\beta_1, \beta_2) = \frac{1}{17\pi} \exp\left[-\tfrac{1}{2}(0.153\beta_1^2 + 0.111\beta_1\beta_2 + 0.111\beta_2^2) \right] }$$

(b) Mechanically, using Eq. 5.56,

$$(\Lambda_y) = (a)(\Lambda_x)(a)^T$$

Now

$$Y_1 = X_1 + 2X_2 - X_3$$
$$Y_2 = X_1 - X_2 + 2X_3$$

$$\therefore \begin{pmatrix} Y_1 \\ Y_2 \end{pmatrix} = \begin{pmatrix} 1 & 2 & -1 \\ 1 & -1 & 2 \end{pmatrix} \begin{pmatrix} X_1 \\ X_2 \\ X_3 \end{pmatrix}$$

$$\therefore (a) = \begin{pmatrix} 1 & 2 & -1 \\ 1 & -1 & 2 \end{pmatrix}$$

and

$$(a)^T = \begin{pmatrix} 1 & 1 \\ 2 & -1 \\ -1 & 2 \end{pmatrix}$$

$$(\Lambda_y) = \begin{pmatrix} 1 & 2 & -1 \\ 1 & -1 & 2 \end{pmatrix} \begin{pmatrix} 2 & 0 & 0 \\ 0 & 1 & 0 \\ 0 & 0 & 2 \end{pmatrix} \begin{pmatrix} 1 & 1 \\ 2 & -1 \\ -1 & 2 \end{pmatrix}$$

$$= \begin{pmatrix} 1 & 2 & -1 \\ 1 & -1 & 2 \end{pmatrix} \begin{pmatrix} 2 & 2 \\ 2 & -1 \\ -2 & 4 \end{pmatrix}$$

$$= \begin{pmatrix} 8 & -4 \\ -4 & 11 \end{pmatrix}$$

as obtained before.

THE CENTRAL LIMIT THEOREM

Apart from the fact that many random variables occurring in nature are approximately gaussian, the gaussian density function is of vital importance in probability theory because of the central limit theorem. This theorem will now be stated without proof.

THEOREM

If $X_1, X_2, \ldots, X_n$ are n independent random variables, each with a mean $\overline{X}$ and a variance σ^2, then the random variable Y defined as

$$Y = \sum_{i=1}^{n} X_i$$

is approximately gaussian, with mean $\overline{Y} = n\overline{X}$ and variance $\sigma_Y^2 = n\sigma^2$.

It is important to realize that the theorem states that the random variables may have different density functions, but only the means and variances must exist and be the same for all the variables.

A satisfying engineering appreciation for the theorem is to demonstrate that the density function for the sum of two random variables is the convolution of the density functions and to realize that continual convolution of a function with itself keeps widening and smoothing it to achieve a gaussian curve.

Example 5.18

(a) If the random variables X and Y are independent with density functions $f_X(\alpha)$ and $f_Y(\beta)$, show that the density function for $Z = X + Y$ is

$$f_Z(\gamma) = \int_{-\infty}^{\infty} f_X(\alpha) f_Y(\gamma - \alpha) \, d\alpha$$

$$= f_X(\gamma) * f_Z(\gamma)$$

where $*$ denotes convolution.

(b) If $f_X(\alpha) = f_Y(\alpha) = 1$, $0 < \alpha < 1$, and is 0 otherwise,
   (1) Find the density function of

   $$Z = X + Y$$

   and note its mean and variance.
   (2) Find the density function of

   $$W = Z + P$$

   where $f_P(\alpha) = 1$, $0 < \alpha < 1$, and P is independent of Z, and note its mean and variance.

SOLUTION

(a) The cumulative distribution function $F_Z(\gamma)$ may be found as indicated in Figure 5.12a.

$$F_Z(\gamma) = P(X + Y < \gamma)$$

$$= \int_{-\infty}^{\infty} \left[\int_{-\infty}^{\gamma - \alpha} f_{XY}(\alpha, \beta) \, d\beta \right] d\alpha$$

$$= \int_{-\infty}^{\infty} f_X(\alpha) \left[\int_{-\infty}^{\gamma - \alpha} f_Y(\beta) \, d\beta \right] d\alpha$$

since X and Y are assumed independent. Using Leibnitz' rule,

$$f_Z(\gamma) = \frac{d}{d\gamma} F_Z(\gamma)$$

$$= \int_{-\infty}^{\infty} f_X(\alpha) f_Y(\gamma - \alpha) \, d\alpha$$

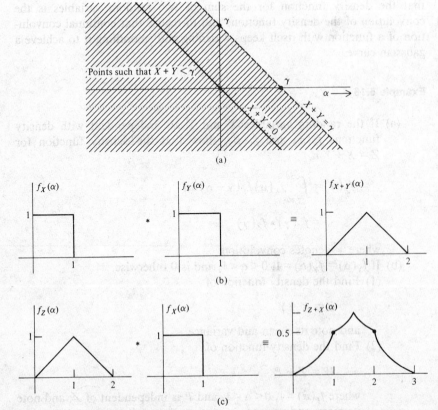

Figure 5.12   (a) Points satisfying $X + Y < \gamma$ on $\alpha\beta$. (b) $f_{X+Y}(\alpha)$ by convolution. (c) f_{Z+X} by convolution for Example 5.18.

(b) (1) If $f_X(\alpha) = f_Y(\alpha) = 1$, $0 < \alpha < 1$, then as is shown in Figure 5.12b we can find $f_Z(\gamma)$ for all γ as follows:

For $-\infty < \gamma < 0$,

$$f_Z(\gamma) = 0$$

For $0 < \gamma < 1$,

$$f_Z(\gamma) = \int_0^\gamma 1 \, d\alpha$$

$$= \gamma$$

For $1 < \gamma < 2$,

$$f_Z(\gamma) = \int_{\gamma-1}^1 d\alpha$$

$$= 2 - \gamma$$

and $f_Z(\gamma) = 0$; $\gamma > 2$. $f_Z(\gamma)$ is shown plotted in Figure 5.12b and it is easy to evaluate $\bar{X} = \bar{Y} = 0.5$, $\bar{Z} = 1$, $\sigma_X^2 = \sigma_Y^2 = \frac{1}{12}$ and $\sigma_Z^2 = \frac{1}{6}$. This demonstrates $\bar{Z} = \bar{X} + \bar{Y}$ and $\sigma_Z^2 = \sigma_X^2 + \sigma_Y^2$.

(2) The density function for $W = Z + P$ may be found as

$$f_W(u) = \int_{-\infty}^\infty f_Z(\alpha) f_P(u - \alpha) \, d\alpha$$

For $0 < u < 1$,

$$f_W(u) = \int_0^u \alpha \, d\alpha$$

$$= \tfrac{1}{2} u^2$$

For $1 < u < 2$,

$$f_W(u) = \int_{-1+u}^1 \alpha \, d\alpha + \int_1^u (2 - \alpha) \, d\alpha$$

$$= -u^2 + 3u - 1.5$$

For $2 < u < 3$,

$$f_W(u) = \int_{-1+u}^2 (2 - \alpha) \, d\alpha$$

$$= 0.5(u^2 - 6u + 9)$$

For $u > 3$,

$$f_W(u) = 0$$

The density function for W is shown plotted in Figure 5.12c, and it can be shown in a straightforward way that $\overline{W} = 1.5$ or $3\overline{X}$ and $\sigma_W^2 = 0.25 = 3\sigma_X^2$. We also notice that the density function is wider and smoother than $f_X(\alpha)$. We should intuitively appreciate that if we define

$$K = \sum_{i=1}^{20} X_i$$

then $f_K(\alpha)$ would be of width 20, have a mean of 20×0.5 and a variance of $\sigma_K^2 = 20\sigma_X^2 = 1.67$. In addition, this density function would approach a gaussian curve in its smoothness.

*5.7 INEQUALITIES INVOLVING RANDOM VARIABLES

In subsequent work and especially in filter theory (Chapter 9) we often are required to calculate bounds on probabilities or on expected values such as the correlation or covariance. A few very important inequalities will be mentioned in this section. All these will be derived using the fact that quantities such as $|X - a|^2$ or $|X - bY|$ or $|X_1 + X_2 + X_3|^2$ must be always positive due to the magnitude operation.

Chebyshev's Inequality

If X is any random variable with a finite variance σ^2 then

$$P\left(|X - \overline{X}| < c\right) \le 1 - \frac{\sigma^2}{c^2} \tag{5.58}$$

for any positive constant c. This is also stated as

$$P\left(|X - \overline{X}| \ge c\right) \le \frac{\sigma^2}{c^2} \tag{5.59}$$

PROOF
In Figure 5.13 is shown a sketch of some density function $f_X(\alpha)$ and the bounds $\overline{X} + c$ and $\overline{X} - c$. Obviously the following inequality holds:

$$\int_{-\infty}^{\infty} (\alpha - \overline{X})^2 f_X(\alpha)\, d\alpha = \int_{\overline{X} - c}^{\overline{X} + c} (\alpha - \overline{X})^2 f_X(\alpha)\, d\alpha$$

$$+ \int_{|\alpha - \overline{X}| > c} (\alpha - \overline{X})^2 f_X(\alpha)\, d\alpha$$

$$\ge c^2 \int_{|\alpha - \overline{X}| > c} f_X(\alpha)\, d\alpha$$

Noting that the left-hand side is the variance and the final right-hand term

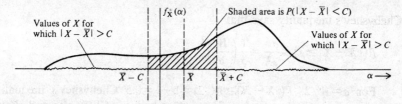

Figure 5.13   Ranges of $|X - \bar{X}|$ used in Chebyshev's inequality.

is $c^2 P(|X - \bar{X}| \geq c])$, we have demonstrated that

$$P(|X - \bar{X}| \geq c) \leq \frac{\sigma^2}{c}$$

which is 5.59.

SPECIALIZATION TO THE CASE WHERE X IS ALWAYS POSITIVE
If $f_X(\alpha) = 0$ for all negative α, then

$$\bar{X} = \int_0^\infty \alpha f_X(\alpha)\, d\alpha$$

$$\geq \int_c^\infty \alpha f_X(\alpha)\, d\alpha$$

$$\geq c P(X > c)$$

and the inequality

$$P(X > c) \leq \frac{\bar{X}}{c} \tag{5.60}$$

results.

Example 5.19

Given that the random variable X has a density function

$$f_X(\alpha) = \frac{1}{2W} \qquad -W < \alpha < W$$

find the probability $|X - \bar{X}| \geq c$ for $c < W/2$ and compare it to its Chebyshev bound.

SOLUTION

$$\bar{X} = 0 \quad \text{and} \quad \sigma^2 = \overline{X^2} = \int_{-W}^W \alpha^2 \frac{1}{2}\, d\alpha = \frac{1}{3}$$

$$P(|X - \bar{X}| \geq c) = 1 - 2\int_0^c \frac{1}{2W}\, d\alpha$$

$$= 1 - \frac{c}{W}$$

Chebyshev's inequality says that

$$P\left(|X - \bar{X}| \ge c\right) \le \frac{\sigma^2}{c^2} = \frac{1}{3}\left(\frac{W}{c}\right)^2$$

For $c = W/2$, $P(|X - \bar{X}| \ge W/2) = 1 - \frac{1}{2} = 0.5$ Chebyshev's inequality says this answer had to be less than $\frac{1}{3}(4)$, which is of little use and shows that Chebyshev's bound in many cases is very imprecise. It would be an instructive exercise to plot $P(|X - \bar{X}| \ge c)$ and the Chebyshev bound versus c as c varies $0 < c < W$.

Cauchy-Schwarz Inequality

An inequality that is of the utmost importance when discussing the correlation of two random variables and properties of autocorrelation and cross-correlation functions is the Cauchy-Schwarz inequality, which says

$$\overline{XY} \le \sqrt{\overline{X^2}\,\overline{Y^2}} \tag{5.61}$$

for any two random variables X and Y, where these moments are defined. The proof once again follows the use of positiveness for squared quantities.

PROOF

$$\overline{(X - cY)^2} \ge 0 \qquad \text{for any real number } c$$

$$\therefore \overline{Y^2}\,c^2 - 2\,\overline{XY}\,c + \overline{X^2} \ge 0$$

Considering this quadratic in c, which is always positive, and differentiating with respect to c to find for what c the quadratic is a minimum we find that

$$2c\,\overline{Y^2} - 2\,\overline{XY} = 0$$

or

$$c = \frac{\overline{XY}}{\overline{Y^2}}$$

Substituting this value of c into the quadratic yields

$$\frac{\left(\overline{XY}\right)^2}{\overline{Y^2}} - 2\frac{\left(\overline{XY}\right)^2}{\overline{Y^2}} + \overline{X^2} \ge 0$$

or

$$\overline{X^2}\,\overline{Y^2} \ge \overline{XY}^2$$

which proves 5.61.

There are many close relatives of the Cauchy-Schwarz inequality that exist for real numbers and real deterministic functions.

Cauchy's Inequality

For any two sets of real numbers,

$$\{x\} = \{x_1, x_2, \ldots, x_n\}$$

and

$$\{y\} = \{y_1, y_2, \ldots, y_n\}$$

$$\therefore \left(\sum_{i=1}^{n} x_i^2 \right) \left(\sum_{i-1}^{n} y_i^2 \right) \geq \left(\sum_{i-1}^{n} x_i y_i \right)^2 \tag{5.62}$$

The proof of this inequality starts with the statement of positiveness

$$\sum_{n} (x_i - cy_i)^2 \geq 0$$

and proceeds as in the proof of the Cauchy-Schwarz inequality.

Schwarz Inequality

For any two real finite-energy time functions $f(t)$ and $g(t)$,

$$\left[\int_{-\infty}^{\infty} f^2(t)\, dt \right] \left[\int_{-\infty}^{\infty} g^2(t)\, dt \right] \geq \left[\int_{-\infty}^{\infty} [f(t)g(t)]^2\, dt \right]$$

PROOF

The proof of this inequality starts with the statement of positiveness

$$\int_{-\infty}^{\infty} [f(t) - cg(t)]^2\, dt \geq 0 \tag{5.63}$$

and proceeds as in the proof of the Cauchy-Schwarz inequality.

Cauchy's inequality and Schwarz's inequality will be the key to deriving a discrete matched filter and continuous matched filter in Chapter 9.

SUMMARY

Two random variables $X(s)$ and $Y(s)$ defined on a sample space are completely characterized by their joint cumulative distribution function $F_{XY}(\alpha, \beta)$, their joint density function $f_{XY}(\alpha, \beta)$, or their joint mass function $p_{XY}(\alpha, \beta)$. The definitions of these functions are:

$$F_{XY}(\alpha, \beta) \triangleq P[(X \leq \alpha) \cap (Y \leq \beta)]$$

$$f_{XY}(\alpha, \beta) \triangleq \frac{\partial^2}{\partial \alpha \partial \beta} F_{XY}(\alpha, \beta)$$

and

$$p_{XY}(\alpha, \beta) \triangleq P[(X = \alpha) \cap (Y = \beta)]$$

Finding $F_{XY}(\alpha, \beta)$ for any two random variables resolves itself into a problem in discrete probability theory utilizing the concept of the intersection of two events and the conditional probability of an event. In practice, finding $f_{XY}(\alpha, \beta) \stackrel{\Delta}{=} f_X(\alpha) f_Y[\beta/(X = \alpha)]$ is more efficient than finding $F_{XY}(\alpha, \beta)$ or for discrete random variables we should first find $p_{XY}(\alpha, \beta)$ as $p_X(\alpha) p_Y[\beta/(X = \alpha)]$ and then $F_{XY}(\alpha, \beta)$ by summation. The properties of distribution, density, and mass functions follow from the axioms of probability theory and were derived in Section 5.2.

Knowing the joint density function of two random variables [or $F_{XY}(\alpha, \beta)$ or $p_{XY}(\alpha, \beta)$] it is possible to answer any probabilistic question concerning an event or conditional event for X and Y, for X, or for Y. The general formulas are

$$P\begin{bmatrix} \text{event } A \text{ involving} \\ X \text{ and } Y \end{bmatrix} = \iint_{\substack{\text{all points} \\ \text{of } \alpha\beta \text{ plane} \\ \text{satisfying } A}} f_{XY}(\alpha, \beta) \, d\alpha \, d\beta$$

$$P[A(X, Y)/B(X, Y)] = \frac{\displaystyle\iint_{\substack{\text{all points} \\ \text{satisfying } A \cap B}} f_{XY}(\alpha, \beta) \, d\alpha \, d\beta}{\displaystyle\iint_{\substack{\text{all points} \\ \text{satisfying } B}} f_{XY}(\alpha, \beta) \, d\alpha \, d\beta}$$

$$f_X(\alpha) = \int_{-\infty}^{\infty} f_{XY}(\alpha, \beta) \, d\beta$$

$$f_X[\alpha/(Y = \beta)] = \frac{f_{XY}(\alpha, \beta)}{f_Y(\beta)}$$

$$f_Y(\beta) = \int_{-\infty}^{\infty} f_{XY}(\alpha, \beta) \, d\alpha$$

and

$$f_Y[\beta/(X = \alpha)] = \frac{f_{XY}(\alpha, \beta)}{f_X(\alpha)}$$

Care must be exercised in applying these formulas with respect to noting the regions for which $f_{XY}(\alpha, \beta)$ is defined and the ranges for $f_X(\alpha)$, $f_Y(\beta)$, $f_X[\alpha/(Y = \beta)]$ and $f_Y[\beta/(X = \alpha)]$.

Statistics, or numbers, associated with a joint density function were defined. The two most important to a communications engineer are the autocorrelation coefficient $R_{XY} \stackrel{\Delta}{=} E(XY)$ and the covariance coefficient L_{XY} or $\text{cov}(X, Y) \stackrel{\Delta}{=} E[(X - \overline{X})(Y - \overline{Y})]$. If two random variables X and Y are independent, which implies $f_{XY}(\alpha, \beta) = f_X(\alpha) f_Y(\beta)$, then

$$E(XY) = \overline{X}\,\overline{Y} \quad \text{and} \quad \text{cov}(X, Y) = 0$$

For these conditions to occur we need not require a condition as strict as independence, and so we define two random variables as being uncorrelated if $\overline{XY} = \overline{X}\overline{Y}$ or if $\text{cov}(X, Y) = 0$. Attention was focused on solving problems involving finding R_{XY} and L_{XY} for random variables found by sampling time waveforms. Since we will see later that noise waveforms are of an infinite-energy type

$$\int_{-\infty}^{\infty} x^2(t)\, dt = \infty$$

or finite-power type

$$\lim_{T \to \infty} \frac{1}{2T} \int_{-T}^{T} x^2(t)\, dt \neq \infty$$

periodic time waveforms were chosen since they are of this type. In Chapter 4 we saw that all first-order statistics defined from a random variable which is defined by sampling a periodic waveform over a period or multiple of periods yielded the same results as the time averages. In Chapter 5 as a result of sampling two values spaced τ s apart we developed some feeling for the "randomness" of our results. If τ approaches 0 we are just defining the first-order statistic $\overline{X^2}$ for R_{XY} and σ_X^2 for L_{XY}. R_{XY} or L_{XY} versus τ must yield a periodic function of τ with the same period as the sampled waveform. The section on second-order statistics concluded with the very powerful fundamental theorem. This theorem states that a second-order statistic for two random variables, say P and θ, may always be obtained on the joint density function of two other random variables X and Y where if $P = g_1(X, Y)$ and $\theta = g_2(X, Y)$, then

$$\overline{h(P, \theta)} = \int_{-\infty}^{\infty} \int_{-\infty}^{\infty} h[g_1(\alpha, \beta) g_2(\alpha, \beta)] f_{XY}(\alpha, \beta)\, d\alpha\, d\beta$$

One example we showed for this was that if we are dealing with a fairly complicated waveform $x(t)$, then $R_{P\theta}$ defined for sampling at two values τ s apart may be evaluated in terms of the random variables X and Y, which sample points on the time axis.

Chapter 5 continued with the system-type problem: Given

and a

| the input $x(t)$ | system with its rule | and the output $y(t)$ |

what is the density function associated with sampling $y(t)$ knowing the density function associated with sampling $x(t)$? Or in the most general form: Given

$$f_{X_1, X_2, \ldots, X_n}(\alpha_1, \alpha_2, \ldots, \alpha_n)$$

find
$$f_{Y_1, Y_2, \ldots, Y_m}(\beta_1, \beta_2, \ldots, \beta_m)$$
where the X random variables describe sampling $x(t)$ at n points and the Y random variables describe sampling $y(t)$ at m points. For many future problems a complete solution may be too difficult, but to find such a solution would be the ultimate for handling any probabilistic question about the output waveform $y(t)$.

Finally, Chapter 5 concluded with the topics of jointly gaussian random variables and important inequalities. The joint gaussian density function was discussed and extended to the n-dimensional case. The central limit theorem was then treated. Four important inequalities were listed: Chebyshev's upper bound on the deviation from the mean, Cauchy-Schwarz' upper bound on the correlation of two random variables and two close relatives, Cauchy's upper bound on the energy product of two sets of numbers (which we will later call the product of the energies of discrete time functions), and Schwarz' upper bound on the product of the energies of two continuous functions. The proofs of all these inequalities depended on a statement of positiveness involving magnitudes of random variables or numbers or functions. These final topics were preceded by an asterisk, which meant they could have been omitted until encountered in practice.

PROBLEMS

1. Consider the phenomenon of sampling the real axis twice. First, a point is chosen uniformly between 0 and 1, and then independently a second point is chosen uniformly between 0 and 2. Define X as the random variable describing the first value chosen and Y as the random variable describing the second value chosen and Z as the random variable describing the sum of the two values obtained. Find and plot
   (a) $f_{XY}(\alpha, \beta)$ and $F_{XY}(\alpha, \beta)$.
   (b) $f_{XZ}(\alpha, \beta) \triangleq f_X(\alpha) f_Z[\beta/(X = \alpha)]$ and $F_{XZ}(\alpha, \beta)$.

2. Consider the random phenomenon which consists of sampling the sawtooth waveform $x(t)$ at a random time t_0 s and at $t_0 + 1$ s. Let X be the random variable describing the sampled value at t_0 by assuming the sampled numerical value and Y be the random variable describing the sampled value at $t_0 + 1$ by assuming the sampled value.
   (a) Find $f_X(\alpha)$ and $f_Y(\beta)$.
   (b) Find $f_Y[\beta/(X = \alpha)]$. This answer will involve a singularity function where the value of β depending on α may or may not be from the same ramp.
   (c) Find $f_{XY}(\alpha, \beta)$.
   (d) Prove that
   $$\int_{-\infty}^{\infty} \left[\int_{-\infty}^{\infty} f_{XY}(\alpha, \beta) \, d\beta \right] d\alpha = 1$$

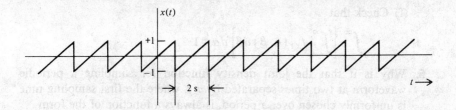

3. Consider the periodic waveform

$$x(t) = \sum_{-\infty}^{\infty} g(t - 3n) \qquad n \text{ an integer}$$

where

$$g(t) = 1 \qquad 0 < t < 1$$

Consider the random phenomenon of sampling this waveform at $t = t$ and $t = t + \frac{1}{2}$, where t is uniformly chosen, $0 < t < 3$. Define the random variable X as describing the waveform value at t and Y as describing the waveform value at $t + \frac{1}{2}$ (with a one-to-one mapping understood).

(a) Sketch $x(t)$ and find and sketch the mass functions $P_X(\alpha)$ and $P_Y(\beta)$.
(b) Find $P_Y[\beta/(X = \alpha)]$ for all α.
(c) Construct the matrix for $P_{XY}(\alpha, \beta)$.

4. Consider the periodic waveform

$$x(t) = \sum_{n = -\infty}^{\infty} g(t - 3n) \qquad n \text{ an integer}$$

where

$$g(t) = 1 - t \qquad 0 < t < 1$$
$$= 0 \qquad \text{otherwise}$$

Let X be a random variable that samples $x(t)$ uniformly over the range $0 < t < 3$ with a one-to-one mapping and let Y be a random variable that samples $x(t)$ simultaneously at $t = t + \tau$. (This problem will be quite challenging.)

(a) Sketch $x(t)$.
(b) What are the following probabilities for a general τ in the range $0 < \tau < 1$?
   (1) Both X and Y are 0.
   (2) X is 0 and Y is not.
   (3) X is not 0 and Y is 0.
   (4) Both X and Y are not 0.
(c) Find $f_X(\alpha), f_Y(\beta)$, and $f_Y[\beta/(X = \alpha)]$ for all α.
(d) Find $f_{XY}(\alpha, \beta)$ for all τ; $0 < \tau < 1$.
(e) Comment on whether your answers for $f_Y[\beta/(X = \alpha)]$ and $f_{XY}(\alpha, \beta)$ make sense for $\tau = 0$ and $\tau = 1$.

(f) Check that

$$\int_{-\infty}^{\infty} \left[\int_{-\infty}^{\infty} f_{XY}(\alpha,\beta)\,d\beta \right] d\alpha = 1$$

5. Why is it that the joint density function for sampling a periodic waveform at two times separated by τ s, where the first sampling time is uniformly chosen over a period, is always a function of the form

$$f_{XY}(\alpha,\beta) = f_X(\alpha)f_Y[\beta/(X=\alpha)] = f_X(\alpha)\delta[\beta - g(\alpha_1)]$$

6. (a) Is it possible for

$$f_{XY}(\alpha,\beta) = A\beta^2 \qquad 0 < 2\alpha < \beta < 2$$
$$= 0 \qquad \text{otherwise}$$

to be a joint density function for any value of A?

(b) If the answer to (a) is yes, find
   (1) $f_X(\alpha)$, $f_Y(\beta)$, $f_X[\alpha/(Y=\beta)]$.
   (2) $P(2X+Y<1)$ and $P[(2X+Y<1)/(X>\tfrac{1}{2})]$.

7. Given

$$f_{XY}(\alpha,\beta) = \frac{1}{\pi^2} \qquad 0 < \alpha^2 + \beta^2 < 1$$

Find
(a) $f_X(\alpha)$, $f_Y(\beta)$, $f_X[\alpha/(Y=\beta)]$ and $f_Y[\beta/(X=\alpha)]$.
(b) $P(\tfrac{1}{4} \le X \le 2)$, $P(|X|>|Y|)$.

8. The value of the joint cumulative distribution function $F_{XY}(\alpha,\beta)$ is given by Table 5.1.

Table 5.1   VALUES OF $F_{XY}(\alpha,\beta)$ AT POINTS IN THE $\alpha\beta$ PLANE

β ╲ α	$\alpha < 2$	$2 \le \alpha < 3$	$3 \le \alpha < 4$	$\alpha \ge 4$
$\beta < 1$	0	0	0	0
$1 \le \beta < 2$	0	0.2	0.4	0.4
$2 \le \beta < 3$	0	0.5	0.6	0.7
$\beta > 3$	0	0.6	0.8	1.0

(a) Find the joint mass function $p_{XY}(\alpha,\beta)$.
(b) Find the marginal mass functions $p_X(\alpha)$, $P_Y(\beta)$.
(c) Find the conditional mass functions $p_X[\alpha/(Y=2)]$ and $P_Y[\beta/(X=3)]$.
(d) Are X and Y independent random variables?

9. In terms of the general joint density function $f_{XYZ}(\alpha,\beta,\gamma)$ write down expressions for the following:
(a) $P[(X \le 3)/(Y=Z) \cap (1<Z<3)]$
(b) $f_{XZ}(\alpha,\gamma)$, $f_{XZ}(3,Y)$
(c) $f_{XY}[(\alpha,3)/(\gamma=2)]$
(d) $f[\alpha/(2<Y<3)\cap(Z>4)]$

(e) $P[(3X+4Y)<2/(X>-2)\cap(Z<4)]$. Be very careful with the limits of integration. For example, what we require here is

$$P[(Y<4)/(Z>3)]$$

$$= \frac{P[(-\infty<X<\infty)\cap(Y<4)\cap(Z>3)]}{P[(-\infty<X<\infty)\cap(-\infty<Y<\infty)\cap(Z>3)]}$$

$$= \frac{\int_{-\infty}^{\infty}\left\{\int_{-\infty}^{4}\left[\int_{3}^{\infty}f_{XYZ}(\alpha,\beta,\gamma)\,d\gamma\right]d\beta\right\}d\alpha}{\int_{-\infty}^{\infty}\left\{\int_{-\infty}^{\infty}\left[\int_{3}^{\infty}f_{XYZ}(\alpha,\beta,\gamma)\,d\gamma\right]d\beta\right\}d\alpha}$$

10. Given
$$f_{XYZ}(\alpha,\beta,Z)=A \qquad 0<\alpha<\beta<2\gamma<2$$
$$=0 \qquad \text{otherwise}$$
Find
   (a) The value of the constant A.
   (b) $f_{XY}(\alpha,\beta), f_{XY}(\alpha,\gamma)$.
   (c) $f_X(\alpha), f_X[\alpha/(Y=0.2)\cap(Z=0.3)]$.

11. For the random variables of Problem 2 evaluate $\bar{X}, \bar{Y}, R_{XY}=\overline{XY}$ and $L_{XY}=\overline{(X-\bar{X})(Y-\bar{Y})}$.

12. For the random variables of Problem 3 evaluate $\bar{X}, \bar{Y}, \overline{XY}$ and $\overline{(X-\bar{X})(Y-\bar{Y})}$

13. Reconsidering Problem 4, let us use the fundamental theorem in two ways.
   (a) First, by defining the random variables P and θ associated with sampling the time axis at $t=t$ and $t=t+\tau$, where t is uniformly chosen, $0<t<3$, we can see that $f_{P\theta}(\alpha,\beta)=\frac{1}{3}\delta(\beta-\alpha); 0<\alpha<3$, $\beta=\alpha$. Now relate X and Y to P and θ and evaluate $\overline{XY}=g(P)M(\theta)$ on the joint density function for P and θ for any general τ; $0<\tau<1$. Here, $g(P)$ and $M(\theta)$ are the functional relations you will find: $X=g(P)$ and $Y=M(\theta)$.
   (b) Second, just consider the random variable P with $f_P(\alpha)=\frac{1}{3}; 0<\alpha<3$, and define $XY=g(P)M(P)$, which is just a function of P, say $l(P)$. Now find the correlation coefficient R_{XY} as

$$\overline{XY}=\overline{l(P)}=\int_{-\infty}^{\infty}l(\alpha)f_X(\alpha)\,d\alpha$$

   (c) As a check on the first two parts find R_{XY} directly based on $f_{XY}(\alpha,\beta)$ from Problem 4d.

14. Consider the output of a summer $z(t)=x(t)+y(t)$. Given that the random variable X describes sampling the waveform $x(t)$ and Y describes sampling $y(t)$ and further that X and Y are independent random variables with X uniformly distributed $0<X<1$ and Y uniformly distributed $0<Y<2$ find $f_Z(\gamma)$. Z is described as the random variable sampling $z(t)$.

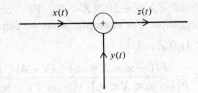

15. Repeat Problem 14 when the device is a subtractor with $z(t) = x(t) - y(t)$.

16. Given

$$f_{XY}(\alpha, \beta) = \tfrac{1}{2} \qquad 0 < \alpha < \beta < 1$$
$$= 0 \qquad \text{otherwise}$$

Find the density functions of the following random variables: $Z = XY$, $Z = X/Y$, $Z = \max(X, Y)$.

17. Given

$$f_{XY}(\alpha, \beta) = \tfrac{1}{2} \qquad 0 < \alpha < \beta < 1$$
$$= 0 \qquad \text{otherwise}$$
$$Z = 6X + 4Y$$

and
$$\times$$
$$W = 7X - 5Y$$

(a) Find $f_{ZW}(\alpha, \beta)$.

(b) If we had only required $\bar{Z}$, $\bar{W}$, and $\overline{ZW}$, find these statistics on the density function of X and Y.

Part III
AN INTRODUCTION
TO RANDOM PROCESSES

Chapter 6
The Definition and Classification of Random Processes

6.1 DEFINITION OF A RANDOM PROCESS

A random process is an ensemble of functions of some parameter (usually time), together with a probability measure, by which it is possible to determine certain statistical properties of a member or a group of members of the process. To appreciate this definition let us consider an ensemble or large collection of identical noise-voltage generators that generate the voltage outputs as depicted in Figure 6.1. We call this collection of waveforms an example of a random process.

Fundamentally we say that the definitions of a random variable and a random process may be related in the following way: A random variable maps the points of a sample description space onto the real axis according to a rule $X(s_i)$, thereby assigning probabilities to points or ranges of the real axis, whereas a random process maps the points of a sample description space S into member waveforms subject to a rule $X(s_i)$, thereby assigning a probability measure to the set of waveforms. In this chapter we will consider some artificial processes wherein the probability measure is evident but in general may be too difficult to specify.

Using the random process $v(t)$ of Figure 6.1 as an example, some nomenclature of random processes will now be explained.

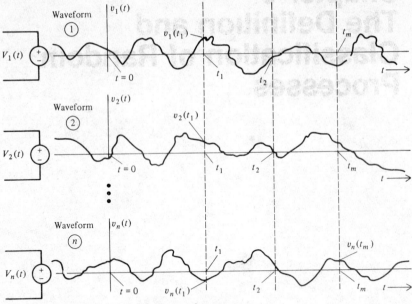

Figure 6.1 Typical members of a random process.

The random process $v(t)$ is said to consist of an infinite number of member waveforms denoted as $v_1(t), v_2(t),\ldots,v_n(t)$, and so on. We define the random variable $v(t_1)$ associated with the phenomenon of uniformly choosing a member of the process, say $v_p(t)$, and sampling it at time $t = t_1$, where we assign the random variable the value of the member waveform at time $t = t_1$, which is $v_p(t_1)$. This definition is very important to appreciate conceptually. Similarly, we define random variables $v(t_2),\ldots,v(t_m)$. For example, in Figure 6.1 we can see that the random variable $v(t_1)$ assumes values $v_1(t_1), v_2(t_1),\ldots,v_m(t_1)$.

Capital letter notation $V_1, V_2, V_3,\ldots,V_m$ is also used for the random variables $v(t_1), v(t_2), v(t_3),\ldots,v(t_m)$, respectively. We can define probabilities associated with the random variable V_1 or $v(t_1)$ on a relative-frequency or ensemble basis. For example, the probability $\alpha < v(t_1) < \beta$ is defined as

$$P(\alpha < V_1 < \beta) \triangleq \lim_{N \to \infty} \frac{N_1}{N}$$

where N_1 is the number of members for which $[\alpha < v_i(t_1) < \beta]$ when N is a large number of ensemble members. This is called an **ensemble average**. Also, we define $\overline{V_1}$ or $v(t_1)$ as

$$\overline{V_1} = \lim_{N \to \infty} \frac{\sum_{i=1}^{N} v_i(t_1)}{N}$$

This is called the mean value of the process at time t_1 defined on an ensemble basis.

Associated with any random variable $v(t_p)$ or V_p we associate a density function $f_{V(t_p)}(\alpha)$ or, more conveniently, $f_{V_p}(\alpha)$. If at $t = t_p$ the process assumes only discrete values, then the mass function notation $p_{v(t_p)}(\alpha_i)$ or $p_{V_p}(\alpha_i)$ may be used. We can denote "ensemble statistics"associated with the process at $t = t_p$ using this density function. For example,

$$\overline{v(t_p)} = \int_{-\infty}^{\infty} \alpha f_{V_p}(\alpha)\, d\alpha$$

$$\overline{v^2(t_p)} = \int_{-\infty}^{\infty} \alpha^2 f_{V_p}(\alpha)\, d\alpha$$

and

$$\overline{g(V_4)} = \int_{-\infty}^{\infty} g(\alpha) f_{V_4}(\alpha)\, d\alpha$$

Associated with any two random variables $v(t_1)$ and $v(t_2)$, which sample the process at $t = t_1$ and $t = t_2$, respectively, there will be an accompanying joint density (or mass) function $f_{V_1 V_2}(\alpha, \beta)$, from which any second-order statistic involving them may be found. Some of the most important of these are

$$R_{vv}(t_1, t_2) \triangleq \overline{v(t_1)v(t_2)}$$

called the autocorrelation function and

$$L_{vv}(t_1, t_2) \triangleq \overline{[v(t_1) - \overline{V_1}][v(t_2) - \overline{V_2}]}$$

called the covariance function.

Associated with any n random variables, $v(t_1), v(t_2), \ldots, v(t_n)$, which sample the process at $t = t_1, t_2, \ldots, t_n$, there will be an accompanying joint density function

$$f_{V_1 V_2 \ldots V_n}(\alpha_1, \alpha_2, \ldots, \alpha_n)$$

Notice that α_1 is now a variable and not a value of the variable α. The vector notation $\mathbf{V}$ and $\boldsymbol{\alpha}$ for arrays is also used. We say that a process is completely known or specified if we can derive the joint density function for any n random variables.

To cement in our minds the definitions and nomenclature of this section we will consider a very contrived example of a random process.

Example 6.1

Consider the random phenomenon of rolling a die and define a random process $x(t)$ as follows: If the result of the die roll is an even number, generate the waveform $x(t)=t$; if the result of the die roll is an odd number, generate a constant waveform with value equal to the number obtained. Sketch some typical members of this process and find the following on an ensemble basis:

(a) $P[1 \leq x(2) \leq 3.6]$
(b) $\overline{x(2)}$
(c) $P[x(2)x(3)<6.6]$

SOLUTION

$$S=\{1,2,3,4,5,6\} \quad \text{and} \quad P(s_i)=\tfrac{1}{6} \qquad 1\leq s_i \leq 6$$

If s_i is even, $X(s_i)=t$ results in the waveform $x(t)=t$ having a probability $P[x_i(t)=t]=\tfrac{1}{2}$. For s_i odd, $X(s_i)=s_i$; $-\infty<t<\infty$, results in

$$P[x_i(t)=1]=P[x_i(t)=3]=P[x_i(t)=5]=\tfrac{1}{6}$$

Typical members of $x(t)$ are shown plotted in Figure 6.2a, b, and c for parts (a), (b), and (c), respectively, the probability measure associated with $x(t)$ is already noted.

(a) The probability that the random variable $x(2)$ assumes a value $1 \leq x(2) \leq 3.6$ is defined on an ensemble basis as

$$P[1 \leq x(2) \leq 3.6]$$

$$= \lim_{N \to \infty} \left[\frac{\text{number of members with } (1 \leq x_i(2) \leq 3.6)}{N} \right]$$

Since $x_i(2)=2$ half the time, $x_i(2)=1$, 3, or 5 one-sixth of the time, then

$$P[1 \leq x(2) \leq 3.6]=P[x(2)=1]+P[x(2)=2]+P[x(2)=3]$$

$$=0.83$$

(b) Observing our process the values of the mass function for $x(2)$ with their probabilities are,

$$P_{x(2)}(1)=P_{x(2)}(3)=P_{x(2)}(5)=\tfrac{1}{6}$$

$$P_{x(2)}(2)=\tfrac{1}{2}$$

$$\therefore \overline{x(2)} = \left(1 \times \tfrac{1}{6}\right)+\left(3 \times \tfrac{1}{6}\right)+\left(5 \times \tfrac{1}{6}\right)+\left(2 \times \tfrac{1}{2}\right)$$

$$=2.5$$

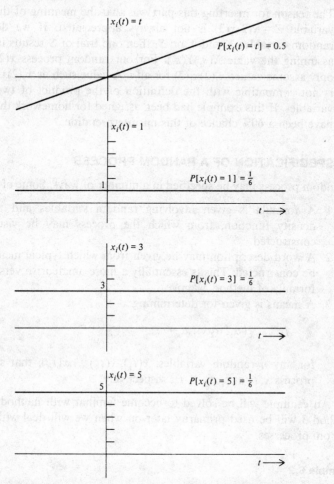

Figure 6.2 Typical members of the random process of Example 6.1.

(c) The different values taken on by $x(2)x(3)$, with their probabilities, are

$$x(2)x(3)=1\times1=1 \quad \text{and} \quad P[x(2)x(3)=1]=\tfrac{1}{6}$$

$$x(2)x(3)=2\times3=6 \quad \text{and} \quad P[x(2)x(3)=6]=\tfrac{1}{2}$$

$$x(2)x(3)=3\times3=9 \quad \text{and} \quad P[x(2)x(3)=9]=\tfrac{1}{6}$$

and

$$x(2)x(3)=5\times5=25 \quad \text{and} \quad P[x(2)x(3)=25]=\tfrac{1}{6}$$

$$\therefore P[x(2)x(3)\leq6.6]=\tfrac{1}{6}+\tfrac{1}{2}=0.67$$

The reason for inserting this part was that the meaning of the random variable $Z = x(2)x(3)$ is not always appreciated. If we define two random variables X and Y on S, then one trial of S results in $Z = XY$ assuming the value $X(s_i)Y(s_i)$. For our random process $x(2)x(3)$ can only assume values $x_i(2)x_i(3)$ for all i. A value such as $x_i(2)x_p(3)$, $p \neq i$ is not compatible with the definition of the product of two random variables. If this example had been assigned for homework there would have been a 60% chance of this misinterpretation.

6.2 SPECIFICATION OF A RANDOM PROCESS

A random process may be specified in a number of ways. Some of these are:

1. A formula is given involving random variables and their joint density function, from which the process may be visualized or constructed.
2. A word description may be given from which typical members may be constructed. This is essentially a more amateurish version of the formula of method 1 above.
3. A means is given for determining

$$f_{X_1 X_2 \ldots X_n}(\alpha_1, \alpha_2, \ldots, \alpha_n)$$

for any n random variables, $x(t_1), x(t_2), \ldots, x(t_n)$, that sample the process $x(t)$ at $t_1, t_2, \ldots, t_n$, respectively.

An example will be solved to become familiar with methods 1 and 2. Method 3 will be used primarily later on when we will deal with gaussian random processes.

Example 6.2

Try to construct how different ensemble members of the following random processes might behave. If possible find $f_X(\alpha)$ at a general time t.

(a) $x(t) = A + 8$, where A is a random variable with the density function

$$f_A(\alpha) = 1 \qquad 0 < \alpha \leq 1$$

$$= 0 \qquad \text{otherwise}$$

(b) $x(t) = A \cos(\omega_c t + \phi)$, where A and ϕ are random variables with the joint density function

$$f_{A\phi}(\alpha, \beta) = \frac{1}{2\pi} \qquad 0 < \alpha \leq 1, 0 < \beta \leq 2\pi$$

$$= 0 \qquad \text{otherwise}$$

(c) A certain random process is defined in the following way: Each member function $x_i(t)$ takes on the values $+1$ or -1 every b s, where each value occurs with equal probability. The amplitude during a time interval b is independent of the previous time interval. Assume that each member has a uniform phase shift ϕ_i to the left, where ϕ_i is chosen uniformly between 0 and b s.

SOLUTION

The main purpose of this problem is to formulate some idea of what a random process is, so the "solution" is somewhat casual.

(a) The understanding of the notation

$$x(t) = A + 8$$

where $f_A(\alpha)$ is given, implies that the different members of the process are

$$x_1(t) = A_1 + 8$$
$$x_2(t) = A_2 + 8$$

$$\vdots$$

$$x_n(t) = A_n + 8$$

where $A_1, A_2, \ldots, A_n$ are generated according to the density function

$$f_A(\alpha) = 1 \qquad 0 < \alpha < 1$$

A_1 could be found by spinning a wheel of fortune infinitely finely calibrated from 0 to 1, and similarly for $A_2, A_3, \ldots, A_n$. The ensemble of waveforms describing $x(t)$ would appear as shown in Figure 6.3a. What we have here is a new concept, and we must visualize the whole ensemble as an entity.

In order to find $f_{X_1}(\alpha)$, the density function for the random variable that samples the process at the time t_1, we can intuitively see that $x(t_1)$ takes on all values between 8 and 9 uniformly and that

$$f_X(\alpha) = 1 \qquad 8 < \alpha < 9$$

$$= 0 \qquad \text{otherwise}$$

(b) The random process now given for perusal is

$$x(t) = A \cos(\omega_c t + \phi)$$

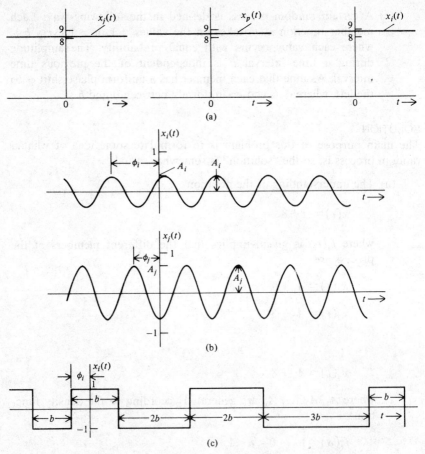

Figure 6.3 Typical members of the random processes of Example 6.2.

where A and ϕ are random variables with a joint density function

$$f_{A\phi}(\alpha, \beta) = \frac{1}{2\pi} \qquad 0 < \alpha < 1, 0 < \beta < 2\pi$$

$$= 0 \qquad \text{otherwise}$$

and ω_c is assumed to be a constant.

The different members of this process might be envisioned as

$$x_1(t) = A_1 \cos(\omega_c t + \phi_1)$$

$$x_2(t) = A_2 \cos(\omega_c t + \phi_2)$$

$$\vdots$$

$$x_n(t) = A_n \cos(\omega_c t + \phi_n)$$

where A_1 and ϕ_1, A_2 and ϕ_2, and so on are generated from the joint density function $f_{A\phi}(\alpha, \beta)$. For this joint density function we can find

$$f_A(\alpha) = \int_{-\infty}^{\infty} f_{A\phi}(\alpha, \beta)\, d\beta \qquad 0 < \alpha < 1$$

$$= \int_0^{2\pi} \frac{1}{2\pi}\, d\beta$$

$$= 1 \qquad 0 < \alpha < 1$$

$$= 0 \qquad \text{otherwise}$$

$$f_\phi(\beta) = \int_{-\infty}^{\infty} f_A(\alpha, \beta)\, d\alpha \qquad 0 < \beta < 2\pi$$

$$= \int_0^1 \frac{1}{2\pi}\, d\alpha$$

$$= \frac{1}{2\pi} \qquad 0 < \beta < 2\pi$$

$$= 0 \qquad \text{otherwise}$$

Also since $f_{A\phi}(\alpha, \beta) = 1/2\pi$ is equal to $f_A(\alpha) f_\phi(\beta) = 1 \times (1/2\pi)$, then the random variables are independent. The member waveforms $x_1(t), x_2(t), \ldots, x_n(t)$ could be found by spinning a wheel of fortune infinitely finely calibrated from 0 to 1 to find A_i and then spinning another wheel of fortune infinitely finely calibrated from 0 to 2π to find ϕ_i. We would carry out this dual experiment to find each member of the process. Two typical members $x_i(t)$ and $x_j(t)$ would look somewhat as shown in Figure 6.3b.

To find $f_{x(t)}(\alpha)$ at a general time $t = t$, we might intuitively feel that it does not depend on the specific time, but at this stage we will pass on trying to find it. If A was a constant and not a random variable, then by the methods of Chapter 4 we could find $f_{x(t)}(\alpha)$, with effort, as

$$f_X(\alpha) = \frac{1}{\pi} \frac{1}{\sqrt{A^2 - \alpha^2}} \qquad -A < \alpha < A$$

(c) In this case the random process is described in words as, "Each member takes on the value $+1$ or -1 every b s, where each value occurs with equal probability and the amplitude during a time interval b is independent of the previous time interval. Assume that each member has a uniform phase shift." This process is called a random binary process. From this description a typical member of $x(t)$ would be as shown in Figure 6.3c.

This member could have been generated by repeatedly tossing a coin. Everytime a head occurs the value 1 is assigned to the waveform for b units, and when a tail occurs the value -1 is assigned. In this way a waveform could be generated from $t = -Nb$ to $t = +Nb$ for a large value of N, and finally the whole waveform could be given a uniform phase shift ϕ_i, where ϕ_i is chosen subject to the density function $f_\phi(\alpha) = 1/b$; $0 < \alpha < b$. It should be clear from the definition of the process that the density function associated with the random variable $x(t)$, sampling the process at $t = t$, is $f_X(\alpha) = \frac{1}{2}\delta(\alpha - 1) + \frac{1}{2}\delta(\alpha + 1)$, which is independent of the time t. This could also be described by the mass function, $p_{x(t)}(-1) = \frac{1}{2}$, $p_{x(t)}(1) = \frac{1}{2}$, and otherwise $p_{x(t)}(\alpha) = 0$. An interesting problem would be to obtain an analytical representation for $x(t)$ utilizing random variables as in parts (a) and (b). This task will now be undertaken.

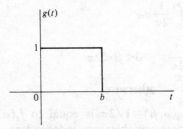

Define the deterministic function

$$g(t) = 1 \qquad 0 < t < b$$
$$\quad\;\; = 0 \qquad \text{otherwise}$$

as is shown in the thumbnail sketch. Now $x(t)$ may be expressed as

$$x(t) = \sum_{n = \infty}^{\infty} A_n g(t - nb + \phi)$$

where A_0, A_1, A_{-1}, and so on are all independent random variables with density function

$$f_{A_i}(\alpha) = \frac{1}{2}\delta(\alpha - 1) + \frac{1}{2}(\alpha + 1)$$

and ϕ is another random variable with density function

$$f_\phi(\alpha) = \frac{1}{b} \qquad 0 < \alpha < b$$

To specify any particular member an infinite number of A's of value $+1$ or -1 must be selected corresponding to repeatedly tossing a coin, and then the whole waveform

$$x_p(t) = \sum_{n=-\infty}^{\infty} A_n g(t - nb)$$

is given a random phase shift ϕ_p. *Students should thoroughly convince themselves that the general formula*

$$x(t) = \sum_{-\infty}^{\infty} A_n g(t - nb - \phi)$$

does indeed describe this random process.

6.3 CLASSIFICATION OF A RANDOM PROCESS

In this section random processes will be divided into a number of categories. There are further subdivisions that can be made, but in this first text we will not go into fine detail. The definitions of these terms will be carefully and systematically considered.

First-Order Stationary Random Process

The random process $x(t)$ is strictly **stationary of order one** if all the first-order statistics associated with the process are independent of time. This implies that

$$f_{x(t_1)}(\alpha) = f_{x(t_2)}(\alpha) \qquad \text{for any } t_1 \text{ and } t_2 \tag{6.1}$$

or, in words, that the density function associated with the random variable describing the values taken on by the different ensemble members at time t_1 is identical to the density function associated with the random variable at any other time t_2.

Some of the more important first-order statistics associated with a first-order stationary random process are

$$\overline{x(t)} = \int_{-\infty}^{\infty} \alpha f_{x(t)}(\alpha)\, d\alpha = \overline{X}$$

called the mean value of the process at $t = t$,

$$\overline{x^2(t)} = \int_{-\infty}^{\infty} \alpha^2 f_{x(t)}(\alpha)\, d\alpha = \overline{X^2}$$

called the mean squared value of the process at $t = t$, and

$$\sigma_X^2(t) = \int_{-\infty}^{\infty} \left[\alpha - \overline{x(t)}\right]^2 f_{x(t)}(\alpha)\, d\alpha$$

called the variance of the process at $t = t$.

Indeed if a process is first-order stationary we write $\bar{X}$, $\overline{X^2}$, and σ_X^2 and omit the t.

Second-Order Stationary Random Process

The random process $x(t)$ is said to be strictly **stationary of order two** if all second-order statistics associated with the random variables $x(t_1)$ and $x(t_2)$ are dependent at most on the time difference defined as $\tau = |t_1 - t_2|$ and not on the two values t_1 and t_2. This implies that

$$f_{x(t)x(t+\tau)}(\alpha, \beta) = f_{x(s)x(s+\tau)}(\alpha, \beta) \qquad \text{for any } t \text{ and } s \tag{6.2}$$

or that the same joint density function is obtained for any two random variables that describe the random process at times separated by τ s.

If a random process is second-order stationary, it must be first-order stationary.

Some of the more important statistics of a second-order stationary process are its **autocorrelation function** $R_{xx}(\tau)$ defined as

$$R_{xx}(\tau) \overset{\triangle}{=} \overline{x(t)x(t-\tau)} \quad \text{or} \quad \overline{x(t)x(t+\tau)}$$

$$= \int_{-\infty}^{\infty} \int_{-\infty}^{\infty} \alpha\beta f_{X_1 X_2}(\alpha, \beta) \, d\alpha \, d\beta \tag{6.3}$$

and its covariance function $L_{xx}(\tau)$ defined as

$$L_{xx}(\tau) = \overline{[x(t) - \bar{X}][x(t+\tau) - \bar{X}]}$$

$$= \int_{-\infty}^{\infty} \int_{-\infty}^{\infty} (\alpha - \bar{X})(\beta - \bar{X}) f_{X_1 X_2}(\alpha, \beta) \, d\alpha \, d\beta \tag{6.4}$$

The notation $\overline{x(t)x(t-\tau)}$, which yields the same result as $\overline{x(t)x(t+\tau)}$ for a second-order stationary process is standard because of its usefulness in some situations not involving stationary processes. However, we will tend to think of $R_{xx}(\tau)$ as being $\overline{x(t)x(t+\tau)}$ where the second random variable $x(t+\tau)$ samples the process at a later time. A somewhat less restrictive random process is defined next.

Wide-Sense Stationary Random Process

A random process $x(t)$ is said to be stationary in the **wide sense** if

$$E[x^2(t)] < \infty$$

and

$$E[x(t_1)x(t_2)] = \overline{x(t_1)x(t_2)}$$

$$\overset{\triangle}{=} \int_{-\infty}^{\infty} \int_{-\infty}^{\infty} f_{X_1 X_2}(\alpha, \beta) \, d\alpha \, d\beta$$

depends only on $|t_1 - t_2|$ or τ for all t_1 and t_2.

If a process is second-order stationary, it is implied that the process is also wide-sense stationary, but a wide-sense stationary process need not be stationary of order two or even order one.

This definition is reminiscent of the distinction we made in Chapter 5 between uncorrelated and independent random variables.

Strictly Stationary Random Process

A random process is said to be **strictly stationary** if for any n random variables, $X_1, X_2, \ldots, X_n$, which sample the process at $t_1, t_2, \ldots, t_n$,

$$f_{X_1 X_2 \cdots X_n}(\alpha_1, \alpha_2, \ldots, \alpha_n) = f_{X_1', X_2', \ldots, X_n'}(\alpha_1, \alpha_2, \ldots, \alpha_n) \tag{6.5}$$

where $X_1', X_2', \ldots, X_n'$, are n other random variables sampling the process at $t_1', t_2', \ldots, t_n'$ and we have the relationship

$$t_j - t_i = t_j' - t_i'$$

for all i and j. If a process is strictly stationary, it must follow that it is stationary of any lower order.

Ergodic Random Processes

A strictly stationary random process is said to be **ergodic** if all ensemble averages or averages based on random variables sampling the process are equal to the corresponding time averages for any specific ensemble member.

A first-order equivalency due to ergodicity is:

$$\lim_{T \to \infty} \frac{1}{2T} \int_{-T}^{T} x_p(t) \, dt \equiv \int_{-\infty}^{\infty} \alpha f_{x(t)}(\alpha) \, d\alpha \tag{6.6}$$

or in shorthand form

$$\widetilde{x(t)} = \overline{x(t)} = \overline{X}$$

where the wavy overbar indicates taking a time average of any specific member and the straight overbar indicates taking an ensemble average. $\overline{x(t)}$ or $\overline{X}$ is called the mean value of the random process.

Another first-order equivalency is

$$\lim_{T \to \infty} \frac{1}{2T} \int_{-T}^{T} x_p^2(t) \, dt \equiv \int_{-\infty}^{\infty} \alpha^2 f_{x(t)}(\alpha) \, d\alpha \tag{6.7}$$

or, in shorthand notation,

$$\widetilde{x^2(t)} = \overline{x^2(t)} = \overline{X^2}$$

In words, this says that the mean square value of any specific member equals the ensemble mean square value.

A second-order equivalency due to ergodicity is given by

$$\lim_{T \to \infty} \frac{1}{2T} \int_{-T}^{T} x_p(t) x_p(t - \tau) \, dt \equiv \int_{-\infty}^{\infty} \alpha \beta f_{x(t)x(t-\tau)}(\alpha, \beta) \, d\alpha \, d\beta \tag{6.8}$$

or, in shorthand notation,

$$\overline{x(t)x(t-\tau)} = \overline{x(t)x(t-\tau)} = R_{xx}(\tau)$$

where $R_{xx}(\tau)$ is called the autocorrelation function of the process. On the left-hand side it is found as a time average for any specific member, whereas on the right-hand side it is the ensemble average for the product of $x(t)$ and $x(t \pm \tau)$.

Another second-order equivalency is as follows:

$$\lim_{T \to \infty} \frac{1}{2T} \int_{-T}^{T} \left[x_p(t) - \overline{X} \right] \left[x_p(t-\tau) - \overline{X} \right] dt$$

$$\equiv \int_{-\infty}^{\infty} \int_{-\infty}^{\infty} (\alpha - \overline{X})(\beta - \overline{X}) f_{X_1 X_2}(\alpha, \beta) \, d\alpha \, d\beta \qquad (6.9)$$

or, in shorthand,

$$\overline{\left[x(t) - \overline{X} \right]\left[x(t-\tau) - \overline{X} \right]} \equiv \overline{\left[x(t) - \overline{X} \right]\left[x(t-\tau) - \overline{X} \right]} = L_{xx}(\tau)$$

where $L_{xx}(\tau)$ is called the covariance function of the process. Rarely do we consider any "statistics" higher than second order, so no others will be defined.

Nonstationary Random Processes

Finally, we define a **nonstationary random process** as one that is not even first-order stationary. We could state this as

$$f_{X_1}(\alpha) \neq f_{X_2}(\alpha) \qquad \text{for some } t_1 \text{ and } t_2 \qquad (6.10)$$

Statistics are in general defined for a nonstationary process and modified for the restrictions or blessings of being first order or second order or strictly stationary or ergodic. The most fundamental definitions of some basic quantities are

$$m_x(t) \triangleq \overline{x(t)} = \int_{-\infty}^{\infty} \alpha f_{x(t)}(\alpha) \, d\alpha \qquad (6.11)$$

which is called the mean function of $x(t)$, and for a nonstationary process it is a function of time.

$$R_{xx}(t, s) = \overline{x(t)x(s)} = \int_{-\infty}^{\infty} \int_{-\infty}^{\infty} \alpha \beta f_{x(t)x(s)}(\alpha, \beta) \, d\alpha \, d\beta \qquad (6.12)$$

and for a nonstationary process it is a function of the times t and s. This is the most general definition of the autocorrelation function.

$$L_{xx}(t, s) \triangleq \overline{\left[x(t) - \overline{x(t)} \right]\left[x(s) - \overline{x(s)} \right]}$$

$$= \int_{-\infty}^{\infty} \int_{-\infty}^{\infty} \left[\alpha - \overline{x(t)} \right]\left[\beta - \overline{x(s)} \right] f_{x(t)x(s)}(\alpha, \beta) \, d\alpha \, d\beta$$

$$\qquad (6.13)$$

and for a nonstationary process it is a function of the times t and s. This is the most general definition of the covariance function of a random process and is sometimes denoted $\phi_{xx}(t, s)$.

All the definitions so far encountered are summarized in Table 6.1. We must keep in mind that they are to be a fluent part of our vocabulary. A few examples will now be solved to help appreciate these definitions in a physical manner. Many of the processes used are extremely trivial from a practical viewpoint, but our intention is to be fluent with vocabulary and definitions to avert being smothered with notation.

Example 6.3

Prove that the random process of Example 6.2, $x(t) = 8 + A$, where $f_A(\alpha) = 1$; $0 < \alpha < 1$, is

(a) First-order stationary.
(b) Second-order stationary.

Table 6.1 A SUMMARY OF SOME NOTATION USED FOR RANDOM PROCESSES

NOTATION	MEANING
$x(t)$	A random process or ensemble of waveforms; a random variable sampling a process at time t, possibly a general specific member.
$x_i(t)$	A specific member of the ensemble (the ith member).
$x_i(t_j)$	The value of the ith member at $t = t_j$.
$f_{X_1}(\alpha)$ or $f_{x(t_1)}(\alpha)$	The density function of the random variable $x(t_1)$.
$\overline{x(t)}$, $\overline{x^2(t)}$, and $\sigma_x^2(t)$	First-order statistics of the random variable $x(t)$: its mean, mean square, and variance.
$f_{X_1 X_2}(\alpha, \beta)$	The joint density function associated with the random variables $x(t_1)$ and $x(t_2)$ of a process.
$R_{xx}(t_1, t_2)$ $L_{xx}(t_1, t_2)$	The autocorrelation and covariance functions, $\overline{x(t_1)x(t_2)}$ and $\overline{(X_1 - \overline{X_1})(X_2 - \overline{X_2})}$ associated with a nonstationary process.
$f_X(\alpha)$	The notation for the density function of a first-order stationary process. This implies $f_{x(t_1)}(\alpha) = f_{x(t_2)}(\alpha)$ for any t_1 and t_2.
$f_{X_1 X_2}(\alpha, \beta, \tau)$	The notation for the joint density function of a second-order stationary process. This implies $f_{X_1 X_2}(\alpha, \beta)$ depends only on $\tau = \lvert t_2 - t_1 \rvert$.
$R_{xx}(\tau), L_{xx}(\tau)$	The autocorrelation and covariance functions for a process that is at least wide-sense stationary.
$\overline{x(t)}, \overline{x(t)x(t + \tau)}$	The notation used for time averages based on a specific member. These may be used in lieu of ensemble averages for ergodic processes.

(c) Strictly stationary.
(d) Not ergodic.

SOLUTION

(a) The solution should be intuitively clear,

$$f_X(\alpha) = 1 \qquad 8 < \alpha < 9$$

$$= 0 \qquad \text{otherwise}$$

as each member at $t = t$ takes on a value $8 + A_i$, which is assigned according to the density function for the random variable A. Since $f_X(\alpha)$ does not involve t, then the process is first-order stationary. Some of its first-order statistics are

$$\overline{x(t)} = \int_8^9 \alpha \, d\alpha = \tfrac{1}{2}(81 - 64) = 8.5$$

$$\overline{x^2(t)} = \int_8^9 \alpha^2 \, d\alpha = \tfrac{1}{3}(729 - 512) = 72.3$$

$$\sigma_x^2(t) = 72.3 - (8.5)^2 = 0.05$$

(b) $x(t)$ is stationary of order 2 if we can show that $f_{X_1 X_2}(\alpha, \beta)$ does not depend on t_1 and t_2 but at most on $\tau = |t_2 - t_1|$ for any t_1 and t_2. Since on any specific trial of the phenomenon for which they are defined, $x(t_1)$ and $x(t_2)$ both must take on the same value $8 + A_i$, as $x_i(t) = 8 + A_i$, then the joint density function is

$$f_{X_1 X_2}(\alpha, \beta) = 1\delta(\beta - \alpha) \qquad 8 < \alpha < 9, \beta = \alpha$$

This obviously yields the same result for t_1 and $t_1 + \tau$ and for t_1' and $t_1' + \tau$. Therefore $x(t)$ is stationary of order two.

(c) $x(t)$ is strictly stationary because we can find

$$f_{X_1 X_2 \ldots X_n}(\alpha, \alpha_2, \ldots, \alpha_n) = f_{X_1' X_2' \ldots X_n'}(\alpha_1, \alpha_2, \ldots, \alpha_n)$$

For example,

$$f_{X_1 X_2 X_3}(\alpha, \beta, \gamma) = 1\delta(\beta - \alpha)\delta(\gamma - \beta) \qquad 8 < \alpha < 9, \beta = \alpha, \gamma = \beta$$

and since it does not involve t_1, t_2, or t_3, $x(t)$ is third-order stationary.

(d) It can readily be noticed that this process is not ergodic, since the different specific members take on different dc values. Obviously, $\widetilde{x(t)}$ is different for each member, whereas $\overline{x(t)} = 8.5$.

Example 6.4

Prove that the random process

$$x(t)=\cos(t+\phi)$$

where ϕ is a random variable with density function

$$f_\phi(\alpha)=\frac{1}{2\pi}\qquad 0<\alpha<2\pi$$

$$=0\qquad \text{otherwise}$$

is

    (a) First-order stationary.
    (b) Second-order stationary.

SOLUTION

    (a) To help visualize the process $x(t)=\cos(t+\phi)$, two typical members
are shown plotted in Figure 6.4a. It is clear that since the phase
shift is uniform (with respect to a period or multiple of periods),
then at any time t the values taken on by the random variable $x(t)$
vary between -1 and $+1$ and do not depend on the specific t.
Obtaining $f_{x(t)}(\alpha)$ becomes an application from Chapter 4 of finding the density function of a function of a random variable whose
density function is known; that is, "Given a known random variable with density function $f_\phi(\alpha)=1/2\pi$; $0<\alpha<2\pi$, find the density function of the random variable $x(t)=Y$, where $Y=\cos(t+\phi)$
for a specific t." We will let $Y=\cos(t+\phi)$ and find $f_Y(\beta)$.
    For the range $\beta<-1$,

$$F_Y(\beta)=0$$

For the range $-1<\beta<0$,

$$F_Y(\beta)=P[\cos(t+\phi)<\beta]$$

Using symmetry and the properties of a sinusoidal waveform as
shown in Figure 6.4b we can state, subject to discipline, that

$$F_Y(\beta)=2P[(\cos^{-1}\beta)-t<\phi<\pi-t]\qquad \text{for }\frac{\pi}{2}<t+\phi<\pi$$

$$=2\int_{\cos^{-1}\beta}^{\pi-t}\frac{1}{2\pi}\,d\alpha$$

$$=1-\frac{1}{\pi}\cos^{-1}\beta$$

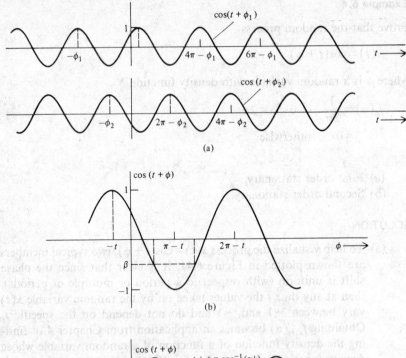

Figure 6.4   (a) Two typical members for the random process of Example 6.4. (b) $\cos(t + \phi)$ versus ϕ for a fixed t. (c) Different possibilities for $x(t) = \alpha$.

We immediately notice that our result is independent of t, so the process is first-order stationary. Now the density function is

$$f_Y(\beta) = \frac{d}{d\beta} F_Y(\beta)$$

$$= \frac{1}{\pi} \frac{1}{\sqrt{1 - \beta^2}} \qquad -1 < \beta < 0$$

For the range $0<\beta<1$, by symmetry $f_Y(\beta)$ must be an even function and

$$f_Y(\beta) = \frac{1}{\pi} \frac{1}{\sqrt{1-\beta^2}}$$

Stated in terms of the notation of our random process, we have found

$$f_{x(t)}(\alpha) = \frac{1}{\pi} \frac{1}{\sqrt{1-\alpha^2}} \qquad -1<\alpha<1$$

and $x(t)$ is a first-order stationary random process. From experience the author considers this a formidable derivation. At this time we should stop and question exactly what is being done as opposed to saying, "I understand the manipulations and algebra, so let's proceed." We are at a stage where notation is piling up and we are relying heavily on the random variable prerequisite material of Chapters 4 and 5, which is philosophically based on discrete probability theory. The topics are tricky and difficult, so the student should not feel inadequate or in a hurry. As an instructor my most frustrating experience is to teach this material to juniors, seniors, and graduate students and to sometimes find graduate students with two previous courses involving communications unable to compete conceptually with juniors because of a lack of appreciating fundamentals.

(b) We are now required to show that $x(t)$ is second-order stationary. For this to be so, the condition is that $f_{X_1 X_2}(\alpha, \beta)$ depends only on α, β and $|t_2 - t_1| = \tau$.

From part (a),

$$f_{X_1 X_2}(\alpha, \beta) = f_{X_1}(\alpha) f_{X_2}\{\beta / [x(t) = \alpha]\}$$

We have found

$$f_{x(t)}(\alpha) = \frac{1}{\pi} \frac{1}{\sqrt{1-\alpha^2}} \qquad -1<\alpha<1$$

$$= 0 \qquad \text{otherwise}$$

and so it remains to find $f_{x(t+\tau)}[\beta/(x(t)=\alpha)]$. If the value of $x(t)$ on a trial of the experiment is α or $\cos(t+\phi)=\alpha$, then in order to find $x(t+\tau)$ uniquely for the same trial we must distinguish between α where the slope is positive $(\alpha\uparrow)$ on the cosine curve and α where the slope is negative $(\alpha\downarrow)$. These situations are shown in Figure 6.4c. If $x(t)=\alpha\uparrow$, then $t+\phi=\cos^{-1}(\alpha\uparrow)$ and $x(t+\tau)$ must take on the value $\cos[\cos^{-1}(\alpha\uparrow)+\tau]$. The joint density function becomes

$$f_{x(t)x(t+\tau)}(\alpha,\beta) = \frac{1}{\pi} \frac{1}{\sqrt{1-\alpha^2}} \delta\{\beta - \cos[\cos^{-1}(\alpha) + \tau]\}$$

where $\alpha\uparrow$ or $\alpha\downarrow$ is used. This proves that $x(t)$ is second-order stationary.

Example 6.5

Consider the random process

$$x(t) = \cos(t + \phi)$$

where ϕ is a random variable with density function

$$f_\phi(\alpha) = \frac{1}{\pi} \qquad -\frac{\pi}{2} < \alpha < \frac{\pi}{2}$$

Prove whether or not the process is stationary.

SOLUTION
If we visualize the members of $x(t)$ we see certain bounds on their behavior as in Figure 6.5. It is obvious that the random variable $x(0)$ takes on only values between 0 and 1, whereas the random variable $x(\pi)$ takes on only

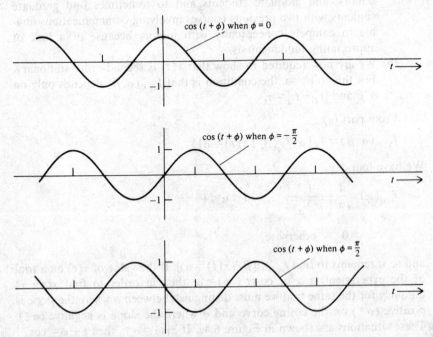

Figure 6.5   Three members for the random process of Example 6.5.

values between -1 and 0.

$$\therefore f_{x(0)}(\alpha) \neq f_{x(\pi)}(\alpha)$$

Hence the process $x(t)$ is not first-order stationary and is indeed nonstationary. As is usual in science, disproving by a counterexample is much easier than proving something.

Example 6.6

Consider the random process

$$x(t) = \sum_{n=-\infty}^{\infty} A_n g(t - 2n + \phi)$$

where the A's and ϕ are independent random variables with density functions

$$f_{A_i}(\alpha) = \tfrac{1}{2}\delta(\alpha) + \tfrac{1}{2}\delta(\alpha - 1)$$

$$f_\phi(\alpha) = 1 \qquad 0 < \alpha < 1$$

$$\qquad\quad = 0 \qquad \text{otherwise}$$

and $g(t)$ is the deterministic function

$$g(t) = 1 \qquad 0 < t < 1$$

$$\quad = 0 \qquad \text{otherwise}$$

Evaluate the density functions $f_{x(0^+)}(\alpha)$ and $f_{x(1^+)}(\alpha)$ and conclude that $x(t)$ is not first-order stationary.

SOLUTION

In order to visualize a typical member of $x(t)$, a plot is shown of $g(t)$ in Figure 6.6a, of

$$\sum_{-\infty}^{\infty} g(t - 2n)$$

in Figure 6.6b, and of two different members of $x(t)$ in Figures 6.6c and d. It is best to visualize a typical member of $x(t)$ as composed of pulses of width 2, which have height 1 or 0 for one unit called the leading edge and then a trailing edge of height 0 for the next unit.

For $f_{x(0^+)}(\alpha)$, if we visualize sampling the members of $x(t)$ at 0^+ then, since the probability of ϕ being exactly 1 is 0, the trailing zero edge of $A_0 g(t + \phi)$ never reaches 0^+. This implies that $p_{x(0^+)}(0) = p_{x(0^+)}(1) = \tfrac{1}{2}$, as indicated in Figure 6.6e.

For $f_{x(1^+)}(\alpha)$, if we visualize sampling the members of $x(t)$ at $t = 1^+$, then the probability of ϕ being exactly 1 is 0 since the forward edge of the pulse $A_1 g(t - 2 + \phi)$ never reaches 1^+, and hence always we have $p_{x(1^+)}(1) = 1$. We have demonstrated that $f_{x(0^+)}(\alpha) \neq f_{x(1^+)}(\alpha)$, and therefore $x(t)$ is a nonstationary random process.

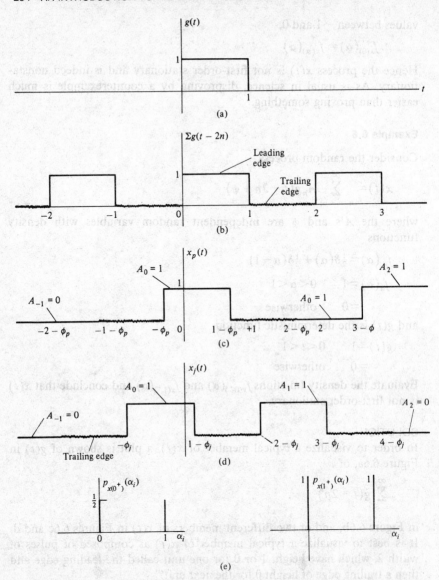

Figure 6.6 (a) The deterministic function $g(t)$ of Example 6.6. (b) $\Sigma g(t-2n)$. (c) A typical member of $\Sigma g(t-2n+\phi)$ for ϕ close to 0. (d) A typyical member of $\Sigma g(t-2n+\phi)$ for ϕ close to 1. (e) The density functions for sampling at 0^+ and 1^+, respectively.

Example 6.7

Consider the random process

$$x(t) = \sum_{n=-\infty}^{\infty} A_n g(t - 2n + \phi)$$

where the A's and ϕ are independent random variables with density functions

$$f_\phi(\alpha) = \tfrac{1}{2}\delta(\alpha) + \tfrac{1}{2}\delta(\alpha - 1)$$

and

$$f_\phi(\alpha) = \tfrac{1}{2} \qquad 0 < \alpha < 2$$
$$= 0 \qquad \text{otherwise}$$

which is a uniform phase shift over the basic pulse length of two and $g(t)$ is the same deterministic function as in Example 6.6. Evaluate the density functions $f_{x(0^+)}(\alpha)$, $f_{x(1^+)}(\alpha)$, and $f_{x(t)}(\alpha)$ for a general t. Conclude that $x(t)$ is first-order stationary.

SOLUTION

Two typical members of $x(t)$ are shown plotted in Figure 6.7a and b. The irregular lines indicate the trailing 0 value edge of each pulse. The different density functions will now be evaluated.

$f_{x(0)^+}(\alpha)$

If $0 < \phi < 1$, as in Figure 6.7a, then the 0 trailing edge of $A_0 g(t - \phi)$ has not arrived at $t = 0^+$ and there is a 50% chance that the leading edge present

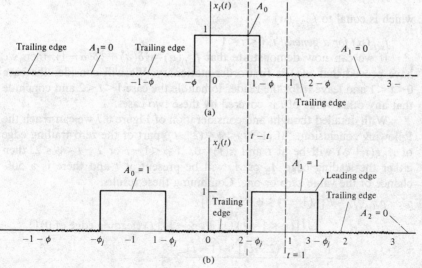

Figure 6.7 Two typical members of the random process of Example 6.7. (a) $0 < \phi < 1$. (b) $1 < \phi < 2$.

there is 0. If $1<\phi<2$, then the 0 trailing edge of $A_0 g(t-\phi)$ is at $t=0^+$ and $x(0^+)=0$. Combining these results we obtain

$$p_{x(0^+)}(0) = P\{(1<\phi<2) \cup [(0<\phi<1) \cap (A_0=0)]\}$$
$$= \tfrac{1}{2} + (\tfrac{1}{2} \times \tfrac{1}{2}) = \tfrac{3}{4}$$

If at $t=0^+$, $0<\phi<1$, then there is a 50% chance that the leading edge A_0 present there is 1.

$$\therefore p_{x(0^+)}(1) = P[(0<\phi<1) \cap (A_0=1)] = \tfrac{1}{4}$$

Summarizing we have found,

$$f_{x(0^+)}(\alpha) = \tfrac{3}{4}\delta(\alpha) + \tfrac{1}{4}\delta(\alpha-1)$$

$f_{x(1^+)}(\alpha)$

At $t=1^+$, if $0<\phi<1$, then the leading edge of $A_1 g(t-2+\phi)$ will not have arrived at 1^+, and therefore $x(1^+)=0$. If $1<\phi<2$, as in Figure 6.7b, then the leading part of the $A_1 g(t-2+\phi)$ pulse will be at 1^+ and there is a 50-50 chance of A_1 being of value one. Combining these results we have

$$p_{x(1^+)}(0) = P\{(0<\phi<1) \cup [(1<\phi<2) \cap (A_1=0)]\}$$
$$= \tfrac{1}{2} + (\tfrac{1}{2} \times \tfrac{1}{2}) = \tfrac{3}{4}$$

and

$$p_{x(1^+)}(1) = P[(1<\phi<2) \cap (A_1=1)] = \tfrac{1}{4}$$

Summarizing, we have found that

$$f_{x(1^+)}(\alpha) = \tfrac{3}{4}\delta(\alpha) + \tfrac{1}{4}\delta(\alpha-1)$$

which is equal to $f_{x(0^+)}(\alpha)$.

$f_{x(t)}(\alpha)$ for a general t, $0<t<1$

If we can now demonstrate that $f_{x(t)}(\alpha) = \tfrac{3}{4}\delta(\alpha) + \tfrac{1}{4}\delta(\alpha-1)$, then we have proved that $x(t)$ is first-order stationary. We will consider the case $0<t<1$ and leave it for the reader to handle the case $1<t<2$ and conclude that any other value of t is covered by these two cases.

With detailed thought and consideration of Figure 6.7 we can reach the following conclusion: "If $(1-t)<\phi<(2-t)$, part of the zero trailing edge of $A_0 g(t+\phi)$ will be at t and $x(t)=0$. If $\phi<1-t$ or $2-t<\phi<2$, then either the leading edge A_0 or A_1 will be present at t and there is a 50% chance of the value zero or one. Combining these results,

$$p_{x(t)}(0) = P\{[(1-t)<\phi<(2-t)]$$
$$\cup [[(\phi<1-t) \cup (\phi>2-t)] \cap (\text{leading edge of } 0)]\}$$
$$= \frac{1}{2} + \frac{[2-(2-t)]+(1-t)}{2} \times \frac{1}{2}$$
$$= \frac{1}{2} + \frac{1}{4} = \frac{3}{4}$$

and

$$p_{x(t)}(1) = P\{[(\phi < 1 - t) \cup (\phi > 2 - t)] \cap (\text{L.E. } 1)\}$$

$$= \tfrac{1}{4}$$

$$f_{x(t)}(\alpha) = \tfrac{3}{4}\delta(\alpha) + \tfrac{1}{4}\delta(\alpha - 1)$$

and $x(t)$ is a first-order stationary process. This derivation is frustrating and requires much thought.

Example 6.8

Prove that the random process

$$x(t) = \sum_{n=-\infty}^{\infty} A_n g(t - 2n + \phi)$$

of Example 6.7 is second-order stationary.

SOLUTION
The random process $x(t)$ is second-order stationary if we can demonstrate that $p_{x(t_1)x(t_2)}(\alpha_i, \beta_j)$ is dependent at most on $|t_2 - t_1|$ and not on t_1 and t_2. Since the process assumes only the discrete values 0 and 1 we can say the joint mass function assumes values for $p(0,0)$, $p(0,1)$ $p(1,0)$ and $p(1,1)$. For convenience we have omitted the subscripts $x(t_1)$ and $x(t_2)$. If we can show that these probabilities are at most dependent on $|t_2 - t_1|$ and not t_1 and t_2, then $x(t)$ is second-order stationary. We will first consider the case $\tau = |t_2 - t_1| < 1$.

$p_{x(t_1)x(t_2)}(1,0); \tau < 1$
$P(1,0) = P\{[x(t_1) = 1] \cap [x(t_2) = 0]\}$ may be found in two ways, either directly or as $p_{X_1}(1)p_{X_2}[0/(x(t_1) = 1)]$. Using the latter we say that the event $(1,0)$ occurs when $X_1 = 1$ and then $X_2 = 0$. We consider as before that a "pulse" of a typical member of $x(t)$ is of width 2 units and that the first half or leading edge is either 0 or 1 and the second half or trailing edge is always 0. Therefore,

$$P(1,0) = P\{[x(t_1) \text{ is from a leading edge of value } 1] \cap [x(t_2) = 0]\}$$

$$= \tfrac{1}{4}P\{(x(t_2) = 0)/[x(t_1) = 1]\}$$

The conditional event of a 0 occurring τ s after a one is the event that $x(t_2)$ is from a trailing edge, and this event has a probability of τ, for $0 < \tau < 1$.

$$\therefore P(1,0) = \tfrac{1}{4}\tau$$

$p_{X_1 X_2}(1,1); \tau < 1$
Considering Figure 6.7,

$$P(1,1) = P\{[x(t_1) = 1] \cap [x(t_2) = 1/x(t_1) = 1]\}$$

$$= \tfrac{1}{4}(1 - \tau) \qquad \text{(Be sure of the factor } 1 - \tau\text{)}$$

$p_{X_1 X_2}(0, 1); \tau < 1$

This probability will be found directly. The value 1 may only be obtained at t_2 if the first value was obtained from a trailing edge 0, since otherwise we always have a 0 at t_2. The event $(0, 1)$ may be written as

$$(0, 1) = \{(\text{a 0 from a trailing edge})$$

$$\cap (\text{a new pulse being generated with value 1})\}$$

$$= \tfrac{1}{2}(\tau) \times \tfrac{1}{2}$$

$$= \tfrac{1}{4}\tau$$

Alternatively, we could find $P(0, 1)$ using conditional probability as

$$P(0, 1) = P(0)P(1/0)$$

$$= \tfrac{3}{4}\left[\left(\tfrac{2}{3}\tau \times \tfrac{1}{2}\right) + \left(\tfrac{1}{3} \times 0\right)\right]$$

$$= \tfrac{1}{4}\tau$$

The factors in brackets may be hard to understand. If a 0 is obtained, then $P(1/0) = 0$ if the 0 is from a leading edge and the chances the 0 was from a leading edge is $\tfrac{1}{3}$ (be sure). If the 0 is from a trailing edge the chances are $\tau/2$ of a 1 at t_2 and the chances the 0 was from a trailing edge is $\tfrac{2}{3}$ (be very sure).

$p_{X_1 X_2}(0, 0); \tau < 1$

The final probability will now be found in two ways. The event $(0, 0)$ may be written as

$$(0, 0) = \{[(\text{a 0 from a leading edge}) \cap (\text{a 0 } \tau \text{ s later})]$$

$$\cup [(\text{a 0 from a trailing edge}) \cap (\text{a 0 } \tau \text{ s later})]\}$$

$$P(0, 0) = \left(\tfrac{1}{4} \times 1\right) + \tfrac{1}{2}\left[(1 - \tau) + \tfrac{1}{2}\tau\right]$$

We notice that the probability a 0 follows a leading-edge 0 τ s later is certain, and the other expressions should now be clear.

$$\therefore P(0, 0) = \tfrac{1}{4} + \tfrac{1}{2} - \tfrac{1}{2}\tau + \tfrac{1}{4}\tau$$

$$= \tfrac{3}{4} - \tfrac{1}{4}\tau$$

Alternatively we can find $P(0, 0)$ using conditional probability as

$$P(0, 0) = P(0)P(0/0)$$

$$= \tfrac{3}{4}\left[\tfrac{2}{3}(1 - \tau) + \tfrac{2}{3}\left(\tfrac{1}{2}\tau\right) + \tfrac{1}{3}\right]$$

$$= \tfrac{1}{2} - \tfrac{1}{2}\tau + \tfrac{1}{4}\tau + \tfrac{1}{4}$$

$$= \tfrac{3}{4} - \tfrac{1}{4}\tau$$

as before.

The joint probability mass matrix for our four points is shown summarized in Figure 6.8. Since the probabilities depend only on $|t_2 - t_1| = \tau$

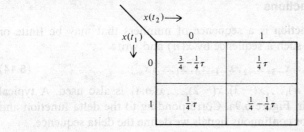

Figure 6.8 The joint mass function for sampling the random process of Example 6.8 when $0 < \tau < 1$.

and not t_1 and t_2 the process is second-order stationary. It is left as an exercise for the reader to form this matrix for $1 < |t_1 - t_2| < 2$ and conclude that for any t_1 and t_2 the probabilities do not depend on both t_1 and t_2 but only on $|t_2 - t_1| = \tau$.

DRILL SET: CLASSIFICATION OF RANDOM PROCESSES

Assume four random processes:

(a) $x(t) = \sum_{-\infty}^{\infty} g(t - 3n + \phi)$ $f_\phi(\alpha) = \frac{1}{3}; \ 0 < \alpha < 3$
(b) $x(t) = \sum_{-\infty}^{\infty} g(t - 3n + \phi)$ $f_\phi(\alpha) = \frac{1}{2}; \ 0 < \alpha < 3$
(c) $x(t) = \sum_{-\infty}^{\infty} A_n g(t - 3n)$
(d) $x(t) = \sum_{-\infty}^{\infty} A_n g(t - 3n + \phi)$

where for (c) and (d) the A_n's are independent random variables with $f_{A_n}(\alpha) = \frac{1}{2}\delta(\alpha - 1) + \frac{1}{2}\delta(\alpha + 1)$, ϕ is also independent with a uniform density function from 0 to 3, and $g(t)$ is defined as

$$g(t) = 1 \quad 0 < t < \tfrac{1}{2}$$

$$= 0 \quad \text{otherwise}$$

Classify each of the random processes as being either

(1) Nonstationary [give an example $f_{X_1}(\alpha) \neq f_{X_2}(\alpha)$].
(2) First-order stationary (find $f_X(\alpha)$ as independent of t).
(3) Second-order stationary.
(4) Ergodic (solve intuitively).

6.4 DISCRETE RANDOM PROCESSES

In this section the concepts and notation of Sections 6.1 to 6.3 for continuous random processes will be treated for discrete random processes. First, discrete time functions will be discussed and then a discrete random process defined as an ensemble of discrete time functions. The nomenclature, specification, and classification of discrete random processes will then be covered.

Discrete Time Functions

A **discrete time function** is a sequence of numbers that may be finite or infinite. We denote such a sequence by $x(n)$ and write

$$x(n) = \{x_{-n}, \ldots, x_{-3}, x_{-2}, x_{-1}, x_0, x_1, \ldots, x_m\} \qquad (6.14)$$

The notation $\{x(-n), \ldots, x(-3), x(-2), \ldots, x(m)\}$ is also used. A typical sequence is shown in Figure 6.9a. Corresponding to the delta function and unit step function for continuous signals we define the delta sequence,

$$\delta(n) = 1 \quad n = 0$$
$$= 0 \quad n \neq 0 \qquad (6.15)$$

and the unit step sequence,

$$u(n) = 1 \quad n \geq 0$$
$$= 0 \quad \text{otherwise} \qquad (6.16)$$

and these are shown plotted in Figures 6.9b and c.

Any discrete time function may conveniently be denoted as

$$x(n) = \sum_{k=-n}^{m} x_k \delta(n-k) \quad \text{or} \quad \sum_{k=-n}^{m} x(k)\delta(n-k) \qquad (6.17)$$

For uniformity when a discrete signal represents the sampled values of an analog signal taken every τ s we still indicate it by $x(n)$, where the sampling rate is unity. We must rescale in time and frequency when we talk about the

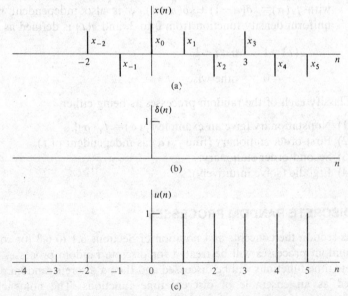

Figure 6.9 (a) A discrete time function. (b) The delta sequence. (c) The unit step sequence.

actual sampled signal or its discrete Fourier transform (introduced in Chapter 8). To become familiar with the notation of indicating a sequence by giving a general formula for x_n, a number of discrete time functions will be visualized and plotted.

Example 6.9

Plot the following discrete time functions:

(a) $x(n)$, where $x_n = 1/4^n u(n)$
(b) $x(n)$, where $x_n = 1/4^{-n} u(-n)$
(c) $x(n)$, where $x_n = 1/4^{|n|}$
(d) $x(n)$, where $x_n = \cos(n/2)$
(e) $x(n)$, where $x_n = \sin(\pi n)/\pi n$

SOLUTION

Considering the definitions, the discrete functions of part (a) to (e) are shown plotted in Figure 6.10a through e, respectively. In part (a), where $x_n = 1/4^n u(n)$, we see that $u(n)$ ensures that $x_n = 0$ for $n < 0$. In part (b), where $x_n = 1/4^{-n} u(-n)$, the term $u(-n)$ ensures that $x_n = 0$ for $n > 0$. In part (c), $x_n = 1/4^{|n|}$ exists for all n, and it is interesting to write $x(n)$ in terms of $u(n)$ and $u(-n)$ as

$$1/4^{|n|} = 1/4^{-n} u(-n) + 1/4^n u(n) - \delta(n)$$

The term $-\delta(n)$ is necessary since both $u(n)$ and $u(-n)$ are defined for $n = 0$, and without it x_0 has value 2 and not 1 as required. In part (d), $x_n = \cos(n/2)$ may be thought of as the sampled values of the analog time function $x(t) = \cos(t/2)$ taken every second. Again, in part (e), $x_n = \sin(\pi n)/\pi n$ are the sampled values of $\sin(\pi t)/\pi t$. Since $\sin(\pi n)$ is 0 for all integers n, $x(n)$ is equivalent to $\delta(n)$.

Most of the operations involving continuous time signals may be defined for discrete time signals. Considering two discrete time functions $x(n)$ and $y(n)$ we define the following terms.

The Sum of Discrete Functions

$$r(n) \triangleq x(n) + y(n) \qquad \text{if } r_k = x_k + y_k \tag{6.18}$$

The Difference of Discrete Functions

$$r(n) \triangleq x(n) - y(n) \qquad \text{if } r_k = x_k - y_k \tag{6.19}$$

A Shifted Function:

$$y(n) \triangleq x(n-l) \qquad \text{if } y_k = x_{k-l}, \text{ for all } k \tag{6.20}$$

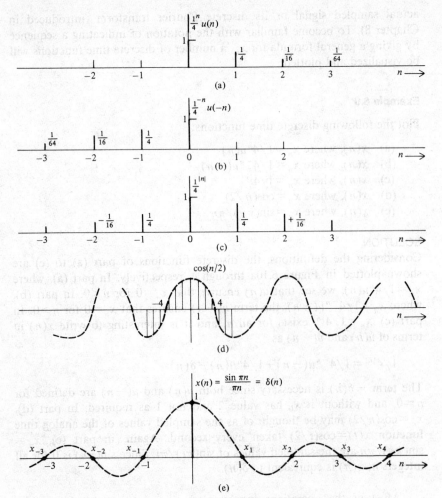

Figure 6.10   The discrete time functions of Example 6.9.

The function $y(n)$ is $x(n)$ shifted l units to the right.

The Product of Discrete Functions

$$r(n) \triangleq x(n)y(n) \qquad \text{if } r_k = x_k y_k \tag{6.21}$$

The Quotient of Discrete Functions

$$r(n) \triangleq \frac{x(n)}{y(n)} \qquad \text{if } r_k = \frac{x_k}{y_k} \tag{6.22}$$

where r_k is not defined if $y_k = 0$.

The Convolution of Discrete Functions

In general the convolution of two discrete time functions is

$$r(n) \stackrel{\triangle}{=} x(n) * y(n) \qquad \text{if } r_n = \sum_{p=-\infty}^{\infty} x_p y_{n-p} \tag{6.23}$$

The symbol $*$ is used to denote convolution. In the case where $x(n)$ and $y(n)$ are causal functions, or $x_k = 0$ and $y_k = 0$ for $k < 0$, then the convolution of $x(n)$ and $y(n)$ is given by

$$\begin{aligned} r_n &= \sum_{k=0}^{n} x_k y_{n-k} \qquad n \geq 0 \\ &= 0 \qquad n < 0 \end{aligned} \tag{6.24}$$

and, as we will see in Chapter 7, due to the commutative property we can also obtain

$$r_n = \sum_{k=0}^{n} y_k x_{n-k} \qquad n \geq 0$$

The convolution of two discrete functions exists if

$$\sum_{-\infty}^{\infty} x_p y_{n-p}$$

is finite for all n, which is the case if both $x(n)$ and $y(n)$ are finite energy functions, which requires

$$\sum_{-\infty}^{\infty} |x_k| < \infty \quad \text{and} \quad \sum_{-\infty}^{\infty} |y_k| < \infty \quad \text{or if} \quad \sum_{-\infty}^{\infty} |r_k| < \infty$$

Discrete Random Processes

DEFINITION OF A DISCRETE RANDOM PROCESS

A discrete random process is an ensemble of discrete time functions, together with a probability measure by which it is possible to determine certain statistical properties of a member or a group of members of the ensemble. Figure 6.11 shows some typical members of a discrete random process.

In order to tie together the definition of a random variable with the definition of a discrete random process or ensemble of number sequences we say that for a random variable X the points of a sample description space map according to a rule $X(s)$ onto the real axis. For a discrete random process, however, the points of S map into discrete time waveforms, thereby assigning a probability measure to the set of discrete waveforms. Since we are a stage where a student needs time to absorb definitions, it is unfair to say that all our definitions are analogous to those in Sections 6.1 to 6.3 for continuous processes and can be instantaneously embraced. We will

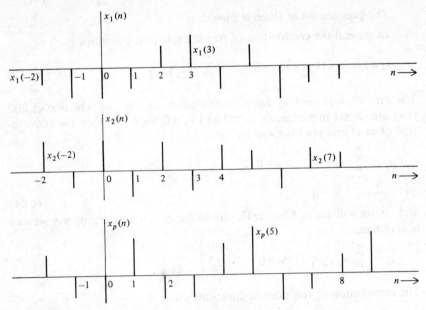

Figure 6.11 Some typical members of a random process.

consider in this section some very artificial discrete processes where the probability measure is evident.

Using the random process $x(n)$ of Figure 6.11 as an example, some nomenclature of discrete random processes will now be defined.

The random process $x(n)$ is said to consist of an infinite number of members $x_1(n), x_2(n), \ldots, x_l(n)$, and so on.

We define the random variable X_k associated with sampling the process at $t = k$, where X_k assumes the value $x_p(k)$ if the member $x_p(n)$ is randomly chosen. X_k is the random variable that describes with a one-to-one mapping the experiment of randomly choosing a member waveform and observing its value at $n = k$. Similarly, we define X_{-2}, X_0, X_4 or X_k for all integer k; $-\infty < k < \infty$. The lowercase notation x_k and $x(k)$ are also used for this random variable, and after a while the $x(k)$ notation will be the most common. To avoid initial confusion between the sampled values of deterministic discrete waveforms and random variables we will use X_k. Later we will use $x(k)$ and be expected to know from the context whether $x(n)$ is a discrete waveform, a random process, or a random variable.

Associated with any random variable X_k we will have a density function $f_{X_k}(\alpha)$, and we will denote the first-order ensemble statistics associated with it as $\overline{X_k}, \overline{X_k^2}, \sigma_{X_k}^2$, etc.

Associated with any two random variables X_n and X_k there will be an accompanying joint density (or mass) function $f_{X_n X_k}(\alpha, \beta)$

$p_{X_n X_k}(\alpha_i, \beta_j)$, from which any second-order statistic involving them may be found. Some of the more important of these are

$$R_{xx}(n, m) \triangleq \overline{X_n X_m}$$

called the autocorrelation function of the process and

$$L_{xx}(n, m) \triangleq \overline{(X_n - \overline{X}_n)(X_m - \overline{X}_m)}$$

called the covariance function of the process.

Associated with any n random variables $X_1, X_2, \ldots, X_n$ there exists an nth-order joint density function

$$f_{X_1 X_2 \ldots X_n}(\alpha_1, \alpha_2, \ldots, \alpha_n) = f_{\mathbf{X}}(\alpha)$$

The general definition of an ensemble "statistic" for a discrete random process is analogous to that used for a continuous process. For example, some ensemble statistics are

$$\overline{x^2(n)} \triangleq \lim_{N \to \infty} \frac{\sum_{p=1}^{N} x_p^2(n)}{N}$$

$$\overline{x(3)x(5)} \triangleq \lim_{N \to \infty} \frac{\sum_{p=1}^{N} x_p(3)x_p(5)}{N}$$

$$\overline{g\{x(n)\}} \triangleq \lim_{N \to \infty} \frac{\sum_{p=1}^{N} g[x_p(n)]}{N}$$

where g indicates some general function. If we know $f_{X_n}(\alpha)$ and $f_{X_3 X_5}(\alpha, \beta)$, the analytic formulas for these three statistics are given by

$$\overline{x^2(n)} = \int_{-\infty}^{\infty} \alpha^2 f_{X_n}(\alpha) \, d\alpha$$

$$\overline{x(3)x(5)} = \int_{-\infty}^{\infty} \int_{-\infty}^{\infty} \alpha\beta f_{X_3 X_5}(\alpha, \beta) \, d\alpha \, d\beta$$

$$\overline{g\{x(n)\}} = \int_{-\infty}^{\infty} g(\alpha) f_{X_n}(\alpha) \, d\alpha$$

From what we have said, this last statistic could be denoted $\overline{g[x(n)]}$ or $\overline{g(X_n)}$ or $\overline{g(x_n)}$; instead of the bar we often use E, as in $E(x_n)$.

SPECIFICATION OF DISCRETE RANDOM PROCESSES

A discrete random process is usually denoted as

$$x(n) = \sum_{k=-\infty}^{\infty} x_k \delta(n-k)$$

where the x_k's are random variables, and a rule is given for obtaining their joint density function $f_{\mathbf{X}}(\boldsymbol{\alpha})$. Here the vector $\mathbf{X}$ may be of any dimension $(X_{-l},\dots,X_{-2}, X_{-1}, X_0, X_1, X_2,\dots,X_m)$ and $\boldsymbol{\alpha}$ is of the same dimension $(\alpha_{-l},\dots,\alpha_{-2},\alpha_{-1},\alpha_0,\alpha_1,\dots,\alpha_m)$. Alternatively, a simple word description may be equivalent to giving the joint density function.

Example 6.10

Try to visualize and sketch typical members of the following:

(a) The random process

$$x(n) = \left[\sum_{k=-\infty}^{\infty} d_k \delta(n-k) \right] - \tfrac{1}{2}^{n} u(n)$$

where the d_k's are independent random variables with

$$f_{d_k}(\alpha) = 1 \qquad -\tfrac{1}{2} < \alpha < \tfrac{1}{2}$$

$$= 0 \qquad \text{otherwise}$$

(b) The random process

$$x(n) = \sum_{k=-\infty}^{\infty} x_k \delta(n-k)$$

where x_k is a discrete random variable that takes on the values 0 and 1 with equal probabilities, and the joint mass function for x_i and x_{i+1} for any i is as shown.

X_i \ X_{i+1}	0	1
0	$\frac{1}{3}$	$\frac{1}{6}$
1	$\frac{1}{6}$	$\frac{1}{3}$

(c) The random process $y(n)$, where $y_k = \tfrac{1}{2}(x_k + x_{k-1})$ given that

$$x(n) = \sum_k \left[\delta(n-k) + A_k \delta(n-k) \right]$$

Here the A's are independent random variables with density functions

$$f_{A_i}(\alpha) = \tfrac{1}{2}[\delta(\alpha - \tfrac{1}{2}) + \delta(\alpha + \tfrac{1}{2})] \qquad \text{for all } i$$

Intuitively, how does $y(n)$ compare to $x(n)$?

SOLUTION

(a) On inspection, $x(n)$ is seen to consist of two parts, $f(n) = \sum \tfrac{1}{2}{}^n \delta(n - k)$, which is a deterministic discrete function, and $d(n) = \sum d_k \delta(n - k)$, which is a discrete random process. A plot of typical members of $f(n)$, $d(n)$, and $x(n)$ is shown in Figure 6.12a. The values of the d_k's in a typical member of $d(n)$ could be obtained by spinning a wheel of fortune (or generating a random number) uniformly distributed from $-\tfrac{1}{2}$ to $+\tfrac{1}{2}$. As n gets large, $f_k \to 0$, and therefore $x(n) = d(n)$ for large values of n.

(b) To visualize

$$x(n) = \sum_k x_k \delta(n - k)$$

where the joint mass function for x_i and x_{i+1} was as given, we first calculate the mass and conditional mass functions

$$p_{x_i}(0) = \tfrac{1}{2} \quad \text{and} \quad p_{x_i}(1) = \tfrac{1}{2} \qquad \text{for any } i$$

$$p_{x_{i+1}}[0/(x_i = 0)] = p_{x_{i+1}}[1/(x_i = 1)] = \tfrac{2}{3}$$

and

$$p_{x_{i+1}}[1/(x_i = 0)] = p_{x_{i+1}}[0/(x_i = 1)] = \tfrac{1}{3}$$

Now we could generate a typical section of a member waveform as follows: "Toss a coin to find x_{-22}, say. If a head is obtained, let $x_{-22} = 1$; otherwise, let $x_{-22} = 0$. Next draw a ducat from an urn with 2 red and 1 black ducats. If a red ducat is drawn, let $x_{-21} = x_{-22}$; otherwise, let x_{-21} take on the other value. Similarly, x_{-20}, x_{-19}, and so on may be generated. A typical section of a member waveform is shown sketched in Figure 6.12b.

(c) The waveform $y(n)$ may be considered as the output of an averager defined as $y_k = \tfrac{1}{2}(x_k + x_{k-1})$ or in system notation as

$$y_n = \sum_{\text{all } p} x_p h_{n-p}$$

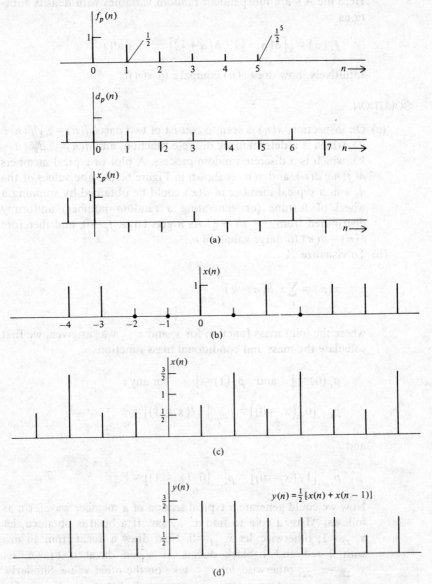

Figure 6.12 Member waveforms of the random processes of Example 6.10.

where $h(n)$ called the system pulse response is defined as

$$h_k = \tfrac{1}{2} \qquad k = 0$$

$$= \tfrac{1}{2} \qquad k = 1$$

$$= 0 \qquad \text{otherwise}$$

and we can see that $y(n)$ is $x(n)$ convolved with $h(n)$.

In order to visualize a member of $y(n)$, a fairly typical member of $x(n)$ as obtained by coin tossing is shown in Figure 6.12c, and then y_k is found for a corresponding member of $y(n)$, as is shown in Figure 6.12d. It can be seen that $y(n)$ is much more similar to $\Sigma\delta(n-k)$ the deterministic part of $x(n)$ than is $x(n)$. The averaging has tended to reduce the noise fluctuations in $x(n)$. A comment should be made about the notation $y(n) = \tfrac{1}{2}(x(n) + x(n-1))$ since it could have two interpretations. The sum of two random processes $z(n) = x(n) + y(n)$ in general means a member of $z(n)$ is formed by adding randomly chosen members of $x(n)$ and $y(n)$. However, when we state that $y(n) = x(n) + x(n-1)$ we mean that a general member of $y(n)$, say $y_p(n)$, is formed by adding $x_p(n)$ to $x_p(n-1)$. This is also implied by writing $y(n) = h(n) * x(n)$ for a deterministic $h(n)$.

CLASSIFICATION OF DISCRETE RANDOM PROCESSES
Discrete random processes may be divided into similar categories as were continuous random processes in Section 6.3.

NONSTATIONARY DISCRETE RANDOM PROCESSES
A random process is nonstationary if

$$f_{x_l}(\alpha) \neq f_{x_k}(\alpha) \tag{6.25}$$

for any l and k. The random variables x_l and x_k describe the amplitudes of the process at $t = l$ and k, and we could also use the notation $x(l)$ and $x(k)$ or X_l or X_k. Statistics associated with a discrete random process are defined in general for nonstationary processes and simplified when different degrees of stationariness are involved. Some of the more general first-order statistics are the mean function $m_x(n)$ defined as

$$m_x(n) = \bar{x}_n \tag{6.26}$$

and it is a function of n, the mean square value $\overline{x_n^2}$ defined as

$$\overline{x_n^2} \triangleq \int_{-\infty}^{\infty} \alpha^2 f_{x_n}(\alpha)\, d\alpha \tag{6.27}$$

and the variance

$$\sigma_{x_n}^2 = \overline{\left(x_n - \overline{x_n}\right)^2} \tag{6.28}$$

Some of the more general second-order statistics are

$$R_{xx}(n, l) \triangleq \overline{x_n x_l}$$

$$= \int_{-\infty}^{\infty} \int_{-\infty}^{\infty} \alpha \beta f_{x_n x_l}(\alpha, \beta) \, d\alpha \, d\beta \tag{6.29}$$

and

$$L_{xx}(n, l) = \overline{(x_n - \bar{x}_n)(x_l - \bar{x}_l)}$$

$R_{xx}(n, l)$, which is a function of both n and l, is called the autocorrelation function of the process, and $L_{xx}(n, l)$, also denoted by $\phi_{xx}(n, l)$, is called the covariance function of the process. It can easily be shown that

$$L_{xx}(n, l) = R_{xx}(n, l) - \bar{x}_n \bar{x}_l$$

FIRST-ORDER STATIONARY DISCRETE RANDOM PROCESS

The random process $x(n)$ is stationary of order one if all the first-order statistics associated with the process at $n = k$ are independent of k. This implies that

$$f_{x_l}(\alpha) = f_{x_k}(\alpha) \tag{6.30}$$

for all l and k. Some of the more important first-order statistics and their notation are

$$\overline{x(n)} = \int_{-\infty}^{\infty} \alpha f_{x_n}(\alpha) \, d\alpha = \bar{x} \tag{6.31}$$

$$\overline{x^2(n)} = \int_{-\infty}^{\infty} \alpha^2 f_{x_n}(\alpha) \, d\alpha = \overline{x^2} \tag{6.32}$$

$$\sigma_x^2 = \int_{-\infty}^{\infty} (\alpha - \bar{x})^2 f_{x_n}(\alpha) \, d\alpha \tag{6.33}$$

$$= \overline{x^2} - \bar{x}^2$$

These are called the mean, mean square value, and variance of the process, respectively.

SECOND-ORDER STATIONARY DISCRETE RANDOM PROCESSES

The random process $x(n)$ is stationary of order two if all second-order statistics associated with the random variables x_p and x_l do not depend on p and l but at most on $|p - l|$. This implies, for any n and l, that

$$f_{x_n x_{n+k}}(\alpha, \beta) \triangleq f_{x_l x_{l+k}}(\alpha, \beta) \tag{6.34}$$

Some of the more important second-order statistics are

$$R_{xx}(k) \triangleq \overline{x_n x_{n+l}} \tag{6.35}$$

called the autocorrelation function and

$$L_{xx}(k) \triangleq \overline{(x_n - \bar{x}_n)(x_{n+k} - \bar{x}_{n+k})} \tag{6.36}$$

called the covariance function.

WIDE-SENSE STATIONARY RANDOM PROCESSES
The random process $x(n)$ is said to be wide-sense stationary if the following conditions hold:

1. $\overline{x(n)} = \bar{x}$ a constant (often 0).
2. $E[(X_n - \bar{X})^2] < \infty$.
3. $E[X_n X_m]$ depends at most on $|n - m|$ for all n and m.

Example 6.11

Consider the two random processes $x(n)$ and $y(n)$, defined as

$$x(n) = \sum_{k=-\infty}^{\infty} x_k \delta(n-k)$$

where the x_k's are independent with uniform density function $f_{x_k}(\alpha) = \frac{1}{2}$; $-1 < \alpha < 1$, and

$$y(n) = \sum_{k=-\infty}^{\infty} y_k \delta(n-k)$$

where for k odd the y's are independent, each with a uniform density function $f_{y_k}(\alpha) = \frac{1}{2}$; $-1 < \alpha < 1$, and for k even the y's are independent, with

$$f_{y_k}(\alpha) = \sqrt{\frac{3}{2\pi}} \, e^{-3\alpha^2/2}$$

Classify these two processes.

SOLUTION
The process $x(n)$ is second-order stationary as

$$f_{x_n x_l}(\alpha, \beta) = f_{x_n}(\alpha) f_{x_l}(\beta)$$

$$= \frac{1}{4} \qquad -1 < \alpha < 1$$

which is independent of n and l. Some first-order statistics are $\bar{x} = 0$, $\overline{x^2} = \frac{1}{3}$, and $\sigma_x^2 = \frac{1}{3}$. Some second-order statistics are $R_{xx}(k) = \frac{1}{3}\delta(k)$ and $L_{xx}(k) = \frac{1}{3}\delta(k)$, since $\bar{x} = 0$.

The process $y(n)$ is obviously not first-order stationary since, for example,

$$f_{y_1}(\alpha) = \frac{1}{2} \qquad -1 < \alpha < 1$$

$$= 0 \qquad \text{otherwise}$$

and

$$f_{y_2}(\alpha) = \sqrt{\frac{3}{2\pi}} \, e^{-3\alpha^2/2}$$

$$f_{y_1}(\alpha) \neq f_{y_2}(\alpha)$$

However, if we evaluate statistics for this process we find that its first-order statistics for n odd, are

$$\overline{y(n_1)} = \overline{y(n_2)} = 0 \qquad \overline{y^2(n_1)} = \overline{y^2(n_2)} = \tfrac{1}{3} \qquad \sigma_y^2 = \tfrac{1}{2}$$

We obtain the same results for n even, since the density function is gaussian with mean 0 and variance $\tfrac{1}{3}$. (Would higher first-order statistics such as $\overline{y^3(n)}$ be the same for odd and even n?) The autocorrelation function for this process is $R_{yy}(n, k) = \overline{y_n y_k}$. On evaluation we find that

$$R_{yy}(0) = \overline{y^2(n)} = \tfrac{1}{3} \qquad n = k$$

$$R_{yy}(n, k) = \overline{y_n y_k} = 0 \qquad n \neq k$$

since y_n and y_k are independent. Therefore, although $y(n)$ is not first-order stationary, we have shown that it is wide-sense stationary since $R_{yy}(n, k)$ does not depend on n and k or even on the magnitude of their difference $|n - k|$. The reader should sketch part of a typical member of $y(n)$.

ERGODIC DISCRETE RANDOM PROCESSES

A discrete random process is said to be ergodic if all corresponding ensemble and time averages are identical. We will denote time averages based on any member $x_p(n)$ by a wavy overbar. Some first-order time averages are

$$\widetilde{x(n)} = \lim_{n \to \infty} \frac{1}{2N+1} \sum_{k=-N}^{N} x_{pk}$$

which is the mean value of the process. We usually delete the p and just write

$$\widetilde{x(n)} = \frac{1}{2N+1} \sum_{k=-N}^{N} x_k$$

The mean square value of the process on a time average basis is

$$\widetilde{x^2(n)} = \lim_{N \to \infty} \frac{1}{2N+1} \sum_{k=-N}^{N} x_k^2$$

and the variance is given by

$$\sigma_x^2 = \lim_{N \to \infty} \frac{1}{2N+1} \left[\sum_{k=-N}^{N} (x_k - \bar{x})^2 \right]$$

The time average autocorrelation and covariance functions are

$$R_{xx}(k) = \lim_{N \to \infty} \left[\frac{1}{2N+1} \left(\sum_{n=-N}^{N} x_{p,n} x_{p,n+k} \right) \right]$$

and

$$L_{xx}(k) = \lim_{N \to \infty} \left[\frac{1}{2N+1} \left(\sum_{n=-N}^{N} (x_{p,n} - \bar{x})(x_{p,n+k} - \bar{x}) \right) \right]$$

Table 6.2 A SUMMARY OF NOTATION FOR DISCRETE PROCESSES

NOTATION	MEANING
$x(n)$	A discrete random process or ensemble of discrete waveforms. Also a random variable sampling $x(n)$ at $t = n$. Sometimes a specific process member.
$x_p(n)$	The pth member of $x(n)$.
$x_p(k)$	The value of the pth member at $n = k$.
$\sum_{\text{all } k} x_k \delta(n-k)$	The most common designation for $x(n)$, where the joint density function is given for the x's.
$f_{X_k}(\alpha)$ or $f_{x_k}(\alpha)$	The density function for the random variable X_k or x_k that describes the values taken on by the process at $n = k$.
$\bar{X}_n,\ \overline{X_n^2},\ \sigma_{x_n}^2$	The mean, mean square value, and variance of a random process at time $t = n$. The notations $\overline{x(n)}$ or $\bar{x}_n$, and $\overline{x^2(n)}$ or $\overline{x_n^2}$, are also used.
$f_{x_n x_k}(\alpha, \beta)$	The joint density function associated with the process at times n and k. If the process is second-order stationary, it depends only on $\|n - k\|$ and not n and k.
$R_{xx}(n, l)$, where $\|n - l\| = k$	The autocorrelation function $\overline{x_n x_l}$, which depends only on k for a wide-sense stationary process; also denoted $R_{xx}(k)$ if stationary. Is given by $$\frac{1}{2N+1} \sum_{n=-N}^{N} x_p(n) x_p(n+k)$$ as $n \to \alpha$ if ergodic.
$L_{xx}(n, l)$, where $\|n - l\| = k$	The covariance function $\overline{(x_n - \bar{x}_n)(x_{n+k} - \bar{x}_{n+k})}$, which depends only on k if at least wide-sense stationary.
$\tilde{x},\quad \overline{x^2}$ $\overline{x(n)x(n+k)}$	Notation for time averages of an ergodic process.

where it is understood that we are referring to any specific member $x_p(n)$. At this stage notation is becoming somewhat burdensome, and we should keep in mind that it must be a fluent part of our vocabulary and not feel ashamed of having to continually go over it and write it out repeatedly.

If $x(n)$ is ergodic, some equivalencies with ensemble averages are

$$\widetilde{x(n)} = \overline{x_k} \qquad \text{for any } k$$

$$\widetilde{x(n)x(n+k)} = \overline{x_n x_{n+k}} \qquad \text{for any } k$$

In Table 6.2 is shown a summary of the definitions and terminology most frequently used for discrete random processes.

SUMMARY

This chapter has introduced the important concept of a random process, which is an ensemble of continuous or discrete waveforms, where there is a probability measure associated with the members. The chapter is essentially an introduction to nomenclature.

Continuous random processes are normally specified by an equation involving random variables from which it is possible to visualize the ensemble members that are generated subject to obeying the joint density function of the random variables. For example,

$$x(t) = A\cos(\omega t + \phi)$$

where A and ϕ are random variables with a joint density function $f_A(\alpha, \beta) = 1/4\pi$; $8 < \alpha < 10$, $0 < \beta \leq 2\pi$, designates a set of sinusoidal waveforms with amplitudes uniformly distributed between 8 and 10 and phases uniformly distributed between 0 and 2π. Discrete random processes are usually described by an equation of the form

$$x(n) = \sum_{k=-\infty}^{\infty} x_k \delta(n-k)$$

where the x's are random variables and a rule is given corresponding to their joint density function.

We define a number of important random variables $X_1, X_2, \ldots, X_n$ for any random process whether continuous or discrete. For a continuous random process, X_i or $x(t_i)$ is the random variable that involves the random phenomenon of choosing a member $x_p(t)$ of the process and observing its value at $t = t_i$. The value $x_p(t_i)$ is assigned to X_i on one trial of the phenomenon. From many trials we can estimate $f_{X_i}(\alpha)$, the density function of X_i. Fairly simple processes were considered in this chapter, from which it was possible to derive $f_{X_i}(\alpha)$ based on symmetry and also from the joint density function for the random variables in terms of which $x(t)$ was defined. Deriving the joint density function for the random variables X_i and X_j involves visualizing the results of sampling the process members at times

t_i and t_j. We then estimated $f_{X_i}(\alpha)$, $f_{X_j}[\beta/(X_i = \alpha)]$ and $f_{X_iX_j}(\alpha, \beta)$ for the process. For the simple analytical processes of Sections 6.2 and 6.3 the joint density functions for $x(t_i)$ and $x(t_j)$ were derived in this fashion.

For discrete random processes we defined x_k as the random variable that describes the amplitudes of the process at $t = k$. The process is usually specified as $x(n) = \Sigma x_k \delta(n - k)$, for all k, where a formula is given for the joint density function of the x's. With this formulation the problem of finding $f_{x_n}(\alpha)$ is trivial. Often, however, we will be interested in finding $f_{y_k}(\alpha)$ where $y(n)$ is the output of linear system defined by $y_n = \Sigma x_p h_{n-p}$, for all p. Finding $f_{X_nX_k}(\alpha, \beta)$ may be trivial, but finding $f_{y_ny_k}(\alpha, \beta)$, where $y_n = \Sigma x_p h_{n-p}$, for all p, will pose challenging problems later in the text.

Associated with a random process are many important first- and second-order statistics which, in lieu of knowing $f_X(\alpha)$ or $f_{X_iX_j}(\alpha, \beta)$, will provide much information about the variation in the amplitude values of the process at a specific time and in the rapidity and variation of values from time t_i to another time t_j. Chapter 7 on the autocorrelation function $R_{xx}(\tau)$ or $R_{xx}(k)$ and Chapter 8 on the power spectral density $S_{xx}(\omega)$, which is either the Fourier transform of $R_{xx}(\tau)$ or the discrete Fourier transform of $R_{xx}(k)$, demonstrate how these functions convey deep information about the "randomness" of a process from a time or frequency point of view. Rarely do engineers have to consider other than first- or second-order statistics.

Random processes are classified as being nonstationary, first-order stationary, wide-sense stationary, second-order stationary, strictly stationary of order n, or ergodic. Important statistics are defined in general for a nonstationary process and simplified as the classification proceeds to a higher order. These definitions were summarized in Tables 6.1 and 6.2.

PROBLEMS

1. Consider the random phenomenon of choosing a ball from an urn containing 3 red balls, 2 blue balls, and 1 black ball. Let us define the following random process. If a red ball is drawn from the urn, we generate the member waveform $x(t) = 1$; if a blue ball is drawn, we generate the member waveform $x(t) = t$; and if a black ball is drawn, we generate $x(t) = t^2$. For this random process obtain the following:
   (a) A sketch indicating different members of $x(t)$ with an indication of their frequency of occurrence.
   (b) The different possible values taken on by different process members at $t = 0$, $t = 1$, and $t = 2$.
   (c) The mass functions of the random variables at $t = 0$, $t = 1$, $t = 2$, and $t = 3$.
   (d) The statistics $\overline{x(2)}$, $\overline{x^2(1)}$, and $\sigma_{x(2)}^2$.
   (e) The joint mass function $p_{x(2)x(3)}(\alpha, \beta)$ and $R_{xx}(2, 3) \triangleq \overline{x(2)x(3)}$.

2. Consider the random phenomenon of spinning a wheel of fortune that is finely calibrated from 0 to 2. Let us define the random process $x(t)$. If the value α is obtained on a trial of the phenomenon then $x(t)=3-\alpha$.

   (a) For this random process, sketch typical members of $x(t)$ from which you can visualize how the process looks.

   (b) Predict the possible values taken on by the random variables $x(1)=X_1$ and $x(2)=X_2$.

   (c) Find the density functions $f_{X_1}(\alpha)$ and $f_{X_2}(\alpha)$.

   (d) Find the conditional density function $f_{x(2)}\{\beta/[x(1)=2.8]\}$.

3. Visualize the following random processes by sketching one, two, or as many members as necessary.

   (a) $x(t)=Atu(t)$, where $u(t)$ is the unit step function defined as

   $$u(t)=0 \qquad t<0$$
   $$=1 \qquad t\geq 0$$

   and A is a random variable defined by

   $$f_A(\alpha)=1 \qquad 0<\alpha<1$$
   $$=0 \qquad \text{otherwise}$$

   (b) $x(t)$ is a pulse-type process with memory defined by

   $$x(t)=\sum_{n=-\infty}^{\infty} A_n g(t-n+\phi) \qquad n \text{ an integer}$$

   where the A's and ϕ are random variables with the following density functions:

   $$f_{A_i}(\alpha)=\tfrac{1}{2}\delta(\alpha)+\tfrac{1}{2}\delta(\alpha-1)$$
   $$f_{A_{i+1}}[\alpha/(A_i=1)]=\tfrac{1}{4}\delta(\alpha)+\tfrac{3}{4}\delta(\alpha-1)$$
   $$f_{A_{i+1}}[\alpha/(A_i=0)]=\tfrac{3}{4}\delta(\alpha)+\tfrac{1}{4}\delta(\alpha-1)$$

   and ϕ is independent of all the A's, with

   $$f_\phi(\alpha)=1 \qquad 0<\phi<1$$

   Now restate this process in words.

   (c) $x(t)=\cos(\omega t+\phi)$, where ω and ϕ are independent random variables with density functions

   $$f_\omega(\alpha)=\tfrac{1}{4} \qquad 8<\omega<12$$
   $$=0 \qquad \text{otherwise}$$

   and

   $$f_\phi(\alpha)=\frac{1}{2\pi} \qquad 0<\alpha<2\pi$$

   (d) $x(t)=A\cos\omega t$, with $f_{A\omega}(\alpha,\beta)=\tfrac{1}{8}$; $0<\alpha<2$, $8<\beta<12$.

(e) $x(t)$ is a "random telegraph" type process composed of pulses of heights $+1$ and -1, respectively. The number of transitions of the t axis in a time 2 is given by

$$P(k \text{ transitions}) = \frac{(4)^k e^{-4}}{k!}$$

You are expected to construct one member where, for each section of 2 or if more convenient a multiple of 2 units, you will decide how many transitions to show. For example, our thumbnail sketch shows six transitions in a time T.

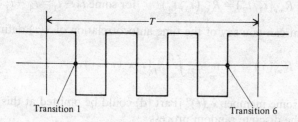

Transition 1   Transition 6

Note: In Problems 4 through 7 we are asked to classify processes as being:
  (a) Nonstationary, by means of showing that an ensemble statistic is different at two times or that $f_{X_1}(\alpha) \neq f_{X_2}(\alpha)$.
  (b) First-order stationary, by using the theory of Chapter 4 on the density function for a function of a random variable and showing that $f_X(\alpha)$ does not depend on t.
  (c) Second-order stationary, by using the theory of Chapter 5 and showing that

$$f_{x(t_1)x(t_1+\tau)}(\alpha, \beta) = f_{x(t_2)x(t_2+\tau)}(\alpha, \beta)$$

  is dependent only on τ.
  (d) Strictly stationary, essentially by intuition plus (c).
  (e) Ergodic, by intuition and the inability to envision a time average that is different than a corresponding ensemble average.
For each process work through the stages (a) through (e) above.
  4. Classify the process of Problem 3a.
  5. Classify the process of Problem 3c.
  6. Classify the process of Problem 3d.
  7. Classify the process of Problem 3e.
  8. Let

$$g(t) = 1 \qquad -\frac{T_0}{2} < t < \frac{T_0}{2}$$

$$= 0 \qquad \text{otherwise}$$

and let A_n's be independent random variables that assume the values

± 1 with equal probability. The random process $x(t)$ is defined by

$$x(t) = \sum_{n=-\infty}^{\infty} A_n g(t - nT_0) \qquad n \text{ an integer}$$

(a) Sketch a typical member of $x(t)$.
(b) Is $x(t)$ first-order stationary?
(c) Show that $x(t)$ is not wide-sense stationary by choosing t_1, t_2, t_3, t_4 and showing that

$$R_{xx}(t_1, t_2) \neq R_{xx}(t_3, t_4) \qquad \text{for some } t_2 - t_1 = t_4 - t_3$$

(d) What can you say of the time autocorrelation of $x(t)$ defined as

$$R_{xx}(\tau) \triangleq \lim_{T \to \infty} \frac{1}{2T} \int_{-T}^{T} x_p(t) x_p(t - \tau)\, dt$$

for some member $x_p(t)$? [Part (d) could be omitted at this time.]

9. Given the discrete random process

$$x(n) = \sum_k A_k \delta(n - k)$$

where the A_k's are found from the joint mass function,

A_k \ $A_k + 1$	0	1
0	$\frac{1}{3}$	$\frac{1}{6}$
1	$\frac{1}{6}$	$\frac{1}{3}$

(a) Plot typical members of
  (1) $x(n)$
  (2) $y(n) = 0.5[x(n) + x(n-1)]$
  (3) $z(n) = 0.5[x(n) - x(n-1)]$
  (4) $y(n)y(n+1)$
  (5) $z(n)z(n+1)$
(b) Evaluate
  (1) $\overline{x(n)}, \overline{x^2(n)}, \sigma_x^2$
  (2) $\overline{y(n)}, \overline{y^2(n)}, \sigma_y^2$
  (3) $\overline{z(n)}, \overline{z^2(n)}, \sigma_z^2$
  (4) $\overline{y(n)y(n+1)}, \overline{z(n)z(n+1)}$

10. Consider

$$x(n) = \sum_k x_k \delta(n - k)$$

where the joint mass function for all k is

x_k \ x_{k+1}	0	1
0	0.2	0.1
1	0.1	0.6

(a) Sketch a typical member of $x(n)$.
(b) Find $p_{x_k}(\alpha_i)$ and $P(X_4 X_5 = 1/X_3 = 0)$.
(c) Is the process stationary?

11. Consider

$$x(n) = \sum_k x(k)\delta(n-k)$$

where the joint mass function for the random variables is

(a)

$x(k)$ \ $x(k+1)$	0	1
0	0.6	0.1
1	0.1	0.2

(b)

$x(k)$ \ $x(k+1)$	0	1
0	0.1	0.2
1	0.6	0.1

Prove that the processes given in (a) and (b) are stationary.

Chapter 7
Ensemble and Time Average Autocorrelation Functions

7.1 AUTOCORRELATION FUNCTIONS FOR CONTINUOUS RANDOM PROCESSES

Chapter 6 introduced the terminology of random processes. Both continuous and discrete processes were defined, specified, and classified. In particular, the classification of random processes was difficult, but it still was restricted to classical problems in random variable theory. Some of these problems and what their solution involves are as follows:

Proving that a random process is first-order stationary entails finding the density function $f_{x(t)}(\alpha)$ and concluding that it is independent of time t. This is an extension of the problem from Chapter 4: "Given $Y = g(X)$, where the density function of X is known, find $f_Y(\beta)$."

Proving that a random process is second-order stationary entails finding the joint density function $f_{X_1 X_2}(\alpha, \beta) = f_{X_1}(\alpha) f_{X_2}[\beta/(X_1 = \alpha)]$, where $X_1 = x(t_1)$ and $X_2 = x(t_2)$, and concluding that it is not dependent on both t_1 and t_2 but only at most on $|t_2 - t_1|$.

Proving that a random process is ergodic is handled in an intuitive manner, and confidence is developed by comparing for artificial processes ensemble averages with time averages based on a specific member.

In this chapter we are embarking on second-moment theory or spectral analysis of linear systems, and it is imperative that we develop a strong physical understanding of the two great second-moment statistics associated with an at least wide-sense stationary random process—the autocorrelation function—and the power spectral density. Sections 7.1 to 7.3 will be devoted to the continuous autocorrelation function $R_{xx}(\tau)$ and its close relatives, the cross-correlation functions $R_{xy}(\tau)$ and $R_{yx}(\tau)$. As a result of evaluating correlation functions an appreciation will be developed as to how $R_{xx}(\tau)$ "measures" or portrays the randomness of a process and how $R_{xy}(\tau)$ compares two different random processes.

The autocorrelation function of a random process $x(t)$ is defined as

$$R_{xx}(t_1, t_2) \triangleq \overline{x(t_1)x(t_2)}$$

$$= \int_{-\infty}^{\infty} \int_{-\infty}^{\infty} \alpha\beta f_{X_1 X_2}(\alpha, \beta)\, d\alpha\, d\beta \tag{7.1}$$

and the covariance function is defined as

$$L_{xx}(t_1, t_2) \triangleq \overline{\left[x(t_1) - \overline{x(t_1)} \right]\left[x(t_2) - \overline{x(t_2)} \right]}$$

$$= \int_{-\infty}^{\infty} \int_{-\infty}^{\infty} \left[\alpha - \overline{x(t_1)} \right]\left[\beta - \overline{x(t_2)} \right] f_{X_1 X_2}(\alpha, \beta)\, d\alpha\, d\beta \tag{7.2}$$

For a process that is at least wide-sense stationary,

$$\overline{x(t_1)x(t_2)} = \overline{x(t)x(t \pm \tau)} = R_{xx}(\tau) \tag{7.3}$$

and

$$\overline{[x(t) - \bar{x}][x(t \pm \tau) - \bar{x}]} = L_{xx}(\tau) \tag{7.4}$$

where $|t_2 - t_1| = \tau$ and $\overline{x(t)} = \bar{x}$, the mean value of the process. In addition,

$$L_{xx}(\tau) = R_{xx}(\tau) - \bar{x}^2 \tag{7.5}$$

If $x(t)$ is an ergodic random process, then $R_{xx}(\tau)$ may be evaluated from any specific time member, say $x_p(t)$, as

$$R_{xx}(\tau) = \widetilde{x(t)x(t - \tau)}$$

$$= \lim_{T \to \infty} \frac{1}{2T} \int_{-T}^{T} x_p(t)x_p(t - \tau)\, dt \tag{7.6}$$

where for time averages it is more conventional to use $x(t)x(t - \tau)$ instead of $x(t)x(t + \tau)$, although they both give the same result. We will now embark on evaluating the autocorrelation function for a number of different processes. In the examples to follow we will consider processes that are at least second-order stationary and evaluate $R_{xx}(\tau)$ in the following two ways:

1. If the process is stationary, the joint density function or joint mass function associated with the random variables $x(t)$ and $x(t + \tau)$,

denoted by X_1 and X_2, is derived as

$$f_{X_1 X_2}(\alpha, \beta) \quad \text{or} \quad p_{X_1 X_2}(\alpha_i, \beta_j)$$

respectively, and $R_{xx}(\tau)$ is next evaluated as

$$R_{xx}(\tau) = \int_{-\infty}^{\infty} \left\{ \int_{-\infty}^{\infty} \alpha \beta f_{X_1}(\alpha) f_{X_2}[\beta/(X_1 = \alpha)] \, d\beta \right\} d\alpha$$

or

$$R_{xx}(\tau) = \sum_{\text{all } i} \sum_{\text{all } j} \alpha_i \beta_j p_{X_1}(\alpha_i) p_{X_2}[\beta_j/(X_1 = \alpha_i)]$$

2. If the process is ergodic, the same procedure as in method 1 may be employed or, alternatively, $R_{xx}(\tau)$ may be found as a time average for one member waveform as

$$R_{xx}(\tau) = \overline{x(t)x(t-\tau)}$$

$$= \lim_{T \to \infty} \frac{1}{2T} \int_{-T}^{T} x_p(t) x_p(t-\tau) \, dt$$

where $x_p(t)$ is any specific member of $x(t)$.

The problems to follow occupy a large portion of text material and throughout their solution, the student should keep in mind the suggested methods 1 and 2 for evaluating autocorrelation functions, try not to be confused with detail, and try to appreciate the necessity for a good background in discrete probability theory. On completion of the problems we will summarize the solutions and develop concrete conclusions about how the autocorrelation function indicates the randomness of a process. Example 7.1 will include five different parts and will be a key in formulating our philosophy about randomness. However, it will take much persistence and perhaps up to 10 hours of study to appreciate it.

Example 7.1

Consider the five binary random processes given in parts (a) through (e). First, sketch a typical member or sufficient members of each of the five processes so that we can visualize them. Second, evaluate $R_{xx}(\tau)$ for each of the five processes and plot all your results on a single graph and then discuss how the autocorrelation function measured randomness.

(a) $x(t) = \sum_{n=-\infty}^{\infty} A_n g(t - nb + \phi)$

where the deterministic function $g(t)$ is defined as

$$g(t) = 1 \qquad 0 < t < b$$

$$= 0 \qquad \text{otherwise}$$

and the A's and ϕ are independent random variables with density functions

$$f_{A_i}(\alpha) = \tfrac{1}{2}\delta(\alpha-1) + \tfrac{1}{2}\delta(\alpha)$$

and

$$f_\phi(\alpha) = \frac{1}{b} \qquad 0 < \alpha < b$$

We could also say that the joint mass function for the A's is as given in Figure 7.1a.

(b) $\qquad x(t) = \sum_{n=-\infty}^{\infty} A_n g(t - nb + \phi)$

where $g(t)$ and ϕ are the same as for part (a). However, the A's are not independent, and their joint mass function is as given in Figure 7.1b.

(c) $x(t)$ is the same as in (a) and (b), except that the joint mass function for the A's is as given in Figure 7.1c.

(d) $x(t)$ is an artificial process that is the same as in (a) and (b), except that the joint mass function is as given in Figure 7.1d.

(e) $x(t)$ is another artificial process that is the same as in (a) and (b), except that the joint mass function is as given in Figure 7.1e.

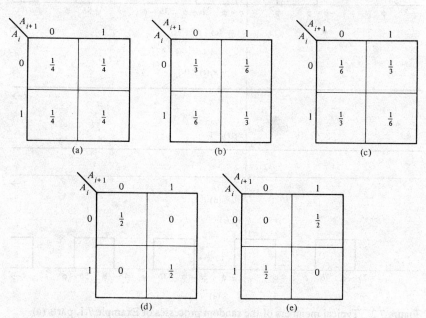

Figure 7.1 The joint mass functions of the random variables defining the processes of Example 7.1, parts (a) through (e), respectively.

SOLUTION

Figure 7.2 shows a typical member for each of parts (a), (b), (c), and (e), whereas part (d) requires two member waveforms. These plots will now be commented on in order.

(a) On consideration, the given definition could be restated in words as, "A source generates a pulse of amplitude 1 or 0 ever b seconds. Both values occur with equal probability, and the amplitude during

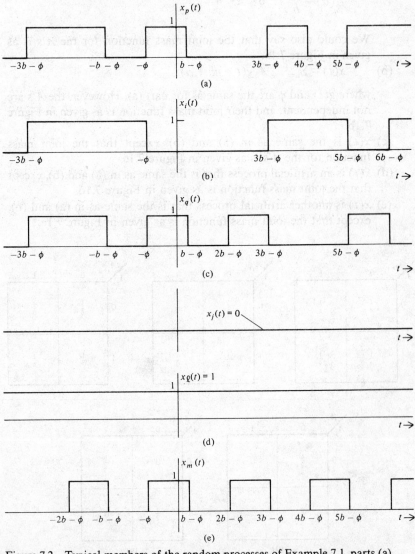

Figure 7.2   Typical members of the random processes of Example 7.1, parts (a) through (e), respectively.

any time interval is independent of previous time intervals. In addition, the whole waveform generated is given a phase shift to the left uniformly distributed from 0 to b seconds." The sample section of the typical member shown was obtained by repeatedly tossing a coin, and the reader is encouraged to generate his or her own typical member.

(b) On examination of the random process we see that it differs from that of part (a) insofar as now we evaluate the conditional mass functions as

$$P_{A_{i+1}}[1/(A_i=1)] = P_{A_{i+1}}[0/(A_i=0)] = \tfrac{2}{3}$$

and

$$P_{A_{i+1}}[1/(A_i=0)] = P_{A_{i+1}}[0/(A_i=1)] = \tfrac{1}{3}$$

These conditional mass functions are reflected in Figure 7.2b by the fact a member tends to retain a value for a longer time than in (a).

(c) The conditional mass functions may be evaluated from Figure 7.2c to yield

$$P_{A_{i+1}}[1/(A_i=1)] = P_{A_{i+1}}[0/(A_i=0)] = \tfrac{1}{3}$$

and

$$P_{A_{i+1}}[0/(A_i=1)] = P_{A_{i+1}}[1/(A_i=0)] = \tfrac{2}{3}$$

We note that the typical member $x_q(t)$ generated indicates the likelihood of changing value every b s. The reader should generate a section of a second member waveform using a coin toss once and continually drawing an object from three objects of which two are similar.

(d) The conditional mass functions may be evaluated from Figure 7.1d as

$$P_{A_{i+1}}[1/(A_i=1)] = P_{A_{i+1}}[0/(A_i=0)] = 1$$

and

$$P_{A_{i+1}}[1/(A_i=0)] = P_{A_{i+1}}[0/(A_i=1)] = 0$$

This implies that a member waveform is always 0 or 1. For this reason two members are shown in Figure 7.2d, each of which occur in the ensemble 50% of the time.

(e) Now, from our joint mass function,

$$P_{A_{i+1}}[1/(A_i=0)] = P_{A_{i+1}}[0/(A_i=1)] = 1$$

and this implies that a member waveform must change amplitude every b s, as is shown in Figure 7.2e. If we consider the five processes shown, we should intuitively feel that those of parts (d) and (e) are the least random or *most predictable* and the process of part (a) is the *most random* or *least predictable*. These terms will be defined in detail later, but for the moment the reader may form mental interpretations.

We will now proceed and evaluate the autocorrelation function $R_{xx}(\tau)$ for the five processes of parts (a) to (e).

(a) On observing Figure 7.2 the reader should immediately conclude that $x(t)$ is first-order stationary, with $f_X(\alpha) = \frac{1}{2}\delta(\alpha) + \frac{1}{2}\delta(\alpha - 1)$. The autocorrelation function will now be evaluated by considering every general range of τ for which a unique answer may be obtained. For each such range the joint mass function $p_{X_1 X_2}(\alpha_i, \beta_j)$, where X_1 is the random variable $x(t)$ and X_2 is the random variable $x(t + \tau)$, will be found. For simplicity the subscripts X_1, X_2 will often be omitted and $P(\alpha_i, \beta_j)$ used instead of $p_{X_1 X_2}(\alpha_i, \beta_j)$.

Intuitively the reader should think about this process and come to the conclusion that it almost certainly seems to be ergodic. To evaluate $R_{xx}(\tau)$ we must consider two different ranges of τ.

For the range $0 < \tau < b$, the joint mass function for $X_1 = x(t)$ and $X_2 = x(t + \tau)$ must now be evaluated for $P(0,0)$, $P(0,1)$, $P(1,0)$, and $P(1,1)$. First we will find $P(0,1)$ and $P(1,0)$. By symmetry,

$$P(0,1) = P(1,0) = P_{X_1}(0) P_{X_2}[1/(X_1 = 0)]$$

The conditional event,

$$[x(t+\tau) = 1/x(t) = 0]$$

$$= \left(\begin{array}{l} \text{The event a new pulse is generated in} \\ \tau \text{ s with value 1 given the present} \\ \text{pulse is 0} \end{array} \right)$$

and

$$P(1/0) = P(0/1)$$

$$= \frac{\tau}{b} \times \frac{1}{2} \qquad \text{(Have absolutely no doubt of this)}$$

Using these results,

$$P(0,1) = P(1,0) = \frac{1}{2}\left(\frac{\tau}{2b}\right) = \frac{\tau}{4b}$$

Now $P(0,0)$ and $P(1,1)$ will be found. Again, by symmetry,

$$P(0,0) = P(1,1) = p_{X_1}(0)p_{X_2}[0/(X_1=0)]$$

The event $[x(t+\tau)=0/x(t)=0]$ is physically defined as

$$(X_2=0/X_1=0) = \left(\begin{array}{l} \text{The event no new pulse is generated in} \\ \tau \text{ s or a new pulse is generated with} \\ \text{value 0} \end{array} \right)$$

$$P(0/0) = \frac{b-\tau}{b} + \frac{\tau}{b}\left(\frac{1}{2}\right)$$

$$= \frac{b-0.5\tau}{b}$$

and

$$P(0,0) = \frac{b-0.5\tau}{2b} = \frac{2b-\tau}{4b}$$

By symmetry, $P(1,1)$ has the same value. This joint mass function for X_1 and X_2 is shown in matrix form in Figure 7.3a.

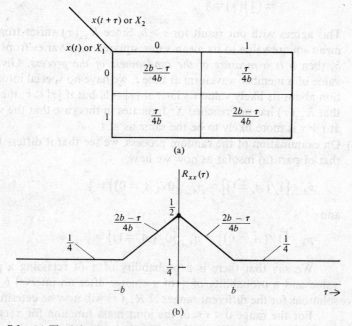

(a)

(b)

Figure 7.3 (a) The joint mass function for $x(t)$ and $x(t+\tau)$; $0<\tau<b$, in Example 7.1a. (b) The autocorrrelation function for Example 7.1a.

The autocorrelation function is now evaluated as

$$R_{xx}(\tau) = \sum \sum \alpha_i \beta_j P(\alpha_i, \beta_j)$$

$$= 0 + 0 + 0 + 1 \left(\frac{2b - \tau}{4b} \right) \qquad 0 < \tau < b$$

For the range $b < \tau < 2b$, the random variable $X_2 = x(t + \tau)$ is independent of $X_1 = x(t)$, and the joint probabilities are $P(0,0) = P(1,1) = P(0,1) = P(1,0) = \frac{1}{4}$. The autocorrelation function is evaluated as

$$R_{xx}(\tau) = 1 \times 1 \times \tfrac{1}{4} = \tfrac{1}{4}$$

$R_{xx}(\tau)$ for the random process of part (a) is now shown plotted in Figure 7.3b. Observing the figure we notice that if $\tau = 0$, $R_{xx}(\tau) = \overline{x^2(t)}$, which is just the mean square value. Moreover, $\overline{x^2(t)} = (1)^2 \frac{1}{2} = \frac{1}{2}$, and this checks with our value for $R_{xx}(0)$. In addition, if $x(t)$ and $x(t + \tau)$ are *independent*,

$$R_{xx}(\tau) = \overline{x(t)x(t + \tau)}$$

$$= \overline{X}\,\overline{X}$$

$$= \left(\tfrac{1}{2}\right)\left(\tfrac{1}{2}\right) = \tfrac{1}{4}$$

This agrees with our result for $\tau > b$. Since $R_{xx}(\tau)$ varies from its mean square value to its mean value squared, as τ varies from 0 to b, then *b is a measure of the randomness of the process.* Given a value of a member waveform at time t, we have no special information about its likely value τ s later if $|\tau| > b$, but if $|\tau| < b$, the fact that $R_{xx}(\tau)$ has not reached $\overline{X}^2$ indicates in this case that the value at $t + \tau$ is more likely to be the same as at t.

(b) On examination of the random process, we see that it differs from that of part (a) insofar as now we have ,

$$p_{A_{i+1}}[1/(A_i = 1)] = p_{A_{i+1}}[0/(A_i = 0)] = \tfrac{2}{3}$$

and

$$p_{A_{i+1}}[1/(A_i = 0)] = p_{A_{i+1}}[0/(A_i = 1)] = \tfrac{1}{3}$$

We say that there is a probability of $\frac{2}{3}$ of retaining a pulse value and a probability of $\frac{1}{3}$ of a change after an interval b. The solutions for the different ranges of $R_{xx}(\tau)$ will now be determined.

For the range $0 < \tau < b$, the joint mass function for $x(t)$ and $x(t + \tau)$ is given in Figure 7.4a. The derivation of the probabilities is left as an exercise as the reasoning is identical to that used for the range $0 < \tau < b$ of the process of part (a). Hence $R_{xx}(\tau)$ may be

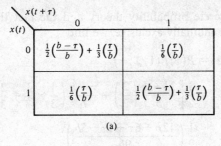

(a)

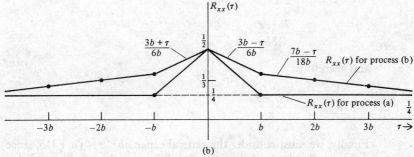

(b)

Figure 7.4   (a) The joint mass function for $x(t)$ and $x(t+\tau)$, $0 < \tau < b$, in Example 7.1b. (b) The autocorrelation function for Example 7.1b compared to Example 7.1a.

evaluated as

$$R_{xx}(\tau) = \overline{x(t)x(t+\tau)}$$

$$= 1 \times 1 \frac{1}{2}\left[\left(\frac{b-\tau}{b}\right) + \frac{2}{3}\left(\frac{\tau}{b}\right)\right]$$

$$= \frac{3b-\tau}{6b}$$

For the range $b < \tau < 2b$, because there is a probability of $\frac{2}{3}$ for retaining a value, we cannot say that $R_{xx}(\tau)$ is the mean squared value. We must evaluate the probabilities $P(0,0)$, $P(0,1)$, $P(1,0)$, and $P(1,1)$. Since $P(1,1)$ is the only one involved in finding $R_{xx}(\tau)$, we will find it but no others. If $b < \tau < 2b$, then either one or two pulse intervals elapse between the two values when sampling a member waveform. (Be sure of what this means.) The event of the value 1 being followed by the value 1, τ s later, is

$$(1/1) = \begin{pmatrix} \text{Event that exactly one new pulse} \\ \text{is generated and its value is 1} \end{pmatrix}$$

$$\cup \begin{pmatrix} \text{Event that exactly two pulses} \\ \text{are generated and the } last \\ \text{pulse is 1} \end{pmatrix}$$

Using discrete probability theory and the fact that the indicated events are mutually exclusive, we find

$$P(1,1) = P(1)P(1/1:\tau)$$

$$= \frac{1}{2}\left[\frac{2b-\tau}{b}\left(\frac{2}{3}\right) + \frac{\tau-b}{b}\left(\frac{1}{3}\times\frac{1}{3} + \frac{2}{3}\times\frac{2}{3}\right)\right]$$

$$= \frac{1}{2}\left(\frac{12b-6\tau+5\tau-5b}{9b}\right)$$

$$= \frac{7b-\tau}{18b}$$

and therefore

$$R_{xx}(\tau) = \frac{7b-\tau}{18b}$$

Finally, we must consider the general range $nb < \tau < (n+1)b$, since theoretically the probability is always more than $\frac{1}{2}$ that a value obtained at t is also obtained at $t+\tau$.

In the range $nb < \tau < (n+1)b$, at least n or at most $n+1$ new pulses are generated.

$$\therefore R_{xx}(\tau) = (1\times1)P_{X_1X_2}(1,1)$$

$$= \frac{1}{2}P\left[\binom{n \text{ new pulses are generated,}}{\text{the last of which is a 1}}\right.$$

$$\left.\cup\binom{n+1 \text{ new pulses are generated,}}{\text{the last of which is a 1}}\right]$$

This is better expressed as

$$R_{xx}(\tau) = \frac{1}{2}\left[P(n \text{ new pulses})P\binom{\text{even number of}}{\text{amplitude changes}}\right.$$

$$\left. + P\binom{n+1 \text{ new}}{\text{pulses}}P\binom{\text{even number of}}{\text{amplitude changes}}\right]$$

$$= \frac{1}{2}\left[\frac{(n+1)b-\tau}{b}\right]\sum_{k \text{ even}}\binom{n}{k}\left(\frac{1}{3}\right)^k\left(\frac{2}{3}\right)^{n-k}$$

$$+ \frac{1}{2}\left[\frac{\tau-nb}{b}\right]\sum_{k \text{ even}}\binom{n+1}{k}\left(\frac{1}{3}\right)^k\left(\frac{2}{3}\right)^{n+1-k}$$

Some algebraic manipulations lead to

$$\sum_{k \text{ even}} \binom{n}{k} \left(\frac{1}{3}\right)^k \left(\frac{2}{3}\right)^{n-k} = \frac{1}{2}\left[\left(\frac{1}{3}+\frac{2}{3}\right)^n + \left(\frac{2}{3}-\frac{1}{3}\right)^n\right]$$

and

$$\sum_{k \text{ even}} \binom{n+1}{k} \left(\frac{1}{3}\right)^k \left(\frac{2}{3}\right)^{n+1-k}$$

$$= \frac{1}{2}\left[\left(\frac{1}{3}+\frac{2}{3}\right)^{n+1} + \left(\frac{2}{3}-\frac{1}{3}\right)^{n+1}\right]$$

$$\therefore R_{xx}(\tau) = \frac{1}{4}\left\{\left[\frac{(n+1)b-\tau}{b}\right]\left[1+\frac{1}{3}^n\right] + \left(\frac{\tau-nb}{b}\right)\left[1+\frac{1}{3}^{n+1}\right]\right\}$$

This calculation for $R_{xx}(\tau)$ in general is intricate and may be omitted and replaced by working out $R_{xx}(\tau)$ for $2b < \tau < 3b$, in order to develop an idea for the behavior of $R_{xx}(\tau)$. In Figure 7.4b, $R_{xx}(\tau)$ is shown plotted along with $R_{xx}(\tau)$ for part (a).

The comparison of the two autocorrelation functions is interesting. $R_{xx}(\tau)$ for part (b) keeps approaching the mean value squared. We can say that random process of part (b) is less random than the process of part (a).

(c) On examination of this random process we see that

$$P_{A_{i+1}}[1/(A_i=1)] = P_{A_{i+1}}[0/(A_i=0)] = \frac{1}{3}$$

and

$$P_{A_{i+1}}[1/(A_i=0)] = P_{A_{i+1}}(0/(A_i=1) = \frac{2}{3}$$

The probability of retaining an amplitude value when a pulse is generated is $\frac{1}{3}$ and that of obtaining a different value is $\frac{2}{3}$. Since the calculation for $R_{xx}(\tau)$ is almost identical as for part (b), we will proceed rapidly.

For the range $0 < \tau < b$,

$$R_{xx}(\tau) = (1 \times 1)P_{X_1}(1)P_{X_2}(1/1)$$

$$= \frac{1}{2}\left[\left(\frac{b-\tau}{b}\right) + \frac{\tau}{b}\left(\frac{1}{3}\right)\right]$$

$$= \frac{3b-2\tau}{6b}$$

For the range $b < \tau < 2b$,

$$R_{xx}(\tau) = (1 \times 1)P_{X_1}(1)P_{X_2}(1/1)$$

$$= \frac{1}{2}\left[\left(\frac{2b-\tau}{b}\right)\left(\frac{1}{3}\right) + \left(\frac{\tau-b}{b}\right)\left(\frac{2}{3} \times \frac{2}{3} + \frac{1}{3} \times \frac{1}{3}\right)\right]$$

$$= \frac{1}{2}\left(\frac{6b - 3\tau + 4\tau - 4b + \tau - b}{9b}\right)$$

$$= \frac{b + 2\tau}{18b}$$

The evaluation of $R_{xx}(\tau)$ for the range $2b < \tau < 3b$ and the general range $nb < \tau < (n+1)b$ is left as an exercise. $R_{xx}(\tau)$ is shown plotted in Figure 7.5 and will be commented on at the conclusion of the problem.

(d) This random process represents an extreme case of a binary process with memory, since now

$$P_{A_{i+1}}[0/(A_i = 0)] = P_{A_{i+1}}[1/(A_i = 1)] = 1$$

The joint mass function is

$$P(0,1) = P(1,0) = 0 \quad \text{and} \quad P(0,0) = P(1,1) = \tfrac{1}{2}$$

The autocorrelation function for all τ is

$$R_{xx}(\tau) = (1 \times 1)p(1,1)$$

$$= \tfrac{1}{2}$$

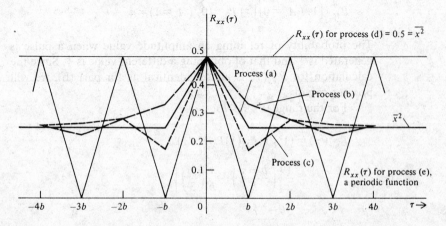

Figure 7.5   The autocorrrelation functions for parts (a) through (e) of Example 7.1.

This autocorrelation function is shown plotted in Figure 7.5, as is the result for part (e).

(e) This random process represents another extreme case in that now

$$p_{A_{i+1}}[0/(A_i=0)] = p_{A_{i+1}}[1/(A_i=1)] = 0$$

and

$$p_{A_{i+1}}[0/(A_i=1)] = p_{A_{i+1}}[1/(A_i=0)] = 1$$

This implies that any specific member alternates in value from 0 to 1 every b s. $R_{xx}(\tau)$ will now be evaluated.

For the range $0 < \tau < b$,

$$R_{xx}(\tau) = (1 \times 1)P_{X_1}(1)P_{X_2}(1/1:\tau)$$

$$= \frac{1}{2}\left[\left(\frac{b-\tau}{b}\right) + \frac{\tau}{b}\left(\frac{1}{3}\right)\right]$$

$$= \frac{b-\tau}{2b}$$

For the range $b < \tau < 2b$,

$$R_{xx}(\tau) = (1 \times 1)P_{X_1}(1)P_{X_2}(1/1:\tau)$$

$$= \frac{2b-\tau}{b}(0) + \frac{\tau-b}{2b}(1)$$

$$= \frac{\tau-b}{2b}$$

For the range $2b < \tau < 3b$ the reader may show $R_{xx}(\tau) = (3b-\tau)/2b$, and for a general range $nb < \tau < (n+1)b$ we can show that

$$R_{xx}(\tau) = \frac{(n+1)b - \tau}{2b} \qquad \text{if } n \text{ is even}$$

and

$$R_{xx}(\tau) = \frac{\tau - nb}{2b} \qquad \text{if } n \text{ is odd}$$

These results are shown sketched in Figure 7.5.

Discussion of the Results

Figure 7.5 shows the autocorrelation functions for processes (a) through (e). The autocorrelation function for (a) varies linearly from its mean square value $\overline{x^2} = 0.5$ at $\tau = 0$, to its mean value squared $\bar{x}^2 = 0.25$ at $\tau = b$ and remains at this value indefinitely. For example, $R_{xx}(0.5b) = (2b - 0.5b)/4b = 0.375$, and this says that if we sample the process at t and $(t + 0.5b)$, the expected product is 0.375; since this is much greater than $\bar{x}^2$, it is much more likely to have the value one at $t + \tau$, given that the value was 1 at t for a specific member, than to have the value 0. A receiver would always predict that a 1 follows a 1 or a 0 follows a 0 for any observation τ s later, for $0 < \tau < b$. Since $R_{xx}(\tau) = \bar{x}^2$, for all $\tau > b$, then a receiver can make no better prediction than the unconditional probabilities $P(1/0) = P(0/1) = 0.5$.

The autocorrelation function for process (b) keeps indefinitely approaching its mean value squared. For any τ, $R_{xx}(\tau)$ is always greater than the mean value squared, and this implies that product terms from the ensemble members occur such as to make $R_{xx}(\tau)$ greater than $\bar{x}^2$. The product terms $(0,0)$ and $(1,1)$ have higher probabilities than 0.25, and the product terms $(0,1)$ and $(1,0)$ have lower probabilities than 0.25. A receiver will always predict that a member waveform retains its value τ s later.

The autocorrelation function for process (c) zigzags back and forth between values greater than and less than $\bar{x}^2$. For example, for $0 < \tau < b$, we found $R_{xx}(\tau) = (3b - 2\tau)/6b$. We can calculate that $R_{xx}(0.25b) = 0.42$, which is greater than 0.25, the mean value squared, that $R_{xx}(0.75b) = 0.25$, and that $R_{xx}(0.9b) = 0.2$, which is less than the mean value squared. When τ equals $0.25b$ s, the product terms $(0,0)$ and $(1,1)$, whose occurrence contributes to $R_{xx}(\tau)$ being greater than the $\bar{x}^2$, have higher probabilities of occurrence than 0.25. When τ equals $0.75b$, $R_{xx}(\tau)$ is equal to the mean value squared and then $(1,1)$, $(1,0)$, $(0,1)$, and $(0,0)$ all occur with probabilities 0.25. When τ equals $0.9b$, since $R_{xx}(\tau)$ is less than the mean value squared, then the product terms $(1,0)$ and $(0,1)$ have higher probabilities of occurrence than 0.25 and of course $(0,0)$ and $(1,1)$ have lower probabilities than 0.25. From the point of view of randomness, (b) and (c) are about equally "random" since we can make similar predictions about the product terms for each one. However, (a) is more random than both processes because if $\tau \gg b$, the random variables $x(t)$ and $x(t + \tau)$ are independent, and from a prediction point of view the statement $P\{[x(t + \tau) = \beta]/[x(t) = \alpha]\}$ equals $P[x(t + \tau) = \beta]$ gives no information because of the condition at $t = t$.

The process in part (d) is the least random of the five processes since a member waveform is either 0 or 1 for all time. This implies that it is entirely predictable, as $P\{[x(t + \tau) = \alpha]/[x(t) = \alpha]\} = 1$ for α equal to both 0 and 1. At this stage we can also surmise that the autocorrelation function does not completely characterize a process, as any random process $x(t) = A$, where A is a random variable with mean square value equal to 0.5, would have the same autocorrelation function as process (d), regardless of $f_{x(t)}(\alpha)$.

Observing the random process (e), we note that due to the fact that a member must change amplitude every b s, then each member is a periodic function with period $2b$. $R_{xx}(\tau)$ is now periodic with period $2b$ and $R(2nb) = \overline{x^2}$ for all integer n values as the random variables $x(t)$ and

$x(t+2nb)$ must jointly take on the same value on a trial of sampling a member waveform. When $\tau = b$, then $x(t)$ and $x(t+\tau)$ must take on different values and $p(0,0) = p(1,1) = 0$ while $p(0,1)$ and $p(1,0) = \frac{1}{2}$. These values yield $R_{xx}(\tau) = 0$. In one sense we say random process (e) is just as predictable as random process (d) because, given that $x(t) = \alpha$ and being told exactly how far from an edge this value occurred, we could exactly predict the value of $x(t+\tau)$. However, knowing that $R_{xx}(0.5b)$ for random process (e) is $(b-0.5b)/2b = 0.25$, we can say only that, on sampling, the ensemble the product terms $(0,0)$, $(0,1)$, $(1,0)$, and $(1,1)$ occur one-quarter of the time.

In summary, from our first detailed and exhaustive problem on evaluating autocorrelation functions, we conclude that process (a) is the most random, processes (b) and (c) are less random and allow us to make predictions about the relative occurrence of the product terms, which are more revealing than the unconditional mass functions when $x(t)$ and $x(t+\tau)$ are independent, and that processes (d) and (e) are the least random or most predictable. The fact that $R_{xx}(\tau)$ rapidly approaches the mean square value indicates randomness only if $R_{xx}(\tau)$ then stays close to this value.

The reader should be relaxed about the foregoing example. If its comprehension tends to create too many difficulties, do not hesitate to forget it for a few days and then return to its detailed consideration after some more experience with autocorrelation functions has been obtained.

Special Drill Problem

Repeat Example 7.1a through e, given that

$$x(t) = \sum_{-\infty}^{\infty} A_i g(t - nb + \phi)$$

where

$$f_{A_i}(\alpha) = \tfrac{1}{2}\delta(\alpha - 1) + \tfrac{1}{2}\delta(\alpha + 1)$$

and otherwise all other conditions are identical. If possible, just intuitively sketch the results graphically as in Figure 7.5.

We will now proceed with more examples illustrating the evaluation of autocorrelation functions.

Example 7.2

Given the random process

$$x(t) = \sum_{n = -\infty}^{\infty} A_n g(t - nb + \phi)$$

where

$$g(t) = 1 \qquad 0 < t < b$$
$$\quad = 0 \qquad \text{otherwise}$$

and the A_n's are independent random variables with density function

$$f_{A_i}(\alpha) = 1 \qquad 0 < \alpha < 1$$
$$\qquad = 0 \qquad \text{otherwise}$$

and ϕ represents a random phase shift with uniform density function $f_\phi(\alpha) = 1/b$; $0 < \alpha < b$, sketch a typical member of $x(t)$ and evaluate the autocorrelation function.

SOLUTION

A typical member of $x(t)$ is shown in Figure 7.6a. This member could have been generated by spinning an infinitely finely calibrated wheel of fortune to obtain each A. We might have done so for $A_{-20}, A_{-19}, \ldots, A_{20}$ to obtain a representative sample. Then ϕ for the waveform could have been chosen by spinning a wheel of fortune calibrated from 0 to b.

For the range $0 < \tau < b$, if we consider the random variables $x(t)$ and $x(t + \tau)$, then there are two possibilities;

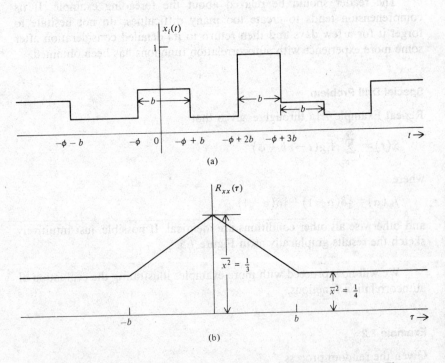

(a)

(b)

Figure 7.6 (a) A typical member of the process of Example 7.2. (b) The autocorrelation function.

1. Either both samples are from the same pulse, or
2. A new pulse is generated during the time τ, and both values are from different pulses (and independent).

The probabilities of these two events are

$$P(\text{both values from same pulse}) = \frac{b-\tau}{b}$$

$$P(\text{a new pulse is generated}) = \frac{\tau}{b}$$

It follows that the joint density function is

$$f_{X_1 X_2}(\alpha, \beta) = f_{X_1}(\alpha) f_{X_2}[\beta/(X_1 = \alpha)]$$

With detailed thought, this yields

$$f_{X_1 X_2}(\alpha, \beta) = 1\left(\frac{b-\tau}{b}\right)\delta(\beta - \alpha) + 1\left(\frac{\tau}{b}\right) \qquad 0<\alpha<1, 0<\beta<1$$

$$R_{xx}(\tau) = \int_0^1 \left[\int_{-\infty}^{\infty} \alpha\beta\left(\frac{b-\tau}{b}\right)\delta(\beta - \alpha)\,d\beta\right] d\alpha + \frac{\tau}{b}\int_0^1\int_0^1 \alpha\beta\,d\alpha\,d\beta$$

$$= \frac{b-\tau}{b}\int_0^1 \alpha^2\,d\alpha + \frac{\tau}{b}\left(\frac{1}{4}\right)$$

$$= \frac{4b-\tau}{12b} \qquad 0<\tau<b$$

If $\tau > b$, then $x(t)$ and $x(t+\tau)$ are independent random variables, and

$$f_{X_1 X_2}(\alpha, \beta) = f_{X_1}(\alpha) f_{X_2}(\beta)$$

$$= 1 \qquad 0<\alpha<1, 0<\beta<1$$

$$R_{xx}(\tau) = \overline{x(t)x(t+\tau)} = \bar{x}^2 = \tfrac{1}{4}$$

$R_{xx}(\tau)$ is shown plotted in Figure 7.6b.

With experience, problems such as Examples 7.1 and 7.2 are solved quite rapidly. For example, Example 7.2 could be solved for $\tau < b$ as

$$R_{xx}(\tau) = \overline{x^2}\,P(\text{no shift}) + \bar{x}^2 P(\text{shift})$$

$$= \frac{1}{3}\left(\frac{b-\tau}{b}\right) + \frac{1}{4}\left(\frac{\tau}{b}\right)$$

$$= \frac{4b-\tau}{12b} \qquad 0<\tau<b$$

and, for $\tau > b$,

$$R_{xx}(\tau) = \left(\overline{x^2}\times 0\right) + \left(\bar{x}^2\times 1\right)$$

$$= \tfrac{1}{4} \qquad \tau > b$$

Such rapid solutions are carried out after thorough understanding of the concepts has been achieved. It should serve as an instructive exercise to resolve Example 7.1 in as rapid a fashion as possible. We notice that the random process of Example 7.2 is ergodic, but once again, since a specific member cannot be analytically represented, the time average autocorrelation function cannot be found. The fact that the autocorrelation function does not uniquely characterize a random process may again be seen, by observing that the processes of Example 7.1a and Example 7.2 gave very similar autocorrelation functions. The student should be able to modify either problem so that they yield the same autocorrelation function.

Example 7.3

Given the random process $x(t) = \sin(\omega_c t + \phi)$, where ϕ is a random variable with the density function

$$f_\phi(\alpha) = \tfrac{1}{2\pi} \qquad 0 < \alpha < 2\pi$$
$$= 0 \qquad \text{otherwise}$$

evaluate the autocorrelation function in two ways:

(a) By the statistical approach,

$$R_{xx}(\tau) = \overline{x(t)x(t+\tau)}$$

(b) By the time average approach,

$$R_{xx}(\tau) = \lim_{T \to \infty} \frac{1}{2T} \int_{-T}^{T} x_p(t)x_p(t-\tau)\, dt$$

where $x_p(t)$ is any member of $x(t)$.

SOLUTION

(a) The statistical approach involves the following operations:

$$R_{xx}(\tau) = \overline{x(t)x(t+\tau)}$$

$$= \overline{\sin(\omega_c t + \phi)\sin[\omega_c(t+\tau)+\phi]}$$

$$= \int_{-\infty}^{\infty} \int_{-\infty}^{\infty} \alpha\beta f_{X_1 X_2}(\alpha, \beta)\, d\alpha\, d\beta$$

This is very difficult to evaluate directly because we need to find $f_{X_1 X_2}(\alpha, \beta)$. However, utilizing the fundamental theorem of Section 5.3, where we consider the random variable Y as defining the phase angle ϕ of $x(t)$ and Z with the phase angle of $x(t+\tau)$, we obtain

$$R_{xx}(\tau) = \int_{-\infty}^{\infty} \int_{-\infty}^{\infty} \sin(\omega_c t + \alpha)$$
$$\times \sin[\omega_c(t+\tau)+\beta] f_{YZ}(\alpha, \beta) \, d\alpha \, d\beta$$

$$= \int_{0}^{2\pi} \int_{-\infty}^{\infty} \left\{ \sin(\omega_c t + \alpha) \cdot \right.$$
$$\left. \times \sin[\omega_c(t+\tau)+\beta] \frac{1}{2\pi} \delta(\beta - \alpha) \, d\beta \right\} d\alpha$$

$$= \frac{1}{2\pi} \int_{0}^{2\pi} \sin(\omega_c t + \alpha)\sin[\omega_c(t+\tau)+\alpha] \, d\alpha$$

Using the trigonometric identity,

$$\cos A - \cos B = 2\sin\left(\frac{A+B}{2}\right)\sin\left(\frac{B-A}{2}\right)$$

we get

$$R_{xx}(\tau) = \frac{1}{2\pi} \int_{0}^{2\pi} \frac{1}{2} \cos \omega_c \tau \, d\alpha$$

$$- \frac{1}{2\pi} \int_{0}^{2\pi} \frac{1}{2} \cos(2\omega_c t + \omega_c \tau + 2\alpha) \, d\alpha$$

$$= \frac{1}{2} \cos \omega_c \tau + 0 \qquad \text{(Be sure the second term is 0)}$$

$$= \frac{1}{2} \cos \omega_c \tau$$

This is a very important derivation, and it should be completely understood. It is worth noting that since $\sin(\omega_c t + \phi)$ and $\sin[\omega_c(t+\tau)+\phi]$ are both functions of the one random variable ϕ, then the product $x(t)x(t+\tau)$ is also a function of the one random variable ϕ, and hence the autocorrelation function could more easily be obtained using the fundamental theorem for one random variable from Chapter 4. Doing this,

$$R_{xx}(\tau) = \overline{x(t)x(t+\tau)}$$

$$= \int_{-\infty}^{\infty} \left\{ \sin(\omega_c t + \alpha)\sin[\omega_c(t+\tau)+\alpha] \right\} f_{\phi}(\alpha) \, d\alpha$$

The density function for ϕ is uniformly distributed between 0 and

2π, and the derivation is now identical to the conclusion of the case for the fundamental theorem using two random variables Y and Z.

(b) Assuming that $x(t)$ is ergodic, any member may be chosen to evaluate $R_{xx}(\tau)$ on a time average basis. Let us decide to let $\phi = 0$ and use the member

$$x_p(t) = \sin \omega_c t$$

Then

$$R_{xx}(\tau) = \widetilde{x(t)x(t-\tau)}$$

$$= \lim_{T \to \infty} \frac{1}{2T} \int_{-T}^{T} \sin \omega_c t \sin \omega_c (t - \tau)\, dt$$

Since $\sin \omega_c t$ is periodic with period $2\pi/\omega_c$, then the product function $\sin \omega_c t \sin \omega_c (t - \tau)$ for any τ is also periodic, with period $2\pi/\omega_c$, and we obtain the same result integrating over one period as for integrating over $2T$ when $T \to \infty$.

$$\therefore R_{xx}(\tau) = \frac{\omega_c}{2\pi} \int_0^{2\pi/\omega_c} \sin \omega_c t \sin \omega_c (t - \tau)\, dt$$

$$= \frac{\omega_c}{2\pi} \int_0^{2\pi/\omega_c} \frac{1}{2} \left[\cos \omega_c \tau - \cos(2\omega_c t - \omega_c \tau) \right] dt$$

$$= \frac{1}{2} \cos \omega_c \tau + 0$$

This is the same result that was obtained by the ensemble average approach of part (a).

It was possible to evaluate the time average autocorrelation function of this random process because we could analytically describe a specific time member. Unfortunately, this is a luxury we will rarely encounter with noise waveforms. For a member of an ergodic random process with period T_0, it should be statistically obvious that $R_{xx}(0)$, $R_{xx}(\pm T_0)$, and $R_{xx}(\pm n T_0)$, for n a positive integer, must all yield the result x^2, and that the resultant autocorrelation function is periodic with period T_0.

Example 7.4

Consider the random process

$$x(t) = \cos(\omega t + \phi)$$

where ω and ϕ are random variables with the joint density function

$$f_{\omega\phi}(\alpha, \beta) = \frac{1}{8\pi} \qquad 8 < \alpha < 12, \, 0 < \beta < 2\pi$$

Find the autocorrelation function $R_{xx}(\tau)$.

SOLUTION

It is easy to show that ω and ϕ are independent random variables with the density functions

$$f_\omega(\alpha) = \tfrac{1}{4} \qquad 8 < \omega < 12$$
$$= 0 \qquad \text{otherwise}$$

and

$$f_\phi(\beta) = \frac{1}{2\pi} \qquad 0 < \beta < 2\pi$$

Two typical members of $x(t)$ are shown in Figure 7.7. Up to now we have always obtained the joint density function $f_{X_1 X_2}(\alpha, \beta)$ and then evaluated $\overline{x(t)x(t+\tau)}$ This is equivalent to proving that the process is second-order stationary and then finding $\overline{x(t)x(t+\tau)}$. In this case, however, it is easy to find $R_{xx}(\tau)$ for a specific member and then find $R_{xx}(\tau)$ as the average of this result over the process, or

$$\boxed{R_{xx}(\tau) = \overline{\widetilde{x_i(\tau)x_i(t+\tau)}}}$$

From example 7.3 for any specific member, say $x_i(t) = \cos(\omega_i t + \phi_i)$, the time autocorrelation function is

$$\widetilde{x_i(t)x_i(t+\tau)} = \frac{\omega_i}{2\pi} \int_0^{2\pi/\omega_i} \cos(\omega_i t + \phi_i)\cos\left[\omega_i(t+\tau) + \phi_i\right] dt$$

$$= \tfrac{1}{2}\cos\omega_i\tau$$

Now, since ω is a random variable, the average of $\tfrac{1}{2}\cos\omega_i\tau$ over the process

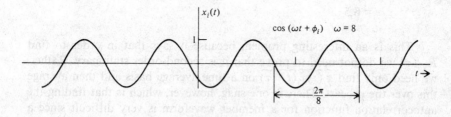

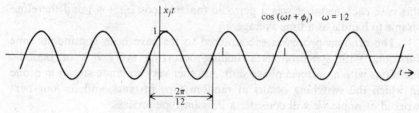

Figure 7.7 Two extreme members of the process of Example 7.4.

is

$$R_{xx}(\tau) = \frac{1}{2}\cos\omega\tau$$

$$= \int_{-\infty}^{\infty} \frac{1}{2}(\cos\alpha\tau)f_\omega(\alpha)\,d\alpha$$

$$= \int_{8}^{12} \frac{1}{2}(\cos\alpha\tau)\frac{1}{4}\,d\alpha$$

$$= \frac{1}{8}\left[\frac{1}{\pi}(\sin\alpha\tau)\Big|_{8}^{12}\right]$$

$$= \frac{\sin 12\tau}{8\tau} - \frac{\sin 8\tau}{8\tau}$$

Students who know their Fourier transforms should have no difficulty showing that $R_{xx}(\tau)$ is even and that $R_{xx}(0) = \frac{12}{8} - 1 = 0.5$. Alternatively, we could expand in a MacLaurin series to find $R_{xx}(0)$ as follows:

$$\frac{\sin 12\tau}{8\tau} = \frac{1}{8\tau}\left[12\tau - \frac{(12\tau)^3}{3!} + \frac{(12\tau)^5}{5!} + \cdots\right]$$

$$\therefore \lim_{\tau \to 0}\left(\frac{\sin 12\tau}{8\tau}\right) = 1.5$$

Similarly,

$$\lim_{\tau \to 0}\left(\frac{\sin 8\tau}{8\tau}\right) = 1$$

and

$$R_{xx}(0) = (1.5) - 1$$
$$= 0.5$$

This is an interesting problem because it says that in order to find $R_{xx}(\tau)$ we do not need to prove that it is second-order stationary. Rather, we need only find $\overline{x_i(t)x_i(t+\tau)}$ on a time average basis and then average this over the process. There is one snag, however, which is that finding the autocorrelation function for a member waveform is very difficult since a member of a random process is normally not specified deterministically. In this case each member was a periodic function $\cos(\omega_i t + \phi_i)$ and therefore simple to handle as a time average.

The random processes encountered so far have been continuous time functions with synchronous switching occurring every b s or periodic functions with a uniform phase shift. Another very common situation is one in which the switching occurs at random time instants, and for our next worked example we will consider a Poisson-type process.

Example 7.5

A random telegraph waveform is defined as follows: Every b s the number of transitions of the waveform from the value 1 to 0 is given by

$$P\left(\begin{array}{c} k \text{ transitions} \\ \text{in } b \text{ s} \end{array}\right) = \frac{e^{-\mu b}(\mu b)^k}{k!}$$

where $\mu b = 3$ and μ is the average number of transitions per second. Let us find the autocorrelation function and covariance function of this process.

SOLUTION

From the Poisson distribution given, the following probabilities may be evaluated for the number of transitions in b s,

$$P(0 \text{ transitions}) = \frac{e^{-3} \times 1}{1} = 0.05$$

$$P(1 \text{ transition}) = \frac{e^{-3} \times 3}{1} = 0.15$$

$$P(2 \text{ transitions}) = \frac{e^{-3} \times 9}{2} = 0.22$$

$$P(3 \text{ transitions}) = \frac{e^{-3} \times 27}{6} = 0.22$$

$$P(4 \text{ transitions}) = \frac{e^{-3} \times 81}{24} = 0.17$$

and so on. Using these probabilities, a typical member of the random process is shown sketched in Figure 7.8a. The autocorrelation function may now be evaluated as

$$R_{xx}(\tau) = \overline{x(t)x(t+\tau)}$$
$$= 1 \times P[x(t)=1] P\{[x(t+\tau)=1]/[x(t)=1]\}$$

With thought the conditional event of $x(t+\tau)$ being 1 given $x(t)$ is 1 is the event of an even number of transitions occurring in τ s. Rescaling the Poisson distribution to τ s we obtain

$$P\left(\begin{array}{c} k \text{ transitions} \\ \text{in } \tau \text{ s} \end{array}\right) = \frac{e^{-3\tau}(3\tau)^k}{k!}$$

and $R_{xx}(\tau)$ is now given by

$$R_{xx}(\tau) = 0.5 P(\text{even number of transitions})$$
$$= 0.5 e^{-3\tau}\left[1 + \frac{1}{2!}(3\tau) + \frac{1}{4!}(3\tau)^4 + \cdots\right]$$

In order to evaluate our series in the brackets we will use the following

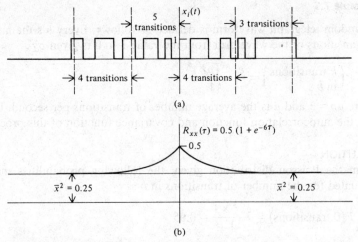

Figure 7.8   (a) A typical member of the Poisson process of Example 7.5. (b) The autocorrelation function.

intuitive mathematical trick as in Chapter 4,

$$1 + \frac{1}{2}x^2 + \frac{1}{4!}x^4 + \frac{1}{6!}x^6 = \frac{1}{2}\left(1 + x + \frac{1}{2}x^2 + \frac{1}{3!}x^3 + \frac{1}{4!}x^4 + \cdots\right)$$

$$+ \frac{1}{2}\left(1 - x + \frac{1}{2}x^2 - \frac{1}{3!}x^3 + \frac{1}{4!}x^4 + \cdots\right)$$

$$= \frac{1}{2}(e^x + e^{-x})$$

Finally,

$$R_{xx}(\tau) = 0.5e^{-3\tau}(0.5e^{3\tau} + 0.5e^{-3\tau})$$

$$= 0.25(1 + e^{-6\tau})$$

This autocorrelation function is shown plotted in Figure 7.8b. The covariance function $L_{xx}(\tau)$ is given by

$$L_{xx}(\tau) = R_{xx}(\tau) - \bar{x}^2$$

$$= 0.25e^{-6\tau}$$

Examples 7.1 through 7.5 were concerned with the mechanics of statistically evaluating autocorrelation functions for stationary or ergodic random processes. The concepts and techniques are initially somewhat of a shock, but the task is easier if the reader can appreciate that in effect she or he is doing nothing more than evaluating expected values of the product of two random variables:

$$E(XY) = \int_{-\infty}^{\infty} \int_{-\infty}^{\infty} \alpha\beta f_{XY}(\alpha, \beta)\, d\alpha\, d\beta$$

or

$$E(XY) = \sum_{\text{all } j} \sum_{\text{all } i} \alpha_i \beta_j P_{XY}(\alpha_i, \beta_j)$$

for the joint density function associated with the random variables that sample the process at two times separated by τ s. If the two random variables are independent, then the answer should be $\bar{x}^2$. The rapidity with which $R_{xx}(\tau)$ approaches its mean value squared is a measure of the randomness of the process. A slowly changing noise waveform would have an autocorrelation function that takes a long time to approach $\bar{x}^2$. The autocorrelation function for the most random-type noise looks like the mathematical model of a weighted delta function. Later we will call such a process a white-noise process. The autocorrelation function for the least random or most predictable random process is $R_{xx}(\tau) = x^2$, where an individual member never varies. We had an example of this process in the last chapter, where $x(t) = A$ and A was a random variable with some density function, and again in Example 7.1d. Knowing the value of a member of such a process at time t, we are able to predict with probability 1 that its value is unchanged τ s later.

DRILL SET: THE EVALUATION OF AUTOCORRELATION FUNCTIONS

1. Consider the random process

$$x(t) = \sum_{-\infty}^{\infty} A_n g(t - 3n)$$

where $g(t) = t$; $0 < t < 1$, and is 0 otherwise, and the A_n's are independent random variables with density functions

$$f_{A_i}(\alpha) = \tfrac{1}{2}\delta(\alpha - 1) + \tfrac{1}{2}\delta(\alpha + 1)$$

Find the mean function $M_x(t) \triangleq \overline{x(t)}$ for this process and plot it as a function of time.

2. Based on the solutions to Examples 7.1 through 7.5, as quickly as possible evaluate the autocorrelation function for

$$x(t) = \sum_{-\infty}^{\infty} A_n g(t - nb + \phi)$$

where $g(t) = 1$; $0 < t < b$, and the A_n's and ϕ are independent with

$$f_{A_i}(\alpha) = \tfrac{1}{10} \qquad -5 < \alpha < 5$$
$$\phantom{f_{A_i}(\alpha)} = 0 \qquad \text{otherwise}$$

and

$$f_\phi(\alpha) = 1/b \qquad 0 < \alpha < b$$

7.2 PROPERTIES OF AUTOCORRELATION AND CROSS-CORRELATION FUNCTIONS

Properties of Autocorrelation Functions

Herewith is listed a number of important properties for the autocorrelation functions of stationary random processes, some of which we have already encountered.

Property 1 $R_{xx}(\tau) = R_{xx}(-\tau)$.

Property 2 $R_{xx}(0) = \overline{x^2}$.

Property 3 $R_{xx}(0) \geq R_{xx}(\tau)$.

Property 4 $R_{xx}(\infty) = \overline{x}^2 **$.[1]

Property 5 $R_{(x+y)(x+y)}(\tau) =$
$R_{xx}(\tau) + R_{yy}(\tau) + \overline{x(t)y(t+\tau)} + \overline{y(t)x(t+\tau)}$.

Property 6 $\int_{-\infty}^{\infty} R_{xx}(\tau)\, d\tau$ exists** or $R_{xx}(\tau)$ is a finite energy function.

Property 7 If $x(t)$ contains a periodic component with period T_0, then $R_{xx}(\tau)$ also contains a periodic component with period T_0.

Property 8 If $R_{xx}(\tau) > \overline{x}^2$, then if $x(t) > \overline{x}$, $x(t+\tau)$ is probably also greater than $\overline{x}$. If the density function is symmetric about $\overline{x}$, we can make the preceding statement with certainty.

As an indication of how to carry out proofs, Properties 3, 5, and 6 will now be proved.

Property 3

Show $R_{xx}(0) \geq R_{xx}(\tau)$. Now $\overline{[x(t) - x(t+\tau)]^2} \geq 0$ for the two random variables $x(t)$ and $x(t+\tau)$.

$$\therefore E[x^2(t) + x^2(t+\tau) - 2x(t)x(t+\tau)] \geq 0$$

$$\therefore 2\overline{x^2(t)} - 2R_{xx}(\tau) \geq 0$$

$$\therefore R(0) \geq R(\tau)$$

The equality sign will be applicable when $x(t)$ is a periodic random process with period T, since then $R_{xx}(0) = R_{xx}(nT)$ for n a positive or negative integer.

---

[1]**The asterisks indicate that Property 4 is true if $R_{xx}(\tau)$ contains no periodic components and Property 6 is true if $R_{xx}(\tau)$ contains no periodic components and if $\overline{x} = 0$.

Property 5

If we form the random process $z(t) = x(t) + y(t)$, then $R_{zz}(\tau)$ may be found as

$$
\begin{aligned}
E[z(t)z(t+\tau)] &= E[(x(t)+y(t))(x(t+\tau)+y(t+\tau))] \\
&= E[x(t)x(t+\tau)] + E[y(t)y(t+\tau)] \\
&\quad + E[x(t)y(t+\tau)] + E[y(t)x(t+\tau)] \\
&= R_{xx}(\tau) + R_{yy}(\tau) + \overline{x(t)y(t+\tau)} + \overline{x(t+\tau)y(t)}
\end{aligned}
$$

which proves Property 5.

If $x(t)$ and $y(t)$ are independent, then

$$
R_{zz}(\tau) = R_{xx}(\tau) + R_{yy}(\tau) + 2\overline{X}\,\overline{Y}
$$

and if, in addition, $\overline{X}$ or $\overline{Y}$ is 0,

$$
R_{zz}(\tau) = R_{xx}(\tau) + R_{yy}(\tau)
$$

We will later give physical interpretation to

$$
R_{xy}(\tau) = \overline{x(t)y(t+\tau)}
$$

which is called the cross-correlation function of $x(t)$ and $y(t)$.

Property 6

Show that

$$
\int_{-\infty}^{\infty} R_{xx}(\tau)\, d\tau < \infty
$$

If $\overline{X} = 0$, $R_{xx}(\tau)$ decreases from $\overline{x^2}$ to 0 when $\tau = \infty$, and therefore $R_{xx}(\tau)$ will be a finite energy function or one whose integral from $-\infty$ to $+\infty$ exists. This is a very significant property, as it implies that although any member of a random process $x_i(t)$ is a nonfinite energy function, the autocorrelation function of $x(t)$ is a finite energy function and therefore must possess a Fourier transform, $F[R_{xx}(\tau)]$. We will refer to the Fourier transform of $R_{xx}(\tau)$ as $S_{xx}(\omega)$, the power spectral density of the random process, and in the next chapter we attach much physical significance to it.

The Definition of Cross-Correlation Functions

A pair of random processes may be jointly classified in a manner similar to a single random process. Let us define, for two random processes $x(t)$ and $y(t)$, the random variables $x(t)$ or X_1, which samples $x(t)$ at time t; $y(t)$ or Y_1, which samples $y(t)$ at time t; $x(t+\tau)$ or X_2, which samples $x(t)$ at $t+\tau$; and $y(t+\tau)$ or Y_2, which samples $y(t)$ at time $t+\tau$. For jointly wide-sense stationary random processes the cross-correlation functions are usually defined as

$$R_{xy}(\tau) \overset{\Delta}{=} \overline{x(t)y(t+\tau)}$$

$$= \int_{-\infty}^{\infty} \int_{-\infty}^{\infty} \alpha\beta f_{X_1 Y_2}(\alpha, \beta) \, d\alpha \, d\beta \tag{7.7}$$

where $R_{xy}(\tau)$ is called the cross correlation of $x(t)$ with $y(t)$, and

$$R_{yx}(\tau) \overset{\Delta}{=} \overline{x(t+\tau)y(t)}$$

$$= \int_{-\infty}^{\infty} \int_{-\infty}^{\infty} \alpha\beta f_{X_2 Y_1}(\alpha, \beta) \, d\alpha \, d\beta \tag{7.8}$$

where $R_{yx}(\tau)$ is called the cross correlation of $y(t)$ with $x(t)$. Occasionally the cross-correlation functions are defined in the reverse manner, with $R_{xy}(\tau)$ being denoted by Eq. 7.8.

If the random processes are jointly ergodic, then these integrals may be evaluated on a time average basis to yield the same results for any $x_p(t)$ or $y_p(t)$. These time average definitions are

$$R_{xy}(\tau) \overset{\Delta}{=} \widetilde{x(t)y(t+\tau)}$$

$$= \lim_{T \to \infty} \frac{1}{2T} \int_{-T}^{T} x_p(t) y_q(t+\tau) \, dt \tag{7.9}$$

or

$$R_{xy}(\tau) = \lim_{T \to \infty} \frac{1}{2T} \int_{-T}^{T} x_p(t-\tau) y_q(t) \, dt$$

and

$$R_{yx}(\tau) \overset{\Delta}{=} \widetilde{x(t+\tau)y(t)}$$

$$= \lim_{T \to \infty} \frac{1}{2T} \int_{-T}^{T} x_p(t+\tau) y_q(t) \, dt \tag{7.10}$$

or

$$\lim_{T \to \infty} \frac{1}{2T} \int_{-T}^{T} x_p(t) y_q(t-\tau) \, dt \tag{7.11}$$

In practice, cross-correlation functions will be useful when $x(t)$ and $y(t)$ are the random input and output waveforms of a system or when a system is defined by adding, subtracting, or multiplying random waveforms. We will now solve a few examples to cement these definitions in our minds. Later, when we encounter systems with random inputs, the cross-correlation integrals will take on much physical significance.

Example 7.6

Given

$$x(t) = \sum_{-\infty}^{\infty} A_n g(t - nb + \phi)$$

and

$$y(t) = \sum_{-\infty}^{\infty} B_n g(t - nb + \theta)$$

where

$$g(t) = 1 \qquad 0 < t < b$$
$$= 0 \qquad \text{otherwise}$$

and the A's, B's, ϕ, and θ are independent random variables with density functions

$$f_\phi(\alpha) = f_\theta(\alpha) = \frac{1}{b} \qquad 0 < \alpha < b$$

and

$$f_{A_i}(\alpha) = f_{B_i}(\alpha) = \tfrac{1}{2}\delta(\alpha) + \tfrac{1}{2}\delta(\alpha - 1)$$

find $R_{xy}(\tau)$ and $R_{yx}(\tau)$ statistically and comment on their time average evaluation.

SOLUTION

$$R_{xy}(\tau) = \overline{x(t)y(t + \tau)}$$
$$= \overline{x(t)} \ \overline{y(t + \tau)}$$
$$= (0.5)^2 = 0.25$$

and, similarly,

$$R_{yx}(\tau) = 0.25$$

If we were to evaluate the cross-correlation function on a time average basis for any specific member over a long time, the product function $x_p(t)y_q(t + \tau)$ would take on the value $+1$ for one-quarter of the time and 0 for three-quarters of the time and again yield the answer 0.25 for the autocorrelation function.

Example 7.7

Given

$$x(t) = 3\cos(\omega_c t + \phi)$$

and

$$y(t) = 2\cos(\omega_c t + \theta)$$

where ϕ and θ are random variables with uniform density functions over the period $T = 2\pi/\omega_c$ but

$$\theta = \phi - \frac{\pi}{2}$$

find $R_{xy}(\tau)$ and $R_{yx}(\tau)$.

SOLUTION
We could describe this relationship as a system that causes an amplification of 0.67 and a phase shift of one-quarter of a period. Since $\theta = \phi - (\pi/2)$, it is implied that we are not allowed to match any $x_p(t)$ with any $y_q(t)$ when finding the cross correlation due to the fact each specific $x_p(t)$ generates a specific mate $y_p(t)$. We now modify our definitions as follows: The cross correlation of $x(t)$ with $y(t)$ is

$$R_{xy}(\tau) = \lim_{N \to \infty} \frac{\displaystyle\sum_{p=1}^{N} x_p(t) y_p(t+\tau)}{N}$$

$$= \overline{x_p(t) y_p(t+\tau)}$$

and

$$R_{xy}(\tau) = \lim_{T \to \infty} \frac{1}{2T} \int_{-T}^{T} x_p(t) y_p(t+\tau)\, dt$$

on a time average basis.

We will evaluate $R_{xy}(\tau)$ statistically and leave it as an exercise for the reader to verify the result as a time average.

$$R_{xy}(\tau) = \overline{x(t) y(t+\tau)}$$

$$= \overline{3\cos(\omega_c t + \phi)2\cos\left[\omega_c(t+\tau) + \phi - \frac{\pi}{2}\right]}$$

Using the fundamental theorem from Chapter 4, this becomes

$$R_{xy}(\tau) = \int_0^{2\pi/\omega_c} 6\cos(\omega_c t + \alpha)\cos\left[\omega_c(t+\tau) + \left(\alpha - \frac{\pi}{2}\right)\right] \frac{\omega_c}{2\pi}\, d\alpha$$

Using the trigonometric identity,

$$\cos A + \cos B \equiv 2\cos\frac{A+B}{2}\cos\frac{A-B}{2}$$

the integral becomes

$$R_{xy}(\tau) = 3\left[\int_0^{2\pi/\omega_c} \cos\left(\omega_c \tau - \frac{\pi}{2}\right)\frac{\omega_c}{2\pi}\, d\alpha + 0\right]$$

where the integral that integrates to 0 is omitted.

$$\therefore R_{xy}(\tau) = 3\cos\left(\omega_c\tau - \frac{\pi}{2}\right)$$

It is interesting to note that $R_{XY}(\tau)$ is not even and that its maximum value occurs at $\tau = \pi/2\omega_c$. The cross correlation of $y(t)$ with $x(t)$ in this case yields

$$R_{yx}(\tau) = 3\cos\left(\omega_c\tau + \frac{\pi}{2}\right)$$

and we note that $R_{yx}(\tau) = R_{xy}(-\tau)$.

Properties of Cross-Correlation Functions

The following are some properties of jointly wide-sense stationary processes,

Property 1 $R_{xy}(\tau) = R_{yx}(-\tau)$.

Property 2 $|R_{xy}(\tau)| \leq \sqrt{R_{xx}(0)R_{yy}(0)}$.

Property 3 If $x(t)$ and $y(t)$ are independent, then $R_{xy}(\tau) = R_{yx}(\tau) = \bar{x}\bar{y}$.

Property 4 If $x(t)$ and $y(t)$ contain a periodic component with period T_0, $R_{xy}(\tau)$ and $R_{yx}(\tau)$ also contain a periodic component with period T_0.

Property 5 If $\bar{x} = 0$ and $\bar{y} = 0$, then $R_{xy}(\tau)$ and $R_{yx}(\tau)$ approach 0 as τ approaches infinity, and the Fourier transforms, which we will call the cross power spectral densities, will exist.

Properties 1, 3, and 4 are intuitively clear, whereas Property 5 will assume most importance in Chapter 8. We will now demonstrate Property 2.

Example 7.8

Demonstrate that $|R_{xy}(\tau)|$ is always less than or at most equal to $\sqrt{R_{xx}(0)R_{yy}(0)}$

SOLUTION
Just as the demonstration that $R_{xx}(0) \geq R_{xx}(\tau)$ was slightly unsatisfactory, so too is this demonstration since we are not skilled with number properties and geometric bounds.[2] The expected value,

$$E\left(\frac{X_1}{\sqrt{R_{xx}(0)}} - \frac{Y_2}{\sqrt{R_{yy}(0)}}\right)^2 \geq 0$$

[2]See Section 5.7 in Chapter 5.

for any two random variables $x(t)$ and $y(t + \tau)$.

$$\therefore \frac{\overline{X_1^2}}{R_{xx}(0)} + \frac{\overline{Y_1^2}}{R_{yy}(0)} - \frac{2\,\overline{X_1 Y_2}}{\sqrt{R_{xx}(0)R_{yy}(0)}} \geq 0$$

$$\therefore 2 \geq \frac{2|R_{xy}(\tau)|}{\sqrt{R_{xx}(0)R_{yy}(0)}}$$

or

$$\sqrt{R_{xx}(0)R_{yy}(0)} \geq |R_{xy}(\tau)|$$

As an example on this bound consider the case from Example 7.7, where

$$x(t) = 3\cos(\omega_c t + \phi) \quad \text{and} \quad y(t) = 2\cos\left(\omega_c t + \phi - \frac{\pi}{2}\right)$$

$$R_{xx}(\tau) = \tfrac{9}{2}\cos\omega_c\tau \qquad R_{yy}(\tau) = 2\cos\omega_c\tau \qquad R_{xy}(\tau) = 3\cos\left(\omega_c\tau - \frac{\pi}{2}\right)$$

We can now find

$$R_{xx}(0) = \tfrac{9}{2} \qquad R_{yy}(0) = 2 \qquad |R_{xy}(\tau)|_{\max} = 3$$

Property 2 states that

$$R_{xx}(0)R_{yy}(0) \geq |R_{xy}(\tau)|$$

$$\therefore \sqrt{(\tfrac{9}{2})(2)} \geq \left|3\cos\left(\omega_c\tau - \frac{\pi}{2}\right)\right| \qquad \text{for all } \tau$$

In this case $R_{xy}(\tau)$ takes on its maximum possible value of 3 at

$$\tau = \frac{\pi}{2\omega_c} + \frac{n2\pi}{\omega_c}$$

for all n since it is periodic and $R_{xx}(0)R_{yy}(0) = |R_{xy}(\tau)|_{\max}^2$.

7.3 CORRELATION FUNCTIONS FOR DISCRETE RANDOM PROCESSES

A discrete random process is specified as

$$x(n) = \sum_{-\infty}^{\infty} x_k \delta(n - k)$$

where the x_k's are random variables whose joint density function is given. The autocorrelation function is defined as

$$R_{xx}(n, k) \triangleq \overline{x_n x_k}$$

$$= \int_{-\infty}^{\infty} \int_{-\infty}^{\infty} \alpha\beta f_{x_n x_k}(\alpha, \beta)\, d\alpha\, d\beta \qquad (7.12)$$

and the covariance function as

$$L_{xx}(n, k) = \overline{(x_n - \bar{x}_n)(x_k - \bar{x}_k)}$$

$$= \int_{-\infty}^{\infty} \int_{-\infty}^{\infty} (\alpha - \bar{x}_n)(\beta - \bar{x}_k) f_{x_n x_k}(\alpha, \beta) \, d\alpha \, d\beta \qquad (7.13)$$

If the process is at least wide-sense stationary,

$$R_{xx}(k) = \overline{x_n x_{n+k}}$$

which depends only on k for any n, and

$$L_{xx}(k) = R_{xx}(k) - \bar{x}^2$$

If the process is ergodic, $R_{xx}(k)$ and $L_{xx}(k)$ may be evaluated on a time average basis as

$$R_{xx}(k) = \overline{x(n)x(n+k)}$$

$$= \lim_{N \to \infty} \frac{1}{2N+1} \sum_{n=-N}^{N} x_p(n)x_p(n+k) \qquad (7.14)$$

and

$$L_{xx}(k) = \overline{[x(n) - \bar{x}][x(n+k) - \bar{x}]}$$

$$= \lim_{N \to \infty} \frac{1}{2N+1} \sum_{n=-N}^{N} [x_p(n) - \bar{x}][x_p(n+k) - \bar{x}]$$

$$= R_{xx}(k) - \bar{x}^2 \qquad (7.15)$$

It is worth noting that we have used $x_p(n)$ here to mean the discrete member function $x_p(n)$ evaluated at $n = n$. We could use the following different notational representations for $R_{xx}(k)$ on a time average basis:

$$\overline{x(n)x(n+k)} = \lim_{N \to \infty} \frac{1}{2N+1} \sum_{n=-N}^{N} x_p(n)x_p(n+k)$$

$$= \lim_{N \to \infty} \frac{1}{2N+1} \sum_{n=-N}^{N} x_{pn}x_{p,n+k}$$

$$= \lim_{N \to \infty} \frac{1}{2N+1} \sum_{n=-N}^{+N} x_n x_{n+k}$$

and be expected to interpret them in the same way from the context. The superscript p very clearly indicates that we are talking about one chosen member of $x(n)$. Although the last formula leaves out p for compactness, it is obviously a time average.

The cross-correlation functions for at least jointly wide-sense stationary random processes are

$$R_{xy}(k) \triangleq \overline{x_n y_{n+k}}$$

$$= \int_{-\infty}^{\infty} \int_{-\infty}^{\infty} \alpha \beta f_{x_n y_{n+k}}(\alpha, \beta) \, d\alpha \, d\beta \qquad (7.16)$$

and

$$R_{yx}(k) \triangleq \overline{x_{n+k}y_n}$$

$$= \int_{-\infty}^{\infty} \int_{-\infty}^{\infty} \alpha\beta f_{x_{n+k}y_n}(\alpha, \beta)\, d\alpha\, d\beta \qquad (7.17)$$

We call $R_{xy}(k)$ the cross correlation of $x(k)$ with $y(k)$ since n is a dummy variable in Eq. 7.16, and we also call $R_{yx}(k)$ the cross correlation of $y(k)$ with $x(k)$. By symmetry, these can also be defined as

$$R_{xy}(k) = \overline{x_{n-k}y_n}$$

and

$$R_{yx}(k) = \overline{x_n y_{n-k}}$$

If the random processes are jointly ergodic, then we may find the cross-correlation functions as time averages

$$R_{xy}(k) \triangleq \lim_{N \to \infty} \frac{1}{2N+1} \sum_{n=-N}^{N} x_{pn} y_{q,n+k}$$

$$R_{yx}(k) \triangleq \lim_{N \to \infty} \frac{1}{2N+1} \sum_{n=-\infty}^{\infty} x_{p,n+k} y_{q,n}$$

where $x_p(n)$ and $y_q(n)$ are any randomly chosen members of $x(n)$ and $y(n)$. In some cases we must be careful with these subscripts. For example, if $y_p(n)$ is the output of a system with $x_p(n)$ as input, then it is implied that we must correlate the member $x_p(n)$ with $y_p(n)$ for any specific p because each is mapped from one point of a sample description space. If $x(n)$ and $y(n)$ are not ergodic but stationary, we could find $R_{xy}(k)$ as the ensemble average of the cross-correlation functions for each $x_p(n)$ and $y_p(n)$.

Most interesting problems with discrete random process will be encountered as examples involving linear systems with random inputs. To cement these definitions and become relaxed with discrete notation, a few examples will be solved.

Example 7.9

Consider the random process

$$x(n) = \sum_{-\infty}^{\infty} x_k \delta(n-k)$$

where the x_k's are characterized by the joint mass function shown in Figure

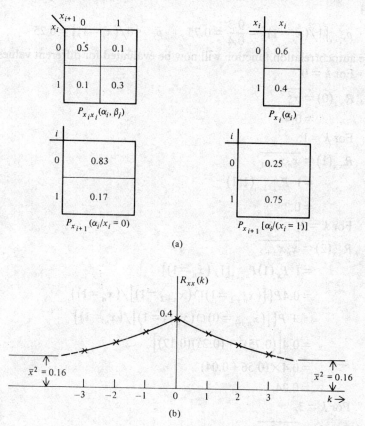

Figure 7.9   (a) The joint, marginal, and conditional mass functions of Example 7.9. (b) The autocorrelation function.

7.9a. Find and plot the autocorrelation function for $k = 0, 1, 2,$ and 3 for this process.

SOLUTION

The probability mass function for any x_i is also shown in Figure 7.9a, and is

$$p_{x_i}(0) = 0.6 \quad \text{and} \quad p_{x_i}(1) = 0.4$$

The mean value of $x(n)$ is

$$\bar{x} = 0(0.6) + 1(0.4)$$
$$= 0.4$$

The conditional mass functions are shown in Figure 7.9a and are

$$p_{x_{i+1}}[1/(x_i = 0)] = \frac{0.1}{0.6} = 0.17 \qquad p_{x_{i+1}}[0/(x_i = 0)] = \frac{0.5}{0.6} = 0.83$$

and

$$p_{x_{i+1}}[1/(x_i=1)] = \frac{0.3}{0.4} = 0.75 \qquad p_{x_{i+1}}[0/(x_i=1)] = 0.25$$

The autocorrelation function will now be evaluated for different values of k.

For $k = 0$,

$$R_{xx}(0) = \overline{x_n^2}$$

$$= 0.4$$

For $k = 1$,

$$R_{xx}(1) = \overline{x_n x_{n+1}}$$

$$= 1 \cdot P_{x_n x_{n+1}}(1,1)$$

$$= 0.3$$

For $k = 2$,

$$R_{xx}(2) = \overline{x_n x_{n+2}}$$

$$= 1 \cdot P_{x_n}(1) P_{x_{n+2}}[1/(x_n=1)]$$

$$= 0.4 P\{[(x_{n+1}=1) \cap (x_{n+2}=1)]/(x_n=1)\}$$

$$\quad + P\{[(x_{n+1}=0) \cap (x_{n+2}=1)]/(x_n=1)\}$$

$$= 0.4[(0.75)^2 + (0.25)(0.17)]$$

$$= 0.4 \times (0.56 + 0.04)$$

$$= 0.24$$

For $k = 3$,

$$R_{xx}(3) = \overline{x_n x_{n+3}}$$

$$= 0.4[(0.75)^3 + (0.75)(0.25)(0.17) + (0.25)(0.17)(0.75)$$

$$\quad + (0.25)(0.83)(0.17)]$$

$$= 0.4[0.42 + (2)(0.03) + 0.04]$$

$$= 0.21$$

For $k = 4$,

$$R_{xx}(4) = \overline{x_n x_{n+4}}$$

$$= 0.4[(0.75)^4 + 3(0.75)^2(0.25)(0.17)$$

$$\quad + 2(0.83)(0.75)(0.25)(0.17) + \cdots]$$

$$= 0.19$$

It is possible to obtain $R_{xx}(k)$ as $R_{xx}(0)$ multiplied by a function of k, but at this stage we will just concentrate on the discrete ensemble averaging. The autocorrelation function is shown plotted in Figure 7.9b. We will summarize our first few results:

For $k = 0$,

$$R_{xx}(0) = (0.4)$$

For $k = 1$,

$$R_{xx}(1) = (0.4)(0.75)^1 = R_{xx}(0) \times 0.75$$

For $k = 2$,

$$R_{xx}(2) = (0.4)(0.75)^2 + (0.4)(0.25)(0.17)$$
$$= R_{xx}(1)(0.75) + R_{xx}(0)$$
$$= 0.75^2 R_{xx}(0) + R_{xx}(0)(0.25)(0.17)$$

Example 7.10

Consider the block diagram of an averager shown in Figure 7.10a. The following discrete random processes are present:

$x(n)$, the input
$y(n) = x(n-1)$, the input delayed by one unit
$s(n) = [x(n) + x(n-1)]$, the summer output
$z(n) = \frac{1}{2}[x(n) + x(n-1)]$, the output of the averager

We may consider our ensemble as consisting of an infinite number of these systems or we may analyze one such system for an ergodic input. Given $x(n) = \Sigma x_k \delta(n-k)$, where the random variables have a joint density function

$$p_{x_i x_{i+1}}(\alpha_i, \beta_j) = 0.25 \qquad i = 0 \text{ or } 1, j = 0 \text{ or } 1$$

(a) Find the cross-correlation function $R_{xy}(k)$.
(b) Find the cross-correlation function $R_{xz}(k)$.
(c) Find the autocorrelation function $R_{zz}(k)$.

SOLUTION
The marginal mass function for x_i is

$$p_{x_i}(0) = p_{x_i}(1) = 0.5$$

and x_n and x_k are independent for all $n \neq k$.

(a) The cross-correlation function $R_{xy}(k)$ is

$$R_{xy}(k) = \overline{x_n y_{n+k}}$$

$$= \overline{x_n (x_{n-1+k})}$$

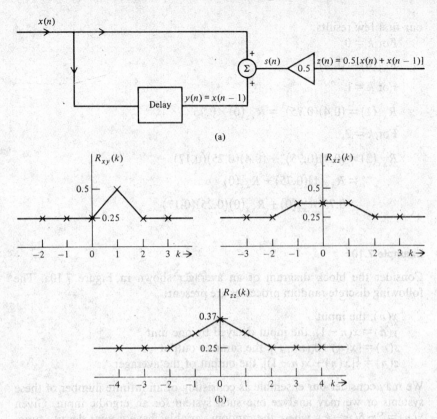

Figure 7.10 (a) Block diagram for the averager of Example 7.10. (b) The cross-correlation and output autocorrelation functions.

If $k = 1$,

$$R_{xy}(k) = \overline{x_n^2}$$

$$= 0.5$$

whereas if $k \neq 1$,

$$R_{xy}(k) = \overline{x_n}^2$$

$$= 0.25$$

This function is shown plotted in Figure 7.10b.

(b) The cross-correlation function $R_{xz}(k)$ is

$$R_{xz}(k) = \overline{x_n\left[\tfrac{1}{2}(x_{n+k} + x_{n-1+k})\right]}$$

$$R_{xz}(k) = \tfrac{1}{2}\,\overline{x_n x_{n+k}} + \tfrac{1}{2}\,\overline{x_n x_{n-1+k}}$$

For $k = 0$,

$$R_{xz}(0) = \tfrac{1}{2}\,\overline{x_n^2} + \tfrac{1}{2}\overline{x}^2$$

$$= 0.375$$

For $k = 1$ we obtain the same numerical result,

$$R_{xz}(1) = \tfrac{1}{2}\,\overline{x^2} + \tfrac{1}{2}\overline{x}^2$$

$$= 0.375$$

For any other k, negative or positive,

$$R_{xz}(k) = \tfrac{1}{2}\overline{x}^2 + \tfrac{1}{2}\overline{x}^2$$

$$= 0.25$$

This cross-correlation function is shown plotted in Figure 7.10b.

(c) The autocorrelation function of the averager output $z(k)$ is

$$R_{zz}(k) = \overline{0.25(x_n + x_{n-1})(x_{n+k} + x_{n-1+k})}$$

$$= 0.25\big(\overline{x_n x_{n+k}} + \overline{x_n x_{n-1+k}} + \overline{x_{n-1} x_{n+k}}$$

$$+ \overline{x_{n-1} x_{n-1+k}}\big)$$

For $k = 0$,

$$R_{zz}(0) = 0.25\big(2\,\overline{x_n^2} + 2\overline{x}_n^2\big)$$

$$= 0.25(1 + 0.5)$$

$$= 0.375$$

For $k = +1$,

$$R_{zz}(1) = 0.25\big(\overline{x^2} + 3\overline{x}^2\big)$$

$$= 0.313$$

For $k = -1$,

$$R_{zz}(-1) = 0.25\big(\overline{x^2} + 3\overline{x}^2\big)$$

$$= 0.313$$

For all other k,

$$R_{zz}(k) = 0.25(4\overline{x}^2)$$

$$= 0.25$$

This autocorrelation function is shown plotted in Figure 7.10b.

A number of detailed comments could now be made about how autocorrelation functions and cross-correlation functions measure randomness. If $R_{xx}(k)$ is greater than $\bar{x}^2$ and the density function is symmetric about the mean, then given $x_{pn} > \bar{x}$, there is a greater than 50% chance that $x_{p,n+k} > \bar{x}$, and the chance becomes closer to 100% as $R_{xx}(k)$ approaches $\bar{x}^2$. It is left for the reader to analyze the situation as was done following Example 7.1.

Properties of Correlation Functions for Discrete Processes

The following are some important properties of correlation functions for at least wide-sense stationary discrete random processes.

Property 1 $R_{xx}(k) = R_{xx}(-k)$.

Property 2 $R_{xx}(0) = \overline{x^2} \geq R_{xx}(k)$.

Property 3 If $x(n)$ is periodic—that is, $x(n+N) = x(n)$—then $R_{xx}(k)$ is also periodic with period N.

Property 4 If $x(n)$ contains no periodic component and $\bar{x} = 0$, then $\Sigma |R_{xx}(k)|$, for all k, is finite and its discrete Fourier transform, called the power spectral density (to be encountered in Chapter 8), exists.

Property 5 $R_{xy}(k) = R_{yx}(-k)$.

Property 6 If x_n and y_k are independent for all n and k,

$$R_{xy}(k) = R_{yx}(k) = \bar{x}\bar{y} \qquad \text{for all } k$$

Property 7 $|R_{xy}(k)| \leq \sqrt{R_{xx}(0)R_{yy}(0)} \qquad$ for all k.

Property 8 If both $x(n)$ and $y(n)$ contain a periodic component with period N, then both $R_{xy}(k)$ and $R_{yx}(k)$ contain a periodic component with period N.

Property 9 If $x(n)$ or $y(n)$ are zero mean and both contain no periodic components, then

$$\sum_{-\infty}^{\infty} |R_{xy}(k)| \quad \text{and} \quad \sum_{-\infty}^{\infty} |R_{yx}(k)|$$

are finite and their discrete Fourier transforms, called the cross spectral densities, exist. We will encounter them in Chapter 8.

Property 10 If $R_{xx}(k) > \bar{x}^2$, then given x_{pn} is greater than $\bar{x}$, $x_{p,n+k}$ is probably greater than $\bar{x}$.

Any of these properties may be demonstrated in an analogous manner to that used in proving the properties of correlation functions for continuous processes.

7.4 EVALUATION OF TIME DOMAIN INTEGRALS

Many important integrals keep recurring throughout system theory. A continuous linear system is characterized in the time domain by its impulse response $h(t)$, and the output for any deterministic finite energy input is the convolution of $x(t)$ with $h(t)$. A discrete system is characterized by its pulse response $h(n)$, and the output for any deterministic finite energy input is the discrete convolution of $x(n)$ with $h(n)$. If the input to a linear system is a member of a continuous random process and it is not a finite energy waveform but a finite power waveform, we characterize the system by $C_{hh}(t)$, which is $h(t)$ correlated with itself, and it will be shown in this section that the output autocorrelation function is the input autocorrelation convolved with $C_{hh}(t)$. If the input to a linear system is a member of a discrete random process which is a finite power waveform, we characterize the system by $C_{hh}(n)$, which is $h(n)$ correlated with itself, and it will be shown in this section that the output autocorrelation function is the discrete convolution of the input autocorrelation function and $C_{hh}(n)$. In addition, many interesting results exist involving convolution and correlation when the cross correlation between input and outputs are found.

Since the conceptual understanding of convolution for deterministic functions and correlation integrals for either deterministic or random functions is almost identical, in this section we will systematically treat their evaluation. The convolution integral, which should already be familiar to the reader, will be presented as a review.

Convolution and Correlation Integrals for Deterministic Functions (A Review)

If we are dealing with finite energy functions—i.e., those for which $\int_{-\infty}^{\infty} |x(t)| \, dt$ exists, we define the following integrals with their respective notations:

$$r_{xy}(t) = x(t) * y(t) \triangleq \int_{-\infty}^{\infty} x(p)y(t-p) \, dp \tag{7.18}$$

$$r_{yx}(t) = y(t) * x(t) \triangleq \int_{-\infty}^{\infty} y(p)x(t-p) \, dp \tag{7.19}$$

$$C_{xx}(t) = x(t) \oplus x(t) \triangleq \int_{-\infty}^{\infty} x(p)x(p-t) \, dp \tag{7.20}$$

$$C_{yx}(t) = y(t) \oplus x(t) \triangleq \int_{-\infty}^{\infty} x(p)y(p-t) \, dp \tag{7.21}$$

$$C_{xy}(t) = x(t) \oplus y(t) \triangleq \int_{-\infty}^{\infty} y(p)x(p-t) \, dp \tag{7.22}$$

where $r_{xy}(t)$ is called the convolution of $x(t)$ with $y(t)$, $C_{xx}(t)$ is called the self-correlation or autocorrelation (the correlation of $x(t)$ with itself), and $C_{xy}(t)$ is called the cross correlation of $x(t)$ with $y(t)$. The symbols $*$ and $\oplus$ are used to indicate convolution and correlation, respectively. Since the mechanics of carrying out these integrals requires discipline and involves much physical interpretation, a number of problems will be solved. For the examples to follow we will use as our two trial functions $x(t)$ and $y(t)$, as shown plotted in Figure 7.11.

Example 7.11

Evaluate the convolution integral

$$r_{xy}(t) = x(t) * y(t) = \int_{-\infty}^{\infty} x(p)y(t - p)\, dp$$

for the two functions shown in Figure 7.11.

SOLUTION
We are being asked to solve an infinite number of problems. For every value of t, $-\infty < t < \infty$, we must evaluate the product of the two functions of p, $x(p)$, and $y(t - p)$ and find the area of their product function. Physically we visualize the function $y(t - p) = y[-(p - t)]$ as being found, by reflecting $y(p)$ and shifting it t units to the right on the p axis (if t is negative, this means $|t|$ units to the left). The normal procedure in evaluating a convolution integral is to start at $t = -\infty$ and work to $t = +\infty$, handling each possible general range of integration. We must very carefully watch the actual limits of integration and the analytic expressions for the product functions. Figure 7.12 shows the different general ranges of t to be considered for this problem. Always $y(t - p)$ is a function extending from $p = t - 1$

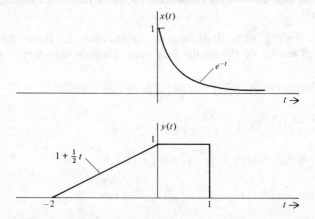

Figure 7.11 The functions used to demonstrate convolution and correlation in Examples 7.11 and 7.12.

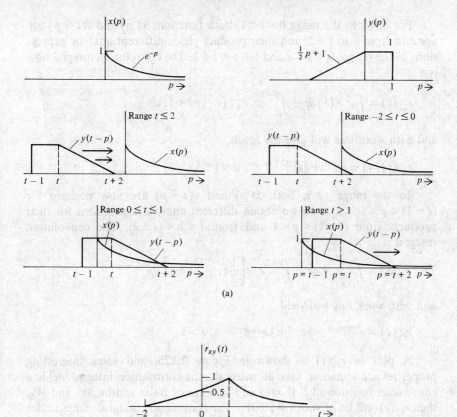

Figure 7.12 (a) The functions in Example 7.11 and the different ranges of t for which the convolution integral must be evaluated. (b) The convolution integral.

to $p = t + 2$, and this reflected version of $y(p)$ is moving to the right as t increases. Referring to Figure 7.12 we now evaluate $r_{xy}(t)$.

For any t in the range $t \le 2$, the product of $x(p)y(t - p) = 0$, since $y(t - p)$ has not reached the lower limit of p for which $x(p)$ is nonzero.

$$\therefore r_{xy}(t) = 0$$

In the range $-2 \le t \le 0$, both functions are nonzero, from $p = 0$ to $p = t + 2$, and the product has one analytic expression. The convolution integral is

$$r_{xy}(t) = \int_0^{t+2} e^{-p} \left[\tfrac{1}{2}(t - p) + 1 \right] dp$$

Treating t as a constant and with substantial work we can find,

$$r_{xy}(t) = \tfrac{1}{2} \left[(t + 1) + e^{-(2+t)} \right]$$

For any t in the range $0 < t < 1$, both functions $x(p)$ and $y(t - p)$ are nonzero from 0 to $t + 2$, and their product yields different analytic expressions in the ranges $0 < p < t$ and $t < p < t + 2$. The convolution integral now is

$$r_{xy}(t) = \int_0^t e^{-p}(1)\, dp + \int_t^{t+2} e^{-p} \left[\tfrac{1}{2}(t - p) + 1 \right] dp$$

and with work this will give the result

$$r_{xy}(t) = 1 - \tfrac{1}{2}e^{-t} + \tfrac{1}{2}e^{-(2+t)} \qquad 0 < t < 1$$

In the range $t > 1$, both $x(p)$ and $y(t - p)$ are now nonzero for $(t - 1) < p < (t + 2)$, and we obtain different analytic expressions for their products from $(t - 1) < p < t$ and from $t < p < (t + 2)$. The convolution integral is

$$r_{xy}(t) = \int_{t-1}^t e^{-p}\, dp + \int_t^{t+2} e^{-p} \left[\frac{1}{2}(t - p) + 1 \right] dp$$

and with work this will yield

$$r_{xy}(t) = e^{-(t-1)} - \tfrac{1}{2}e^{-t} + \tfrac{1}{2}e^{-(2+t)} \qquad t > 1$$

A plot of $r_{xy}(t)$ is shown in Figure 7.12b, and some interesting properties are apparent. Like all integrals, the convolution integral yields a continuous function of t. If $x(t)$ and $y(t)$ are of finite widths, W_1 and W_2, then $r_{xy}(t)$ will be of width $W_1 + W_2$. The convolution integral comes under many different names: "the superposition integral," "the folding integral," "the smoothing integral," and so on, and we will meet it again when summarizing results for linear systems with deterministic inputs and deriving results for linear systems with random inputs in the next section.

Example 7.12

Evaluate the autocorrelation function

$$C_{xx}(t) = x(t) \oplus x(t) = \int_{-\infty}^{\infty} x(p)x(p - t)\, dp$$

where $x(t)$ is as shown in Figure 7.11.

SOLUTION
The mechanics of carrying out a correlation integral for finite energy functions is similar to convolution, though somewhat simpler because no reflection is involved in $x(p - t)$. The function $x(p - t)$ for any fixed value of t represents $x(p)$ shifted t units to the right along the p axis (or $|t|$ units to the left if t is a negative number). Figure 7.13a shows the different general ranges of t to be considered in this correlation integral. We notice that

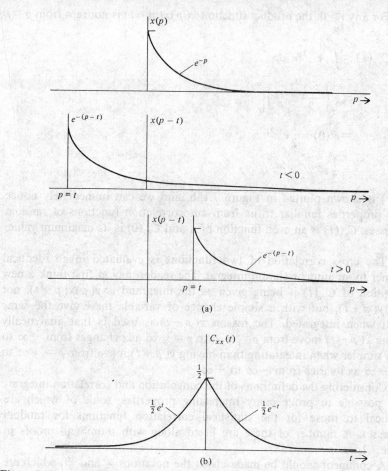

Figure 7.13 (a) Different ranges of t for which the correlation integral must be evaluated in Example 7.12. (b) The correlation integral.

$x(p-t)$ always extends from $p=t$ to $p=+\infty$. Figure 7.13a shows $x(p-t)$ plotted for negative and positive t.

For any $t<0$, the product function $x(p)x(p-t)$ is nonzero from $p=0$ to $p=\infty$.

$$C_{xx}(t) = \int_0^\infty e^{-p} e^{t-p} \, dp$$

$$= e^t \int_0^\infty e^{-2p} \, dp$$

$$= -\frac{1}{2} e^t \left(e^{-2p} \big|_0^\infty \right)$$

$$= \frac{1}{2} e^t \qquad t<0$$

For any $t > 0$, the product function $x(p)x(p-t)$ is nonzero from $p = t$ to $p = \infty$.

$$C_{xx}(t) = \int_t^\infty e^{-2p}e^t \, dp$$

$$= e^t \left(\frac{-1}{2} e^{-2p} \right) \Big|_t^\infty$$

$$= e^t(0) + \frac{1}{2} e^{-2t}$$

$$= \frac{1}{2} e^{-t} \qquad t > 0$$

$C_{xx}(t)$ is shown plotted in Figure 7.13b, and we can immediately notice some properties familiar to us from autocorrelation functions of random processes. $C_{xx}(t)$ is an even function of t, and $C_{xx}(0)$ is its maximum value.

The cross correlation of two functions is evaluated in an identical manner to an autocorrelation integral. The reader may at first think a new definition of $C_{xy}(t)$ is being given as the integrand is $y(p)x(p-t)$, not $x(p)y(p+t)$, but with a simple change of variable these give the same result when integrated. The reason $x(p-t)$ is used is that analytically having $x(p-t)$ move from $p = -\infty$ to $p = +\infty$ as t changes from $-\infty$ to ∞ is simpler when integrating than having $y(p+t)$ move from $p = +\infty$ to $p = -\infty$ as t varies from $-\infty$ to $+\infty$.

Considering the definitions of the convolution and correlation integrals it is possible to prove many interesting properties, some of which are identical to those for the statistical correlation functions for random processes. A number of these are listed along with thumbnail proofs in Table 7.1.

A comment should be made about the notations $*$ and $\oplus$, which are used for convolution and correlation, respectively. Notation is good only up to a point, and one should keep in mind that no amount of it actually evaluates an integral or solves a problem. For example, an expression such as $x(t)*[y(t)*z(t)]$ is strictly notational for convolving $y(t)*z(t)$ and obtaining an answer $r_{yz}(t)$ as a function of t, and then convolving $x(t)$ with $r_{yz}(t)$ and obtaining the answer as a function of t. This is why we write $x(t)*[y(t)*z(t)]$ and use the variable t only. Writing a double integral for this takes some practice:

$$y(t)*z(t) = \int_{-\infty}^\infty y(l)z(t-l) \, dl$$

and

$$x(t)*r_{yz}(t) = \int_{-\infty}^\infty x(p)r_{yz}(t-p) \, dp$$

$$= \int_{-\infty}^\infty x(p) \left[\int_{-\infty}^\infty y(l)z(t-p-l) \, dl \right] dp$$

OPERATION	PROPERTIES	AN ILLUSTRATED MATHEMATICAL PROOF
CONVOLUTION $$x*y \triangleq \int_{-\infty}^{\infty} x(p)y(t-p)\,dp$$ $$r_{xy}(t) = x(t)*y(t)$$	1. $x(t)*y(t) = y(t)*x(t)$ 2. $x(t)*[y(t)+h(t)]$ $= x(t)*y(t) + x(t)*h(t)$ 3. $x(t)*[y(t)*h(t)]$ $= [x(t)*y(t)]*h(t)$	Proof of 1: $$x*y = \int_{-\infty}^{\infty} x(p)y(t-p)\,dp$$ Let $s = t-p$; then $$x*y = \int_{-\infty}^{\infty} x(t-s)y(s)(-ds)$$ $$= \int_{-\infty}^{\infty} x(t-s)y(s)\,ds$$ $$= y(t)*x(t)$$
CORRELATION $$C_{xx}(t) = \int_{-\infty}^{\infty} x(p)x(p-t)\,dp$$ $$C_{xx}(t) = x(t) \oplus x(t)$$	1. $C_{xx}(t) = C_{xx}(-t)$ 2. $C_{xx}(0) \geq C_{xx}(t)$ for any t	Proof of 1: $$C_{xx}(t) = \int_{-\infty}^{\infty} x(p)x(p-t)\,dp \quad\text{(a)}$$ Let $p-t = u$; then $$C_{xx}(t) = \int_{-\infty}^{\infty} x(u+t)x(u)\,du$$ Now $$C_{xx}(-t) = \int_{-\infty}^{\infty} x(p)x[p-(-t)]\,dp$$ $$= \int_{-\infty}^{\infty} x(p+t)x(p)\,dp \quad\text{(b)}$$ (a) = (b)
CROSS CORRELATION $$C_{yx}(t) = \int_{-\infty}^{\infty} x(p)y(p-t)\,dp$$ $$= y(t) \oplus x(t)$$ and $$C_{xy}(t) = x(t) \oplus y(t)$$ $$= \int_{-\infty}^{\infty} y(p)x(p-t)\,dp$$	1. $x \oplus y \neq y \oplus x$ 2. $x \oplus (y+z) = (x \oplus y) + (x \oplus z)$ 3. $x \oplus (y \oplus z) = (x \oplus y) \oplus z$ 4. $C_{xy}(t) = C_{yx}(-t)$	Proof of 4: $$C_{yx}(t) = \int_{-\infty}^{\infty} x(p)y(p-t)\,dp \quad\text{(a)}$$ Let $p-t = s$; then $$C_{yx}(t) = \int_{-\infty}^{\infty} x(s+t)y(s)\,ds$$ $$C_{xy}(-t) = \int_{-\infty}^{\infty} y(u)x[u-(-t)]\,du$$ $$= \int_{-\infty}^{\infty} x(u+t)y(u)\,du \quad\text{(b)}$$ (a) = (b)

Here l and p are dummy variables, and for any specific t we obtain a numerical answer. This skill will be useful when deriving input-output system relations. When we write integral expressions for all the convolution and correlation operations in Table 7.1, we should keep the physical meaning of what we are doing in mind. Notation should be our slave and not vice versa.

DRILL SET: CONVOLUTION AND CORRELATION INTEGRALS FOR DETERMINISTIC FUNCTIONS

Consider

$$x(t) = t - 1 \qquad 0 < t < 1$$
$$= 0 \qquad \text{otherwise}$$

(a) Sketch and give analytical formulas for $x(p - t)$ versus p for $t = -3$, $t = 1$, $t = 4$, and a general t.
(b) Sketch and give analytical formulas for $x(t - p)$ versus p for $t = -3$, $t = 1$, $t = 4$, and a general t.
(c) For what values of t are $x(p)x(p - t)$ and $x(p)x(t - p)$ equal to 0?
(d) Evaluate and sketch $x(t) * x(t)$ and $x(t) \oplus x(t)$.
(e) If

$$y(t) = 2 \qquad 1 < t < 3$$
$$= 0 \qquad \text{otherwise}$$

evaluate and sketch $C_{xy}(t)$ and $C_{yx}(t)$.

Correlation Integrals for Periodic Functions

The preceding subsection defined correlation integrals for finite energy waveforms. A periodic waveform

$$x(t) = \sum_{-\infty}^{\infty} g(t - nT)$$

where $g(t)$ only exists $0 < t < T$, is not a finite energy waveform because

$$\int_{-\infty}^{\infty} x^2(t) \, dt = \infty$$

and so all the definitions for convolution and correlation integrals given in Eqs. 7.18 through 7.22 would blow up or not exist. We consider a periodic waveform to be a finite power waveform as

$$\lim_{T \to \infty} \frac{1}{2T} \int_{-T}^{T} x^2(t) \, dt$$

is finite and may be obtained by averaging over one period. For periodic or general finite power waveforms we make the following definitions for correlation integrals:

$$R_{xx}(\tau) \triangleq \lim_{T \to \infty} \frac{1}{2T} \int_{-T}^{T} x(t)x(t-\tau)\,dt \tag{7.23}$$

is called the time autocorrelation function, and we have encountered it previously for member ensemble waveforms;

$$R_{yx}(\tau) \triangleq \lim_{T \to \infty} \frac{1}{2T} \int_{-T}^{T} x(t)y(t-\tau)\,dt \tag{7.24}$$

is called the cross correlation of $y(\tau)$ with $x(\tau)$; and

$$R_{xy}(\tau) = \lim_{T \to \infty} \frac{1}{2T} \int_{-T}^{T} y(t)x(t-\tau)\,dt \tag{7.25}$$

is called the cross correlation of $x(\tau)$ with $y(\tau)$.

These are general definitions for any finite power waveform. The only finite power waveforms we can handle analytically are periodic waveforms, but if $x(t)$ or $y(t)$ represent nonperiodic members of a random process, we will approximate Eqs. 7.23 through 7.25 based on a large segment of a waveform (see next section). In Example 7.3 we evaluated a time average autocorrelation function for the random process $x(t) = \sin(\omega_c t + \phi)$, where $f_\phi(\alpha) = 1/2\pi$, $0 < \alpha < 2\pi$. One more example will be solved here before proceeding to nonperiodic noise waveforms.

Example 7.13

Given the periodic waveform

$$x(t) = \sum_{n=-\infty}^{\infty} g(t-2n)$$

where

$$\begin{aligned} g(t) &= 1 & 0 < t < 1 \\ &= -1 & 1 < t < 2 \\ &= 0 & \text{otherwise} \end{aligned}$$

evaluate the autocorrelation function

$$R_{xx}(\tau) = \lim_{T \to \infty} \frac{1}{2T} \int_{-T}^{T} x(t)x(t-\tau)\,dt$$

SOLUTION

Figure 7.14 shows a plot of $x(t)$ and of $x(t-\tau)$ for two ranges of τ: $-1 < \tau < 0$ and $0 < \tau < 1$. The actual evaluation of $R_{xx}(\tau)$ is very similar to the case of $C_{xx}(\tau)$ for a finite energy function. In this case, since $x(t)$ is periodic with period 2 and the product function $x(t)x(t-\tau)$ is also a

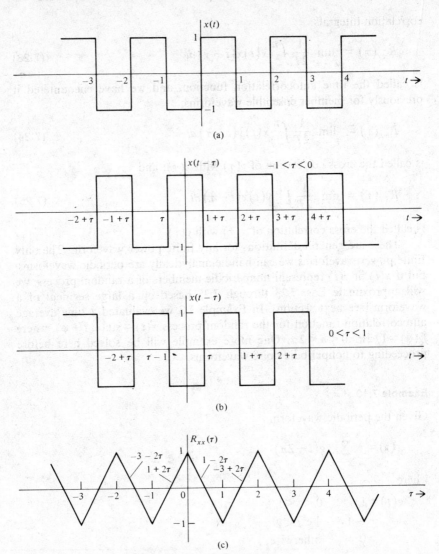

Figure 7.14 (a) The periodic function $x(t)$ for Example 7.13. (b) $x(t-\tau)$ for two different ranges of τ. (c) The autocorrelation function.

periodic function of τ with period 2,

$$\therefore R_{xx}(\tau) = \lim_{T \to \infty} \frac{1}{2T} \int_{-T}^{T} x(t)x(t-\tau)\, dt$$

$$= \frac{1}{2} \int_{t_0}^{t_0+2} x(t)x(t-\tau)\, dt$$

where we may choose any value for t_0. In our case we will use $t_0 = 0$.

For the range $0 < \tau < 1$,

$$R_{xx}(\tau) = \frac{1}{2}\left[\int_0^\tau (1)(-1)\,dt + \int_\tau^1 (1)(+1)\,dt \right.$$
$$\left. + \int_1^{1+\tau}(1)(-1)\,dt + \int_{1+\tau}^2 (-1)(-1)\,dt\right]$$
$$= \frac{1}{2}\{-\tau + (1-\tau) - (1+\tau-1) + [2-(1+\tau)]\}$$
$$= \frac{1}{2}(2-4\tau)$$
$$= 1 - 2\tau$$

For the range $-1 < \tau < 0$, if the integration is carried out the result is
$$R_{xx}(\tau) = 1 + 2\tau$$

By inspection we see that $R_{xx}(\tau)$ is even. Since $R_{xx}(\tau)$ is periodic, it may be plotted as shown in Figure 7.14c. With experience an engineering student should be able to sketch this result by simple inspection.

The mechanics of evaluating a cross-correlation function for two periodic functions is identical to that used for autocorrelation and so an example will not be included. The fact that constructing a random process by assigning a uniform phase shift over a period to a periodic function yields an ergodic random process should be apparent. Statistically saying the resulting autocorrelation function is periodic because obviously

$$\overline{x(t)x(t+\tau)} = \overline{x(t)x(t+\tau+nT_0)}$$

is equivalent to saying, in the time domain analysis of one waveform $x_i(t)$ with period T_0, "Obviously,

$$R_{x_ix_i}(\tau) = R_{x_ix_i}(\tau + nT_0)$$

since the product time function $x(t)x(t-\tau)$ is left unchanged when $x(t-\tau)$ shifts $x(t)$ a multiple of periods to the right or left."

DRILL SET: CORRELATION INTEGRALS FOR PERIODIC FUNCTIONS

Consider the deterministic periodic functions

$$x(t) = \sum_{n=-\infty}^{+\infty} g(t-3n) \quad \text{and} \quad y(t) = \sum_{n=-\infty}^{\infty} g\left(t - 3n + \frac{1}{2}\right)$$

where

$$g(t) = 1 \quad 0 < t < 1$$
$$= 0 \quad \text{otherwise}$$

(a) Find and plot $R_{xx}(\tau)$ and $R_{yy}(\tau)$.
(b) Without doing any unnecessary work, plot $R_{xy}(\tau)$ and $R_{yx}(\tau)$. Is $R_{xy}(\tau) = R_{yx}(-\tau)$, as it should be?

Correlation Integrals for Discrete Waveforms or Quantized Analog Waveforms

This section will concentrate on correlation and convolution integrals for discrete waveforms or the techniques used when continuous waveforms are quantized and analyzed by a digital computer. Formulas will be developed for two classes of waveforms:

1. Finite energy discrete waveforms or approximated continuous waveforms.
2. Finite power discrete waveforms or finite power continuous waveforms approximated by a set of sampled values. In practice these are probably member waveforms of random processes.

The formulas given here will be those that will represent the required quantities in noise analysis. In practice, however, we try to estimate them as accurately as possible. There are numerous special algorithms using fast transforms available for this purpose. The whole study of algorithms is a special topic in its own right, and the combination of good software knowledge and an understanding of noise analysis for systems is an enviable achievement for a modern graduate.

FINITE ENERGY WAVEFORMS

Figure 7.15a and b shows two functions $f(u)$ and $g(u)$ and pulse-type approximations to them, $f_A(u)$ and $g_A(u)$, respectively. Over each little range Δ —say, from $i\Delta + (i+1)\Delta$ — $f(u)$ is approximated by a pulse function with constant value f_i, where f_i is either the value at $f(i\Delta + \frac{1}{2}\Delta)$, or $f(i\Delta)$ or is chosen with some other criterion in mind. Consider that we want to approximately evaluate

$$C_{ff}(t) = \int_{-\infty}^{\infty} f(u)f(u-t)\, du$$

at $t = 0$, $\pm \Delta$, $\pm 2\Delta$, and so on. Instead of using $f(u)$ we will use $f_A(u)$, and the following results will be obtained:

$$C_{ff}(0) = \Delta\{f_{-M_1}^2 + f_{-M_1+1}^2 + \cdots + f_0^2 + \cdots + f_{N_1}^2\}$$

$$C_{ff}(\Delta) = \Delta\{f_{-M_1+1}f_{-M_1} + \cdots + f_0 f_{-1} + f_1 f_0 + \cdots + f_{N_1} f_{N_1-1}\}$$

$$C_{ff}(-\Delta) = \Delta\{f_{-M_1}f_{-M_1+1} + \cdots + f_0 f_1 + f_{N_1-1} f_{N_1}\}$$

$$C_{ff}(2\Delta) = \Delta\{f_{-M_1+2}f_{M_1} + \cdots + f_0 f_{-2} + f_1 f_{-1} + \cdots + f_{N_1-2} f_{N_1}\}$$

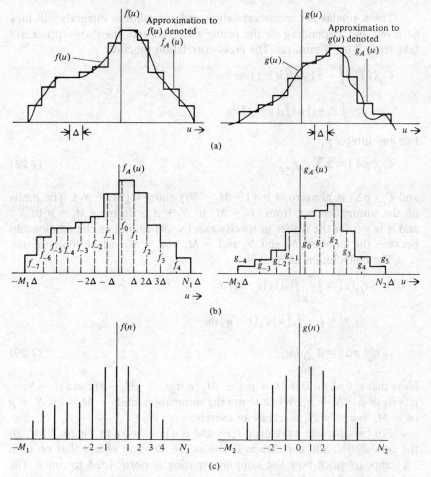

Figure 7.15 (a) Two continuous finite energy functions $f(u)$ and $g(u)$. (b) Pulse or quantized approximations $f_A(u)$ and $g_A(u)$. (c) Discrete functions $f(n)$ and $g(n)$.

These results are written out longhand so that the student may visualize them. In general, we can write

$$C_{ff}(p\Delta) = \Delta \sum_{p=-M_1}^{N_1+p} f_i f_{i-p} \qquad \text{if } -N_1 - M_1 < p < 0 \qquad (7.26)$$

$$= \Delta \sum_{p=-M_1+p}^{N} f_i f_{i-p} \qquad \text{if } 0 \le p < N_1 + M_1 \qquad (7.27)$$

Expressions such as Eqs. 7.26 and 7.27 are very suitable for computer evaluation when N_1 and M_1 are numbers in the hundreds.

The formulas for cross-correlation and convolution integrals will now be given and, depending on the reader's background, their absorption will take from 3 to 50 minutes. The cross-correlation function

$$C_{fg}(t) = \int_{-\infty}^{\infty} g(u)f(u-t)\,du$$

$$\approx \int_{-\infty}^{\infty} g_A(u)f_A(u-t)\,du$$

For any integer p,

$$C_{fg}(p\Delta) = \Delta \sum_i g_i f_{i-p} \qquad (7.28)$$

and $C_{fg}(p\Delta)$ is nonzero if $p > (-M_2 - N_1)$ and $p < (M_1 + N_2)$. The limits on the summation are from $i = -M_2$ to $N_2 + p$ or from $-M_2 + p$ to N_2, and it is left for the reader to specify exactly, depending on the relationship between the numbers N_1 and N_2 and $-M_1$ and $-M_2$, which limits to use.

The convolution integral is

$$r_{fg}(t) = \int_{-\infty}^{\infty} f(u)g(t-u)\,du$$

$$\approx \int_{-\infty}^{\infty} f_A(u)g_A(t-u)\,du$$

$$\therefore r_{fg}(p\Delta) = \Delta \sum_i f_i g_{p-i} \qquad (7.29)$$

Note that $r_{fg}(p\Delta)$ is 0 if $M_2 + p < -M_1$ or if $p < -M_1 - M_2$ and if $-N_1 + p > N_2$ or $p > N_1 + N_2$. When to use the summation limits $-M_1 < i < N_1 + p$ or $-M_1 + p < i < N_1$ is left as an exercise.

For two discrete waveforms $f(n)$ and $g(n)$ as shown in Figure 7.15c, all the convolution and correlation formulas are identical except that no term "Δ" appears since now the sampling spacing is normalized to unity. The discrete convolution is

$$r_{fg}(n) = y(n)$$

where

$$y_p = \sum_{\text{all } i} f_i g_{p-i} \quad \text{or} \quad \sum_{\text{all } i} g_i f_{p-i} \qquad (7.30)$$

The discrete autocorrelation and cross-correlation functions are

$$C_{ff}(n) = y(n)$$

where

$$y_p = \sum_{\text{all } i} f_i f_{i-p} \quad \text{or} \quad \sum_{\text{all } i} f_i f_{i+p} \qquad (7.31)$$

and

$$C_{fg}(n) = y(n)$$

where

$$y_p = \sum_{\text{all } i} g_i f_{i-p} \quad \text{or} \quad \sum_{\text{all } i} f_i g_{i+p} \tag{7.32}$$

At this stage it should be noticed that many interesting random processes are finite energy waveforms and of finite duration. In biomedicine the three classic waveforms associated with the heart, brain, and muscle are the EKG, the EEG, and the EMG, respectively. We could clearly say that they are not stationary random processes, but it would be easy to make them stationary by considering each pattern as one period of a periodic random process. In practice, whenever we analyze a segment of a member of a random process we essentially assume it to be one period from a periodic process.

Correlation Integrals for Finite Power Noise Waveforms

It now remains to adapt the formulas for finite energy functions to infinite energy or finite power-type waveforms.

Figure 7.16a shows a typical member of a continuous random process $f(t)$. Figure 7.16b shows a section of this waveform of length T_1 (the criterion for choosing this section is that it must be statistically representative of the whole waveform, and such a choice is of the utmost importance[3] in signal processing). Figure 7.16c shows representative values that are either sampled from Figure 7.16b or part of a discrete noise waveform. We will denote these values by $f_{-M}, \ldots, f_0, \ldots, f_N$ and say there are $N + M + 1$ such pulses, or sampled values. The choice of pulse width or spacing between sampled values is also very important.[3]

We could now obtain the following approximate values for the autocorrelation function

$$R_{ff}(\tau) = \lim_{T \to \infty} \frac{1}{2T} \int_{-T}^{T} f(t) f(t - \tau)\, dt$$

$$\approx \frac{1}{W} \int_{-M\Delta}^{N\Delta} f_T(t) f_T(t - \tau)\, dt \tag{7.33}$$

where W represents the common range of $f_T(t)$ and $f_T(t - \tau)$, which is $T_1 - \tau$ or $(N + M)\Delta - \tau$, and the formula is useful only when $|\tau| \ll (N + M)$. (Why?)

In order to evaluate $R_{ff}(\tau)$ using a computer,

$$R_{ff}(p\Delta) = \frac{\Delta}{(N + M - |p|)} \sum_{i=-M}^{N} f_i f_{i-p} \tag{7.34}$$

where we use $f_i = 0$, $i < -M$ or $i > N$, and this formula is only applied over

[3] Ralph B. Blackman and John W. Tukey, *The Measurement of Power Spectra*, (New York: Dover Publications, 1958).

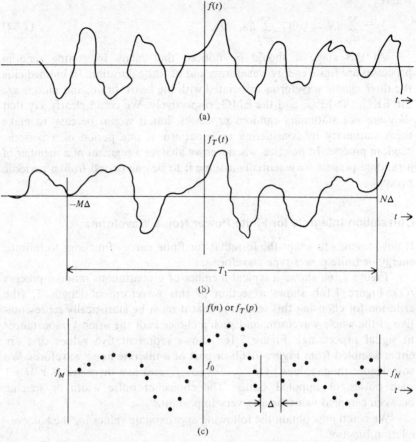

Figure 7.16   (a) A finite power noise waveform $f(t)$. (b) A truncated section of length T_1. (c) Its sampled values or a discrete noise waveform $f(n)$.

the range $-k < p < k$, where $k \ll N + M$. In practice, $N + M$ could be a number of the order of a few hundred and k would be of the order of 20 or 30.

The cross-correlation function $R_{fg}(\tau)$ for $f(t)$ and another function $g(t)$ may be approximated in the same way as for $R_{ff}(p\Delta)$; that is,

$$R_{fg}(p\Delta) = \frac{\Delta}{(N+M-p)} \sum_{i=-M+p}^{N} g_i f_{i-p} \qquad p \text{ positive} \qquad (7.35)$$

and

$$R_{fg}(p\Delta) = \frac{\Delta}{(N+M-|p|)} \sum_{i=-M}^{N-|p|} g_i f_{i-p} \qquad p \text{ negative} \qquad (7.36)$$

and we assume for convenience that both $f_T(t)$ and $g_T(t)$ extend from $-M\Delta$ to $+N\Delta$, where both M and N are positive integer numbers.

If Figure 7.16c represents a section of a discrete member waveform of a random process $f(n)$ and we have another section of a discrete member waveform from a process $g(n)$ also extending from $-M$ to $+N$, then if $f(n)$ and $g(n)$ are jointly ergodic, the formulas for estimations of the correlation functions are

$$R_{ff}(p) = \frac{1}{N+M-p} \sum_{i=-M+p}^{N} f_i f_{i-p} \qquad p \geq 0 \qquad (7.37)$$

and

$$R_{ff}(P) = \frac{1}{N+M+p} \sum_{i=-M}^{N+p} f_i f_{i-p} \qquad p < 0 \qquad (7.38)$$

and

$$R_{fg}(p) = \frac{1}{N+M-|p|} \sum_{\text{all } i} g_i f_{i-p} \qquad (7.39)$$

where the limits are identical as for the autocorrelation function.

At first we might look at these formulas and conclude that the more sample values we use, the more accurate our estimations for the correlation functions should be. Unfortunately in the next chapter we will see that things are not quite this simple. This complicated problem of estimates will be considered further in Chapter 9. We will conclude our coverage of correlation functions by intuitively discussing how a time average correlation integral measures randomness.

TUTORIAL TIME AVERAGE DISCUSSION OF RANDOMNESS

In the first part of this chapter we developed feelings for the randomness of a process by statistically evaluating

$$R_{xx}(\tau) = \overline{x(t)x(t+\tau)}$$

and noticing the rapidity with which the function decreases from its mean square value to its mean value squared. We could develop the same physical feelings about $R_{xx}(\tau)$ in the time domain. Evaluating

$$R_{xx}(\tau) \triangleq \lim_{T \to \infty} \frac{1}{2T} \int_{-T}^{T} x(t)x(t-\tau)\, dt$$

for a finite power waveform is said to be a measure of the similarity of $x(t)$ to itself. In Figure 7.17 is shown a waveform $x(t)$ with zero mean and a shifted version of the waveform $x(t-\tau)$. It should be clear that if the ordinates of the waveforms are multiplied together, the result for $\tau = 0$ is a large positive number, whereas for any other τ there is a tendency for some products to be positive and some negative, so the final sum of the products should be much less than when $\tau = 0$. If $R_{xx}(\tau)$ is plotted for any waveform, it must show a maximum at $\tau = 0$ and decrease to a minimum at $\tau \to \infty$.

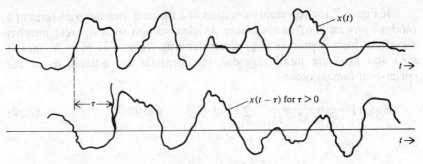

Figure 7.17 A finite power waveform and a shifted version of it.

$R_{xx}(0)$ is the mean square value and $R_{xx}(\infty)$ the mean value squared. The rapidity with which $R_{xx}(\tau)$ approaches its mean value squared is again a measure of the similarity or randomness of $x(t)$. Obviously, a rapidly changing waveform is much more random than a slowly changing one, although they both may assume the same range of values and have the same density function for the random variable of sampling the waveform.

7.5 INPUT-OUTPUT RELATIONS FOR SYSTEMS WITH RANDOM INPUTS

This section will develop formulas for relating the output autocorrelation function of a linear time-invariant causal system to its input autocorrelation function and the cross-correlation function of the input with the output to the input autocorrelation function. Applications of these relations will occur later, when we will discuss the design of linear systems or filters to achieve desired results such as maximizing the output signal to noise at a specified time (the matched filter) or minimizing the mean square error fluctuations in a signal (Wiener filter). As a prelude to these derivations a number of important results from an introductory systems course will be summarized as will certain properties of correlation integrals for functions possessing even symmetry.

Linear Systems with Deterministic Inputs (A Review)

In a first or second course in circuits a linear system with input $x(t)$ and output $y(t)$ is characterized by one of three methods.

The System Differential Equation:

$$a_n \frac{d^n y}{dt^n} + a_{n-1} \frac{d^{n-1} y}{dt^{n-1}} + \cdots + a_0 y(t) = b_0 x(t) + \cdots + b_m \frac{d^m x}{dt^m} \quad (7.40)$$

The System Function H(s):

$$H(s) = \frac{\text{Forced response when } x(t) = e^{st}}{e^{st}}$$

or

$$H(s) = \frac{Y(s)}{X(s)} \tag{7.41}$$

where $Y(s)$ and $X(s)$ are the Laplace transforms of the output and input, respectively. For a system with a differential equation as in Eq. 7.40,

$$H(s) = \frac{b_0 + b_1 s + \cdots + b_m s^m}{a_0 + a_1 s + \cdots + a_n s^n}$$

The Impulse Response h(t):

$$h(t) = y(t) \tag{7.42}$$

when $x(t) = \delta(t)$ and the initial system energy is 0. We may determine $h(t)$ by finding it as the inverse Laplace transform of $H(s)$ or by solving Eq. 7.40. The delta function is probably the most fundamental function in system theory, and theoretically any input $x(t)$ may be written as a string of weighted delta functions. This is shown in stages in Figure 7.18.

$$x(t) \approx \Delta \sum_{-\infty}^{\infty} x(n\Delta) \delta(t - n\Delta)$$

For a linear, causal, time-invariant system, the response due to $x(n\Delta)\delta(t - n\Delta)$ is $x(n\Delta)h(t - n\Delta)$ and the response due to $x(t)$ is

$$y(t) = \sum_{\text{all } n} \Delta x(n\Delta) h(t - n\Delta)$$

$$= \int_{-\infty}^{\infty} x(p) h(t - p) \, dp \tag{7.43}$$

We have demonstrated this result in Figure 7.19.

$$y(t) = x(t) * h(t)$$

or

$$y(t) = h(t) * x(t) \tag{7.44}$$

A very simple problem will be solved to illustrate these results.

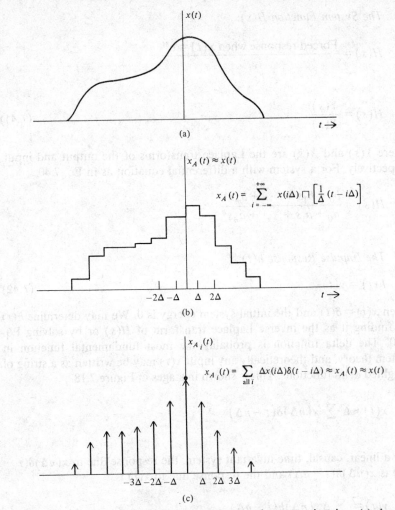

Figure 7.18   (a) A continuous function. (b) A pulse approximation. (c) An approximation by a string of delta functions.

Example 7.14

For the *RC* filter shown:

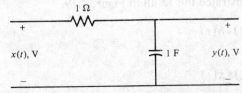

(a) Find $H(s)$, $h(t)$ and the system differential equation.
(b) Find the output when $x(t) = u(t)$, the unit step function.

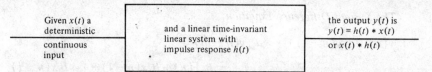

Given $x(t)$ a deterministic	and a linear time-invariant linear system with impulse response $h(t)$	the output $y(t)$ is $y(t) = h(t) * x(t)$
continuous input		or $x(t) * h(t)$

Figure 7.19   The output of a continuous linear system with impulse response $h(t)$.

SOLUTION

(a) For this circuit,

$$H(s) = \frac{1/s}{1 + (1/s)} = \frac{1}{s+1}$$

and

$$h(t) = e^{-t}u(t)$$

The system differential equation is

$$\frac{dy}{dt} + y(t) = x(t)$$

(b) The system output is

$$y(t) = \int_{-\infty}^{\infty} x(t-r)h(r)\,dr$$

where $x(t-r)$ and $h(r)$ are shown here in a thumbnail sketch.

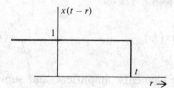
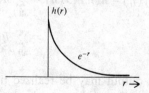

Now $y(t) = 0$ for $t < 0$; and for $t > 0$,

$$y(t) = \int_{0}^{t} e^{-r}\,dr$$

$$= -e^{-r}\big|_{0}^{t}$$

$$= 1 - e^{-t}$$

DISCRETE SYSTEMS

A linear, time-invariant, discrete system with input $x(n)$ and output $y(n)$ is characterized by one of three methods:

The System Difference Equation:

$$a_0 y(n) + a_1 y(n-1) + \cdots a_m y(n-m)$$
$$= b_0 x(n) + b_1 x(n-1) + \cdots b_l x(n-l)$$

$$(7.45)$$

The System Function H(z):

$$H(z) = \frac{Y(z)}{X(z)} \qquad\qquad (7.46)$$

where $Y(z)$ and $X(z)$ are the z transforms of the output and input, respectively. For a system with a difference equation, as in Eq. (7.45),

$$H(z) = \frac{b_0 + b_1 z^{-1} + \cdots + b_m z^{-m}}{a_0 + a_1 z^{-1} + \cdots + a_l z^{-l}}$$

The Pulse Response:

$$h(n) = y(n) \qquad \text{when } x(n) = \delta(n) \qquad (7.47)$$

The response $h(n)$ may be found as the inverse z transform of $H(z)$ or by solving Eq. 7.45 with $x(0) = 1$ and $x(n) = 0$ otherwise.

For a linear time invariant system with impulse response $h(n)$ the output $y(n)$ is

$$y(n) = h(n) * x(n) = \sum_{\text{all } k} h(k)x(n-k)$$
$$= \sum_{\text{all } k} h(n-k)x(k) \qquad (7.48)$$

This output may be realized by delay elements, amplifiers, and summing devices. An analog signal may be approximated by discrete versions of the input and impulse response sampled every τ s. In the latter case the approximate values of $y(n)$ must be scaled by τ. This result is demonstrated in Figure 7.20.

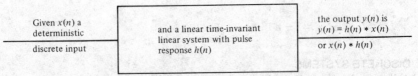

Given $x(n)$ a deterministic discrete input	and a linear time-invariant linear system with pulse response $h(n)$	the output $y(n)$ is $y(n) = h(n) * x(n)$ or $x(n) * h(n)$

Figure 7.20 The output of a discrete linear system with pulse response $h(n)$.

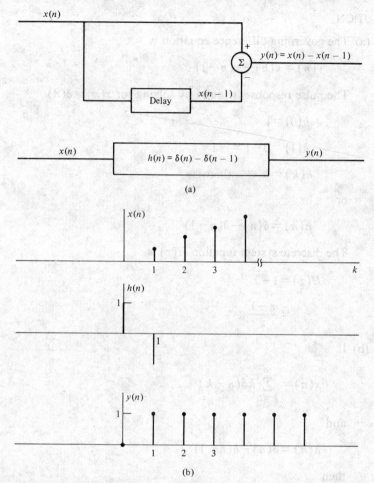

(a)

(b)

Figure 7.21 (a) The discrete system for Example 7.15. (b) The input, pulse response, and output.

Example 7.15

Consider the discrete system shown in Figure 7.21a, where the input $x(n)$ is delayed by one unit and subtracted from itself.

(a) Find the difference equation relating $y(n)$ and $x(n)$, the pulse response $h(n)$, and the discrete system function $H(z)$.

(b) Find $y(n)$ if

$$x(n) = \sum_0^\infty x(k)\delta(n-k)$$

where $x(k) = k$.

SOLUTION

(a) The governing difference equation is

$$y(n) = x(n) - x(n-1)$$

The pulse response is found by solving for $x(n) = \delta(n)$.

$$\therefore h(0) = 1$$

$$h(1) = 0 - 1 = -1$$

$$h(k) = 0 \qquad \text{otherwise}$$

or

$$h(n) = \delta(n) - \delta(n-1)$$

The discrete system function $H(z)$ is

$$H(z) = 1 - z^{-1}$$

$$= \frac{z-1}{z}$$

(b) If

$$x(n) = \sum_{k=0}^{\infty} k\delta(n-k)$$

and

$$h(n) = \delta(n) - \delta(n-1)$$

then

$$y(n) = \sum_{k=0}^{\infty} h(n-k)x(k)$$

Term by term, this yields

$$y(0) = 0$$

$$y(1) = h(1)0 + h(0)1$$

$$= 1$$

$$y(2) = 0 + h(1)1 + h(0)2$$

$$= -1 + 2$$

$$= 1$$

$$y(n) = h(1)(n-1) + h(0)n$$

$$= -(n-1) + n$$

$$= 1$$

$$\therefore y(n) = 1 \qquad n \geq 1$$

Our system is just a differentiator, and since the input is a ramp function, its derivative is always 1. Figure 7.21b shows $x(n)$, $h(n)$, and $y(n)$ graphically.

Correlation Functions for Systems with Random Inputs

The general problem to be considered in this section is, "Given the input to a system is a member of an at least second-order stationary continuous or discrete random process, find the cross-correlation function $R_{xy}(\tau)$ or $R_{xy}(n)$ between the input and the output and, in particular, the output autocorrelation function $R_{yy}(\tau)$ or $R_{yy}(n)$ in terms of $R_{xx}(\tau)$ and $h(\tau)$ or in terms of $R_{xx}(n)$ and $h(n)$." We should be crystal clear on exactly how the random processes are defined. When we talk about $y(t)$ or $y(n)$, we visualize an ensemble of identical continuous or discrete systems as shown in Figure 7.22a and b, where the input to each system is a member of $x(t)$ or $x(n)$. Before we proceed with the derivations, a number of results from convolu-

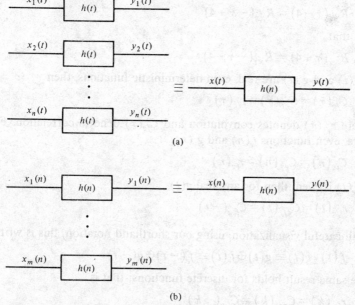

Figure 7.22 (a) An ensemble of continuous system inputs and outputs. (b) An ensemble of discrete system inputs and outputs.

tion and correlation which keep recurring will be stated, as will theorems involving integrating random processes and taking expected values of random processes.

SOME RESULTS FROM CONVOLUTION AND CORRELATION OF REAL FUNCTIONS

Often the properties

and

$$R_{xx}(\tau) = R_{xx}(-\tau) \\ R_{xx}(n) = R_{xx}(-n)$$ for random processes

or

and

$$C_{ff}(\tau) = C_{ff}(-\tau) \\ C_{ff}(n) = C_{ff}(-n)$$ for deterministic functions

are used in not so obvious a fashion. For example, saying that as a function of τ, for any t, r, and l,

$$R_{xx}(t - \tau + r - l) = R_{xx}(\tau - t - r + l)$$

and

$$R_{xx}(t - \tau + r - l) = R_{xx}(\tau - t + r - l)$$

is conceptually equivalent to noting that

$$R_{xx}(\tau - 4) = R_{xx}(-\tau + 4)$$

and that

$$R_{xx}(\tau + 4) \neq R_{xx}(-\tau - 4)$$

If $f_e(t)$ and $g_e(t)$ are real, even deterministic functions, then

$$C_{fg}(\tau) = C_{gf}(\tau) = r_{fg}(\tau)$$

where $r_{gf}(\tau)$ denotes convolution and $C_{fg}(\tau)$ denotes correlation. For discrete, even functions $f_e(n)$ and $g_e(n)$,

$$C_{fg}(n) = C_{gf}(n) = r_{fg}(n)$$

If $f_e(t)$ is even, then for any real $g(t)$,

$$r_{f_e g}(t) = C_{gf_e}(t) = C_{f_e g}(-t)$$

With careful visualization, using our shorthand notation, this is written

$$f_e(t) * g(t) \equiv g(t) \oplus f_e(t) \equiv f_e(-t) \oplus g(-t)$$

The same result holds for discrete functions; that is,

$$r_{f_e g}(k) = C_{gf_e}(k) = C_{f_e g}(-k)$$

INTEGRALS OF RANDOM PROCESSES AND EXPECTED VALUES OF INTEGRALS

Two important theorems on random processes will now be stated. With thought, these theorems should be intuitively acceptable. Indeed, it is imperative to dwell on simple examples, so this will be the case. In subsequent graduate-level courses much time will be devoted to a more rigorous development.

THEOREM 1 (CONTINUOUS CASE)

(a) The integral of a stationary zero-mean random process is a stationary random process.
(b) A weighted integral of a stationary zero-mean random process is a
stationary random process.

We say that the random process $y(t)$ is the integral of $x(t)$ if for each member $x_p(t)$ we form the corresponding member

$$y_p(t) = \int_{-\infty}^{t} x_p(\alpha)\,d\alpha$$

for all t; $-\infty < t < \infty$.

The random process $y(t)$ is a weighted integral of $x(t)$ if, for every $x_p(t)$,

$$y_p(t) = \int_{-\infty}^{t} m(\alpha)x_p(\alpha)\,d\alpha$$

for all t; $-\infty < t < \infty$, where $m(t)$ is a real deterministic function of t.

THEOREM 1 (DISCRETE VERSION)

If $x(n)$ is a zero-mean stationary random process, then $y(n)$ is also zero-mean stationary. Corresponding to $x_p(n)$, $y_p(n)$ is defined by

$$y_p(n) = \sum_{k=-\infty}^{n} x_p(k)$$

for all n; $-\infty < n < \infty$.

If $y(n)$ is a weighted sum of a zero-mean stationary random process $x(n)$, then $y(n)$ is also zero-mean stationary. Corresponding to $x_p(n)$, $y_p(n)$ is defined by

$$y_p(n) = \sum_{k=-\infty}^{n} l(k)x_p(k)$$

where $l(k)$ is a deterministic discrete waveform.

THEOREM 2

The operation of taking an expected value and integration (or summation) may be interchanged. For example, if $x(t)$ is a stationary random process

and

$$y(t) = \int_{-\infty}^{t} m(\alpha)x(\alpha)\, d\alpha \quad \text{or} \quad \int_{-\infty}^{\infty} m(\alpha)x(\alpha, t)\, d\alpha$$

then

$$\overline{y(t)} = \int_{-\infty}^{\infty} m(\alpha)\,\overline{x(\alpha, t)}\, d\alpha$$

$$\overline{y(t)y(t+\tau)} = E\left\{ \left[\int_{-\infty}^{\infty} m(\alpha)x(\alpha, t)\, d\alpha \right] \int_{-\infty}^{\infty} m(r)x(r, t+\tau)\, dr \right\}$$

$$= \int_{-\infty}^{\infty}\int_{-\infty}^{\infty} m(\alpha)m(r)\,\overline{x(\alpha, t)x(r, t+\tau)}\, d\alpha\, dr$$

In the discrete case, if $x(n)$ is a stationary random process,

$$y(n) = \sum_{-\infty}^{n} l(k)x(k) \quad \text{or} \quad \sum_{-\infty}^{\infty} l(k)x(k, n)$$

then

$$\overline{y(n)} = \sum_{-\infty}^{\infty} l(k)\,\overline{x(k, n)}$$

and

$$\overline{y(k)y(k+n)} = E\left[\sum_{-\infty}^{\infty} l(r)x(r, k) \sum_{-\infty}^{\infty} l(m)x(m, k+n) \right]$$

$$= \sum_{-\infty}^{\infty} \left\{ \sum_{-\infty}^{\infty} \left[l(r)l(m)\,\overline{x(r, k)x(m, k+n)} \right] \right\}$$

LINEAR SYSTEMS WITH RANDOM INPUTS

If we consider a linear system with impulse response $h(t)$, whose input is a member of a zero-mean stationary random process with autocorrelation function $R_{xx}(\tau)$, then we can evaluate the following functions.

The Cross-Correlation Function $R_{xy}(\tau)$

$$R_{xy}(\tau) = \overline{x(t)y(t+\tau)}$$

$$= E\left[x(t) \int_{-\infty}^{\infty} h(t+\tau-r)x(r)\, dr \right]$$

$$= E\left[\int_{-\infty}^{\infty} x(t)x(r)h(t+\tau-r)\, dr \right]$$

$$= \int_{-\infty}^{\infty} R_{xx}(r-t)h(t+\tau-r)\, dr \quad \text{(by Theorem 2)}$$

Letting $t - r = s$ and noting $R_{xx}(-s) = R_{xx}(s)$, this becomes

$$R_{xy}(\tau) = \int_{-\infty}^{\infty} R_{xx}(s)h(\tau-s)\, ds$$

$$= R_{xx}(\tau) * h(\tau)$$

and, since $R_{xx}(\tau)$ is even, we could also obtain

$$R_{xy}(\tau) = h(\tau) \oplus R_{xx}(\tau) \quad \text{or} \quad R_{xx}(-\tau) \oplus h(-\tau)$$

In summary,

$$
\begin{aligned}
R_{xy}(\tau) &= R_{xx}(\tau) * h(\tau) \\
&= h(\tau) \oplus R_{xx}(\tau) \quad \text{or} \\
&= R_{xx}(-\tau) \oplus h(-\tau)
\end{aligned}
\tag{7.49}
$$

The Output Autocorrelation Function $R_{yy}(\tau)$

$$
\begin{aligned}
R_{yy}(\tau) &= \overline{y(t)y(t+\tau)} \\
&= E\left[\int_{-\infty}^{\infty} h(t-r)x(r)\,dr \int_{-\infty}^{\infty} h(t+\tau-m)x(m)\,dm \right] \\
&= \int_{-\infty}^{\infty} h(t-r)\left[\int_{-\infty}^{\infty} R_{xx}(r-m)h(t+\tau-m)\,dm \right] dr
\end{aligned}
$$

Interchanging the order of integration, this becomes

$$
R_{yy}(\tau) = \int_{-\infty}^{\infty} h(t+\tau-m)\left[\int_{-\infty}^{\infty} R_{xx}(r-m)h(t-r)\,dr \right] dm
$$

Let $r - m = u$,

$$
R_{yy}(\tau) = \int_{-\infty}^{\infty} h(t+\tau-m) \int_{-\infty}^{\infty} R_{xx}(u)h(t-m-u)\,du\,dm
$$

and interchanging the order of integration again,

$$
\begin{aligned}
R_{yy}(\tau) &= \int_{-\infty}^{\infty} R_{xx}(u) \int_{-\infty}^{\infty} h(t+\tau-m)h(t-m-u)\,dm\,du \\
&= \int_{-\infty}^{\infty} R_{xx}(u)C_{hh}(u+\tau)\,du \quad \text{(Be sure)} \\
&= R_{xx}(\tau) \oplus C_{hh}(\tau)
\end{aligned}
$$

Since $R_{xx}(\tau)$ and $C_{hh}(\tau)$ are both even functions, we could obtain the following equivalent results:

$$
\begin{aligned}
R_{yy}(\tau) &= R_{xx}(\tau) \oplus C_{hh}(\tau) \\
&= C_{hh}(\tau) \oplus R_{xx}(\tau) \\
&= R_{xx}(\tau) * C_{hh}(\tau)
\end{aligned}
\tag{7.50}
$$

This result is called the **Wiener-Khinchine theorem**. The derivations involving manipulations for expected values and interchanging the order of integration keep recurring in system theory and are often handled very poorly by students and even by professors when in a hurry. We should be confident and able to obtain these results in different ways. The reader should repeat the derivation for Eq. 7.50 immediately, proceeding as follows:

$$R_{yy}(\tau) = \overline{y(t)y(t+\tau)}$$

$$= E\left\{\left[\int_{-\infty}^{\infty} h(p)x(t-p)\,dp\right]\left[\int_{-\infty}^{\infty} h(l)x(t+\tau-l)\,dl\right]\right\}$$

and so forth, to obtain the same result.

DISCRETE VERSION OF A LINEAR SYSTEM WITH RANDOM INPUTS

If we consider a discrete system with pulse response $h(n)$ and a zero mean random input with autocorrelation function $R_{xx}(n)$, then we can evaluate $R_{xy}(n)$ as follows:

$$R_{xy}(n) = \overline{x(k)y(k+n)}$$

$$= E\left[x(k)\sum_{\text{all }l} h(k+n-l)x(l)\right]$$

$$R_{xy}(n) = \sum_{l} \overline{x(k)x(l)}\,h(k+n-l)$$

$$= \sum_{l} R_{xx}(l-k)h(k+n-l)$$

Substituting $l-k=r$,

$$R_{xy}(n) = \sum_{r} R_{xx}(r)h(n-r)$$

Allowing for the fact $R_{xx}(r)$ is an even function,

$$R_{xy}(n) = R_{xx}(n) * h(n)$$

$$= h(n) \oplus R_{xx}(n)$$

$$= R_{xx}(-n) \oplus h(-n) \tag{7.51}$$

which is Eq. 7.49 in discrete form.

Now we will derive the discrete version of the Wiener-Khinchine theorem.

$$R_{yy}(n) = \overline{y(k)y(k+n)}$$

$$= E\left[\sum_{l} h(k-l)x(l)\sum_{m} h(n+k-m)x(m)\right]$$

$$= \sum_{m}\left[\sum_{l} \overline{x(l)x(m)}\,h(k-l)\right]h(n+k-m)$$

$$= \sum_{m}\left[\sum_{l} R_{xx}(l-m)h(k-l)\right]h(n+k-m)$$

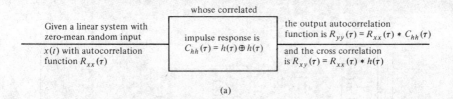

(a)

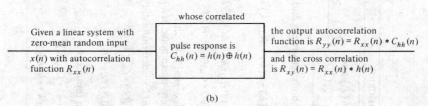

(b)

Figure 7.23   (a) A continuous system with random input. (b) A discrete system with random input.

Substituting $l - m = p$ we obtain

$$R_{yy}(n) = \sum_m \left[\sum_p R_{xx}(p)h(k-m-p) \right] h(n+k-m)$$

Now interchanging the order of summation we obtain

$$R_{yy}(n) = \sum_p R_{xx}(p) \sum_m h(k-m-p)h(n+k-m)$$

$$= \sum_p R_{xx}(p)C_{hh}(p+n)$$

Allowing for the fact both $R_{xx}(n)$ and $C_{hh}(n)$ are even discrete functions we obtain the following equivalent results:

$$R_{yy}(n) = R_{xx}(n) \oplus C_{hh}(n)$$

$$= C_{hh}(n) \oplus R_{xx}(n)$$

$$= C_{hh}(n) * R_{xx}(n) \tag{7.52}$$

Again the reader is exhorted to develop facility, confidence, and courage by repeating these derivations in an alternative manner. Figure 7.23a and b show system diagram summaries of these results. We will now extend this theory to two very important communication problems.

A LINEAR SYSTEM WITH A DETERMINISTIC SIGNAL PLUS NOISE INPUT
In Table 7.2 is shown a linear system with input

$$x(t) = f(t) + n(t)$$

where $f(t)$ is a known deterministic signal and $n(t)$ is a member of an

Table 7.2 LINEAR SYSTEMS WITH SIGNAL PLUS NOISE INPUTS

CONTINUOUS

$$x(t) = f(t) + n(t) \quad \boxed{\begin{array}{c} h(t) \text{ and} \\ C_{hh}(\tau) \end{array}} \quad y(t) = g(t) + m(t)$$

CASE 1 $f(t)$ and $n(t)$ both zero-mean random and uncorrelated

$$R_{xy}(\tau) = R_{ff}(\tau) * h(\tau) + R_{nn}(\tau) * h(\tau)$$

or

$$h(\tau) \oplus R_{xx}(\tau) + h(\tau) \oplus R_{nn}(\tau)$$

$$R_{yy}(\tau) = R_{ff}(\tau) * C_{hh}(\tau) + R_{nn}(\tau) * C_{hh}(\tau)$$

CASE 2 A deterministic signal $f(t)$ plus uncorrelated zero-mean noise $n(t)$

$$g(t) = f(t) * h(t) - \text{deterministic output}$$

$$R_{mm}(\tau) = R_{nn}(\tau) * C_{hh}(\tau)$$

$$R_{nm}(\tau) = R_{nn}(\tau) * h(\tau)$$

CASE 3 Purely random input $x(t)$

$$R_{yy}(\tau) = R_{xx}(\tau) * C_{hh}(\tau)$$

$$R_{xy}(\tau) = R_{xx}(\tau) * h(\tau)$$

DISCRETE

$$x(k) = f(k) + n(k) \quad \boxed{\begin{array}{c} h(k) \text{ and} \\ C_{hh}(k) \end{array}} \quad y(k) = g(k) + m(k)$$

CASE 1 $f(k)$ and $n(k)$ both zero-mean and uncorrelated

$$R_{xy}(k) = R_{ff}(k) * h(k) + R_{nn}(k) * h(k)$$

or

$$h(k) \oplus R_{ff}(k) + h(k) \oplus R_{nn}(k)$$

$$R_{yy}(k) = R_{ff}(k) * C_{hh}(k) + R_{nn}(k) * C_{hh}(k)$$

CASE 2 A deterministic signal $f(k)$ plus uncorrelated zero-mean noise $n(k)$

$$g(k) = f(k) * h(k) - \text{deterministic output}$$

$$R_{mm}(k) = R_{nn}(k) * C_{hh}(k)$$

$$R_{nm}(k) \doteq R_{nn}(k) * h(k)$$

CASE 3 Purely random input $x(k)$

$$R_{yy}(k) = R_{xx}(k) * C_{hh}(k)$$

$$R_{xy}(k) = R_{xx}(k) * h(k)$$

uncorrelated zero-mean random process, which implies that $R_{fn}(\tau) = R_{nf}(\tau) = 0$. We wish to find the system output $y(t)$, the cross correlation $R_{xy}(\tau)$ between $x(t)$ and $y(t)$, and the autocorrelation $R_{mm}(\tau)$ of the output noise. If the impulse response is $h(t)$, then

$$
\begin{aligned}
y(t) &= [f(t) + n(t)] * h(t) \\
&= f(t) * h(t) + n(t) * h(t) \\
&= g(t) + m(t)
\end{aligned}
$$

Obviously, $g(t)$ is the deterministic output signal, and if $n(t)$ is a member of a stationary zero-mean random process, so is $m(t)$. The correlation functions are

$$
\begin{aligned}
R_{xy}(\tau) &= E\left\{ [f(t) + n(t)] \left[\int_{-\infty}^{\infty} h(t + \tau - l)(f(l) + n(l))\, dl \right] \right\} \\
&= \int_{-\infty}^{\infty} R_{ff}(l - t)h(t + \tau - l)\, dl \\
&\quad + \int_{-\infty}^{\infty} R_{nn}(l - t)h(t + \tau - l)\, dl + 0 \\
&= R_{ff}(\tau) * h(\tau) + R_{nn}(\tau) * h(\tau)
\end{aligned}
\tag{7.53}
$$

When using shorthand notation we must be careful; that is,

$$
R_{xy}(\tau) = \overline{\{ [f(t) + n(t)][f(t + \tau) * h(t + \tau) + n(t + \tau) * h(t + \tau)] \}}
$$

could easily be misinterpreted without detailed attention.

The output autocorrelation is as follows:

$$
\begin{aligned}
R_{yy}(\tau) &= \overline{\{ [f(t) + n(t)] * h(t) \}\{ [f(t + \tau) + n(t + \tau)] * h(t + \tau) \}} \\
&= R_{ff}(\tau) * C_{hh}(\tau) + R_{nn}(\tau) * C_{hh}(\tau) \\
&= R_{gg}(\tau) + R_{mm}(\tau)
\end{aligned}
\tag{7.54}
$$

An important problem involving this result, to be encountered in Chapter 9, is to design a "matched filter" with impulse response $h(t)$ so that the signal-to-noise ratio is maximized at the output at some specific time.

The output signal-to-noise (S/N) ratio is defined as

$$
(S/N)_{out} = \frac{g^2(t_0)}{\overline{m^2(t_0)}} = \frac{g^2(t_0)}{R_{mm}(0)}
$$

In the case of deterministic signals, Eqs. 7.53 and 7.54 would probably not be utilized as such, and we just say that the output signal is

$$
g(t) = f(t) * h(t)
\tag{7.55}
$$

and the noise output autocorrelation function is

$$
R_{mm}(\tau) = R_{nn}(\tau) * C_{hh}(\tau)
\tag{7.56}
$$

where $n(t)$ is zero-mean noise uncorrelated with $f(t)$.

For a discrete system with pulse response $h(k)$ and a deterministic input $f(k)$ plus uncorrelated noise $n(k)$, the discrete deterministic output is

$$g(k) = f(k) * h(k) \tag{7.57}$$

and the autocorrelation function of the output noise is

$$R_{mm}(k) = R_{nn}(k) * C_{hh}(k) \tag{7.58}$$

and the cross-correlation function between the input and output is

$$R_{xy}(k) = R_{ff}(k) * h(k) + R_{nn}(k) * h(k) \tag{7.59}$$

The variable k is used to avoid confusion with n in the noise notation $n(k)$.

A LINEAR SYSTEM WITH RANDOM SIGNAL PLUS RANDOM NOISE INPUT

The most general situation that can arise is when the input to a system with impulse response $h(t)$ is

$$x(t) = f(t) + n(t)$$

where both $f(t)$ and $n(t)$ are random with correlation functions $R_{ff}(\tau)$, $R_{nn}(\tau)$, $R_{fn}(\tau)$ and $R_{nf}(\tau)$. The expressions for $R_{xy}(\tau)$ and $R_{yy}(\tau)$ are

$$R_{yy}(\tau)$$

$$= \overline{[f(t)*h(t) + n(t)*h(t)][f(t+\tau)*h(t+\tau) + n(t+\tau)*h(t+\tau)]}$$
$$= R_{ff}(\tau)*C_{hh}(\tau) + R_{nn}(\tau)*C_{hh}(\tau) + R_{fn}(\tau)*C_{hh}(\tau)$$
$$+ R_{nf}(\tau)*C_{hh}(\tau) \tag{7.60}$$

If $f(t)$ and $n(t)$ are uncorrelated and either has zero mean, then

$$R_{yy}(\tau) = R_{ff}(\tau)*C_{hh}(\tau) + R_{nn}(\tau)*C_{hh}(\tau) \tag{7.61}$$

and, as derived previously,

$$R_{xy}(\tau) = R_{ff}(\tau)*h(\tau) + R_{nn}(\tau)*h(\tau) \tag{7.62}$$

An important problem to be discussed in Chapter 9 is that of the Wiener filter, which is to design $h(t)$ such that the error signal

$$\overline{\varepsilon^2(t)} = \overline{[y(t) - f(t)]^2}$$

is minimized, where $R_{ff}(\tau)$ and $R_{nn}(\tau)$ are known. The formulas for a discrete system with pulse response $h(k)$ and a random signal $f(k)$ plus random noise $n(k)$ as input are

$$R_{yy}(k) = R_{ff}(k)*C_{hh}(k) + R_{nn}(k)*C_{hh}(k) + R_{fn}(k)*C_{hh}(k)$$
$$+ R_{nf}(k)*C_{hh}(k) \tag{7.63}$$

and

$$R_{xy}(k) = R_{ff}(k)*h(k) + R_{nn}(k)*h(k) + R_{fn}(k)*h(k)$$
$$+ R_{nf}(k)*h(k) \tag{7.64}$$

This completes the derivations for linear system with signal plus noise inputs. The results for the different possible situations are shown in summary form in Table 7.2. We have now completed the time domain analysis of random processes, and in Chapter 8 we consider their frequency interpretation and the frequency relations for linear systems with signal plus noise inputs.

Example 7.16

Consider the linear system shown in Table 7.2, where $h(t) = e^{-t}u(t)$, and the input consists of the deterministic signal $f(t) = 4\cos 2t$ plus uncorrelated zero mean noise with an autocorrelation function $R_{nn}(\tau) = 2e^{-10|\tau|}$. For analytical purposes assume the noise to be approximately white and hence find the output signal, the output noise autocorrelation function, the input and output signal-to-noise ratios, and the noise figure.

SOLUTION
The system function is $H(s) = 1/(s + 1)$ and the output signal is found most easily by using phasors from a basic circuits course; that is,

$$g(t) = \text{Re}\left[\left(\frac{1}{1 + j2}\right)4\angle 0°e^{j2t}\right]$$

where Re denotes the real part

$$\therefore g(t) = \frac{4}{\sqrt{5}}\cos\left(2t - \tan^{-1}\frac{2}{1}\right)$$

$$= 1.78\cos(2t - 63°)$$

The impulse response is $h(t) = e^{-t}u(t)$, and the correlated impulse response $C_{hh}(\tau)$ is

$$C_{hh}(\tau) = \int_{\tau}^{\infty} e^{-(r-\tau)}e^{-r}\,dr \qquad \tau > 0$$

$$= e^{\tau}\left[-\frac{1}{2}e^{-2r}\Big|_{\tau}^{\infty}\right]$$

$$= \frac{1}{2}e^{-\tau}$$

Since $C_{hh}(\tau)$ is even, then

$$C_{hh}(\tau) = \tfrac{1}{2}e^{-|\tau|} \qquad \text{for all } \tau$$

The output autocorrelation function is given by

$$R_{mm}(\tau) = R_{nn}(\tau) * C_{hh}(\tau)$$

$$= 2e^{-10|\tau|} * \tfrac{1}{2}e^{-|\tau|}$$

Since $2e^{-10|\tau|}$ is very narrow compared to $0.5e^{-|\tau|}$, we may replace it with a delta function weighted by its area.

$$\text{Area of } R_{nn}(\tau) = 2\int_0^\infty 2e^{-10\tau}\,d\tau$$

$$= \frac{4}{-10}\left[e^{-10\tau}\Big|_0^\infty\right]$$

$$= 0.4$$

For this system,

$$R_{nn}(\tau) \approx 0.4\delta(\tau)$$

and

$$R_{mm}(\tau) \approx 0.4\delta(\tau) * 0.5e^{-|\tau|}$$

$$= 0.2e^{-|\tau|}$$

The input signal-to-noise ratio is found as follows:

$$\overline{f^2(\tau)} = \overline{(4\cos 2t)^2}$$

$$= \left(\frac{4}{\sqrt{2}}\right)^2$$

$$= 8$$

$$\overline{n^2(t)} = R_{nn}(0) = 2$$

$$S/N = \frac{\overline{f^2(t)}}{\overline{n^2(t)}} = 4$$

The output signal-to-noise ratio is found as follows:

$$\overline{g^2(t)} = \left(\frac{1.78}{\sqrt{2}}\right)^2 = 1.5$$

$$\overline{m^2(t)} = R_{mm}(0) = 0.2$$

$$S/N = \frac{1.5}{0.2} = 7.5$$

In this case our signal-to-noise ratio is higher at the output than at the input.

The noise figure in decibels is normally defined as

$$NF = -10\log_{10}\frac{S/N|_{\text{out}}}{S/N|_{\text{in}}}$$

$$= 10\log_{10}\frac{7.5}{4}$$

$$= 2.7 \text{ dB}$$

Example 7.17

Consider a discrete system with pulse response $h(n) = \frac{1}{2}\delta(n) + \delta(n-1) + \frac{1}{2}\delta(n-2)$ has an input consisting of the discrete waveform $f(n) = u(n)$ plus uncorrelated, approximately white noise with autocorrelation function $R_{nn}(n) = 2\delta(n)$ and a mean square value of 4.

(a) Find the output signal and the output signal-and-noise autocorrelation function.

(b) Find the input and output signal-to-noise ratios and the noise figure.

SOLUTION

(a) To find the output signal,

$$g(k) = h(k) * x(k) = \{0.5, 1, 0.5\} * \{1, 1, 1, \ldots, 1\}$$

$$g(0) = 0.5$$

$$g(1) = 1.5$$

$$g(k) = 2, \qquad k \geq 2$$

The output autocorrelation function is found as follows:

$$C_{hh}(k) = h(k) \oplus h(k) = \{0.5, 1, 0.5\} \oplus \{0.5, 1, 0.5\}$$

$$C_{hh}(-2) = C_{hh}(2) = 0.25$$

$$C_{hh}(-1) = C_{hh}(1) = 1$$

$$C_{hh}(0) = 1.5$$

otherwise,

$$C_{hh}(k) = 0$$

$$R_{mm}(k) = R_{nn}(k) * C_{hh}(k)$$

$$= 2\delta(k) * C_{hh}(k)$$

$$= 2C_{hh}(k)$$

(b) The input signal-to-noise ratio is found as follows:

$$\overline{f^2(n)} = 1$$

$$\overline{n^2(k)} = 4$$

$$S/N = 0.25$$

The output signal-to-noise ratio is found as follows:

$$\overline{g^2(n)} = 2 \qquad \text{if } n > 2$$

$$\overline{m^2(k)} = R_{mm}(0)$$

$$= 3$$

Hence

$$S/N = 0.67$$

In this case the noise figure (NF) is

$$NF = 10 \log_{10} \frac{0.67}{0.25}$$

$$= 4.3 \text{ dB}$$

SUMMARY

Chapter 7 was devoted to the autocorrelation and cross-correlation functions of continuous and discrete random processes. The analysis problems were treated in depth. These are as follows:

Given a second-order stationary random process, find its autocorrelation function as

$$R_{xx}(\tau) = \overline{x(t)x(t + \tau)}$$

or

$$R_{xx}(k) = \overline{x(n)x(n + k)}$$

from the joint density function.

Given an assumed ergodic process, find its autocorrelation function on a time average basis as

$$R_{xx}(\tau) = \lim_{T \to \infty} \frac{1}{2T} \int_{-T}^{T} x_p(t) x_p(t - \tau) \, dt$$

or

$$R_{xx}(k) = \lim_{N \to \infty} \frac{1}{2N + 1} \sum_{n = -N}^{N} x_p(n) x_p(n + k)$$

where $x_p(t)$ or $x_p(n)$ are any member of the appropriate process, and if the sampling rate is Δ we must rescale by Δ.

Find an estimate of the autocorrelation function of a continuous ergodic random process at $t = (k\Delta)$ or of a discrete random process at k

$$R_{xx}(k) = \frac{1}{2N+1-k} \sum_{\text{all } n} x_p(n)x_p(n+k) \qquad k \ll N$$

where we assume we have $2N+1$ samples or values from $x(-N)$ to $x(+N)$ and for convenience the sampling rate is $\Delta t = 1$.

Given two jointly second-order stationary random processes, find the cross-correlation function as

$$R_{xy}(\tau) = \overline{x(t)y(t+\tau)}$$

or

$$R_{xy}(k) = \overline{x(n)y(n+k)}$$

from the joint density function.

Given assumed jointly ergodic random processes, find the cross correlation on a time average basis as

$$R_{xy}(\tau) = \lim_{T \to \infty} \frac{1}{2T} \int_{-T}^{T} x_p(t-\tau)y_p(t)\, dt$$

or

$$R_{xy}(k) = \lim_{N \to \infty} \frac{1}{2N+1} \sum_{n=-N}^{N} x_p(n-k)y_p(n)$$

Find an estimate of the cross-correlation function of jointly ergodic random processes from $2N+1$ samples of a continuous process or values of a discrete process as

$$R_{xy}(k) = \frac{1}{2N+1-k} \sum_{N} x_p(n)y_p(n+k)$$

where the result is useful $k \ll n$.

The concept of how the autocorrelation function measures randomness was developed intuitively by observing the results of solved problems. Unfortunately in practice we must consider the reverse synthesis problem. Knowing only estimates of the autocorrelation function, how much can we say about the process? The statistical statements we make from observed data is the basis of much of statistical communication theory.

The analysis of linear systems with random inputs in the chapter merged together the disciplines of system analysis and random process theory. The deterministic input-output system relation

$$y(t) = x(t) * h(t)$$

or

$$y(k) = x(k) * h(k)$$

leads to the Wiener-Khinchine relationship of the output and input autocorrelation functions for a linear system with a stationary random input as

$$R_{yy}(\tau) = C_{hh}(\tau) * R_{xx}(\tau)$$

and the relationship of the input-output cross-correlation function to the input autocorrelation function as

$$R_{xy}(\tau) = h(\tau) \oplus R_{xx}(\tau)$$

or

$$R_{xy}(\tau) = R_{xx}(\tau) * h(\tau)$$

$C_{hh}(\tau)$, called the "power pulse response" by the author, is the impulse response correlated with itself, and it is the output autocorrelation function when $R_{xx}(\tau) = \delta(\tau)$ (white noise). The symbols $\oplus$ and $*$ are used to denote correlation and convolution, respectively, for finite energy deterministic functions; that is,

$$f(t) \oplus g(t) \triangleq \int_{-\infty}^{\infty} f(u) g(u+t) \, du$$

$$\triangleq \int_{-\infty}^{\infty} f(u-t) g(u) \, du$$

and

$$f(t) * g(t) \triangleq \int_{-\infty}^{\infty} f(u) g(t-u) \, du$$

Detailed consideration was given to the general system problem of a linear system with signal plus noise input for all cases of correlation between the signal and noise such as whether or not the signal is deterministic or random. The relations developed serve as a starting point to many great classical problems in communication filter theory, and the manipulations and derivations should be practiced until they become second nature.

PROBLEMS

1. Consider the random process

$$x(t) = \sum_{n=-\infty}^{\infty} g(t - 3n + \phi) \qquad n \text{ an integer}$$

where

$$g(t) = 1 \qquad 0 < t < 1$$
$$= 0 \qquad \text{otherwise}$$

and ϕ is a random variable with a density function

$$f_\phi(\alpha) = \tfrac{1}{2} \qquad 0 < \alpha < 2$$
$$= 0 \qquad \text{otherwise}$$

(a) Prove that $x(t)$ is not a first-order stationary process.

(b) Find and plot $m_x(t) \overset{\triangle}{=} \overline{x(t)}$ for all t, $-\infty < t < \infty$.

2. Repeat the questions of Problem 1 for the random process

$$x(t) = \sum_{n=-\infty}^{\infty} g(t - 3n + \phi) \qquad n \text{ an integer}$$

where

$$g(t) = 2t \qquad 0 < t < 1$$
$$= 0 \qquad \text{otherwise}$$

and

$$f_\phi(\alpha) = \tfrac{1}{2} \qquad 0 < \alpha < 2$$
$$= 0 \qquad \text{otherwise}$$

3. Is $m_x(t)$, the mean function of a random process, an even function? If your answer is no, give a specific counterexample.

4. Consider the random process

$$x(t) = A \sin(\omega_0 t + \phi)$$

where A and ϕ are random variables with a joint density function

$$f_{A\phi}(\alpha, \beta) = \frac{1}{8\pi} \qquad 8 < \alpha < 12, 0 < \phi < 2\pi$$

(a) Sketch a few typical members of $x(t)$.

(b) Prove that $x(t)$ is stationary of order 1 by finding $f(\alpha)$. Evaluate $\overline{X}$, $\overline{X^2}$, and σ_x^2.

(c) Prove that $x(t)$ is stationary or order 2 and evaluate $R_{xx}(\tau)$ and $L_{xx}(\tau)$.

(d) Find the time autocorrelation function for any specific member and hence find the ensemble average of all the different time autocorrelation functions. Is your result the same as in part (c)?

(e) Based on your result for part (d), try to develop a theorem relating ensemble autocorrelation functions and the ensemble of time autocorrelation functions for stationary random processes.

5. Consider the Poisson random process composed of pulses of height ± 1 generated in a time T, where

$$P\left(\begin{array}{c} k \text{ transitions} \\ \text{in } T \text{ s} \end{array}\right) = \frac{(\mu T)^k e^{-\mu T}}{k!}$$

Evaluate $R_{xx}(t)$.

6. This problem is a simple extension of the not-too-simple example in the text. Consider the random process with memory where every b s a pulse is generated with value $+1$ or -1. If the value of a pulse is $+1$, then the probability of the next pulse being $+1$ is $\frac{2}{3}$; whereas if the value of a pulse is -1, then the value of the next pulse being -1 is $\frac{2}{3}$.
   (a) Sketch a typical member of $x(t)$ using some mechanism to derive the different pulses.
   (b) Evaluate the autocorrelation function for the ranges $0 < |\tau| < b$ and $b < |\tau| < 2b$ and sketch your result.
   (c) Noticing your result, can you sketch on the graph for part (b) the result for $R_{xx}(\tau)$ when the memory is such that there is a probability of $\frac{1}{3}$ of retaining a pulse value and also the no-memory case where the values $+1$ and -1 are always generated with equal probabilities.

7. (a) Think of a number of random processes that have

$$R_{xx}(\tau) = a\left(1 - \frac{|\tau|}{T}\right) + b\frac{|\tau|}{T} \qquad |\tau| < T$$

$$= b \qquad |\tau| > T$$

   for some constants a, b, and T, for their autocorrelation function.
   (b) Repeat part (a) for

$$R_{xx}(\tau) = (a - b)e^{-n|\tau|} + b \qquad \text{all } \tau$$

   where a, b, and α are constants. Relate all constants to statistics of $f_x(\beta)$ or to pulse widths or some other physical statistic of the process $x(t)$.

8. Consider the random process

$$x(t) = \sum_{-\infty}^{\infty} A_n g(t - n + \phi) \qquad n \text{ an integer}$$

where the A's and ϕ are independent random variables with

$$f_{A_i}(\alpha) = \tfrac{1}{2}\delta(\alpha) + \tfrac{1}{2}\delta(\alpha - 1)$$

and

$$f_\phi(\alpha) = 1 \qquad 0 < \phi < 1$$

If a new random process is formed by synchronously adding three members of $x(t)$ (this means adjusting ϕ to the same value before adding) to obtain

$$z_i(t) = x_i(t) + x_j(t) + x_k(t)$$

or, in general notation (which we must be careful with), the process $z(t)$ is defined as

$$z(t) = x(t) + x(t) + x(t)$$

find

(a) $p_Z(\alpha)$, $\overline{Z}$, $\overline{Z^2}$, and σ_Z^2.

(b) $p_{Z_1 Z_2}(\alpha_i, \beta_j)$ and $R_{ZZ}(\tau)$.

9. Repeat Problem 8 for $z_i(t) = x_i(t) + x_j(t) - x_k(t)$.

10. If $x(t)$ and $y(t)$ are independent zero-mean random processes with autocorrelation functions $R_x(\tau)$ and $R_y(\tau)$, respectively, and we form two new processes

$$z(t) = x(t) + y(t)$$

and

$$u(t) = x(t) - y(t)$$

find in terms of $R_{xx}(\tau)$ and $R_{yy}(\tau)$:

(a) $R_{zz}(\tau)$ and $R_{uu}(\tau)$.

(b) $R_{xy}(\tau)$, $R_{zu}(\tau)$, and $R_{uz}(\tau)$, where $R_{xy}(\tau) \triangleq \overline{x(t)y(t-\tau)}$.

11. For $x(t)$ and $y(t)$, as defined in Problem 10, define

$$z(t) = x(t)y(t)$$

and find $R_{zz}(\tau)$ in terms of $R_{xx}(\tau)$ and $R_{yy}(\tau)$.

12. The derivative of a random process is defined by

$$y(t) = \frac{d}{dt}x(t)$$

which means that each member $x_i(t)$ is differentiated to form $y_i(t)$. Prove that

$$R_y(\tau) = \frac{d^2}{d\tau^2} R_x(\tau)$$

if $R_y(\tau)$ exists. What are the conditions for $R_y(\tau)$ to exist?

13. Give examples of specific random processes that possess and do not possess derivatives.

14. The integral of a random process is defined as

$$y(t) = \int_{-\infty}^{t} x(p)\, dp$$

where we mean that a specific member is formed as

$$y_i(t) = \int_{-\infty}^{t} x_i(p)\, dp$$

What can you say about $R_{yy}(\tau)$?

The following problems are some review problems on convolution and correlation integrals for deterministic finite energy, time functions, plus extensions to periodic and noise waveforms.

15. Given

$$p(t) = 2 \qquad 3 < t < 5$$
$$\quad = 0 \qquad \text{otherwise}$$

and

$$q(t) = 1 \qquad 2 < t < 3$$
$$\quad = 0 \qquad \text{otherwise}$$

(a) Find and plot $p(t-3)$, $p(4-t)$, and $p(-2t-7)$.
(b) Find and plot $p(s-t)$, $p(t-s)$ versus s on the s axis, indicating the different locations of the pulse as the specific number t varies in the range $-\infty < t < \infty$.
(c) For what values of t are

$$p(s)q(t-s)$$
$$p(s)p(s-t)$$

and

$$p(s)q(s-t)$$

nonzero functions of s?
(d) Evaluate

$$r_{pq}(t) = p(t) * q(t) = \int_{-\infty}^{\infty} p(s)q(t-s)\, ds$$

$$C_{pp}(t) = p(t) \oplus p(t) = \int_{-\infty}^{\infty} p(s)p(s-t)\, ds$$

$$C_{pq}(t) = p(t) \oplus q(t) = \int_{-\infty}^{\infty} q(s)p(s-t)\, ds$$

$$C_{qp}(t) = q(t) \oplus p(t) = \int_{-\infty}^{\infty} p(s)q(s-t)\, ds$$

After finding $r(t)$ it is hoped that the other three functions will be obtained by graphically sketching the result by inspection and then obtaining the appropriate analytic expressions.

16. Repeat Problem 15d for

$$p(t) = 2 \qquad 3 < t < 5$$
$$\quad = 0 \qquad \text{otherwise}$$

and

$$q(t) = e^{-t} \qquad t > 0$$
$$\quad = 0 \qquad \text{otherwise}$$

Make sketches to find the ranges of t for which $r(t)$, $C_{pp}(t)$, and $C_{pq}(t)$ exist and sketch the final results.

17. If a function $p(t)$ of width W_1 (that is, $p(t)$ is nonzero only for a continuous range of points $t_1 < t < t_2$, where $t_2 - t_1 = W_1$) and another

function $q(t)$ of width W_2 are being convolved, correlated, and cross-correlated, find the widths of $r(t)$, $C_{pp}(t)$, $C_{pq}(t)$, $C_{qp}(t)$, and $C_{qq}(t)$.

18. Consider the periodic waveforms

$$x(t) = \sum_{-\infty}^{\infty} g_1(t-3n) \qquad n \text{ an integer}$$

where

$$g_1(t) = 1 \qquad 0 < t < 1$$
$$= 0 \qquad \text{otherwise}$$

and

$$y(t) = \sum_{-\infty}^{\infty} g_2(t-3n) \qquad n \text{ an integer}$$

where

$$g_2(t) = t \qquad 0 < t < 1$$
$$= 0 \qquad \text{otherwise}$$

Evaluate

(a) $R_{xx}(\tau) = \lim\limits_{T \to \infty} \dfrac{1}{2T} \displaystyle\int_{-T}^{T} x(t)x(t-\tau)\, dt$

Use the fact that your answer is an even function of τ to do the problem as quickly as possible.

(b) $R_{yy}(\tau)$ and $R_{xy}(\tau)$.

19. Consider the two finite energy functions $x(t)$ and $y(t)$, where $x(t)=0$ if $t<0$ or $t>2$ and $y(t)=0$ if $t<0$ and $t>\frac{3}{2}$ given that we know only the sampled values of these functions—that is, $x(\frac{1}{4})=2$, $x(\frac{1}{2})=1$, $x(\frac{3}{4})=3$, $x(1)=3$, $x(\frac{5}{4})=5$, $x(\frac{3}{2})=0$, $x(\frac{7}{4})=2$ and $y(\frac{1}{4})=3$, $y(\frac{1}{2})=2$, $y(\frac{3}{4})=1$, $y(1)=2$, $y(\frac{5}{4})=1$—and the fact that they are otherwise 0. Find $r_{xy}(t)$, $C_{xx}(t)$ and $C_{xy}(t)$.

20. Given $x(n) = \sum x(k)\delta(n-k)$, where the joint mass function for the random variables is as shown find and plot the autocorrelation and covariance functions $R_{xx}(k)$ and $L_{xx}(k)$.

(a)

$x(k)$	$x(k+1)$ 0	1
0	0.4	0.1
1	0.1	0.4

(b)

$x(k)$	$x(k+1)$ 0	1
0	0.1	0.4
1	0.4	0.1

21. (a) Prove that $R_{xy}(k) \leq [R_{xx}(0)R_{yy}(0)]^{0.5}$.

   (b) If $f(t)$ is even, prove that for another real function, $g(t)$,

$$C_{gf}(\tau) = r_{fg}(\tau) = C_{gf}(-\tau)$$

   where C and r indicate correlation and convolution, respectively.

   (c) If $f(n)$ is even, then for another real discrete function $g(n)$, prove that

$$C_{gf}(n) = r_{g(n)}(n) = C_{fg}(-n)$$

22. Given that the input to a discrete system with

$$h(k) = 2\delta(k) + \delta(k-1)$$

   is a deterministic signal $f(k) = ku(k)$ mixed with zero-mean uncorrelated noise with $R_{nn}(k) = 2\delta(k)$, find

   (a) $\overline{R_{mm}(k)}$, $R_{xy}(k)$, and $g(k)$.

   (b) $\overline{m^2(k)}$.

   where $x(k) = f(k) + n(k)$ and $y(k) = g(k) + m(k)$.

23. Assume that $x(t)$ is a random input to a linear system with $h(t) = e^{-2t}u(t)$ and $R_{xx}(\tau) = 2\delta(\tau)$.

   (a) Find and sketch $R_{xy}(\tau)$, $R_{yx}(\tau)$, and $R_{yy}(\tau)$.

   (b) Do these functions possess the predicted properties?

24. Assume that $x(n)$ is a random input to a linear system with $h(k) = 2\delta(k)$ and $R_{xx}(k) = (0.5)^k u(k)$.

   (a) Find and sketch $R_{xy}(k)$, $R_{yx}(k)$, and $R_{yy}(k)$.

   (b) Do these functions possess the predicted properties?

25. Derive the formulas for a discrete linear system, with pulse response $h(k)$ where

   (a) $x(k) = f(k) + n(k)$, in which $f(k)$ is deterministic and $n(k)$ is zero-mean uncorrelated noise.

   (b) $x(k) = f(k) + n(k)$, in which both signal and noise are zero-mean uncorrelated random processes.

26. Two filters, $h_1(\tau)$ or $h_1(k)$ and $h_2(\tau)$ or $h_2(k)$, have the same random inputs, as shown.

   (a) Derive R_{xy}, R_{yx}, R_{yz}, R_{zy}, R_{yy}, and R_{zz} in terms of R_{xx} and h_1 and h_2, for both the discrete and continuous cases.

   (b) If $R_{xx}(k) = \delta(k)$ and $h_1(k) = u(k)$ and $h_2(k) = (0.5)^k u(k)$ solve for all the quantities listed in part (a).

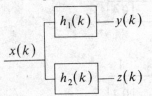

Chapter 8
The Power Spectral Density and Input-Output Relations for a Linear System with Random Inputs

8.1 A SUMMARY OF LAPLACE AND z TRANSFORMS

The time domain interpretation of random processes and of input-output correlation function results for a linear system with signal plus random inputs were developed in Chapter 7. In Chapter 8 we will consider the frequency interpretation of random processes via spectral density functions, and the frequency version of input-output linear-system spectral function relations will be developed.

Even a beginning student of modern system theory and communication theory is expected to be versatile in the use and interrelation of many transforms. Chief among these are:

The one-sided or conventional Laplace transform
The two-sided or bilateral Laplace transform
The one-sided or conventional z transform
The two-sided or bilateral z transform
The Fourier transform
The discrete Fourier transform
Fast transform algorithms

This is a far cry from the not too distant past, when a complete course was often devoted to the one-sided Laplace transform plus a few simple circuit applications. Now much material is presented rapidly and in such a general manner that the expectations from a student are often unrealistic. In addition, faculty members are often so involved with specific current applications that they tend to neglect fundamentals in their desire to land students in an immediate usage situation.

Sections 8.1 and 8.2 will summarize results from z, Laplace, and Fourier transforms, with an emphasis toward their applications in communications. It is assumed the reader is already familiar with the one-sided Laplace and z transforms and with Fourier series theory and the Fourier transform. If such is not the case, Chapters 8 and 9 should motivate a desire for such familiarity. It is further assumed that the reader has a slight acquaintance with complex variables. Some of the more important results from complex Integration and Laurent series are included in Appendix B at the end of the text.

The Two-Sided z Transform

The two-sided z transform associated with a sequence of numbers or a discrete waveform

$$f(n) = \sum_{-\infty}^{\infty} f(k)\delta(n-k)$$

is defined as

$$F(z) = \sum_{-\infty}^{\infty} f(k)z^{-k} \tag{8.1}$$

if $F(z)$ converges for $\rho_1 < |z| < \rho_2$. The motivation for assigning a series to a discrete function is the realization that convolving discrete functions and multiplying series term by term are equivalent. For example, by discrete convolution,

$$\{f(-1), f(0), f(1)\} * \{h(0), h(1)\} = \{g\}$$

where $g(-1) = h(0)f(-1)$, $g(0) = h(0)f(0) + h(1)f(-1)$, $g(1) = h(0)f(1) + h(1)f(0)$, and $g(2) = h(1)f(1)$. Now if we define the z transform as

$$Z[f(n)] = F(z) = f(-1)z + f(0) + f(1)z^{-1} \quad \text{(per Eq. 8.1)}$$

and

$$Z[h(n)] = H(z) = h(0) + h(1)z^{-1}$$

then multiplying the two series,

$$f(-1)z + f(0) + f(1)z^{-1} \quad \text{and} \quad h(0) + h(1)z^{-1}$$

together yields

$$G(z) = \underbrace{f(-1)h(0)z}_{g(-1)} + \underbrace{[f(-1)h(1) + f(0)h(0)]}_{g(0)}$$

$$+ \underbrace{[f(1)h(0) + f(0)h(1)]z^{-1}}_{g(1)} + \underbrace{f(1)h(1)z^{-2}}_{g(2)}$$

which results in the same values for $g(k)$ as before. If $f(n)$ and $h(n)$ are of finite duration, there is obviously no advantage in using the z transform over convolution, but if it is possible to find a closed-form expression for the series for $F(z)$ and also for the series for $H(z)$, then convolution may be avoided by expressing $G(z)$ in a Laurent series to obtain $g(n)$. For this reason the z transform is useful only when

$$F(z) = \sum_{-\infty}^{\infty} f(k)z^{-k}$$

may be expressed as the ratio of two polynomials in z (or in z^{-1}) and the series converges. The convergence limitation implies $F(z)$ will always exist for $\rho_1 < |z| < \rho_2$, which is called the annulus of convergence, as illustrated in the following example.

Example 8.1

Find the two-sided z transform of the following functions:

(a) $f_1(n) = \sum_{-\infty}^{\infty} f(k)\delta(n-k)$ where $f(k) = 1;\ k \geq 0$
                                                    $= 0;$ otherwise

(b) $f_2(n) = \sum_{-\infty}^{\infty} f(k)\delta(n-k)$ where $f(k) = 0;\ k > 0$
                                                    $= 1;$ otherwise

(c) $f_3(n) = \sum_{-\infty}^{\infty} f(k)\delta(n-k)$ where $f(k) = \frac{1}{2}^k;\ k > 0$
                                                    $= 1;$ $k < 0$

SOLUTION

The functions $f_1(n)$, $f_2(n)$, and $f_3(n)$ are shown plotted in Figure 8.1.

(a) By definition,

$$F_1(z) = 1 + z^{-1} + z^{-2} + \cdots + z^{-n} + \cdots$$

$$= \frac{1}{1 - z^{-1}} = \frac{z}{z - 1}$$

and the series converges if $|z^{-1}| < 1$ or $|z| > 1$. A pole-zero diagram of $F_1(z)$ is shown in Figure 8.1a.

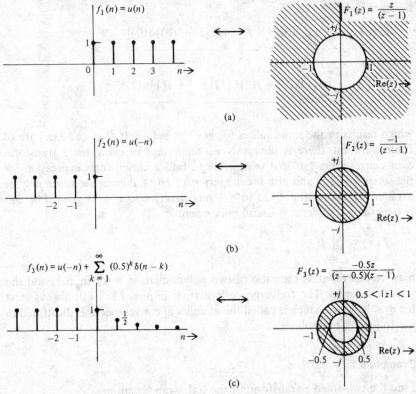

Figure 8.1 Three discrete time functions and their two-sided z transforms in Example 8.1.

(b) By definition,

$$F_2(z) = 1 + z + z^2 + \cdots + z^n + \cdots$$

$$= \frac{1}{1-z} = \frac{-1}{z-1}$$

and this series converges if $|z| < 1$. A pole-zero diagram of $F_2(z)$ is shown in Figure 8.1b.

(c) By definition,

$$F_3(z) = 1 + \tfrac{1}{2}z^{-1} + \tfrac{1}{4}z^{-2} + \cdots + \tfrac{1}{2}^k z^{-k} + \cdots$$

$$+ z + z^2 + z^3 + \cdots + z^k + \cdots$$

$$= \frac{1}{1-\tfrac{1}{2}z^{-1}} + z\left(\frac{1}{1-z}\right)$$

The series of negative powers of z converges if $\left|\tfrac{1}{2}z^{-1}\right| < 1$ or $z > 0.5$, whereas the series of positive powers of z converges if $|z| < 1$. Therefore

$$F_3(z) = \frac{z}{z - \frac{1}{2}} + \frac{-z}{z-1}$$

$$= \frac{z(z-1) + (-z)(z-\frac{1}{2})}{(z-\frac{1}{2})(z-1)}$$

$$= \frac{-0.5z}{(z-\frac{1}{2})(z-1)} \qquad 0.5 < |z| < 1$$

A pole-zero diagram of $F_3(z)$ is shown in Figure 8.1c. We note that the coefficients for negative powers of z converge outside the pole at $z = \frac{1}{2}$ and those for positive powers of z converge inside the pole at $z = 1$.

THE INVERSE TWO-SIDED z TRANSFORM
Given

$$F(z) = \frac{N(z)}{D(z)} \qquad \rho_1 < |z| < \rho_2$$

we may find

$$F(z) = \sum_{-\infty}^{\infty} f(n)z^{-n}$$

or, in Laurent series form,

$$\sum_{-\infty}^{\infty} A_n z^n$$

(see Appendix B) as

$$f(n) = A_{-n} = \frac{1}{2\pi j} \oint_C \frac{F(z)}{z^{-n+1}} \, dz$$

$$= \frac{1}{2\pi j} \oint_C z^{n-1} F(z) \, dz$$

where $C = \rho e^{j\theta}$ and $\rho_1 < \rho < \rho_2$. Hence

$$f(n) = \sum \begin{bmatrix} \text{residues of the poles of} \\ z^{n-1} F(z) \text{ inside } C \end{bmatrix} \qquad (8.2)$$

One of the most powerful theorems—one that is much neglected in traditional complex variable texts—is that for any $K(z) = N_1(z)/D_1(z)$, where $D_1(z)$ is a polynomial of order at least one higher than $N(z)$:

$$\frac{1}{2\pi j} \oint_C \frac{N_1(z)}{D_1(z)} \, dz = \sum \begin{bmatrix} \text{residues of poles} \\ \text{of } K(z) \text{ inside } C \end{bmatrix} \qquad (8.3a)$$

or

$$\frac{1}{2\pi j}\oint_C \frac{N_1(z)}{D_1(z)}\, dz = -\Sigma \left[\begin{array}{l}\text{residues of poles}\\ \text{of } K(z) \text{ outside } C\end{array}\right] \tag{8.3b}$$

The theorem is illustrated in Appendix B, and we will call it the inside-outside theorem.

Returning to the inverse z transform given by Eq. 8.2, we can say that if $F(z) = N(z)/D(z)$, where the order of $D(z)$ is at least one higher than $N(z)$, then

$$f(n) = \frac{1}{2\pi j}\oint z^{n-1}\frac{N(z)}{D(z)}\, dz$$

depending on whether n is positive or negative, can be found from Eq. 8.4 or 8.5.

For $n > 0$, z^{n-1} contributes at most a zero at the origin; hence

$$f(n) = \Sigma \left[\begin{array}{l}\text{residues of poles of}\\ z^{n-1}F(z) \text{ inside } C\end{array}\right] \tag{8.4}$$

For $n \le 0$, z^{n-1} contributes a simple or higher pole of order $|n| + 1$ (if $N(z)$ has no zero at the origin), and finding the residue of this pole involves differentiation for all n. We therefore will use Eq. 8.3b to yield

$$f(n) = -\Sigma \left[\begin{array}{l}\text{residues of poles of}\\ z^{n-1}F(z) \text{ outside } C\end{array}\right] \tag{8.5}$$

A simple inverse two-sided z transform will now be found.

Example 8.2

Given the z transforms

(a) $F_1(z) = \dfrac{z}{(z-1)(z-2)} \qquad 0 < |z| < 1$

(b) $F_2(z) = \dfrac{z}{(z-1)(z-2)} \qquad 1 < |z| < 2$

(c) $F_3(z) = \dfrac{z}{(z-1)(z-2)} \qquad |z| > 2$

find and plot the discrete waveforms $f_1(n)$, $f_2(n)$, and $f_3(n)$.

SOLUTION

(a) $f_1(n) = \dfrac{1}{2\pi j}\oint_{C_1} \dfrac{z^n}{(z-1)(z-2)}\, dz$

If $n > 0$, then $f_1(n) = 0$ since there are no poles inside C_1 for any C defined by $|z| = \rho$, where $0 < \rho < 1$.

If $n \le 0$, then

$$f_1(n) = \frac{1}{2\pi j} \oint_{C_1} \frac{1}{z^{|n|}(z-1)(z-2)} dz$$

In order to avoid the higher-order pole at $z = 0$, the inside-outside theorem will be used to yield

$$f_1(n) = -\Sigma \left(\begin{array}{c} \text{residues of poles at} \\ z = 1 \text{ and } z = 2 \end{array} \right)$$

$$\therefore f_1(n) = -\frac{1}{(-1)} - \frac{1}{2^{|n|}}$$

$$= 1 - \frac{1}{2^{|n|}}$$

and $f_1(n)$ is shown plotted in Figure 8.2a.

(b) $f_2(n) = -\dfrac{1}{2\pi j} \oint_{C_2} \dfrac{z^n}{(z-1)(z-2)} dz$

where C_2 is defined by $|z| = \rho$ and $1 < \rho < 2$.
   If $n > 0$, then

$$f_2(n) = \text{residue of pole at } z = 1$$

$$= \frac{1}{(-1)} = -1$$

   If $n \le 0$, the inside-outside theorem will be used in order to avoid the higher-order pole at $z = 0$.

$$f_2(n) = -(\text{residue of pole at } z = 2)$$

$$= -\frac{1}{2^{|n|}}$$

and $f_2(n)$ is shown plotted in Figure 8.2b.

(c) $f_3(n) = \dfrac{1}{2\pi j} \oint_{C_3} \dfrac{z^n}{(z-1)(z-2)} dz$

where now the region of convergence is outside all the poles as in Figure 8.2c.
   If $n \ge 0$, then

$$f_3(n) = \Sigma \left(\begin{array}{c} \text{residue of the poles at} \\ z = 1 \text{ and } z = 2 \end{array} \right)$$

$$= -1 + 2^n$$

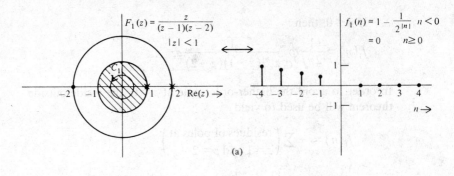

(a)

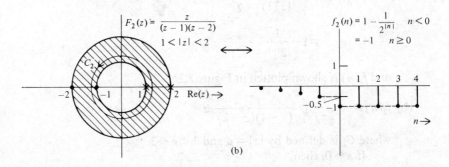

(b)

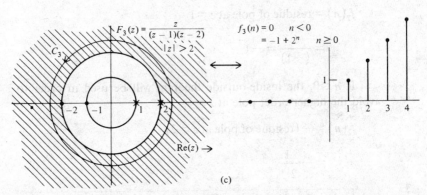

(c)

Figure 8.2   Three two-sided z transforms and their inverses in Example 8.2.

If $n < 0$, in order to avoid the higher-order pole at $z = 0$ the inside-outside theorem is used to yield

$$f_3(n) = -\sum (\text{residues of poles outside})$$

$$= 0$$

since there are no poles outside. The function $f_3(n)$ is shown plotted in Figure 8.2c.

Observing our results we notice that $f_3(n)$ is an unstable function, $f_2(n)$ is borderline stable, and $f_1(n)$ is stable. In general, if

$$F(z) = \frac{N(z)}{D(z)} \qquad \rho_1 < |z| < \rho_2$$

$f(n)$ is stable if $\rho_1 < 1$ and $\rho_2 > 1$.

The One-Sided z Transform

The one-sided z transform of

$$f(n) = \sum_{k=0}^{\infty} f(k)\delta(n-k)$$

is defined as

$$F(z) = \sum_{k=0}^{\infty} f(k)z^{-k} \tag{8.6}$$

if $F(z)$ converges for $|z| > \rho_1$. In the case in which $f(k) = 0$ for all $k \geq n$, then $F(z)$ is a finite polynomial and converges for all z. If $F(z) = N(z)/D(z)$, then $F(z)$ converges for all $|z| > \rho_1$, where ρ_1 is the distance to the pole farthest from the origin or $|z| > \rho_1$ is the annulus outside all the poles of $F(z)$. The theory of the inverse has already been treated in the discussion of the two-sided transform, and its application now yields

$$f(n) = \frac{1}{2\pi j} \oint_C z^{n-1} F(z)\, dz$$

If $n \geq 0$,

$$f(n) = \sum \left[\begin{array}{l} \text{residue of all the} \\ \text{poles of } z^{n-1}F(z) \end{array} \right] \tag{8.7}$$

If $n < 0$, then $f(n) = 0$ since there are no poles of $F(z)$ outside C.

Example 8.3

Find the z transform of

(a) $f(n) = a^n u(n)$.
(b) $f(n) = nu(n)$.

SOLUTION

(a) $F(z) = 1 + az^{-1} + a^2 z^{-2} + \cdots + a^n z^{-n} + \cdots$

$$= \frac{1}{1 - az^{-1}} = \frac{z}{z - a} \qquad |z| > a$$

(b) $F(z) = 0 + z^{-1} + 2z^{-2} + 3z^{-3} + \cdots + nz^{-n}$

$$= \sum_n nz^{-n} \tag{1}$$

Now

$$\frac{z}{z-a} = 1 + az^{-1} + a^2 z^{-2} + \cdots + a^n z^{-n}$$

and

$$\frac{d}{da}\frac{z}{z-a} = \frac{z}{(z-a)^2} \qquad \text{or differentiating term by term}$$

$$= z^{-1} + 2az^{-2} + 3az^{-3} + \cdots + na^{n-1}z^{-n}$$

If $a = 1$, then

$$\frac{z}{(z-1)^2} = \sum nz^{-n} \tag{2}$$

From (1) and (2),

$$z[nu(n)] = \frac{z}{(z-1)^2} \qquad |z| > 1$$

Example 8.4

Given

$$F(z) = \frac{z^{-m}}{(z+0.5)} \qquad |z| > 0.5$$

find and plot $f(n)$ and discuss what operation is implied by multiplying a transform by z^{-m}, where m is a positive integer.

SOLUTION

$$f(n) = \frac{1}{2\pi j}\oint_C \frac{z^{n-1}z^{-m}}{(z-0.5)}\,dz$$

$$= \frac{1}{2\pi j}\oint \frac{z^{n-(m+1)}}{z-0.5}\,dz$$

Now if $n \le m$, there is a pole at $z = 0$ and, by the inside-outside theorem, $f(n)$ is 0.

If $n > m$,

$$f(n) = \sum \left(\begin{array}{c} \text{residue of the pole} \\ \text{at } z = +0.5 \end{array} \right)$$

$$= (0.5)^{n-(m+1)} \qquad n > m$$

$F(z)$ and $f(n)$ are shown plotted in Figure 8.3.

If we define

$$g(n) = (0.5)^n u(n) \quad \text{and} \quad G(z) = \frac{z}{z-0.5}$$

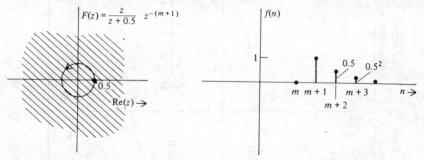

Figure 8.3   Illustration of multiplying by $z^{-(m+1)}$ and shifting to the right $(m+1)$ units in Example 8.4.

then

$$f(n) = g[n-(m+1)] = (0.5)^{n-(m+1)}u[n-(m+1)]$$

and $F(z) = G(z)z^{-(m+1)}$. So shifting $g(n)$ by $m+1$ units to the right in the time domain corresponds to multiplying by $z^{-(m+1)}$ in the z domain, and vice versa.

This section has stressed the use of residue theory in evaluating inverse z transforms as a student should know some complex variable theory but it is also possible to clearly evaluate inverse z transforms using a few basic transforms and partial fractions. Table 8.1 gives a brief summary of some z transforms, and Table 8.2 lists some of the more important theorems for one-sided transforms. Finally, to finish this review we will discuss when the two-sided and one-sided z transforms will be used. (As an exercise, the reader should fill in the blanks in Table 8.1.)

Applications of z Transforms

THE TWO-SIDED z TRANSFORM

The most common discrete functions that exist for both positive and negative time, which we have encountered are:

1. Discrete autocorrelation and cross-correlation functions of random processes $R_{xx}(k)$, $R_{xy}(k)$, and $R_{yx}(k)$.
2. The correlation function $C_{hh}(n)$ of the pulse response $h(n)$ with itself, called the "correlation transfer function."

$$C_{hh}(n) = \sum_{k=n}^{\infty} h(k)h(k-n) \qquad n \geq 0$$

$$= \sum_{k=0}^{\infty} h(k)h(k-n) \qquad n < 0$$

Table 8.1 TWO-SIDED AND ONE-SIDED z TRANSFORM PAIRS

TWO-SIDED z TRANSFORM $X(z)$	REGION OF CONVERGENCE	TIME FUNCTION $x(n)$ $= \Sigma$ RESIDUES OF $z^{n-1}X(z)$	CAUSAL TIME FUNCTION $x(n)$	ONE-SIDED z TRANSFORM
$z^k \quad k>0$	all z	$\delta(n+k)$	$\delta(n)$	1
$z^{-k} \quad k>0$	all z	$\delta(n-k)$	$a^n u(n)$	$\dfrac{z}{z-a}$
$\dfrac{z}{z-1}$	$|z|>1$	$u(n)$	$na^{n-1}u(n)$	$\dfrac{d}{da}\dfrac{z}{z-a}=\dfrac{z}{(z-a)^2}$
	$|z|<1$	$-u(-n-1)$	$n(n-1)a^{n-2}u(n)$	$\dfrac{d}{da}\dfrac{z}{(z-a)^2}=\dfrac{2z}{(z-a)^3}$

$$\frac{z}{z-a}, \quad 0<a<1 \qquad |z|>|a|$$

$$\frac{z}{(z-a)^2}, \quad 0<a<1 \qquad |z|<|a|$$

$$a^n u(n)$$

$$-\left(\frac{1}{a}\right)^n u(-1-n)$$

$$na^{n-1}u(n)$$

$$nu(n)$$

$$n(n-1)u(n)$$

$$\frac{z}{(z-1)^2}$$

$$\frac{2z}{(z-1)^3}$$

$$\frac{Az+B}{(z-a)(z-b)}$$

?

Table 8.2   SOME ONE-SIDED z TRANSFORM THEOREMS

THEOREM	TIME FUNCTION	ONE-SIDED TRANSFORM
Linearity	$ax(n) + by(n)$	$aX(z) + bY(z)$
Shifting	$x(n+1)u(n)$ $x(n-1)u(n)$ $x(n-1)u(n-1)$	$zX(z) - x(0)$ $z^{-1}X(z) + x(-1)$ $z^{-1}X(z)$
	$x(n+k)u(n)$	$z^k X(z)$
	$x(n-k)u(n)$	$z^{-m}X(z) + \displaystyle\sum_{p=1}^{k} x(-p)z^{k-p}$
	$x(n-k)u(n-k)$	$z^{-k}X(z)$
Convolution	$\displaystyle\sum_{p=0}^{n} f(p)g(n-p)$ or $\displaystyle\sum_{p=0}^{n} f(n-p)g(p)$	$F(z)G(z)$

or

$$C_{hh}(n) = h(n) \oplus h(n)$$

3. A member waveform $x_p(n)$ of a random process $x(n)$.

If $R_{xx}(k)$, $R_{xy}(k)$, and $R_{yx}(k)$ are zero mean, then they will possess a two-sided z transform. $C_{hh}(n)$ the correlation transfer function will always possess a two-sided z transform if $h(n) \to 0$ for $n \to \infty$. In addition, since these time functions are real, even functions, their z transforms will have interesting symmetry with respect to the poles inside and outside the annulus of convergence. For a member waveform of a random process we will have to discuss very carefully whether it possesses a z transform since it is an infinite energy function.

THE ONE-SIDED z TRANSFORM
If a discrete system with input $x(n)$ and output $y(n)$ is governed by a difference equation,

$$a_0 y(k) + a_1 y(k-1) + \cdots + a_n y(k-n)$$
$$= b_0 x(k) + \cdots + b_m x(k-m) \tag{8.8}$$

[1] This may also be written

$$a_n y(k+m) + a_{n-1} y(k+n-1) + \cdots + a_0 y(k) = b_0 x(k) + \cdots + b_m x(k+m)$$

then its discrete system function is

$$H(z) = \frac{Z[y(n)]}{Z[x(n)]}$$

$$= \frac{b_0 + b_1 z^{-1} + \cdots + b_m z^{-m}}{a_0 + a_1 z^{-1} + \cdots + a_n z^{-n}} \tag{8.9}$$

$$= \frac{N(z)}{D(z)}$$

The z transform of the output is

$$Y(z) = H(z)X(z)$$

and the discrete output is

$$y(n) = Z^{-1}[H(z)X(z)]$$

$$= \frac{1}{2\pi j} \oint_C z^{n-1} H(z) X(z) \, dz \tag{8.10}$$

where C is a closed contour enclosing all the poles of $Y(z)$. This yields, for $n \geq 0$,

$$y(n) = \sum \begin{bmatrix} \text{residues of poles} \\ \text{of } z^{n-1} H(z)X(z) \end{bmatrix} \tag{8.11}$$

and, for $n < 0$,

$$y(n) = 0$$

If the z transform of the deterministic input $x(n)$ yields the ratio of two polynomials in z then Eq. 8.11 is always utilized to find the deterministic output $y(n)$. This theory is demonstrated schematically in Figure 8.4, where the results obtained by time domain analysis in Chapter 7 are also included for completeness.

The Two-Sided Laplace Transform

The two-sided Laplace transform of a time function $f(t)$ is

$$F_B(s) \triangleq \int_{-\infty}^{\infty} f(t) e^{-st} \, dt \tag{8.12}$$

where $F_B(s)$ is defined for all complex s for which the integral exists. The strict condition for $F_B(s)$ to exist is that

$$\int_{-\infty}^{\infty} |f(t) e^{-st}| \, dt < \infty$$

which leads to an acceptable region of convergence for s. If s is denoted as $s = \sigma + j\omega$, then this condition leads to the strip of convergence $\sigma_1 < \sigma < \sigma_2$.

<table>
<tr><td></td><td>to a discrete system
with difference equation
$a_0 y(k) + \cdots + a_n y(k-n) =$
$b_0 x(k) + \cdots + b_m x(k-m)$</td><td>then the output z transform</td></tr>
<tr><td>Given a deterministic</td><td>or</td><td>of the output is</td></tr>
<tr><td>input $x(n)$</td><td>$H(z) = \dfrac{b_0 + b_1 z^{-1} + \cdots + b_m z^{-m}}{a_0 + a_1 z^{-1} + \cdots + a_n z^{-n}}$</td><td>$Y(z) = H(z) X(z)$
or
$y(n) = \dfrac{1}{2\pi j} \oint z^{n-1} H(z) X(z) \, d(z)$</td></tr>
</table>

Transform Domain (Chapter 8)

$Y(z) = H(z) X(z) \qquad |z| > \zeta_y$

$y(n) = \sum \left[\begin{array}{c} \text{residues of poles of} \\ z^{n-1} H(z) X(z) \end{array} \right] \quad n \geq 0$

$\quad\quad = 0 \quad n \leq 0$

Time Domain (Chapter 7)

$h(n) = z^{-1} [H(z)]$

or the solution to the governing difference equation with
$x(0) = 1, x(n) = 0; n \neq 0.$

$y(n) = h(n) * x(n)$

$\quad = \sum_{k=0}^{n} x(n-k) h(k) \quad n \geq 0$

$\quad = 0 \quad\quad \text{otherwise}$

Figure 8.4 The one-sided z transform in discrete system analysis.

Example 8.5

Evaluate the bilateral Laplace transforms of the following functions:

(a) $f_1(t) = e^{3t} u(-t) + e^{2t} u(t)$.
(b) $f_2(t) = (e^{2t} - e^{3t}) u(t)$.
(c) $f_3(t) = (-e^{2t} + e^{3t}) u(-t)$.

SOLUTION

(a) For $f_1(t) = e^{3t} u(-t) + e^{2t} u(t)$,

$$F_{1B}(s) = \int_{-\infty}^{0} e^{(3-s)t} \, dt + \int_{0}^{\infty} e^{(2-s)t} \, dt$$

$$= \frac{e^{(3-s)t}}{3-s} \bigg|_{-\infty}^{0} + \frac{e^{(2-s)t}}{2-s} \bigg|_{0}^{\infty}$$

$$= \frac{1}{3-s} - \frac{1}{2-s} \qquad 2 < \sigma < 3$$

$$\therefore F_{1B}(s) = \frac{-1}{(s-3)(s-2)} \qquad 2 < \mathrm{Re}(s) < 3$$

It should be noticed that the region of convergence is found from the facts that

$$\frac{e^{(3-s)t}}{3-s} \bigg|_{t=-\infty} = 0 \qquad \text{if } \mathrm{Re}(s) < 3$$

and

$$\left.\frac{e^{(2-s)t}}{2-s}\right|_{t=\infty} = 0 \qquad \text{if } \mathrm{Re}(s) > 2$$

The function $f_1(t)$ and the pole-zero diagram for $F_{1B}(s)$ are shown sketched in Figure 8.5a.

(b) For $f_2(t) = (e^{2t} - e^{3t})u(t)$,

$$F_{2B}(s) = \int_0^\infty e^{(2-s)t}\,dt - \int_0^\infty e^{(3-s)t}\,dt$$

$$= \frac{1}{s-2} - \frac{1}{s-3} \qquad \text{if } \sigma > 2 \text{ and } \sigma > 3$$

$$= \frac{-1}{(s-2)(s-3)} \qquad \text{if } \sigma > 3$$

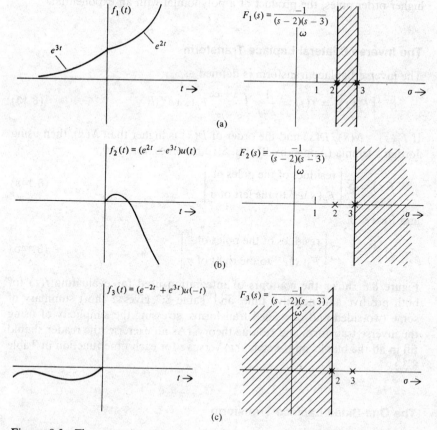

(a)

(b)

(c)

Figure 8.5 Three continuous time functions and their two-sided Laplace transforms in Example 8.5.

The function $f_2(t)$ and the pole-zero diagram for $F_{2B}(s)$ are shown sketched in Figure 8.5b.

(c) For $f_3(t) = (-e^{2t} + e^{3t})u(-t)$ it can easily be shown that

$$F_{3B}(s) = \frac{-1}{(s-2)(s-3)} \qquad -\infty < \sigma < 2$$

The function $f_3(t)$ and the pole-zero diagram for $F_{3B}(s)$ are shown sketched in Figure 8.5c.

It is interesting to note that all three time functions had the same transform but, of course, different regions of convergence. If $F(s)$ is the ratio of two polynomials, then the poles to the right of the region of convergence are determined by $f(t)$ for negative time and the poles to the left are determined by $f(t)$ for positive time. Also, to yield the ratio of two polynomials the time function must be composed of exponential functions (this includes trigonometric and damped sinusoidal functions) and, for higher-order poles, the product of a polynomial with an exponential.

The Inverse Bilateral Laplace Transform

The inverse Laplace transform is defined as

$$\mathcal{L}^{-1}[F_B(s)] = f(t) = \frac{1}{2\pi j} \int_{\sigma - j\infty}^{\sigma + j\infty} F_B(s) e^{st}\, ds \qquad (8.13)$$

If $F_B(s) = N(s)/D(s)$ and the order of $D(s)$ is higher than $N(s)$, then using Jordan's lemma (Appendix B), Eq. 8.13 becomes, for $t > 0$,

$$f(t) = \Sigma \begin{bmatrix} \text{residues of the poles of} \\ F_B(s)e^{st} \text{ to the left of } \sigma \end{bmatrix} \qquad (8.14a)$$

and for $t < 0$,

$$f(t) = -\Sigma \begin{bmatrix} \text{residues of the poles of} \\ F_B(s)e^{st} \text{ to the right of } \sigma \end{bmatrix} \qquad (8.14b)$$

Figure 8.6 shows the contours of integration used for evaluating $f(t)$ for both positive and negative time, and Table 8.3 gives a short summary of some two-sided and one-sided transforms, stressing the simplicity of using the inverse transform and residue theory. (As an exercise, the reader should fill in all the blanks and sketch $x(t)$ versus t for each time function in Table 8.3.)

The One-Sided Laplace Transform

It is assumed that one-sided Laplace transforms, as well as their usage and applications, are of second nature to the reader. For completeness the

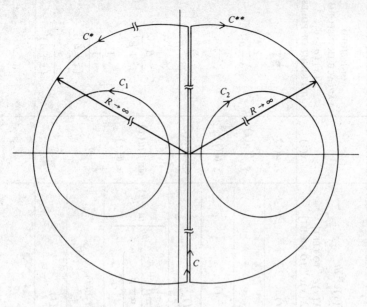

Contour definitions

C: $S(\rho) = \sigma + j\rho$ $-\infty < \rho < \infty$

C^*: $S(\theta) = \sigma + jRe^{j\theta}$ as $R \to \infty$ $\dfrac{\pi}{2} < \theta < \dfrac{3\pi}{2}$

C^{**}: $S(\theta) = \sigma + jRe^{j\theta}$ as $R \to \infty$ $-\dfrac{\pi}{2} < \theta < \dfrac{\pi}{2}$

C_1: $C + C^*$ or any closed path enclosing all the poles to the left of σ
C_2: $C + C^{**}$ or any closed path enclosing all the poles to the right of σ

Figure 8.6   The contours of integration for the inverse Laplace transform.

definitions will be reiterated. The one-sided Laplace transform of $f(t)$ is
defined as

$$F(s) = \int_{0^-}^{\infty} f(t) e^{-st} dt \qquad (8.15)$$

where, if $F(s)$ exists, it does so for an allowable range of $\text{Re}(s) = \sigma$ where
$\sigma > \sigma_1$. If $F(s) = N(s)/D(s)$, then σ is to the right of all the poles.
The inverse transform is defined as

$$\mathcal{L}^{-1}[F(s)] = \frac{1}{2\pi j} \int_{C, \sigma - j\infty}^{\sigma + j\infty} F(s) e^{+st} ds \qquad (8.16)$$

Using Jordan's lemma (Appendix B) the following results are obtained
for positive and negative time:

For $t > 0$, the contour C may be closed to the left as in Figure 8.6 as
$F(s)e^{st} \to 0$, $t > 0$, on C^*.

$$\therefore f(t) = \frac{1}{2\pi j} \oint_{C + C^*} F(s) e^{st} ds$$

$$= \text{residues of poles of } F(s) e^{st}$$

Table 3.3 TWO-SIDED AND ONE-SIDED LAPLACE TRANSFORM

TWO-SIDED LAPLACE TRANSFORM $X(s)$	REGION OF CONVERGENCE	$x(t) = \dfrac{1}{2\pi j}\oint_C X(s)e^{st}\,dt$ $t>0 = \Sigma[\text{RESIDUES OF } X(s)e^{st} \text{ TO LEFT}]$ $t<0 = -\Sigma[\text{RESIDUES OF } X(s)e^{st} \text{ TO RIGHT}]$	$x(t)$	ONE-SIDED LAPLACE TRANSFORM $X(s)$ ($\sigma > \sigma_1$, σ_1 TO RIGHT OF POLES)
$\dfrac{A}{s}$	$\sigma < 0$	$-Au(-t)$	$u(t)$	$\dfrac{1}{s}$
	$\sigma > 0$	$Au(t)$	$e^{-at}u(t)$	$\dfrac{1}{s+a}$
$\dfrac{A}{s+\alpha}$	$\sigma < -\alpha$	$-Ae^{-\alpha t}u(-t)$	$tu(t)$	$\dfrac{1}{s^2}$
	$\sigma > -\alpha$	$Ae^{-\alpha t}u(t)$	$\cos \beta t\,u(t)$	$\dfrac{s}{s^2+\beta^2}$
$\dfrac{A_1s+A_2}{(s+\alpha)(s+\beta)}\quad -\alpha > -\beta$	$\sigma < -\beta$	$-\left[\underbrace{\dfrac{A_1(-\alpha)+A_2}{-\alpha+\beta}}_{r_1}e^{-\alpha t}+\underbrace{\dfrac{A_1(-\beta)+A_2}{\alpha-\beta}}_{r_2}e^{-\beta t}\right]u(-t)$	$\sin \beta t\,u(t)$	$\dfrac{\beta}{s^2+\beta^2}$
	$-\beta < \sigma < -\alpha$	$-r_1u(-t)+r_2u(t)$	$\delta(t)$	1
	$\sigma > -\alpha$	$(r_1+r_2)u(t)$	?	$\dfrac{1}{(s+a)^2}$
$\dfrac{A_1s+A_2}{s^2+\beta^2}$	$\sigma < 0$	?	?	$\dfrac{s}{(s+a)^2}$
	$\sigma > 0$	?		

For $t < 0$, the contour C may be closed to the right as in Figure 8.6 as $F(s)e^{st} \to 0$; $t < 0$, on C^{**}.

$$\therefore f(t) = \frac{1}{2\pi j} \oint_{C + C^{**}} F(s)e^{st}\,ds$$
$$= 0$$

since no poles are included.

Table 8.4 lists some well-known one-sided Laplace transform theorems, which should already be familiar to the reader.

Applications of Laplace Transforms

TWO-SIDED LAPLACE TRANSFORM APPLICATIONS
The most common continuous time functions that exist for both positive and negative time in system theory are as follows:

1. The continuous autocorrelation and cross-correlation functions $R_{xx}(\tau)$, $R_{xy}(\tau)$, and $R_{yx}(\tau)$.
2. The correlation function of the impulse response $h(t)$ with itself,

Table 8.4 SOME ONE-SIDED LAPLACE TRANSFORM THEOREMS

THE LINEARITY THEOREM

$$ax(t) + by(t) \longleftrightarrow aX(s) + bY(s)$$

THE SCALING THEOREM

$$y(at) \longleftrightarrow \frac{1}{a} Y\left(\frac{s}{a}\right) \qquad \text{if } a > 0$$

This theorem cannot be defined for $a < 0$. (Why?)

THE SHIFTING THEOREM

$$y(t - a)u(t - a) \longleftrightarrow e^{-as}F(s) \qquad \text{if } a > 0$$

THE CONVOLUTION THEOREM

$$x(t) * y(t) = \int_0^t x(r)y(t - r)\,dr \longleftrightarrow X(s)Y(s)$$

THE DERIVATIVE THEOREM

$$y'(t) \longleftrightarrow sY(s) - y(0^+)$$

and

$$y^n(t) \longleftrightarrow s^n Y(s) - \sum_{p=0}^{n-1} s^{p-n-1} y^p(0^+)$$

s DERIVATIVE THEOREM

$$ty(t) \longleftrightarrow -F'(s)$$
$$t^n y(t) \longleftrightarrow (-1)^n F^n(s)$$

called the "correlation transfer function," given by

$$C_{hh}(\tau) = \int_{-\infty}^{\infty} h(p)h(p \pm \tau)\, dp$$

which is designated as

$$C_{hh}(\tau) = h(\tau) \oplus h(\tau)$$

3. A member waveform $x_p(t)$ of a random process $x(t)$.

If $R_{xx}(\tau)$, $R_{xy}(\tau)$, and $R_{yx}(\tau)$ are of zero mean, then they will possess two-sided Laplace transforms. $C_{hh}(\tau)$, the correlation time transfer function, will always possess a two-sided Laplace transform if the impulse response approaches zero for infinite positive time. (Does this always happen?) In addition, since all correlation functions are real, even functions of time, the Laplace transforms will possess interesting symmetry properties with respect to the poles to the left and right of σ. For a member waveform of a random process we will have to be very careful in deciding whether or not a bilateral transform can be defined.

ONE-SIDED LAPLACE TRANSFORM APPLICATIONS

The one-sided Laplace transform is used to find the response of linear time invariant systems to continuous inputs. If a system is governed by the linear differential equation

$$a_n \frac{d^n y}{dt^n} + a_{n-1} \frac{d^{n-1}}{dt^{n-1}} + \cdots + a_0 y(t) = b_0 x(t) + \cdots + b_m \frac{d^m x}{dt^m} \quad (8.17)$$

where $x(t)$ is the input and $y(t)$ the output, then the system function $H(s)$ is defined as

$$H(s) = \frac{\mathcal{L}[y(t)]}{\mathcal{L}[x(t)]}$$

$$= \frac{b_0 + b_1 s + \cdots + b_m s^m}{a_0 + a_1 s + \cdots + a_n s^n} \quad (8.18)$$

For a passive network composed of resistors, inductors, and capacitors (plus transformers and gyrators), $H(s)$ is very easily found. (How?) Given $H(s)$, which is always the ratio of two polynomials in s, and $x(t)$ the input, the zero initial energy output is

$$y(t) = \frac{1}{2\pi j} \int_{\sigma - j\infty}^{\sigma + j\infty} H(s) X(s) e^{st}\, ds$$

For $t > 0$, this becomes

$$y(t) = \sum \left[\begin{array}{l} \text{residues of the poles} \\ \text{of } H(s)X(s)e^{st} \end{array} \right] \quad (8.19)$$

	to a system with a differential equation, $$a_0 y(t) + a_1 \frac{dy}{dt} + \cdots + a_n \frac{d^n y}{dt^n} =$$ $$b_0 x(t) + \cdots + b_m \frac{d^m x}{dt^m}$$ or system function $$H(s) = \frac{b_0 + b_1 s + \cdots + b_m s^m}{a_0 + a_1 s + \cdots + a_n s^n}$$	
Given a continuous		then the output transform is
input $x(t)$		$Y(s) = H(s)X(s)$ or the output is $y(t) = \dfrac{1}{2\pi j} \displaystyle\oint H(s)X(s)e^{st}\, ds \quad t > 0$ $\qquad\quad = 0 \quad t < 0$

Transform Domain (Chapter 8)

$Y(s) = H(s) X(s) \qquad \text{Re}(s) = \sigma > \sigma_y$

$$y(t) = \sum \begin{bmatrix} \text{residues of poles of} \\ H(s)X(s)e^{st} \end{bmatrix} \quad t > 0$$

$\qquad = 0 \quad t < 0$

Time Domain (Chapter 7)

The impulse response $h(t) = \mathcal{L}^{-1}[H(s)]$, or it is the solution to the governing differential equation with zero initial energy and input $\delta(t)$

$y(t) = h(t) * x(t)$

$$\qquad = \int_{-\infty}^{\infty} x(t - p)h(p)\, dp \quad t > 0$$

$\qquad = 0 \quad \text{otherwise}$

Figure 8.7 The one-sided Laplace transform in continuous system analysis.

and for $t < 0$,

$$y(t) = 0$$

The one-sided Laplace transform is always utilized when $X(s)$ is the ratio of two polynomials in s and the denominator of $Y(s)$ is of higher order than the numerator.

The reader should be already familiar with the use of the one-sided Laplace transform in circuit theory and control theory. A summary for the use of the one-sided Laplace transform in finding the response of a linear system to a deterministic input is given in Figure 8.7. For completeness the results obtained by time domain analysis in Chapter 7 are also included.

8.2 A SUMMARY OF FOURIER AND DISCRETE FOURIER TRANSFORMS

Section 8.1 treated the easiest analytical transforms used in electrical engineering—namely, the z and Laplace transforms. However, when visualizing the randomness of a random process or discussing linear systems with signal plus random inputs from a frequency point of view, the Fourier transform and discrete Fourier transform are used. For time functions that are analytically simple we will see that the Fourier transform is very similar to the two-sided Laplace transform and that the discrete Fourier transform is very similar to the two-sided z transform. If the time functions do not result in manageable Laplace or z transforms, then the language of "frequency signal content" is used and the Fourier transform or discrete Fourier transform is approximated using algorithms employing "fast Fourier transforms." We will now give a review of Fourier transforms, which should also be acceptable as a first offering to a serious student.

The Definition of a Fourier Transform Pair

The Fourier transform $Y(f)$ of a time function $y(t)$ is

$$Y(f) \triangleq \int_{-\infty}^{\infty} y(t)e^{-2\pi jft}\,dt \tag{8.20}$$

$$\triangleq \int_{-\infty}^{\infty} y(t)\cos 2\pi ft\,dt + j\int_{-\infty}^{\infty} y(t)\sin 2\pi ft\,dt$$

where $Y(f)$ is defined for all real f. The Fourier transform for a waveform is said to exist if Eq. 8.20 can be evaluated, which is the case if $y(t)$ is a finite energy waveform satisfying

$$\int_{-\infty}^{\infty} |y(t)|^2\,dt < \infty \tag{8.21}$$

Unfortunately, there are three almost equally used notations for the Fourier transform—$Y(f)$, $Y(\omega)$, and $Y(j\omega)$—and they are all treated as the same symbol for the right-hand side of Eq. 8.20. The most common is probably $Y(\omega)$, and we will use both $Y(\omega)$ and $Y(f)$.

Some shorthand notations for the Fourier transform and the inverse Fourier transform are as follows:

$$y(t) \longleftrightarrow Y(f) \quad \text{or} \quad Y(\omega) \quad \text{or} \quad Y(j\omega)$$
$$F[y(t)] = Y(f) \quad \text{or} \quad Y(\omega)$$
$$\overline{y(t)} = Y(f) \quad \text{or} \quad Y(\omega)$$
$$\overline{Y(\omega)} \quad \text{or} \quad \overline{Y(f)} = y(t)$$
$$F^{-1}[Y(\omega)] \quad \text{or} \quad F^{-1}[Y(f)] = y(t)$$

The meanings of these are evident: $y(t) \longleftrightarrow Y(f)$ says that $y(t)$ and $Y(f)$ are a Fourier transform pair and $F[y(t)] = Y(f)$ says that the Fourier transform of $y(t)$ is $Y(f)$.

The inverse Fourier transform of $Y(f)$ is defined as

$$y(t) = \int_{-\infty}^{\infty} Y(f)e^{+2\pi jft}\,df$$

or

$$y(t) = \frac{1}{2\pi} \int_{-\infty}^{\infty} Y(\omega)e^{j\omega t}\,d\omega \tag{8.22}$$

where $y(t)$ is defined for all real t; $-\infty < t < \infty$. $y(t)$ and $Y(f)$ form a unique pair of functions, which implies that

$$\int_{-\infty}^{\infty} \left[\int_{-\infty}^{\infty} y(\alpha)e^{-2\pi jf\alpha} d\alpha \right] e^{2\pi jft} df = y(t) \tag{8.23}$$

or, in shorthand notation, $F^{-1}\{F[y(t)]\} = y(t)$.

A student normally becomes familiar with the definitions given by Eqs. 8.20 and 8.22 by actually evaluating the transforms and inverse transforms of a number of functions. Some good cases to practice with are the following pairs:

$$e^{-at}u(t) \longleftrightarrow \frac{1}{a + j\omega} \quad \text{if } a > 0 \text{ (Why?)}$$

$$= \frac{a}{a^2 + 4\pi^2 f^2} - j\frac{2\pi f}{a^2 + 4\pi^2 f^2}$$

$$\sqcap(t) \longleftrightarrow \frac{\sin \pi f}{\pi f} \quad \text{or} \quad 2\frac{\sin(\omega/2)}{\omega}$$

$$e^{-a|t|} \longleftrightarrow \frac{2a}{a^2 + 4\pi^2 f^2} \quad \text{or} \quad \frac{2a}{a^2 + \omega^2}$$

$$\delta(t) \longleftrightarrow 1$$

where we have used the notation, $\sqcap(t)$ and $u(t)$ for the functions defined as

$$\sqcap(t) = 1 \quad -\tfrac{1}{2} < t < \tfrac{1}{2}$$
$$= 0 \quad \text{otherwise}$$

and

$$u(t) = 1 \quad t > 0$$
$$= 0 \quad \text{otherwise}$$

$\sqcap(t)$ or $P_T(t)$ is called the rectangle or unit pulse function and $u(t)$ is the unit step function. These transforms are shown plotted in Figure 8.8.

One stumbling block at this stage to our whole concept of frequency is the restriction that $y(t)$ must be a finite energy function to possess a Fourier transform, and this excludes such functions as $\cos \omega_c t$, dc waveforms, and periodic functions that obviously contain specific frequencies. To include these functions we relent on the strictness of Eq. 8.21 and define a "Fourier transform in the limit." This will now be done somewhat loosely.

FOURIER TRANSFORM IN THE LIMIT
Equation 8.21 gave the strict condition for the existence of a Fourier transform as

$$\int_{-\infty}^{\infty} |y(t)|^2 dt < \infty$$

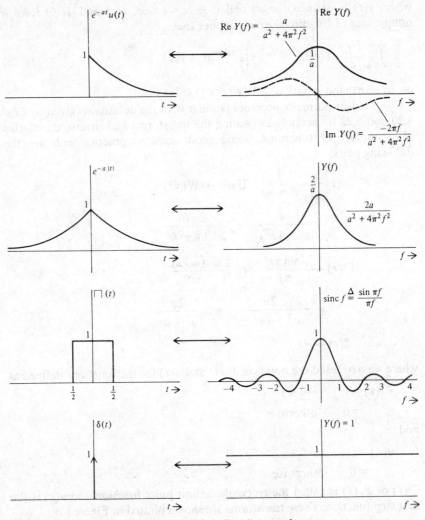

Figure 8.8 Four time functions and their Fourier transforms.

However, a number of functions that do not obey this condition are said to possess a **Fourier transform in the limit**. These functions will all obey the following conditions:

1. $\int_{-\infty}^{\infty} |y(t)| \, dt$ does not exist.
2. $\int_{-\infty}^{\infty} \lim_{a \to 0^+} |y(t)e^{-a|t|}| \, dt$ does exist. In this condition we are evaluating the integral for some positive value of $a > 0$.

If $y(t)$ obeys conditions 1 and 2 we say that

$$F\left[\lim_{a \to 0^+} \left[y(t)e^{-a|t|} \right] \right] = \lim_{a \to 0^+} Y_1(f) = Y(f)$$

where $Y_1(f)$ denotes the Fourier transform of $y(t)e^{-a|t|}$ and we denote its limit as $a \to 0^+$ by $Y(f)$. The important mathematical condition involved here is the interchange of the operations of integration and taking a limit. As an example, let us find the Fourier transform of $y(t) = 1$.

Example 8.6

Evaluate the Fourier transform of $y(t) = 1$.

SOLUTION

$$\int_{-\infty}^{\infty} 1 \, dt = \infty$$

$$\int_{-\infty}^{\infty} e^{-|a|t} \, dt = 2 \int_{0}^{\infty} e^{-at} \, dt = 2\left[\frac{1}{a} (1-0) \right]$$

$$= \frac{2}{a}$$

and this exists for any a not equal to 0.

$$F\left[(1) e^{-|a|t} \right] = \frac{2a}{a^2 + 4\pi^2 f^2}$$

$$\therefore F(1) = \lim_{a \to 0^+} \frac{2a}{a^2 + 4\pi^2 f^2} = Y(f)$$

On examining $Y(f)$ we see that $Y(0) = \infty$, $Y(f) = 0$ when $f \neq 0$, and

$$\int_{-\infty}^{\infty} Y(f) \, df = 1$$

which defines the mathematical model of a delta function.

$$\therefore F(1) = \delta(f) \quad \text{or} \quad 2\pi \delta(\omega)$$

using the delta function property that

$$\delta(au + b) = \frac{1}{|a|} \delta\left(u + \frac{b}{a} \right)$$

For functions that possess Fourier transforms in the limit we will always obtain delta functions in their transforms. Some other important transforms in the limit are

$$F(\cos \omega_c t) = \frac{1}{2} \left[\delta(f + f_c) + \delta(f - f_c) \right]$$

$$F(\sin \omega_c t) = \frac{j}{2} \left[\delta(f + f_c) - \delta(f - f_c) \right]$$

and

$$F\left[\sum_{-\infty}^{\infty} g(t - nT) \right] = \sum_{-\infty}^{\infty} A_n \delta(f - nf_0)$$

where

$$\sum_{-\infty}^{\infty} g(t - nT)$$

is a periodic function with period T and the A_n's are its exponential Fourier coefficients. When singularity functions occur, we must be careful when transferring from $X(f)$ to $X(\omega)$. For example,

$$F(1) = 2\pi\delta(\omega)$$
$$F(\cos \omega_c t) = \pi[\delta(\omega - \omega_c) + \delta(\omega + \omega_c)]$$
$$F(\sin \omega_c t) = j\pi[\delta(\omega + \omega_c) - \delta(\omega - \omega_c)]$$

and

$$\sum_{-\infty}^{\infty} g(t - nT) = 2\pi \sum_{-\infty}^{\infty} A_n\delta(\omega - n\omega_0)$$

To a beginning student the use of both $X(f)$ and $X(\omega)$ is confusing. However, there is still a lack of conformity. Such noted authors as Bracewell, Davenport, and Root use f, whereas most others, such as Papoulis, Schwartz, and Lathai use ω.

Finally, a short table of Fourier transforms is given in Table 8.5 for most of the functions mentioned in this subsection.

SOME IMPORTANT FOURIER TRANSFORM THEOREMS INVOLVING TIME AND FREQUENCY

Table 8.6 lists some of the more important theorems involving Fourier transforms. These theorems are useful when deriving Fourier transforms from a small basic set such as in Table 8.5 or at different stages of solving linear systems, filter, distortionless transmission, modulation, sampling, or other problems involved in electrical engineering. In a systems course much effort is placed on proving, interpreting, and physically understanding these theorems, and as an illustration we will prove three of them.

Example 8.7

Prove and comment on the reciprocal spreading theorem:

$$F[y(at)] = \frac{1}{|a|} Y\left(\frac{f}{a}\right) \quad \text{or} \quad \frac{1}{|a|} Y\left(\frac{\omega}{a}\right)$$

SOLUTION
By definition,

$$F[y(t)] = \int_{-\infty}^{\infty} y(t)e^{-2\pi jft} \, dt$$

Table 8.5 A BRIEF TABLE OF SOME FOURIER TRANSFORM PAIRS

$x(t) \longleftrightarrow X(f)$ or $X(\omega)$

$e^{-a|t|} \longleftrightarrow \dfrac{2a}{a^2 + 4\pi^2 f^2}$ or $\dfrac{2a}{a^2 + \omega^2}$

$e^{-at}u(t) \longleftrightarrow \dfrac{a}{a^2 + 4\pi^2 f^2} - j\dfrac{2\pi f}{a^2 + 4\pi^2 f^2}$ $a > 0$

$\qquad = \dfrac{1}{a + j2\pi f}$ or $\dfrac{1}{a + j\omega}$

$\Pi(t) \longleftrightarrow \dfrac{\sin \pi f}{\pi f} = \mathrm{sinc}\, f$ or $\dfrac{\sin \omega/2}{\omega/2}$

$\lambda(t) \longleftrightarrow \mathrm{sinc}^2 f$ or $\dfrac{4 \sin \omega/2}{\omega}$

$1 \longleftrightarrow \delta(f)$ or $2\pi\delta(\omega)$

$\delta(t) \longleftrightarrow 1$

$\cos t \longleftrightarrow \dfrac{1}{2}\left[\delta\left(f + \dfrac{1}{2\pi}\right) + \delta\left(f - \dfrac{1}{2\pi}\right)\right]$ or $\pi[\delta(\omega + 1) + \delta(\omega - 1)]$

$\sin t \longleftrightarrow \dfrac{j}{2}\left[\delta\left(f + \dfrac{1}{2\pi}\right) - \delta\left(f - \dfrac{1}{2\pi}\right)\right]$ or $j\pi[\delta(\omega + 1) - \delta(\omega - 1)]$

Notation

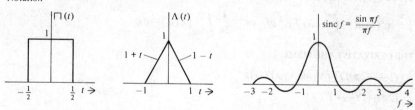

When $a > 0$,

$$F[y(at)] \triangleq \int_{-\infty}^{\infty} y(at)e^{-2\pi jft}\, dt$$

Now let $at = p$.

$$\therefore dt = \frac{1}{a}\, dp$$

$$F[y(at)] = \int_{-\infty}^{\infty} y(p)e^{-2\pi jfp/a}\left(\frac{1}{a}\right) dp$$

$$= \frac{1}{a}\int_{-\infty}^{\infty} y(p)e^{-2\pi j(f/a)p}\, dp$$

$$= \frac{1}{a}Y\left(\frac{f}{a}\right)$$

Table 8.6 SOME IMPORTANT THEOREMS INVOLVING
FOURIER TRANSFORMS

THE LINEARITY THEOREM

$$ax(t) + by(t) \longleftrightarrow aX(f) + bY(f)$$

$$\text{or} \quad aX(\omega) + bY(\omega)$$

THE RECIPROCAL SPREADING THEOREM

$$y(at) \longleftrightarrow \frac{1}{|a|} Y\left(\frac{f}{a}\right) \quad \text{or} \quad \frac{1}{|a|} Y\left(\frac{\omega}{a}\right)$$

THE SHIFTING THEOREMS

$$y(t-a) \longleftrightarrow e^{-2\pi jfa} y(t) \quad \text{or} \quad e^{-j\omega a} Y(\omega)$$

$$e^{j2\pi at} y(t) \longleftrightarrow Y(f-a) \quad \text{(modulation)}$$

$$\text{or} \quad Y(\omega - 2\pi a)$$

THE CONVOLUTION THEOREMS

$$x(t) * y(t) \longleftrightarrow X(f)Y(f) \quad \text{or} \quad X(\omega) * Y(\omega)$$

$$x(t)y(t) \longleftrightarrow X(f) * Y(f) \quad \text{or} \quad X(\omega) * Y(\omega)$$

THE CORRELATION THEOREMS

$$x(t) \oplus y(t) \longleftrightarrow X(f)Y(-f) \quad \text{or} \quad X(\omega)Y(-\omega)$$

$$x(t) \oplus x(t) \longleftrightarrow |X(f)|^2 \quad \text{or} \quad |X(\omega)|^2$$

$$\int_{-\infty}^{\infty} x^2(t)\, dt = \int_{-\infty}^{\infty} |X(f)|^2\, df \quad \text{or} \quad \frac{1}{2\pi} \int_{-\infty}^{\infty} |X(\omega)|^2\, d\omega$$

THE DERIVATIVE THEOREMS

$$x'(t) \longleftrightarrow 2\pi jf Y(f) \quad \text{or} \quad j\omega Y(\omega)$$

$$-j2\pi t y(t) \longleftrightarrow Y'(f)$$

THE RECIPROCITY THEOREM

If $y(t) \longleftrightarrow Y(f)$ or $Y_1(\omega)$, then

$$Y(t) \longleftrightarrow y(-f) \quad \text{or} \quad \text{(be careful)}$$

THE MOMENT THEOREM

$$\int_{-\infty}^{\infty} t^n y(t)\, dt = \frac{1}{(-2\pi j)^n} Y^n(0)$$

When $a < 0$, let $at = p$; then

$$dt = \frac{1}{a} dp = -\frac{1}{|a|} dp$$

and, as t varies over the range $-\infty < t < \infty$, p varies over the range $\infty > p > -\infty$,

$$F[y(at)] = \frac{1}{-|a|} \int_{\infty}^{-\infty} Y(p) e^{-2\pi j(f/a)p} dp$$

$$= \frac{1}{-(-|a|)} \int_{-\infty}^{\infty} y(p) e^{-2\pi j(f/a)p} dp$$

$$= \frac{1}{|a|} Y\left(\frac{f}{a}\right)$$

We should appreciate that $Y(f/a)$ is an entirely different function—actually it is a reflected version of $Y(f/|a|)$—when $a < 0$ rather than when $a > 0$. This theorem states that a contracted function of time [$y(at)$ when $a > 1$] has values of $Y(f)$ for f large, whereas a wide function of time [$y(at)$ when $a < 1$] has essentially only values of $Y(f)$ for small f. Figure 8.9 shows a plot of the three transform pairs

$$\sqcap(t) \longrightarrow \frac{\sin \pi f}{\pi f}$$

$$\sqcap(2t) \longrightarrow \frac{1}{2} \frac{\sin(\pi/2)f}{(\pi/2)f}$$

and

$$\sqcap(\tfrac{1}{2}t) \longrightarrow 2 \frac{\sin 2\pi f}{2\pi f}$$

This figure shows the classical "fat and thin" relationship between a function and its Fourier transform.

Example 8.8

Prove the convolution theorem. This says that the Fourier transform of the convolution of two functions is the product of the individual transforms. The theorem may be proved in two ways:

(a) First, we show that

$$\int_{-\infty}^{\infty} \left[\int_{-\infty}^{\infty} x(p) y(t-p) \, dp \right] e^{-2\pi j f t} \, dt = X(f) Y(f) \qquad (1)$$

or

(b) Second, we take the inverse transform and show that

$$\int_{-\infty}^{\infty} X(f) Y(f) e^{2\pi j f t} \, df = \int_{-\infty}^{\infty} x(p) y(t-p) \, dp \qquad (2)$$

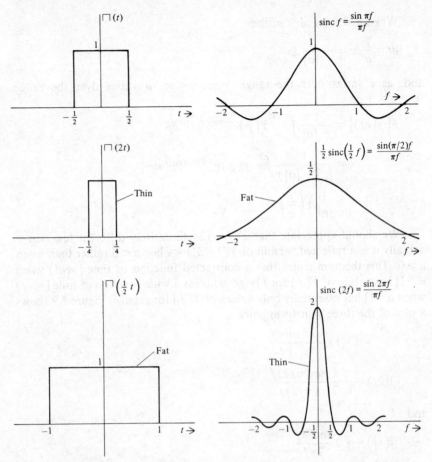

Figure 8.9 Illustration of the reciprocal spreading theorem in Example 8.7.

SOLUTION

(a) Interchanging the order of integration of the left-hand side of (1) yields

$$\int_{-\infty}^{\infty} x(p) \left[\int_{-\infty}^{\infty} y(t-p) e^{-2\pi jft} \, dt \right] dp$$

Now substitute $t - p = m$ and this becomes

$$\int_{-\infty}^{\infty} x(p) \left[\int_{-\infty}^{\infty} y(m) e^{-2\pi jf(m+p)} \, dm \right] dp$$

$$= \int_{-\infty}^{\infty} x(p) e^{-2\pi jfp} \left[\int_{-\infty}^{\infty} y(m) e^{-2\pi jfm} \, dm \right] dp$$

$$= Y(f) \int_{-\infty}^{\infty} x(p) e^{-2\pi jfp} \, dp$$

$$= X(f) Y(f)$$

(b) It is left as an exercise for the reader to prove this theorem by the inverse transform method.

The convolution theorem is the most powerful theorem of system analysis. Instead of evaluating the response of a linear system characterized by an impulse response $h(t)$ as

$$y(t) = h(t) * x(t)$$

the convolution theorem allows us to evaluate

$$Y(f) = H(f)X(f)$$

and then find $y(t)$ by taking the inverse transform of $Y(f)$.

Example 8.9

(a) Prove that the Fourier transform of the correlation function of $x(t)$ with itself is equal to the magnitude of the Fourier transform squared.

(b) Hence show that the energy of any signal equals the energy of its spectrum (this is called **Parseval's theorem**).

SOLUTION

(a) We wish to show that

$$F[x(t) \oplus x(t)] = |X(f)|^2 \quad \text{or} \quad |X(\omega)|^2$$

By definition,

$$F[x(t) \oplus x(t)] = \int_{-\infty}^{\infty} \left[\int_{-\infty}^{\infty} x(p)x(p-t)\, dp \right] e^{-2\pi j f t}\, dt$$

Interchanging the order of integration yields, for the right-hand side (denoted RHS),

$$\text{RHS} = \int_{-\infty}^{\infty} x(p) \left[\int_{-\infty}^{\infty} x(p-t)e^{-2\pi j f t}\, dt \right] dp$$

If we let $p - t = s$ (remembering that p is like a constant), then

$$F[x(t) \oplus x(t)] = \int_{-\infty}^{\infty} x(p) \left[\int_{\infty}^{-\infty} x(s)e^{-2\pi j f(p-s)}(-ds) \right] dp$$

$$= \int_{-\infty}^{\infty} x(p)e^{-2\pi j f p} \left[\int_{-\infty}^{\infty} x(s)e^{+2\pi j f s}\, ds \right] dp$$

By definition the term in brackets is $X^*(f)$, where the asterisk stands for the complex conjugate of $X(f)$ and

$$\therefore F[x(t)\oplus x(t)] = X(f)X^*(f)$$

$$= |X(f)|^2$$

(b) Now it is asked to prove Parseval's theorem,

$$\int_{-\infty}^{\infty} x^2(t)\,dt = \int_{-\infty}^{\infty} |X(f)|^2\,df \quad \text{or} \quad \frac{1}{2\pi}\int_{-\infty}^{\infty} |X(\omega)|^2\,d\omega$$

From part (a), using the definition of the inverse Fourier transform,

$$\int_{-\infty}^{\infty} |X(f)|^2 e^{2\pi jft}\,df = \int_{-\infty}^{\infty} x(p)x(p-t)\,dp$$

The value of each side when $t = 0$ is

$$\int_{-\infty}^{\infty} |X(f)|^2\,df = \int_{-\infty}^{\infty} x^2(p)\,dp$$

which is Parseval's theorem for a real time function $x(t)$.

A PHYSICAL INTERPRETATION OF THE FOURIER TRANSFORM AND THE ENERGY SPECTRAL DENSITY

Reconsidering the last two subsections, we have defined Fourier transforms, evaluated them, and then summarized some important theorems concerning real-time waveforms and their spectra or Fourier transforms. Now let us consider how we think of Fourier transforms. A Fourier transform measures the relative-frequency content of a waveform. If we find the Fourier transform of a dc waveform of value A we obtain $A\delta(f)$, which tells us that the relative frequency content at zero frequency to any other frequency is infinite. The Fourier transform of $\cos\omega_c t$ is $\frac{1}{2}\delta(f-f_c)+\frac{1}{2}\delta(f-f_c)$, which says that the waveform contains only the discrete frequencies f_c and $-f_c$ in equal proportions and that the relative content of these compared to any other frequency is infinite. Similarly a periodic function, which contains a specific fundamental frequency and its harmonics, will consist of a string of delta functions with weighting factors proportional to the relative strength of the harmonics, and the delta functions indicate that the relative content of the harmonic frequencies to nonharmonic frequencies is infinite. If we consider a nonperiodic function $x(t)$, we will always find that $|X(f)|$ is a continuous function of f, which indicates there is no meaning to specific frequency content but only to the frequency content of a band. Having studied random variables we can see at this stage that a Fourier transform is somewhat like a density function. If the random variable is continuous, the density function is continuous, and thus there is meaning only to the probability of taking on a value in a range and not to the probability of a point. If the random variable is discrete, then the density function consists of a string of delta functions whose weighting factors indicate the probabilities of assuming the value of each point.

A function that is used to indicate the frequency content of a waveform is the **energy spectral density** function $E(f)$. For any waveform $x(t)$ with Fourier transform $X(f)$, we have shown that the total signal energy is as follows:

$$\text{signal energy} = \int_{-\infty}^{\infty} x^2(t)\, dt$$

Also, from Table 8.6,

$$\int_{-\infty}^{\infty} x^2(t)\, dt = \int_{-\infty}^{\infty} |X(f)|^2\, df \quad \text{or} \quad \frac{1}{2\pi} \int_{-\infty}^{\infty} |X(\omega)|^2\, d\omega$$

Therefore the signal energy contained from $f = f_0$ to $f = f_0 + \Delta f$ is $|X(f_0)|^2 \Delta f$. We define the energy spectral density as

$$E(f) = |X(f)|^2 \quad \text{or} \quad |X(\omega)|^2 \tag{8.24}$$

We say that $E(f) = |X(f)|^2$ truly represents the distribution of energy with frequency and indicates the relative-frequency content of the waveform $x(t)$ with respect to its energy. It is worth noting that the energy spectral density is not defined for sinusoidal or periodic functions, although we have loosely interpreted their transforms. When we soon return to random processes we will discuss the power spectral density for infinite energy signals such as member waveforms of a random process, which like sinusoidal or periodic waveforms are finite power waveforms.

Applications of Fourier Transforms

The Fourier transform is a two-sided transform, and its definition is identical to the two-sided Laplace transform with s replaced by $j\omega$. Indeed, if $\sigma = 0$ is an acceptable value for the existence of $F_B(s)$, then both transforms exist and are identical, with

$$F(\omega) = F_B(j\omega)$$

where the s of the Laplace transform is replaced by $j\omega$. The Fourier transform will be utilized for functions of both positive and negative time, such as

1. Autocorrelation and cross-correlation functions of continuous random processes. These are $R_{xx}(\tau)$, $R_{xy}(\tau)$, and $R_{yx}(\tau)$.
2. The system correlation transfer function, $C_{hh}(\tau)$, which is defined as the impulse response $h(\tau)$ correlated with itself.
3. A member waveform of a random process $x_p(t)$. Such a member waveform will be treated with extreme care because for a stationary process each member is of infinite energy.

We might pose the question, "If the Fourier and bilateral Laplace transforms are almost identical, why bother with both?" The answer is that Laplace and Fourier transforms are like two different languages. The Laplace transform deals with a pole-zero interpretation of signals, whereas the Fourier transform deals with a frequency magnitude and phase interpretation of signals. Many devices are approximated using Fourier transform language. Some of these are low-pass, band-pass, and high-pass ideal filters. These are very simple when plotted as $|H(\omega)|$ versus f or $\angle H(\omega)$ versus f, but the corresponding Laplace transform or time function is analytically somewhat dreadful. Similarly, the system function $H(s)$ of any passive network is a very simple ratio of two polynomials with left-half-plane poles when Laplace transforms are used.

For spectral analysis of random signals the Fourier transform is always used, and data are presented and interpreted in terms of frequency magnitude and phase. In a first graduate course much time is spent on such interpretation.

The Fourier transform may be used to obtain the deterministic output $y(t)$ of a linear system with deterministic input $x(t)$. If we denote the Fourier transform of $x(t)$ by $X(\omega)$ and that of the impulse response $h(t)$ by $H(\omega)$, since

$$y(t) = x(t) * h(t)$$

then $Y(\omega)$ the Fourier transform of $y(t)$ is

$$Y(\omega) = X(\omega)H(\omega)$$

and $y(t)$ may be found as $F^{-1}[Y(\omega)]$.

THE SAMPLING THEOREM

We will conclude our summary of results from Fourier transform theory by giving, without proof, the sampling theorem. Two very important applications of Fourier transforms in communication theory are:

1. The interpretation of sampled data.
2. The choice of a sampling rate to obtain the autocorrelation function or cross-correlation functions of a member of an ergodic random process or noise waveform.

There are certain waveforms that may be recovered from a set of sampled values and a specific recovery formula exists for doing this. This result is called the **sampling theorem**.

THE SAMPLING THEOREM

If a time function $x(t)$ is band limited, with a cutoff frequency f_c (i.e., $X(f) = 0$; $f > f_c$), then this function may be recovered from its values sampled every $\tau = 1/2f_c$ s. The recovery formula for any time t is

$$x(t) = \sum_{-\infty}^{\infty} x(n\tau) \frac{\sin 2\pi f_c(t - n\tau)}{2\pi f_c(t - n\tau)} \tag{8.25}$$

The proof of the sampling theorem will not be given, but it can be found in any book dealing at all seriously with the Fourier transform. We will now interpret the theorem. Figure 8.10a shows the sampled values of a function that is assumed to be such that $X(f)=0$, $f>f_c$. Figure 8.10b shows the reconstruction of the continuous function $x(t)$ by adding together

$$\frac{\sin 2\pi f_c(t-n\tau)}{2\pi f_c(t-n_\tau)}$$

functions, each weighted by the sampled value $x(n\tau)$. For instance, at a specific time t_1,

$$x(t_1) = x(0)\frac{\sin 2\pi f_c t_1}{2\pi f_c t_1} + x(\tau)\frac{\sin 2\pi f_c(t_1-\tau)}{2\pi f_c(t_1-\tau)}$$

$$+ x(-\tau)\frac{\sin 2\pi f_c(t_1+\tau)}{2\pi f_c(t_1+\tau)} + \cdots$$

Applying the recovery formula to find $x(t)$ at a value other than at a sampled point is a very tall order. In practice, a researcher tries to use an interpolation formula to give a close approximation to Eq. 8.25. The simplest (if clumsy), but sometimes sufficient, approximation would be at a midpoint between $t=p\tau$ and $t=(p+1)\tau$, where we assume that

$$x\left[\left(p+\tfrac{1}{2}\right)\tau\right] = \tfrac{1}{2}\left[x(p+1)\tau\right]+x(p\tau)\right]$$

The next simplest approximation might be to give some weight to $x[(p-2)\tau]$ and $x[(p+2)\tau]$ so that the approximation for $x[(p+\tfrac{1}{2})\tau]$ would be closer to the result for actually using the weighted sinc functions. This whole

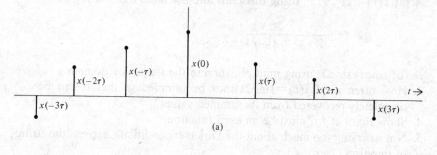

(a)

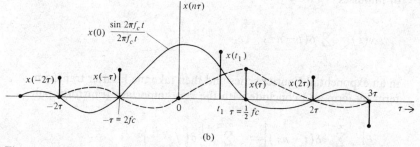

(b)

Figure 8.10 Theoretically using the sampling theorem to interpolate a function.

problem of reconstructing a function from sampled data is a topic in its own right, which we are just hinting at.

In conclusion, it should be said that there are certain idealizations in the sampling theorem. From the reciprocal spreading theorem of the Fourier transform we can see that if a function is band limited, then it must exist for all time, $-\infty < t < \infty$. In practice, we always deal with a finite width T_1 of data, which implies that the assumption of a cutoff frequency f_c and a sampling rate $\tau = 1/2f_c$ is somewhat in error. When we do recover $x(t)$, we have essentially ignored high-frequency components ($f > f_c$) or fluctuations more rapid than can be detected by $1/2f_c$-s samples. We again iterate that the choice of τ is of extreme importance and may involve certain preconceived assumptions about the function or noise waveforms being analyzed.

The purpose of this section on the sampling theorem was informational as opposed to demanding theoretical understanding. The fact that a waveform can be almost completely recovered from a discrete set of numbers or sampled values is a powerful conceptual idea and should make us aware of the basis for the choices of T_1 and τ when we are approximately evaluating an autocorrelation function for a member of an ergodic random process.

DRILL SET: THE FOURIER TRANSFORM

1. Given that the density function of a random variable is $f_X(\alpha) = \frac{1}{2}e^{-|\alpha|}$ use the moment theorems to evaluate $\overline{X}$ and $\overline{X^2}$.
2. Find and graph the Fourier transforms of
    (a) $y(t) = 2e^{-4|t-2|}$ using theorems and the basic transform pair

$$e^{-a|t|} \longleftrightarrow \frac{2a}{a^2 + 4\pi^2 f^2}$$

   (b) $\operatorname{sinc} t * \operatorname{sinc} 2t$ using multiplication in the frequency domain.
3. How often need $y(t) = \operatorname{sinc} 2t \operatorname{sinc} t$ be sampled so that it can be completely recovered from its sampled values?
4. Show that $|X(f)|$ must be an even function.
5. Not worrying too much about the Dirichlet conditions, express the string of impulses

$$x(t) = \sum_{-\infty}^{\infty} \delta(t - n\tau)$$

in an exponential Fourier series and then take the Fourier transform term by term and conclude with the very important transform pair

$$F\left[\sum_{n=-\infty}^{\infty} \delta(t - n\tau) \right] = \sum_{n=-\infty}^{\infty} \frac{1}{\tau} \delta\left(f - \frac{n}{\tau} \right)$$

The Fourier Series and the Discrete Fourier Transform

Associated with a continuous finite energy time function is a Fourier transform. We will now define, in association with a discrete finite energy time function, a discrete Fourier transform. A key to philosophically accepting this definition is an understanding of the theory and concepts of exponential Fourier series. For this reason we will first rapidly review and summarize some results from Fourier series.

FOURIER SERIES

The Fourier series representation of a periodic time function of period T s in trigonometric form is

$$x(t) = \tfrac{1}{2}a_0 + \sum_{n=1}^{\infty} a_n \cos n\omega_0 t + \sum_{n=1}^{\infty} b_n \sin n\omega_0 t \qquad (8.26a)$$

for integer n. The exponential form is

$$x(t) = \sum_{-\infty}^{\infty} A_n e^{jn\omega_0 t} \qquad (8.26b)$$

for integer n, where $\omega_0 = 2\pi/T$. The exponential coefficients may be found as

$$A_n = \frac{1}{T} \int_{\alpha}^{\alpha+T} x(t) e^{-jn\omega_0 t} dt \qquad -\infty < n < \infty \qquad (8.27)$$

for any conveniently chosen α. We note from symmetry that $A_n = A_{-n}^*$ where * denotes the complex conjugate. The trigonometric coefficients can be found directly as

$$a_n = \frac{1}{T} \int_{\alpha}^{\alpha+T} x(t) \cos n\omega_0 t \, dt \qquad n \geq 0 \qquad (8.28)$$

and

$$b_n = \frac{1}{T} \int_{\alpha}^{\alpha+T} x(t) \sin n\omega_0 t \, dt \qquad (8.29)$$

or, from observing the formula for A_n in Eq. 8.27,

$$a_0 = 2A_0$$
$$a_n = 2\mathrm{Re}[A_n]$$
$$b_n = -2\mathrm{Im}[A_n]$$

where "Re" and "Im" are used to denote the real and imaginary parts, respectively. It is assumed the reader is familiar with the Dirichlet conditions for the coefficients to exist and how the series approximates the time function in the best mean square error sense. In this section we will focus on the exponential series. Observing the definition we say that the exponential

Fourier series associates with a periodic time function

$$x(t) = \sum_{-\infty}^{\infty} x_T(t - nT)$$

a set of complex numbers $\{A_n\}$:

$$x(t) \longleftrightarrow \{\ldots, A_{-2}, A_{-1}, A_0, A_1, A_2, \ldots\}$$

which could be considered as a complex discrete function of frequency.

Relationship Between Fourier Transforms and Fourier Series

A periodic time function $x(t)$ does not have a Fourier transform in the strict sense since

$$\int_{-\infty}^{\infty} |x(t)| \, dt$$

is infinite, but as we have seen it is possible to associate with $x(t)$ a Fourier transform in the limit, which we still denote by $X(f)$ or $X(\omega)$. If a periodic function is expressed in an exponential Fourier series,

$$x(t) = \sum_{-\infty}^{\infty} A_n e^{jn\omega_0 t}$$

$$= + \cdots A_{-1} e^{-j\omega_0 t} + A_0 + A_1 e^{j\omega_0 t} + \cdots$$

then using the shifting theorem its Fourier transform is

$$X(f) = \cdots + A_{-1}\delta(f + f_0) + A_0\delta(f) + A_1\delta(f - f_0) + \cdots$$

$$= \sum_{-\infty}^{\infty} A_n\delta(f - nf_0) \tag{8.30}$$

This says that the Fourier transform of a periodic function is a string of delta functions, each weighted by the appropriate exponential coefficient.

We will now obtain another expression for the Fourier transform of a periodic function by writing it as the sum of shifted versions of one period, $x_T(t)$.

$$x(t) = \sum_{-\infty}^{\infty} x_T(t - nT)$$

and this may be rewritten as

$$x(t) = x_T(t) * \sum_{-\infty}^{\infty} \delta(t - nT)$$

using the property of a singularity function, that $f(t) * \delta(t - a) = f(t - a)$. An important Fourier transform pair is that

$$\sum_{-\infty}^{\infty} \delta(t - n\tau) \longleftrightarrow \sum_{-\infty}^{\infty} \frac{1}{\tau}\delta\left(f - \frac{n}{\tau}\right)$$

and using this theorem and the fact that the transform of convolved functions is the product of the transforms, we obtain

$$F\left[x_T(t) * \sum_{-\infty}^{\infty} \delta(t - nT)\right] = X_T(f)\frac{1}{T}\sum_{-\infty}^{\infty}\delta\left(f - \frac{n}{T}\right) \qquad (8.31)$$

where $X_T(f)$ or $X_T(\omega)$ denotes the Fourier transform of $x_T(t)$. Now Eqs. 8.30 and 8.31 give two alternative formulas for the transform of a periodic function. Reiterating, these are

$$F\left[\sum_{-\infty}^{\infty} x_T(t - nT)\right] = \sum_{-\infty}^{\infty} A_n\delta(f - nf_0)$$

or

$$F\left[\sum_{-\infty}^{\infty} x_T(t - nT)\right] = \sum_{-\infty}^{\infty}\left[\frac{1}{T}X_T(nf_0)\right]\delta(f - nf_0)$$

This implies that the exponential Fourier coefficients may be related to sampled values of the Fourier transform of one period $X_T(f)$ by the formula,

$$A_n = \frac{1}{T}X_T(nf_0) \qquad (8.32)$$

where $f_0 = 1/T$. Theoretically this is a very powerful result, which says that the Fourier coefficients of any periodic function for which the Fourier transform of one period (does it matter which period?) is known may be found by Eq. 8.32. In practice, we will use this relationship in the opposite manner. The Fourier transform of a "difficult to handle" analytical function will be found by assuming that it is essentially periodic with some period T and applying Eq. 8.27 to find its Fourier coefficients and, from these, then predicting $X(\omega)$ as being TA_n, where $\omega = n/T$. We will now illustrate the use of Eq. 8.32 by an example.

Example 8.10

Given

$$x_T(t) = 1 \qquad -\tfrac{1}{2} < t < \tfrac{1}{2}$$
$$= 0 \qquad \text{otherwise}$$

and the periodic function

$$x(t) = \sum_{n=-\infty}^{\infty} x_T(t - 4n)$$

express $x(t)$ in an exponential Fourier series.

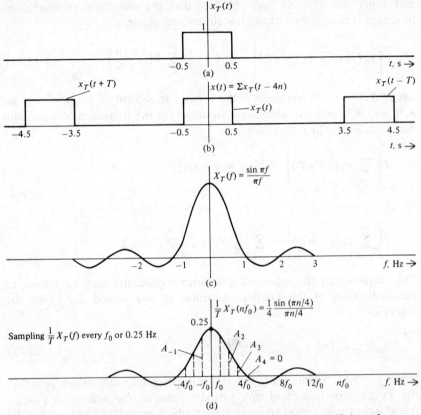

Figure 8.11   The periodic function and its exponential Fourier coefficients from Example 8.10. (a) One period $x_T(t)$ of a periodic function. (b) The periodic function. (c) The Fourier transform of one period. (d) The exponential Fourier coefficients.

SOLUTION

Both $x_T(t)$ and $x(t)$ are shown sketched in Figures 8.11a and b. Now the Fourier transform of $x_T(t)$ is

$$X_T(f) = \frac{\sin \pi f}{\pi f}$$

By Eq. 8.32, the exponential coefficients are

$$A_n = \frac{1}{T} \frac{\sin \pi n f_0}{\pi n f_0}$$

where $T = 4$ and $f_0 = \frac{1}{4}$. This gives

$$A_n = \frac{1}{4} \frac{\sin \pi (n/4)}{\pi (n/4)} = \frac{\sin(\pi n/4)}{\pi n}$$

This yields, for some of the coefficients,

$$A_0 = \frac{1}{4}, \quad A_1 = A_{-1} = \frac{1}{4}\left(\frac{0.707}{0.785}\right) = 0.23, \quad A_2 = A_{-2} = \frac{1}{4}\left(\frac{1}{1.57}\right) = 0.16,$$

$$A_3 = A_{-3} = 0.08, \quad A_4 = A_{-4} = 0$$

and so forth. Figures 8.11c and d show $X_T(f)$, the envelope of the exponential coefficients, $(1/T)X_T(f)$, and the A_n's. In this case, since $x(t)$ is even, the A_n's are real and even.

THE DISCRETE FOURIER TRANSFORM

The Fourier series associates with a periodic time function $x(t)$ a set of complex numbers, $\{\ldots, A_{-2}, A_{-1}, A_0, A_1, \ldots\}$, via the formula

$$x(t) = \sum_{-\infty}^{\infty} A_n e^{jn\omega_0 t}\, dt$$

These complex coefficients are found as

$$A_n = \frac{1}{T}\int_{\alpha}^{\alpha+T} x(t)e^{-jn\omega_0 t}\, dt$$

This motivates us to associate with a set of numbers in the time domain or with a discrete time function $x(n)$ a periodic function of frequency $X(\omega)$, which we call the discrete Fourier transform of $x(n)$. The definitions are

$$X(\omega) = \sum_{-\infty}^{\infty} x(n)e^{-jn\omega/\tau} \tag{8.33}$$

and

$$x(n) = \frac{\tau}{2\pi}\int_{-\pi/\tau}^{\pi/\tau} X(\omega)e^{jn\omega/\tau}\, d\omega \tag{8.34}$$

if the values of $x(n)$ are spaced τ s apart. It is conventional, as we have seen, to assume unity spacing for our discrete time function and to define

$$X(\omega) = \sum_{-\infty}^{\infty} x(n)e^{-jn\omega} \tag{8.35}$$

and

$$x(n) = \frac{1}{2\pi}\int_{-\pi}^{\pi} X(\omega)e^{jn\omega}\, d\omega \tag{8.36}$$

It is immediately apparent that the discrete Fourier transform equals the two-sided z transform with $z^{-n} = e^{-j\omega}$, and $x(n)$ may be found as

$$x(n) = \frac{1}{2\pi j}\oint X(z)z^{n-1}\, dz$$

$$= \sum \begin{bmatrix} \text{residues of poles} \\ \text{of } X(z)z^{n-1} \text{ inside } C \end{bmatrix} \quad n > 0$$

$$= -\sum \begin{bmatrix} \text{residues of poles} \\ \text{of } X(z)z^{n-1} \text{ outside } C \end{bmatrix} \quad n \leq 0$$

if $X(z) = N(z)/D(z)$ and $D(z)$ is of at least order one higher than $N(z)$, and $X(z) = X(\omega)$ with every $e^{j\omega} = z$. Sometimes the notation $X(e^{j\omega})$ is used for the discrete Fourier transform to indicate this substitution. For the moment we will consider a frequency interpretation of $X(\omega)$. A number of discrete Fourier transforms will be evaluated and we will endeavor to

1. Appreciate the definitions given by Eqs. 8.35 and 8.36 and realize that we are just using our background in finding Fourier series in a reverse manner.
2. Develop an appreciation of how we should synthesize discrete time functions so that $X(\omega)$ is readily obtained analytically.

Example 8.11

Evaluate the discrete Fourier transforms of the following functions:

(a) $x(n) = 1;\ n = 0$
$\quad = 0;$ otherwise
(b) $x(n) = \frac{1}{2};\ n = \pm 1$
$\quad = 0;$ otherwise
(c) $x(n) = \frac{1}{2};\ n = \pm 2$
$\quad = 0;$ otherwise
(d) $x(n) = \frac{1}{2};\ n = +1$
$\quad = -\frac{1}{2};\ n = -1$
$\quad = 0;$ otherwise

SOLUTION

(a) From the definition of a discrete Fourier transform,

$$X(\omega) = 1$$

which is analogous to saying in Fourier series theory that a dc waveform is already in series form. This transform pair is shown plotted in Figure 8.12a.

(b) $X(\omega) = \frac{1}{2}e^{-j\omega} + \frac{1}{2}e^{j\omega}$
$\quad = \cos \omega$

We notice that the discrete Fourier transform of an even discrete function of time, is a real, even continuous periodic function of ω. This transform pair is shown plotted in Figure 8.12b.

(c) $X(\omega) = \frac{1}{2}e^{-j2\omega} + \frac{1}{2}e^{j2\omega}$
$\quad = \cos 2\omega$

This transform pair is shown plotted in Figure 8.12c. We observe that $x(n)$ is changing less rapidly than in part (b) and that its discrete transform contains a larger low-frequency content.

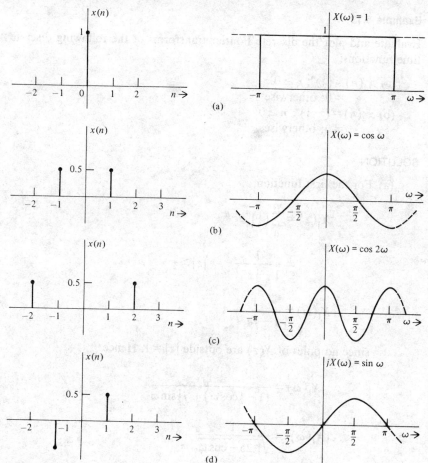

Figure 8.12 Four discrete functions and their discrete Fourier transforms from Example 8.11.

(d) $X(\omega) = \frac{1}{2}e^{-j\omega} - \frac{1}{2}e^{-j\omega}$

$\qquad = -j\sin\omega$

This transform pair is shown plotted in Figure 8.12d. We notice that the discrete Fourier transform of an odd sequence is an imaginary odd function of frequency. This time function changes no more rapidly than in part (c). (Are you sure?)

As in the case of z transforms, an important representation for discrete signals is when $x(n) = a^n$, where a is a constant. $X(\omega)$ will exist for the infinite sequence if $|a| < 1$, since $|e^{j\omega n}| = 1$. Another example will be solved to illustrate the use of already mastered z transforms as a vehicle to develop more feelings concerning low- and high-frequency sequences.

Example 8.12

Evaluate and plot the discrete Fourier transforms of the following discrete time functions:

(a) $x_1(n) = (\frac{1}{2})^n$; $n \geq 0$
　　$= 0$; otherwise
(b) $x_2(n) = (-\frac{1}{2})^n$; $n \geq 0$
　　$= 0$; otherwise

SOLUTION

(a) For the first function,

$$X_1(z) = \sum_0^\infty (\tfrac{1}{2})^n z^{-n}$$

$$= \frac{1}{1 - \frac{1}{2}z^{-1}} \qquad |z| > \tfrac{1}{2}$$

$$\therefore X_1(\omega) = \frac{1}{1 - \frac{1}{2}e^{-j\omega}}$$

since no poles of $X(z)$ are outside $|z| = 1$. Hence

$$X_1(\omega) = \frac{1}{(1 - \frac{1}{2}\cos\omega) + j\frac{1}{2}\sin\omega}$$

$$\therefore |X_1(\omega)| = \frac{1}{\sqrt{1.25 - \cos\omega}}$$

and

$$\angle X_1(\omega) = -\tan^{-1}\left(\frac{\sin\omega}{2 - \cos\omega}\right)$$

Figure 8.13a shows plots of $|X_1(\omega)|$ and $\angle X_1(\omega)$ versus ω. We will comment about these plots after solving part (b).

(b) For the second function,

$$X_2(z) = \sum_0^\infty (-\tfrac{1}{2})^n z^{-n}$$

$$= \frac{1}{1 + \frac{1}{2}z^{-1}} \qquad |z| > \tfrac{1}{2}$$

The discrete Fourier transform exists, since all the poles of $X_2(z)$ lie

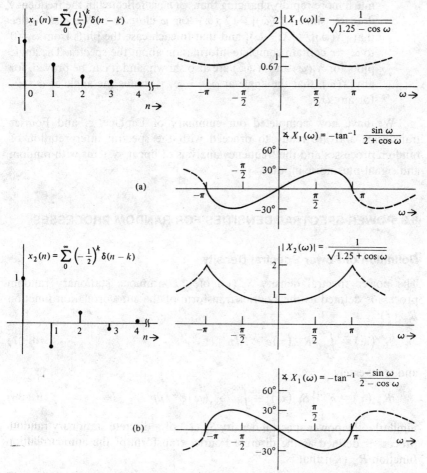

Figure 8.13   The discrete functions of Example 8.12 and their discrete Fourier transforms.

in the annulus $|z| < 1$, and

$$|X_2(\omega)| = \frac{1}{\sqrt{1.25 + \cos \omega}}$$

and

$$\angle X_2(\omega) = -\tan^{-1} \frac{-\sin \omega}{2 + \cos \omega}$$

Figure 8.13b shows a plot of $|X_2(\omega)|$ and $\angle X_2(\omega)$ versus ω. On observing the plots for parts (a) and (b), we can see that both $X_1(\omega)$ and $X_2(\omega)$ are periodic with period 2π. The fact that $x_2(n)$ is much more rapidly changing than $x_1(n)$ is reflected in the frequency domain, since $|X_2(\omega)| \gg |X_1(\omega)|$ for ω close to π. We also notice that $|X(\omega)| = |X(-\omega)|$ and that in each case the plots from $\omega = 0$ to $\omega = \pi$ contain complete information about the spectra. The angle plots for $X_1(\omega)$ and $X_2(\omega)$ are also shown, and it can be proved, for any $X(\omega)$, that the angle at $\omega = -\omega_1$ is minus the angle at $\omega = \omega_1$ for any ω_1.

We have now completed our summary of Laplace, z, and Fourier transforms and are ready to proceed with the spectral interpretation of random processes and the frequency analysis of linear systems with random and signal-plus-noise inputs.

8.3 POWER SPECTRAL DENSITIES FOR RANDOM PROCESSES

Definition of Power Spectral Density

The power spectral density $S_{xx}(\omega)$ of a continuous stationary random process is defined as the Fourier transform of the autocorrelation function $R_{xx}(\tau)$:

$$S_{xx}(\omega) \triangleq \int_{-\infty}^{\infty} R_{xx}(\tau) e^{-j\omega\tau} d\tau \tag{8.37}$$

and, conversely,

$$R_{xx}(\tau) = F^{-1}[S_{xx}(\omega)] = \int_{-\infty}^{\infty} S_{xx}(\omega) e^{j\omega\tau} df \tag{8.38}$$

Similarly, the power spectral density $S_{xx}(\omega)$ of a discrete stationary random process is defined as the discrete Fourier transform of the autocorrelation function $R_{xx}(k)$; that is,

$$S_{xx}(\omega) = \sum_{-\infty}^{\infty} R_{xx}(k) e^{-jk\omega} \tag{8.39}$$

and, conversely,

$$R_{xx}(k) = \int_{-0.5}^{0.5} S_{xx}(\omega) e^{jk\omega} df \quad \text{or} \quad \frac{1}{2\pi} \int_{-\pi}^{\pi} S_{xx}(\omega) e^{jk\omega} d\omega \tag{8.40}$$

These are formal definitions and all the properties of power spectral densities may be derived from a frequency interpretation of the autocorrelation function properties. Some of these were:

1. $R_{xx}(\tau)$ and $R_{xx}(k)$ are real, even functions and
2. $R_{xx}(0) \geq R_{xx}(\tau)$ and $R_{xx}(0) \geq R_{xx}(k)$.

In order to appreciate calling $S_{xx}(\omega)$ the power spectral density a few physical interpretations of $S_{xx}(\omega)$ will be discussed.

A PHYSICAL INTERPRETATION OF $S_{xx}(\omega)$

Consider the problem of passing a member of a continuous stationary random process, with autocorrelation function $R_{xx}(\tau)$ and power spectral density $S_{xx}(\omega)$, through a narrow-band filter as is shown in Figure 8.14. The expected value of the total average power output is

$$P_{av} = \overline{y^2(t)} = R_{yy}(0)$$

and

$$R_{yy}(\tau) = C_{hh}(\tau) * R_{xx}(\tau)$$

The Fourier transform of this is

$$S_{yy}(\omega) = |H(\omega)|^2 S_{xx}(\omega)$$

and

$$R_{yy}(0) = \int_{-\infty}^{\infty} S_{xx}(\omega)|H(\omega)|^2 \, df$$

For our filter,

$$R_{yy}(0) = \int_{-f_0 - \Delta f/2}^{-f_0 + \Delta f/2} S_{xx}(\omega) \, df + \int_{f_0 - \Delta f/2}^{f_0 + \Delta f/2} S_{xx}(\omega) \, df$$

$$= S_{xx}(-\omega_0)\Delta f + S_{xx}(\omega_0)\Delta f$$

The Fourier transform of a real, even function of time [such as $R_{xx}(\tau)$] is a real, even function of frequency [such as $S_{xx}(\omega)$]. Therefore the total average power output from the narrow-band filter is

$$P_{av} = 2S_{xx}(\omega_0)\Delta f$$

This result says that $S_{xx}(\omega_0)\Delta f$ is the power contained in the frequency band Δf about ω_0. Therefore $S_{xx}(\omega)$ truly gives the distribution of average power with frequency.

A SECOND INTERPRETATION OF POWER SPECTRAL DENSITY

There is another physical interpretation of $S_{xx}(\omega)$ that is often given. If a typical member of a continuous stationary random process is considered as is shown in Figure 8.15, it is seen that $x_i(t)$ does not possess a Fourier

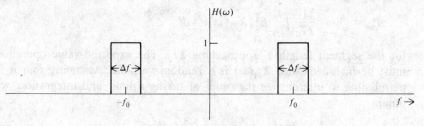

Figure 8.14 The system function of an ideal narrow-band filter.

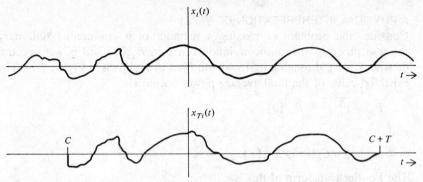

Figure 8.15   A member of a continuous random process and a segment of length T from it.

transform because

$$\int_{-\infty}^{\infty} |x_i(t)|^2 \, dt$$

is infinite. However, if a truncated version of $x_i(t)$ is defined as

$$x_{Ti}(t) = x_i(t) \qquad c < t < c + T$$
$$= 0 \qquad \text{otherwise}$$

then, for any c, $x_{Ti}(t)$ is a finite energy function and must possess a Fourier transform. We call its Fourier transform $X_{Ti}(\omega)$ and

$$X_{Ti}(\omega) = \int_{c}^{c+T} x_i(t) e^{-j\omega t} \, dt$$

This Fourier transform is a random variable, as a different result is obtained for any c or for any chosen member $x_i(t)$. From Parseval's theorem the signal energy of $x_{Ti}(t)$ must equal the spectrum energy; that is,

$$\int_{-\infty}^{\infty} x_{Ti}^2(t) \, dt = \int_{-\infty}^{\infty} |X_{Ti}(\omega)|^2 \, df$$

The total average power of $x_i(t)$ is

$$P_{av} = \lim_{T \to \infty} \frac{1}{2T_0} \int_{-T_0}^{T_0} x_i^2(t) \, dt$$

and the total average power for the process is

$$P_{av} = \lim_{T_0 \to \infty} \frac{1}{2T_0} \int_{-T_0}^{T_0} E\big[|X_{Ti}(\omega)|^2\big] \, df$$

for the segment length T approaching $2T_0$. The expected-value operation must be included since $X_{Ti}(\omega)$ is a random variable. Assuming that it is permissible to interchange the order of taking a limit and integration, we obtain

$$P_{av} = \int_{-\infty}^{\infty} \left\{ \lim_{T \to \infty} \frac{1}{2T} E\big[|X_{Ti}(\omega)|^2\big] \right\} df \qquad (8.41)$$

We have already defined power spectral density $S_{xx}(\omega)$ to mean

$$P_{av} = \int_{-\infty}^{\infty} S_{xx}(\omega)\, df$$

Therefore, observing Eq. 8.41, we obtain a new definition that

$$S_{xx}(\omega) = \lim_{T \to \infty} \frac{1}{2T} E\big[|X_T(\omega)|^2\big] \tag{8.42}$$

A fine discussion of this derivation is given by Cooper and McGillem.[2] A formula often used for power spectral density, which is in classical error but is normally demonstrated for cases in which it gives a correct result, is

$$S_{xx}(\omega) = \lim_{T \to \infty} \left[\frac{1}{2T}|X_T(\omega)|^2\right]$$

The omission of the expected value is incorrect. A reference for this formula will not be given.

Example 8.13

The power spectral density has been defined as follows:

1. Formally, $S_{xx}(\omega) \overset{\Delta}{=} F[R_{xx}(\tau)]$.

2. Physically, $S_{xx}(\omega) \overset{\Delta}{=} \lim_{T \to \infty} \left\{ \frac{1}{2T} E\big[|X_T(\omega)|^2\big] \right\}$.

In order to become more familiar with these definitions, find the power spectral densities of the following random processes, using both definitions.

(a) $x(t) = A$, where $f_A(\alpha) = 1$; $3 < \alpha < 4$.
(b) $x(t) = \cos(t + \phi)$, where $f_\phi(\alpha) = 1/2\pi$; $0 < \alpha < 2\pi$.

SOLUTION

(a) We will first find $S_{xx}(\omega)$ as the Fourier transform of the autocorrelation function. Now $R_{xx}(\tau) = \overline{x^2}$, since a member waveform never changes value. We find that

$$\overline{x^2} = \int_3^4 \alpha^2 \, d\alpha$$

$$= 12.33$$

$$\therefore S_{xx}(\omega) = 12.33\delta(f) \quad \text{or} \quad 74.47\delta(\omega)$$

We will now interpret

$$S_{xx}(\omega) = \lim_{T \to \infty} \left\{ \frac{1}{2T} E\big[|X_T(\omega)|^2\big] \right\}$$

[2] See George R. Cooper and Clare D. McGillem, *Probabilistic Methods of Signal and System Analysis* (New York: Holt, Rinehart and Winston, 1971), pp. 133–135.

For a specific waveform, $x_i(t) = A_i$, we wish to find $X_T(\omega)$ or $X_T(f)$ for a segment from c to $c + T$. In this case we will choose $c = -\frac{1}{2}T$. Recalling that $\sqcap(t)$ is the pulse time function with value one from $t = -0.5$ to $t = 0.5$ and 0 value otherwise, from Fourier transforms, we have

$$\sqcap(t) \longleftrightarrow \frac{\sin \pi f}{\pi f}$$

Now

$$x_{Ti}(t) = A \sqcap \left(\frac{1}{T} t \right)$$

and using the reciprocal spreading theorem of Fourier transforms,

$$F[x_{Ti}(t)] = AT \frac{\sin \pi Tf}{\pi Tf} = AT \frac{\sin T\omega/2}{T\omega/2}$$

$$= X_T(f) \quad \text{or} \quad X_T(\omega)$$

We will first interpret $S_{xx}(f)$ for the member where $A = A_i$ and then find:

$$S_{xx}(f) = \lim_{T \to \infty} \left\{ \frac{1}{2T} E\left[|X_T(f)|^2 \right] \right\}$$

$$S_{xx}(f) = \lim_{T \to \infty} \left\{ \frac{1}{2T} A_i^2 T^2 \frac{\sin^2 \pi Tf}{\pi^2 T^2 f^2} \right\}$$

If we closely examine the function,

$$\lim_{T \to \infty} \frac{T \sin^2 \pi Tf}{\pi^2 T^2 f^2}$$

which we call $P(f)$, we can show that $P(f) = \infty$ if $f = 0$, $P(f) = 0$ if $f \neq 0$, and that

$$\int_{-\infty}^{\infty} P(f) \, df = 1$$

This means that $P(f)$ is a mathematical model of a delta function. Finally, we conclude that

$$S_{xx}(f) = A_i^2 \delta(f)$$

based on a specific member $x_i(t) = A_i$. If we say that the ensemble power spectral density is the ensemble average of the power spectral densities, then

$$S_{xx}(f) = \overline{S_{xx}(f)}$$

$$= \overline{A^2}\,\delta(f)$$

$$= \left(\int_3^4 \alpha^2\, d\alpha\right)\delta(f)$$

$$= 12.33\delta(f) \quad \text{or} \quad 74.47\delta(\omega)$$

as before.

(b) If $x(t) = \cos(t + \phi)$, where $f_\phi(\alpha) = 1/2\pi$; $0 < \alpha < 2\pi$, we can assume that $x(t)$ is ergodic and find $R_{xx}(\tau)$ based on any member to yield $R_{xx}(\tau) = 0.5\cos\tau$. (The reader is urged to do this.) Hence

$$S_{xx}(f) = F(\tfrac{1}{2}\cos\tau)$$

$$= 0.25\big[\delta(f - \tfrac{1}{2\pi}) + \delta(f + \tfrac{1}{2\pi})\big]$$

or

$$S_{xx}(\omega) = 0.5\pi\big[\delta(\omega - 1) + \delta(\omega + 1)\big]$$

As a check, the total average power is

$$P_{av} = \int_{-\infty}^{\infty} 0.25\big[\delta(f - \tfrac{1}{2}\pi) + \delta(f + \tfrac{1}{2}\pi)\big]\,df$$

$$= 0.5$$

This result agrees with that from a basic circuits course; that is, if a current of $i(t) = \cos(t + \phi)$ A flows through a 1-Ω resistor, then

$$P_{av} = I_{rms}^2(1)$$

$$= 0.5 \text{ W}$$

It is left as an exercise for the reader to derive the same result for $S_{xx}(\omega)$ using the definition

$$S_{xx}(\omega) = \lim_{T \to \infty}\left\{\frac{1}{2T}E\big[|X_T(\omega)|^2\big]\right\}$$

SOME COMMENTS ABOUT DISCRETE PROCESSES

For a normalized discrete process or one where the time spacing is unity, the power spectral density is defined as

$$S_{xx}(\omega) = \sum_{-\infty}^{\infty} R_{xx}(k)e^{-j\omega k} \tag{8.43}$$

$S_{xx}(\omega)$ is a periodic function of ω with period 2π. The total average power of the signal is

$$R_{xx}(0) = \frac{1}{2\pi} \int_{-\pi}^{\pi} S_{xx}(\omega)\, d\omega \quad \text{or} \quad \int_{-0.5}^{0.5} S_{xx}(f)\, df$$

ALTERNATE INTERPRETATION OF $S_{xx}(\omega)$

Since $x(n)$ is a stationary random process, the discrete Fourier transform (or two-sided z transform) of a member waveform $x(n)$ will not exist, but as in the continuous case we can define $x_{Ni}(n)$ as

$$x_{Ni}(n) = x_i(n) \quad p < n < p + N$$
$$= 0 \quad \text{otherwise}$$

$x_i(n)$ and $x_{Ni}(n)$ are shown for a typical member waveform of a random process in Figure 8.16. The discrete Fourier transform of $x_{Ni}(n)$ will always exist and is

$$X_N(\omega) = \sum_{n=p}^{p+N} x_i(n) e^{-j\omega n}$$

and this discrete Fourier transform is a random variable because a different result is obtained for every p. From Parseval's theorem the signal energy must equal the spectral energy, or

$$\sum_{n=p}^{P+N} x_{Ni}^2(n) = \frac{1}{2\pi} \int_{-\pi}^{\pi} |X_N(\omega)|^2\, d\omega$$

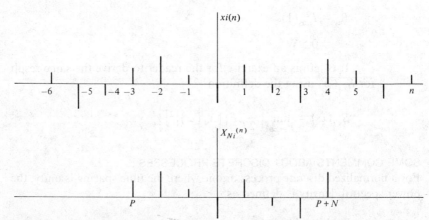

Figure 8.16   A member of a discrete random process and a segment of N values from it.

The total average power of $x_i(n)$ is

$$P_{av} = \lim_{N \to \infty} \frac{1}{2N+1} \sum_{-N}^{N} x_i^2(n)$$

$$= \lim_{N \to \infty} \frac{1}{2N+1} \int_{-1/2}^{1/2} E\big[|X_N(\omega)|^2\big] \, df$$

$$= \frac{1}{2\pi} \int_{-\pi}^{\pi} S_{xx}(\omega) \, d\omega$$

Corresponding to

$$S_{xx}(\omega) = \lim \frac{1}{2T} E\big[|X_T(\omega)|^2\big]$$

in the continuous case we obtain an analogous discrete-form physical definition of $S_{xx}(\omega)$ as

$$S_{xx}(\omega) = \lim_{N \to \infty} \left\{ \frac{1}{2N+1} E\big[|X_N(\omega)|^2\big] \right\} \tag{8.44}$$

The expected-value symbol is essential due to the fact that a definition of $X_N(\omega)$ based on samples from p to $p + N$ pertains to a random variable, and so the exclusion of the expected-value symbol yields one value that the random variable

$$\frac{1}{2N+1} |X_N(\omega)|^2$$

could assume.

Properties of Power Spectral Densities and Power Transfer Functions

In Chapter 7 the autocorrelation functions $R_{xx}(\tau)$ and $R_{xx}(k)$ were studied for continuous and discrete processes and their properties noted. Some of these were as follows:

Property 1 $R_{xx}(\tau)$ and $R_{xx}(k)$ are even functions.

Property 2 $R_{xx}(0) \geq R_{xx}(\tau)$ and $R_{xx}(0) \geq R_{xx}(k)$ for all τ and k.

Property 3 If $x(t)$ or $x(n)$ contain a periodic component, then $R_{xx}(\tau)$ and $R_{xx}(k)$ contain a similar periodic component.

Since the power spectral density is the Fourier transform of the autocorrelation function, it too possesses many interesting properties. Among these are the following:

Property 1 $S_{xx}(\omega) = S_{xx}(-\omega)$ for both continuous and discrete processes.

Property 2 $S_{xx}(\omega) \geq 0$ for all ω, for both continuous and discrete processes.

Property 3 $S_{xx}(\omega) = S_{xx}(\omega + 2\pi n)$ for n integer, for a discrete process. This states that the power spectral density of a discrete process is periodic, with a period of 2π radians per second.

Property 4 For a continuous process,

$$\int_{-\infty}^{\infty} S_{xx}(\omega)\, df = \overline{x^2}$$

and for a discrete process,

$$\frac{1}{2\pi} \int_{-\pi}^{\pi} S_{xx}(\omega)\, d\omega = \overline{x^2}$$

A form of power spectral density that often occurs and is particularly suitable is the ratio of two polynomials in ω. For a continuous random process the form of such an expression is

$$S_{xx}(\omega) = \frac{A(\omega^{2n} + a_{2n-2}\omega^{2n-2} + \cdots + a_2\omega^2 + a_0)}{\omega^{2m} + b_{2m-2}\omega^{2m-2} + \cdots + b_2\omega^2 + b_0} \tag{8.45}$$

If $x(t)$ has no dc component and no periodic component, then $n < m$. In this case it is easy to find the bilateral Laplace transform of the autocorrelation function, inasmuch as the region of convergence will include $\sigma = 0$ since the Fourier transform exists. The bilateral Laplace transform is found by replacing $j\omega$ by s or, if the bilateral transform is known, $S_{xx}(\omega)$ is found by replacing s by $j\omega$. Symbolically, we can write

$$\mathcal{L}[R_{xx}(\tau)] = S_{xx}(\omega)\big|_{j\omega=s}$$

$$= S_{xx}(s)$$

or

$$F[R_{xx}(\tau)] = S_{xx}(s)\big|_{s=j\omega}$$

and this is denoted as $S_{xx}(\omega)$ or $S_{xx}(f)$ or $S_{xx}(j\omega)$.

The advantage of the Laplace transform representation is of course, that it leads to pole-zero interpretations and the use of residue theory. It is worth investigating the poles-and-zeros configuration of $S_{xx}(s)$. Before doing this another function with all the properties of a power spectral density will be defined.

THE POWER TRANSFER FUNCTION $|H(\omega)|^2$ OR $H(s)H(-s)$

We have seen in Chapter 7 that the output autocorrelation function of a linear system with random input is

$$R_{yy}(\tau) = C_{hh}(\tau) * R_{xx}(\tau)$$

Taking the Fourier transform, we find that the output power spectral density is

$$S_{yy}(\omega) = F[C_{hh}(\tau)] S_{xx}(\omega)$$

The Fourier transform of $C_{hh}(\tau)$, which is the impulse response $h(\tau)$ correlated with itself, is $|H(\omega)|^2$, which we will denote by $T(\omega)$ and call the pulse transfer function. Since $C_{hh}(\tau)$ is a real, even function with all the properties of $R_{xx}(\tau)$, then $T(\omega)$ has all the same properties as $S_{xx}(\omega)$ and will always occur in the rational form of $S_{xx}(\omega)$ given by Eq. 8.45. We will give a very simple example before postulating the properties of $S_{xx}(s)$ and $T(s)$.

Example 8.14

Find the system function $H(\omega)$ or $H(j\omega)$ and the pulse transfer function $T(\omega)$ or $T(s)$ for the system shown.

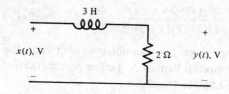

SOLUTION

$$H(s) = \frac{2}{3s+2} = \frac{0.67}{s+0.67}$$

$$H(\omega) = \frac{0.67}{j\omega+0.67}$$

and

$$T(\omega) = |H(\omega)|^2 = \frac{(0.67)}{0.67+\omega^2}$$

$$= \frac{0.67}{(j\omega+0.67)(-j\omega+0.67)}$$

Relating the Fourier and bilateral Laplace transforms we obtain

$$T(s) = |H(\omega)|^2|_{j\omega=s}$$

$$= \frac{0.67}{(s+0.67)(-s+0.67)} = \frac{-0.67}{(s+0.67)(s-0.67)}$$

and the region of convergence is $-0.67 < \sigma < 0.67$.

Example 8.15

If

$$H(s) = \frac{s+2}{(s+1)(s+3)}$$

find $T(\omega)$ and $T(s)$.

SOLUTION

$$H(\omega) = \frac{j\omega + 2}{(j\omega + 1)(j\omega + 3)}$$

$$T(\omega) = |H(\omega)|^2 = \frac{(j\omega + 2)(-j\omega + 2)}{(j\omega + 1)(-j\omega + 1)(j\omega + 3)(-j\omega + 3)}$$

$$= \frac{\omega^2 + 4}{(\omega^2 + 1)(\omega^2 + 9)}$$

$$T(s) = H(s)H(-s)$$

$$= \frac{(s + 2)(-s + 2)}{(s + 1)(-s + 1)(s + 3)(-s + 3)}$$

$$= \frac{-(s + 2)(s - 2)}{(s + 1)(s - 1)(s + 3)(s - 3)} \qquad -1 < \text{Re}(s) < +1$$

As a result of these two examples we will state the general rules for relating a power spectral density $S_{xx}(\omega)$ or power transfer function $|H(\omega)|^2$ with its bilateral Laplace transform.

Case 1

If

$$S_{xx}(\omega) = \frac{A_0 \left(\omega^{2n} + a_{n-2}\omega^{2n-2} + \cdots + a_2\omega^2 + a_0 \right)}{\omega^{2m} + b_{m-2}\omega^{2m-2} + \cdots + b_2\omega^2 + b_0}$$

$S_{xx}(s)$ is found by replacing ω^2 by $-s^2$.

Case 2

If a system function $H(s)$ is given, then the power transfer function $T(s)$ is found as

$$T(s) = H(s)H(-s)$$

and

$$|H(\omega)|^2 = T(s)|_{s=j\omega}$$

For the case $H(s) = N(s)/D(s)$, where $D(s)$ is at least of order one higher than $N(s)$, $T(s)$ will always have poles symmetrically located in the left and right half planes and the region of convergence will include $\text{Re}(s) = 0$.

THE DISCRETE CASE

For discrete random processes a form of power spectral density that often occurs and is particularly suitable for analytical manipulation is

$$S_{xx}(\omega) = S_{xx}(z)\Big|_{z=e^{j\omega}} = \frac{N(z)}{D(z)}\Big|_{z=e^{j\omega}}$$

where $S_{xx}(z)$ is the two-sided z transform of the autocorrelation function and its annulus of convergence $|a| < |z| < (b)$ is such that $|z| = 1$ is included. If $S_{xx}(\omega)$ is known, then the two-sided z transform is found by

$$S_{xx}(z) = S_{xx}(\omega)|_{e^{j\omega} = z}$$

We have seen in Chapter 7 that the output autocorrelation function of a linear system with a random input is

$$R_{yy}(n) = C_{hh}(n) * R_{xx}(n)$$

Now, by discrete Fourier transform theory,

$$S_{yy}(\omega) = F[C_{hh}(n)] S_{xx}(\omega)$$

$C_{hh}(n)$ is the pulse response $h(n)$ correlated with itself, and it has a Fourier transform $|H(\omega)|^2$, where $H(\omega)$ is the discrete Fourier transform of $h(n)$. Since $C_{hh}(n)$ is a real, even function of n, then $|H(\omega)|^2$ has all the identical properties of a power spectral density and must be a real, nonnegative, even function of ω.

Example 8.16

Given a discrete system with system function $H(z) = 1 - z^{-1}$, find

$$|H(\omega)|^2 = T(\omega) \quad \text{and} \quad T(z)$$

SOLUTION

$$H(\omega) = 1 - e^{-j\omega}$$

$$|H(\omega)| = \sqrt{(1 - \cos\omega)^2 - \sin^2\omega}$$

$$= \sqrt{1 - 2\cos\omega + 1}$$

$$= \sqrt{2(1 - \cos\omega)}$$

$$= \sqrt{2\left(1 - \cos^2\frac{\omega}{2}\right)}$$

$$= \sqrt{4\sin^2\frac{\omega}{2}}$$

$$= 2\sin\frac{\omega}{2}$$

and

$$|H(\omega)|^2 = 4\sin^2\frac{\omega}{2} = 2(1 + \cos\omega)$$

The pulse response is given by

$$h(n) = \{1, -1\}$$

and

$$C_{hh}(n) = \{-1, 2, -1\}$$

where

$$C_{hh}(-1) = C_{hh}(+1) = -1$$

The two-sided z transform is

$$z[C_{hh}(n)] = -z + 2 - z^{-1} = T(z)$$

Since

$$H(z) = 1 - z^{-1} \quad \text{and} \quad H(z^{-1}) = 1 - z$$

then the two-sided z transform of $C_{hh}(n)$ may also be expressed as the product of the two series $H(z)$ and $H(z^{-1})$. This is a general relationship, which should be obvious after some thought. If

$$h(n) = [h(0), h(1), \ldots, h(n)]$$

and

$$H(z) = h(0) + h(1)z^{-1} + \cdots + h(n)z^{-n}$$

then $C_{hh}(n)$ may be found by correlation or by multiplying the series

$$H(z) = h(0) + h(1)z^{-1} + \cdots + h(n)z^{-n}$$

and

$$H\left(\frac{1}{z}\right) = h(n)z^n + h(n-1)z^{n-1} + \cdots + h(0)$$

term by term. In our simple case we found,

$$1 - z^{-1}$$

multiplied by

$$\frac{1-z}{-z+2-z^{-1}} = C_{hh}(-1)z + C_{hh}(0) + C_{hh}(1)z^{-1} = T(z)$$

Example 8.17

Given a discrete system with system function

$$H(z) = \frac{1}{1 - 0.5z^{-1}} \qquad |z| > 0.5$$

find $T(z) = Z[C_{hh}(n)]$, and hence $C_{hh}(n)$.

SOLUTION

$$H(z) = \frac{z}{z - 0.5}$$

$$H(z^{-1}) = \frac{z^{-1}}{z^{-1} - 0.5} = \frac{1}{1 - 0.5z}$$

$$= \frac{-2}{(z-2)} \qquad |z| < 2$$

$T(z)$, the two-sided z transform of $C_{hh}(n)$, is found as follows:

$$T(z) = H(z)H(z^{-1})$$

$$= \frac{-2z}{(z-0.5)(z-2)} \qquad 0.5 < |z| < 2$$

The self-correlation function of $h(n)$ is

$$C_{hh}(n) = Z^{-1}\left[\frac{-2z}{(z-0.5)(z-2)}\right]$$

$$= \frac{1}{2\pi j}\oint \frac{-z^{n-1}2z}{(z-0.5)(z-2)}\,dz$$

On evaluation by the residue theorem, we find

For $n > 0$,

$$C_{hh}(n) = \text{residue of the pole at } z = 0.5$$

$$C_{hh}(n) = \frac{-2(0.5)^n}{-1.5}$$

$$= +1.33(0.5)^n \qquad n > 0$$

For $n \leq 0$,

$$C_{hh}(n) = \frac{1}{2\pi j}\oint \frac{-2z^n}{(z-0.5)(z-2)}\,dz$$

Since this expression contains a higher-order pole at $z = 0$, we use the inside-outside theorem to obtain

$$C_{hh}(n) = -\text{residue of the pole at } z = 2$$

$$= -\frac{-2^{n+1}}{1.5}$$

$$= +1.33(2)^{-|n|}$$

Figure 8.17 shows a plot of $H(z)$, $T(z) = H(z)H(z^{-1})$, and $C_{hh}(n)$.

GENERAL POLE-ZERO PROPERTIES OF $T(z) = H(z)H(z^{-1})$ AND $S_{xx}(z)$

If

$$T(z) = \frac{N(z)}{D(z)} \qquad \rho_1 < |z|\rho_2$$

is the z transform of $C_{hh}(n)$, then

$$T(z) = \sum_{-\infty}^{\infty} C_{hh}(n)z^{-n}$$

will converge for $|z| = 1$, and if $T(z)$ has a pole at $|z| = a(|a| < 1)$, it will also have a pole at $|z| = 1/a$. All of these properties also hold for $S_{xx}(z)$, the two-sided z transform of $R_{xx}(n)$, or the power spectral density with argu-

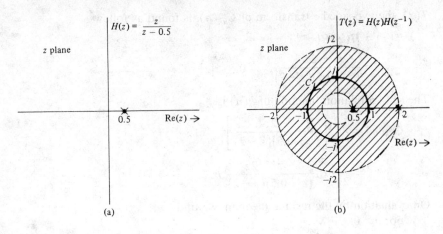

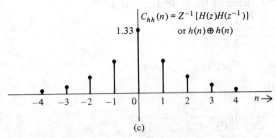

Figure 8.17   (a) The system function $H(z)$ of Example 8.17. (b) The power transfer function $T(z) = H(z)H(z^{-1})$. (c) The correlation transfer function.

ment z. Figure 8.18 shows a plot of poles for a typical $T(z)$ or $S_{xx}(z)$ function.

FREQUENCY PROPERTIES OF $|H_{xx}(\omega)|^2$ AND $S_{xx}(\omega)$ FOR A DISCRETE SYSTEM

As we have seen, $T(z) = H(z)H(z^{-1})$ contains factors of the form $(z - z_0)$ and $(z^{-1} - z_0)$ in the numerator and denominator.

$$(z - z_0)\left(\frac{1}{z} - z_0\right) = 1 - z_0\left(z + \frac{1}{z}\right) + z_0^2$$

and this is a function of $z + (1/z)$ for every pair of poles or zeros at $z = z_p$ and $1/z_p$. The system function $H(\omega)$ is found by replacing z by $e^{j\omega}$ and z^{-1} by $e^{-j\omega}$. Therefore, each $z + (1/z)$ term may be replaced by $2\cos\omega$, and $|H(\omega)|^2$ will be a rational function of $\cos\omega$. Similarly, a convenient analytical form for $S_{xx}(\omega)$, which has the same properties as $T(\omega) = |H(\omega)|^2$, is a rational function of $\cos\omega$. This result could have been obtained directly from the definition of the discrete Fourier transform and the fact that $S_{xx}(\omega)$ must be real, even, and positive.

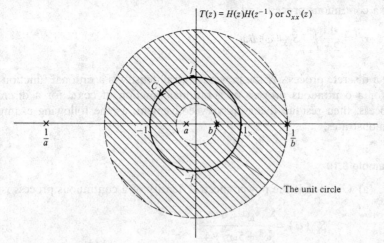

Figure 8.18 A typical pole-zero diagram for a power transfer function $T(z)$ or power spectral density $S_{xx}(z)$.

Example 8.18

Given

$$H(z) = \frac{z - 0.5}{z - 0.25}$$

find $|H(\omega)|^2$ as a rational function of $\cos \omega$.

SOLUTION

$$T(z) = H(z)H(z^{-1})$$

$$= \frac{(z - 0.5)(z^{-1} - 0.5)}{(z - 0.25)(z^{-1} - 0.25)}$$

$$= \frac{1 - 0.5(z^{-1} + z) + 0.25}{1 - 0.25(z^{-1} + z) + 0.0625}$$

$$T(\omega) = |H(\omega)|^2$$

$$= \frac{1 - 0.5(2\cos \omega) + 0.25}{1 - 0.25(2\cos \omega) + 0.0625}$$

$$= \frac{20 - 16\cos \omega}{17 - 8\cos \omega}$$

THE EVALUATION OF MEAN SQUARE VALUES BY RESIDUES

Knowing the power spectral density $S_{xx}(\omega)$, it is often required to find the mean square value of a random process; that is,

$$\overline{x^2} = \int_{-\infty}^{\infty} S_{xx}(\omega) \, df$$

for a continuous process and

$$\overline{x^2} = \frac{1}{2\pi} \int_{-\pi}^{\pi} S_{xx}(\omega) \, d\omega$$

for a discrete process. If the power spectral density is a rational function of ω for a continuous process or a rational function of $\cos \omega$ for a discrete process, then residue theory is very convenient, as the following example demonstrates.

Example 8.19

(a) Given that the power spectral density of a continuous process is

$$S_{xx}(\omega) = \frac{\omega^2 + 9}{\omega^4 + 5\omega^2 + 4}$$

find the mean square value of the process.

(b) Given that the power spectral density of a discrete random process is

$$S_{xx}(\omega) = \frac{4 - 2\cos \omega}{10 - 6\cos \omega}$$

find the mean square value of the process.

SOLUTION

(a) For the continuous process,

$$S_{xx}(s) = \frac{-s^2 + 9}{s^4 - 5s^2 + 4}$$

$$= \frac{(s+3)(s-3)}{(s^2-4)(s^2-1)}$$

$$= \frac{-(s+3)(s-3)}{(s-2)(s+2)(s-1)(s+1)} \qquad -1 < \mathrm{Re}(s) < 1$$

$$\overline{x^2} = \frac{1}{2\pi j} \int_{C\,\sigma - j\infty}^{\sigma + j\infty} S_{xx}(s) \, ds$$

$$= \Sigma \begin{bmatrix} \text{residues of the poles of} \\ S_{xx}(s) \text{ to the left of } \sigma \end{bmatrix}$$

$$= \frac{-(-1)(-5)}{(-4)(-3)(-1)} + \frac{-(2)(-4)}{(-3)(1)(-2)}$$

$$= 0.42 + 1.33 = 1.75$$

We could also have found $\overline{x^2}$ as being minus the residues of the poles to the right of an acceptable σ.

(b) For the discrete process,

$$S_{xx}(z) = \frac{4-2[0.5(z+1/z)]}{10-6[0.5(z+1/z)]}$$

$$= \frac{4-z-z^{-1}}{10-3z+3z^{-1}}$$

$$= \frac{z^2-4z+1}{3z^2-10z+3}$$

$$= \frac{(z-0.27)(z-3.73)}{3(z-0.34)(z-2.98)} \qquad 0.34<|z|<2.98$$

$$\overline{x^2} = \frac{1}{2\pi j}\oint_C S_{xx}(z)\,dz$$

$$= \text{residue of the pole at } z=0.34$$

$$= \frac{(0.34-0.27)(0.34-3.73)}{3(0.34-2.98)}$$

$$= 0.03$$

THE DEFINITION OF CROSS SPECTRAL DENSITIES

When comparing two random processes such as velocity and current fluctuations in a diode or the input and output waveforms of a system, **cross spectral densities** are defined. For continuous processes,

$$S_{xy}(\omega) \triangleq F[R_{xy}(\tau)]$$

$$= \int_{-\infty}^{\infty} R_{xy}(\tau)e^{-j\omega\tau}\,d\tau \tag{8.46}$$

and

$$S_{yx}(\omega) \triangleq F[R_{yx}(\tau)]$$

$$= \int_{-\infty}^{\infty} R_{yx}(\tau)e^{-j\omega\tau}\,d\tau \tag{8.47}$$

$S_{xy}(\omega)$ is called the cross spectral density of the process $x(t)$ and $y(t)$, and similarly $S_{yx}(\omega)$ is the cross spectral density of $y(t)$ and $x(t)$.

For discrete stationary (or at least correlation stationary) processes,

$$S_{xy}(\omega) \triangleq F[R_{xy}(k)]$$

$$= \sum_{-\infty}^{\infty} R_{xy}(k)e^{-j\omega k}$$

and

$$S_{yx}(\omega) = \sum_{-\infty}^{\infty} R_{yx}(k)e^{-j\omega k}$$

Some properties of cross spectral densities that follow from the properties of cross-correlation functions and Fourier transforms are:

1. $S_{xy}(\omega)$ and $S_{yx}(\omega)$ are not usually real, even, positive functions of ω.
2. If $S_{xy}(\omega) = R(\omega) + jI(\omega)$ then $S_{yx}(\omega)$ is its complex conjugate; that is, $S_{yx}(\omega) = R(\omega) - jI(\omega)$.
3. If $S_{xy}(\omega) = R(\omega) + jI(\omega)$, then $R(\omega)$ is an even function of ω and $I(\omega)$ is an odd function of ω.

8.4 INPUT-OUTPUT SPECTRAL RELATIONS FOR SYSTEMS WITH RANDOM INPUTS

This section will develop formulas for relating the output power spectral density of a linear time-invariant causal system to its input power spectral density and the cross power spectral density of the input with the output to the input power spectral density. The cases of pure random inputs, deterministic signal plus an independent random input, and a random signal plus independent random noise input will all be considered, as was done in Section 7.5. These results will be found by appropriately transforming the results from Section 7.5, but we will derive a number of them directly in the frequency, s, or z domains. At this stage we are assumed to have meshed together both random signal theory and transform theory and to be in the process of feeling comfortable with both disciplines.

Spectral Functions for Systems with Random Inputs

The general problem to be considered in this section is, "Given the input to a system is a member of an at least second-order stationary continuous or discrete random process, find the cross spectral density $S_{xy}(\omega)$ between the input and output and in particular find the output power spectral density $S_{yy}(\omega)$ in terms of the system function $H(\omega)$ and the input power spectral density $S_{xx}(\omega)$." Before we proceed with these derivations a number of constantly recurring results involving properties of Fourier or Laplace or z transforms will be stated.

SOME TRANSFORM RESULTS FOR SPECTRAL FUNCTIONS OR FUNCTIONS WITH SYMMETRY

For a real, continuous, stationary random process $x(t)$,

$$S_{xx}(\omega) = S_{xx}(-\omega)$$

when the Fourier transform is used, and

$$S_{xx}(s) = S_{xx}(-s)$$

when it is possible to use the bilateral Laplace transform.

For a real, discrete random process $x(n)$,

$$S_{xx}(\omega) = S_{xx}(-\omega)$$

when the discrete Fourier transform is used, and

$$S_{xx}(z) = S_{xx}(z^{-1})$$

when it is possible to use the two-sided z transform. These properties also apply to the power transfer function $T(\omega) = |H(\omega)|^2$, which is the Fourier transform of the real, even function $C_{hh}(\tau) = h(\tau) \oplus h(\tau)$ or, in the discrete case, $C_{hh}(n) = h(n) \oplus h(n)$, where $h(\tau)$ and $h(n)$ are the impulse response and the discrete pulse response, respectively.

When dealing with two real functions $f(t)$ and $g(t)$, one of which is even—say, $f(t) = f_e(t)$—then the following relations hold for the Fourier transforms $F(\omega)$ and $G(\omega)$. Using the symbol F to denote the Fourier transform and $C_{f_e g}(\tau)$ to mean $f(t) \oplus g(t)$,

$$F\left[C_{f_e g}(\tau)\right] = G(\omega)F(-\omega) = G(\omega)F(\omega)$$

This is also the result obtained by convolution and

$$\therefore F\left[C_{f_e g}(\tau)\right] = F\left[r_{f_e g}(\tau)\right]$$

where r denotes convolution. By symmetry,

$$F\left[C_{g f_e}(\tau)\right] = F(\omega)G(-\omega)$$

If $F[C_{f_e g}(\tau)] = K(\omega)$, then

$$F\left[C_{g f_e}(\tau)\right] = K(-\omega)$$

since $F(\omega)$ is real and even.

These same results hold for two discrete functions, one of which, $f_e(n)$, is even; that is,

$$F\left[C_{f_e g}(n)\right] = F\left[r_{f_e g}(n)\right] = F(\omega)G(\omega)$$

$$= K(\omega)$$

$$F\left[C_{g f_e}(n)\right] = F(\omega)G(-\omega) = K(-\omega)$$

Keeping these results in mind we will proceed with our derivations.

CASE OF A LINEAR SYSTEM WITH A PURE RANDOM NOISE INPUT
CONTINUOUS CASE

Let us consider a continuous linear system with impulse response $h(\tau)$ or system function $H(\omega)$ and a zero mean random noise input $x(t)$ with autocorrelation function $R_{xx}(\tau)$ and power spectral density $S_{xx}(\omega)$. From Chapter 7 the output autocorrelation function is

$$R_{yy}(\tau) = C_{hh}(\tau) * R_{xx}(\tau)$$

By taking the Fourier transform on both sides, the output power spectral density is

$$S_{yy}(\omega) = |H(\omega)|^2 S_{xx}(\omega) \tag{8.48}$$

As a physical exercise we can find this result directly in the frequency domain using the definition

$$S_{xx}(\omega) \triangleq \lim_{T \to \infty} \left\{ \frac{1}{2T} E\big[|X_T(\omega)|^2 \big] \right\}$$

Let $y_{iT}(t)$ be the output, when a section of the ith member of the random process $x_i(t)$ from $t = c$ to $c + T$ is the input to a linear system with system function $H(\omega)$. We will use $X_{iT}(\omega)$ and $Y_{iT}(\omega)$ to denote the Fourier transforms of $x_{iT}(t)$ and $y_{iT}(t)$ respectively.

$$Y_{iT}(\omega) = X_{iT}(\omega)H(\omega)$$

$$|Y_{iT}(\omega)|^2 = [X_{iT}(\omega)H(\omega)][X_{iT}(-\omega)H(-\omega)]$$

$$= |X_{iT}(\omega)|^2 |H(\omega)|^2$$

$$S_{yy}(\omega) = \lim_{T \to \infty} \left\{ \frac{1}{2T} \big[E|Y_T(\omega)|^2 \big] \right\}$$

$$= |H(\omega)|^2 \lim_{T \to \infty} \left\{ \frac{1}{2T} \big[E|X_T(\omega)|^2 \big] \right\}$$

$$= |H(\omega)|^2 S_{xx}(\omega)$$

as obtained earlier in Eq. 8.48.

The cross power spectral densities $S_{xy}(\omega)$ and $S_{yx}(\omega)$ may be found by transforming the results from Chapter 7; that is,

$$R_{xy}(\tau) = h(\tau) \oplus R_{xx}(\tau) \quad \text{or} \quad R_{xx}(\tau) * h(\tau) \quad \text{(from Eq. 8.47)}$$

$$\therefore S_{xy}(\omega) = S_{xx}(\omega)H(\omega) \tag{8.49}$$

$$R_{yx}(\tau) = R_{xx}(\tau) \oplus h(\tau)$$

$$\therefore S_{yx}(\omega) = S_{xx}(\omega)H(-\omega) \tag{8.50}$$

The form of these results is almost identical if the two-sided Laplace transform is used. The great analytical advantage is in working from the transform to the time domain by residue theory. For example,

$$S_{yy}(s) = S_{xx}(s)T(s)$$

where $T(s) = H(s)H(-s)$ and

$$R_{yy}(\tau) = \frac{1}{2\pi j} \oint S_{xx}(s)T(s)e^{s\tau} \, ds$$

For $\tau > 0$,

$$R_{yy}(\tau) = \sum \left[\begin{array}{l} \text{residues of the poles of} \\ S_{xx}(s)T(s)e^{s\tau} \text{ to the left of } \sigma \end{array} \right] \tag{8.51}$$

For $\tau < 0$,

$$R_{yy}(\tau) = -\sum \left[\begin{array}{l} \text{residues of the poles of} \\ S_{xx}(s)T(s)e^{s\tau} \text{ to the right of } \sigma \end{array} \right] \tag{8.52}$$

In Eqs. 8.51 and 8.52, σ is an acceptable value of the real part of s in the region of convergence for $S_{yy}(s)$ and, of course, $R_{xx}(\tau) = R_{xx}(-\tau)$ for any τ.

Utilizing residue theory for the cross spectral densities, we have

$$S_{xy}(s) = S_{xx}(s)H(s) \qquad \sigma_1 < \text{Re}(s) < \sigma_2$$

For $\tau > 0$,

$$R_{xy}(\tau) = \sum \left[\begin{array}{l} \text{residues of poles of } S_{xy}(s)e^{s\tau} \\ \text{to the left of } \sigma \end{array} \right] \tag{8.53}$$

For $\tau < 0$,

$$R_{xy}(\tau) = -\sum \left[\begin{array}{l} \text{residues of poles of } S_{xy}(s)e^{s\tau} \\ \text{to the right of } \sigma \end{array} \right] \tag{8.54}$$

Similarly,

$$S_{yx}(s) = S_{xx}(s)H(-s)$$

For $\tau > 0$,

$$R_{yx}(\tau) = \sum \left[\begin{array}{l} \text{residues of poles of } S_{yx}(s) \\ \text{to the left of } \sigma \end{array} \right] \tag{8.55}$$

For $\tau < 0$,

$$R_{yx}(\tau) = -\sum \left[\begin{array}{l} \text{residues of poles of } S_{yx}(s) \\ \text{to the right of } \sigma \end{array} \right] \tag{8.56}$$

These results are shown summarized in Figure 8.19a.

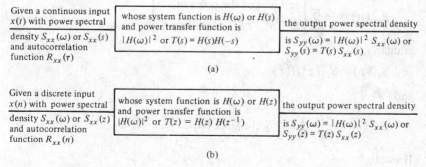

Given a continuous input $x(t)$ with power spectral density $S_{xx}(\omega)$ or $S_{xx}(s)$ and autocorrelation function $R_{xx}(\tau)$	whose system function is $H(\omega)$ or $H(s)$ and power transfer function is $|H(\omega)|^2$ or $T(s) = H(s)H(-s)$	the output power spectral density is $S_{yy}(\omega) = |H(\omega)|^2 S_{xx}(\omega)$ or $S_{yy}(s) = T(s) S_{xx}(s)$

(a)

Given a discrete input $x(n)$ with power spectral density $S_{xx}(\omega)$ or $S_{xx}(z)$ and autocorrelation function $R_{xx}(n)$	whose system function is $H(\omega)$ or $H(z)$ and power transfer function is $|H(\omega)|^2$ or $T(z) = H(z)\,H(z^{-1})$	the output power spectral density is $S_{yy}(\omega) = |H(\omega)|^2 S_{xx}(\omega)$ or $S_{yy}(z) = T(z) S_{xx}(z)$

(b)

Figure 8.19 (a) A continuous system with a random input. (b) A discrete system with a random input.

CASE OF A DISCRETE LINEAR SYSTEM

Let us consider a discrete linear system with pulse response $h(n)$, whose discrete Fourier transform $H(\omega)$ is the system function. From Chapter 7, when the input is a discrete zero-mean random process with autocorrelation function $R_{xx}(n)$ the output autocorrelation function is

$$R_{yy}(n) = R_{xx}(n) * C_{hh}(n)$$

and the cross correlation of the input with the output is

$$R_{xy}(n) = R_{xx}(n) * h(n)$$

Taking the discrete Fourier transforms of both sides yields

$$S_{yy}(\omega) = S_{xx}(\omega)|H(\omega)|^2 \qquad (8.57)$$

$$S_{xy}(\omega) = S_{xx}(\omega)H(\omega) \qquad (8.58)$$

and

$$S_{yx}(\omega) = S_{xx}(\omega)H(-\omega) \qquad (8.59)$$

The form of these results is almost identical if the two-sided z transform is used.

$$S_{yy}(z) = S_{xx}(z)T(z)\cdots \qquad \rho_1 < |z| < \rho$$

where $T(z) = H(z)H(z^{-1})$ and

$$R_{yy}(n) = \frac{1}{2\pi j}\oint_C z^{n-1}S_{xx}(z)T(z)\,dz$$

If $n \geq 0$,

$$R_{yy}(n) = \Sigma \begin{bmatrix} \text{residues of the poles of} \\ z^{n-1}S_{xx}(z)T(z) \text{ inside } C \end{bmatrix} \qquad (8.60)$$

If $n \leq 0$,

$$R_{yy}(n) = -\Sigma \begin{bmatrix} \text{residues of the poles of} \\ z^{n-1}S_{xx}(z)T(z) \text{ outside } C \end{bmatrix} \qquad (8.61)$$

Similarly,

$$S_{xy}(z) = S_{xx}(z)H(z)$$

and

$$R_{xy}(n) = \frac{1}{2\pi j}\oint_C z^{n-1}S_{xx}(z)H(z)\,dz$$

If $n \geq 0$,

$$R_{xy}(n) = \Sigma \begin{bmatrix} \text{residues of the poles} \\ \text{of } S_{xy}(z) \text{ inside } C \end{bmatrix} \qquad (8.62)$$

If $n \leq 0$,

$$R_{xy}(n) = -\sum \begin{bmatrix} \text{residues of the poles} \\ \text{of } S_{xy}(z) \text{ outside } C \end{bmatrix} \quad (8.63)$$

and an almost identical set of formulas hold for $S_{yx}(z)$ and $R_{yx}(n)$. In addition, we know that $R_{yx}(n) = R_{xy}(-n)$.

All of these results are shown summarized in Table 8.7.

A Linear System with a Signal plus Uncorrelated Noise Input

Corresponding to the time domain results developed in Chapter 7 for a linear system with an input $x(t)$ or $x(k)$ consisting of a signal $f(t)$ or $f(k)$ plus uncorrelated noise $n(t)$ or $n(k)$ we will develop the frequency domain version of these formulas. We will again consider the two cases where the signal may be deterministic or a zero-mean noise signal.

DETERMINISTIC SIGNAL PLUS UNCORRELATED ZERO-MEAN NOISE (CONTINUOUS CASE)

Table 8.7 shows a linear system with input $x(t) = f(t) + n(t)$. If the function $f(t)$ is deterministic and causal and the noise is assumed to be uncorrelated and of zero mean, then from Chapter 7 the output is denoted as

$$y(t) = g(t) + m(t)$$

where $g(t)$ is the deterministic function

$$g(t) = f(t) * h(t)$$

and $m(t)$ is an uncorrelated random process with autocorrelation function

$$R_{mm}(\tau) = R_{nn}(\tau) * C_{hh}(\tau)$$

and the cross correlation between $n(t)$ and $m(t)$ is

$$R_{nm}(\tau) = R_{nn}(\tau) * h(\tau)$$

The frequency ω domain or Laplace s domain results are as follows:

The transform of the deterministic output is

$$G(\omega) = H(\omega)F(\omega) \quad \text{or} \quad G(s) = H(s)F(s) \quad (8.64)$$

where the capital letter denotes the Fourier or Laplace transform.

For the spectral density of output noise,

$$S_{mm}(\omega) = |H(\omega)|^2 S_{nn}(\omega) \quad \text{or} \quad S_{mm}(s) = S_{nn}(s)H(s)H(-s) \quad (8.65)$$

For the cross spectral density of input and output noise,

$$S_{nm}(\omega) = S_{nn}(\omega)H(\omega) \quad \text{or} \quad S_{nm}(s) = S_{nn}(s)H(s) \quad (8.66)$$

DISCRETE CASE

In the case of a discrete system with a deterministic input $x(k)$ plus zero-mean uncorrelated noise $n(k)$, where we use k so as not to be slightly upset with the notation $n(n)$, the frequency domain or z transform results

Table 8.7 SUMMARY OF DIGITAL OR ANALOG SYSTEMS WITH DIFFERENT SIGNAL-PLUS-NOISE-INPUTS

CONTINUOUS		DISCRETE	
$x(t) = f(t) + n(t)$ with $S_{nn}(\omega)$ or $S_{nn}(s)$	$H(\omega), T(\omega) = \lvert H(\omega)\rvert^2$ or $H(s), T(s) = H(s)H(-s)$ — $y(t) = g(t) + m(t)$	$x(k) = f(k) + n(k)$	$H(\omega), T(\omega) = \lvert H(\omega)\rvert^2$ $H(z), T(z) = H(z)H(z^{-1})$ — $y(k) = g(k) + m(k)$

CONTINUOUS

CASE 1 $f(t)$ and $g(t)$ both zero-mean random and uncorrelated

$S_{xy}(\ ) = S_{ff}(\ )H(\ ) + S_{nn}(\ )H(\ )$
$S_{yy}(\ ) = S_{ff}(\ )T(\ ) + S_{nn}(\ )T(\ )$
Arguments ω or s

CASE 2 $f(t)$ deterministic plus zero-mean uncorrelated noise

$G(\ ) = F(\ )H(\ )$
$S_{mm}(\ ) = S_{nn}(\ )T(\ )$
$S_{nm} = S_{nn}(\ )H(\ )$
Arguments ω or s

CASE 3 Pure random $x(t)$

$S_{yy}(\ ) = S_{xx}(\ )T(\ )$
$S_{xy}(\ ) = S_{xx}(\ )H(\ )$
Arguments ω or s

DISCRETE

CASE 1 $f(k)$ and $g(k)$ both zero-mean random and uncorrelated

$S_{xy}(\ ) = S_{ff}(\ )H(\ ) + S_{nn}(\ )H(\ )$
$S_{yy}(\ ) = S_{ff}(\ )T(\ ) + S_{nn}(\ )T(\ )$
Arguments ω or z

CASE 2 $f(k)$ deterministic plus zero-mean uncorrelated noise

$G(\ ) = F(\ )H(\ )$
$S_{mm}(\ ) = S_{ff}(\ )T(\ )$
$S_{nm}(\ ) = S_{ff}(\ )H(\ )$
Arguments ω or z

CASE 3 Pure random $x(k)$

$S_{yy}(\ ) = S_{xx}(\ )T(\ )$
$S_{xy}(\ ) = S_{xx}(\ )H(\omega)$
Arguments ω or z

are

$$G(\omega) = H(\omega)F(\omega) \quad \text{or} \quad G(z) = H(z)F(z) \tag{8.67}$$

$$S_{mm}(\omega) = |H(\omega)|^2 S_{nn}(\omega) \quad \text{or} \quad S_{mm}(z) = [H(z)H(-z)]S_{nn}(z)$$

and

$$S_{nm}(\omega) = S_{nn}(\omega)H(\omega) \quad \text{or} \quad S_{nm}(z) = S_{nn}(z)H(z)$$

These results for the continuous and discrete cases are shown in Table 8.7. One of their main applications will occur in the design of a matched filter to indicate the presence of a deterministic signal buried in noise.

A RANDOM SIGNAL INPUT PLUS UNCORRELATED ZERO-MEAN NOISE

The most general situation that can arise is when a system input consists of a random signal $f(t)$ or $f(k)$ plus noise $n(t)$ or $n(k)$. If we assume that both signal and noise are uncorrelated zero-mean members of stationary random processes, then from Chapter 7 we have

$$R_{yy}(\tau) = R_{ff}(\tau) * C_{hh}(\tau) + R_{nn}(\tau) * C_{hh}(\tau)$$

and

$$R_{xy}(\tau) = R_{xx}(\tau) * h(\tau) + R_{nn}(\tau) * h(\tau)$$

These results expressed in the ω or s domain are

$$S_{yy}(\omega) = S_{ff}(\omega)|H(\omega)|^2 + S_{nn}(\omega)|H(\omega)|^2$$

or

$$S_{yy}(s) = S_{ff}(s)[H(s)H(-s)] + S_{nn}(s)[H(s)H(-s)]$$

and

$$S_{xy}(\omega) = S_{ff}(\omega)H(\omega) + S_{nn}(\omega)H(\omega)$$

or

$$S_{xy}(s) = S_{ff}(s)H(s) + S_{nn}(s)H(s)$$

DISCRETE CASE

The corresponding results for a discrete system with input

$$x(k) = f(k) + n(k)$$

which are uncorrelated zero-mean members of random processes are

$$S_{yy}(\omega) = S_{ff}(\omega)|H(\omega)|^2 + S_{nn}(\omega)|H(\omega)|^2$$

or

$$S_{yy}(z) = S_{ff}(z)[H(z)H(z^{-1})] + S_{nn}(z)[H(z)H(z^{-1})]$$

and

$$S_{xy}(\omega) = S_{ff}(\omega)H(\omega) + S_{nn}(\omega)H(\omega)$$

or

$$S_{xy}(z) = S_{ff}(z)H(z) + S_{nn}(z)H(z)$$

In these formulas ω refers to discrete Fourier transforms and z to two-sided z transforms, respectively. Table 8.7 summarizes the results of this section in order of the less general results:

> Case 1 is given for both a random signal and a random noise waveform.
> Case 2 gives the results when the signal is deterministic.
> Case 3 describes a purely random signal input.

One of our main applications in practice of a random signal plus random noise input is the design of a Wiener filter to recover the best mean square error approximation of the signal. If such a filter is too difficult to achieve, we may consider a close relative, called a Kalman filter. The name dropping of the terms "matched," "Wiener," and "Kalman" filter serves to indicate how close we are to dealing with their design.

We have now completed the frequency interpretation of random processes and the transform analysis of linear systems with signal plus noise inputs. A few examples will be solved so that the reader will become familiar and comfortable with these linear system results.

Example 8.20

Consider a first-order system with system function $H(\omega) = 1/(j\omega + 4)$, which has an input

$$x(t) = \cos 5t + n(t)$$

where $n(t)$ behaves like white noise with a power spectral density $S_{nn}(\omega) = 10$ and actual mean square fluctuations $\overline{n^2(t)} = 50$. Assuming that the noise and signal are uncorrelated, find the output noise power spectral density, the cross power spectral densities of input and output noise, and the output signal-to-noise ratio.

SOLUTION
The phasor of the input signal $f(t)$ is $1 \angle 0°$ and the output signal phasor is

$$G_{ph} = 1 \angle 0° \frac{1}{j5 + 1}$$

$$\therefore G_{ph} = \frac{1 \angle 0°}{5.1 \angle 79°}$$

$$= 0.2 \angle -79°$$

and the output signal is

$$g(t) = 0.2\cos(5t - 79°)$$

The output power spectral density is

$$S_{mm}(\omega) = 10 S_{nn}(\omega)$$

$$= 10 \left| \frac{1}{j\omega + 4} \right|^2$$

$$= \frac{10}{16 + \omega^2}$$

and the cross spectral densities are

$$S_{nm}(\omega) = S_{nn}(\omega) H(\omega)$$

$$= \frac{10}{j\omega + 4}$$

$$= \frac{10}{\omega^2 + 16} - \frac{j\omega 10}{\omega^2 + 16}$$

and

$$S_{mn}(\omega) = S_{nn}(\omega) H(-\omega)$$

$$= \frac{10}{4 - j\omega}$$

$$= \frac{10}{\omega^2 + 16} + j\frac{\omega 10}{\omega^2 + 16}$$

We notice that the real and imaginary part of the cross spectral densities have the desired properties of Section 8.3.

The output noise fluctuations are found as follows:

$$S_{mm}(\omega) = \frac{10}{\omega^2 + 16}$$

or

$$S_{mm}(s) = \frac{10}{-s^2 + 16}$$

$$= \frac{10}{-(s-4)(s+4)} \qquad -4 < \text{Re}(s) < 4$$

and

$$\overline{m^2(t)} = \frac{1}{2\pi} \int_{-\infty}^{\infty} S_{mm}(\omega)\, d\omega$$

$$= \frac{1}{2\pi j} \int_{\sigma - j\infty}^{\sigma + j\infty} \frac{-10}{(s-4)(s+4)}\, ds \qquad -4 < \sigma < +4$$

$$= \sum \left(\begin{array}{l} \text{residue of the pole} \\ \text{to the left of } \sigma \end{array} \right)$$

$$= \frac{-10}{-8}$$

$$= 1.25$$

The output signal-to-noise ratio is

$$(S/N)_{\text{out}} = \frac{\overline{g^2(t)}}{\overline{m^2(t)}}$$

$$= \frac{\left(0.2/\sqrt{2}\right)^2}{1.25}$$

$$= 0.016$$

This compares with the input signal-to-noise ratio

$$(S/N)_{\text{in}} = \frac{\left(1/\sqrt{2}\right)^2}{50} = 0.010$$

Example 8.21

Consider a discrete linear system with the system function

$$H(z) = \frac{1}{1 - 0.6z^{-1}}$$

which has as its input a discrete deterministic signal

$$f(k) = \sum_{n=0}^{\infty} \delta(k - n)$$

plus uncorrelated zero mean noise $n(k)$, with power spectral density

$$S_{nn}(z) = \frac{2z}{(z + 0.5)(z + 2)}$$

(a) Find the deterministic output $g(k)$ and the output noise power spectral density.

(b) Find the mean square noise fluctuations at the output and the output signal-to-noise ratio.

SOLUTION

(a) The output signal is found as follows:

$$G(z) = H(z)X(z)$$

$$= \frac{z}{z - 1} \sum_{k=0}^{\infty} (0.6z^{-k})$$

$$= \frac{z}{z - 1} \left(\frac{z}{z - 0.6} \right)$$

$$\therefore g(n) = \frac{1}{2\pi j} \oint_C \frac{z^2 z^{n-1}}{(z - 1)(z - 0.6)} dz$$

where C encloses all the poles. Hence

$$g(n) = \frac{1}{2\pi j} \oint_C \frac{z^{n+1}}{(z-1)(z-0.6)} dz$$

For $n < 0$, $g(n) = 0$ because the denominator polynomial is at least of order two higher than the numerator. For $n \geq 0$,

$$g(n) = \sum (\text{residues of the poles at } z = 0.6 \text{ and } z = 1)$$

$$= 2.5 - 2.5(0.6)^{n+1}$$

The output noise power spectral density is determined as follows:

$$S_{mm}(z) = S_{nn}(z)H(z)H(z^{-1})$$

$$= \frac{2z}{(z+0.5)(z+2)} \frac{1}{1-0.6z^{-1}} \frac{1}{1-0.6z}$$

$$= \frac{2z}{(z+0.5)(z+2)} \frac{-0.6}{(z-0.6)[z-(1/0.6)]}$$

$$= \frac{-3.2z}{(z-0.6)(z+0.5)(z+2)(z-1.67)} \qquad 0.6 < |z| < 1.67$$

(b) The noise fluctuations are formed as follows:

$$\overline{m^2(t)} = \frac{1}{2\pi j} \oint_C S_{mm}(z) dz \qquad 0.6 < |z| < 1.67$$

$$= \sum (\text{residues of the poles inside } C)$$

or

$$= -\sum (\text{residues of the poles outside } C)$$

Evaluating the sum of the residues inside, we obtain

$$\overline{m^2(t)} = \frac{-3.2(0.6)}{1.1(2.6)(-1.07)} + \frac{1.6}{-1.1(1.5)(-2.17)}$$

$$= 0.63 + 0.45$$

$$= 1.08$$

Since the output signal is $g(n) = 2.5(1 - 0.6^{n+1})$ and $g^2(n)$ is a function of n, then the signal-to-noise ratio at $n = n$ is as follows:

$$(\text{S/N})_{\text{out}} = \frac{2.5(1 - 0.6^{2n+2})}{1.08}$$

Table 8.8 THE STAGES OF STATIONARY RANDOM PROCESSES

TRANSFORM	DEFINITION	INVERSE	SPECTRAL DENSITIES
Two-sided z transform $F(z)$ or $F_B(z)$ or $Z(f(n))$	$F(z) = \sum_{-\infty}^{\infty} f(n) z^{-n}$ $\rho_1 < \rho < \rho_2$ exists if $\sum_{-\infty}^{\infty} \lvert f(n) \rvert \rho^{-n} < \infty$	$f(n) = \dfrac{1}{2\pi j} \oint_C z^{n-1} F(z)\, dz$ $n > 0$ $f(n) = \sum \left(\begin{array}{c} \text{residues of poles} \\ \text{inside } C \end{array} \right)$ $n \le 0$ $f(n) = -\sum \left(\begin{array}{c} \text{residues of} \\ \text{poles outside } C \end{array} \right)$	$S_{xx}(z) = Z[R_{xx}(n)]$ $S_{xy}(z) = Z[R_{xy}(n)]$ $S_{yx}(z) = Z[R_{yx}(n)]$
Discrete Fourier transform $F(\omega)$ or $F[f(n)]$	$F(\omega) = \sum f(n) e^{-jn\omega}$ exists if $\sum \lvert f(n) \rvert < \infty$ $F(\omega) = F(z)$ with $z^{-n} = e^{-jn\omega}$ $F(\omega) = F(\omega + 2\pi n)$	$f(n) = \dfrac{1}{2\pi} \int_{-\pi}^{\pi} F(\omega) e^{j\omega n}\, d\omega$	$S_{xx}(\omega) = F[R_{xx}(n)]$ $S_{xy}(\omega) = F[R_{xy}(n)]$ $S_{yx}(\omega) = F[R_{yx}(n)]$
Fourier transform $F(\omega)$ or $F[f(t)]$	$F(\omega) = \int_{-\infty}^{\infty} f(t) e^{-j\omega t}\, dt$** exists if $\int_{-\infty}^{\infty} \lvert f(t) \rvert^2\, dt < \infty$ exists in limit if $\int_{-\infty}^{\infty} \lvert f(t) \rvert e^{-a\lvert t \rvert}\, dt < \infty$ for $a \to 0^{+}$	$f(t) = \int_{-\infty}^{\infty} F(\omega) e^{j\omega t}\, df$	$S_{xx}(\omega) = F[R_{xx}(\tau)]$ $S_{xy}(\omega) = F[R_{xy}(\tau)]$ $S_{yx}(\omega) = F[R_{yx}(\tau)]$
Bilateral Laplace transform $F(s)$ or $F_B(s)$ or $\mathcal{L}[f(t)]$	$F(s) = \int_{-\infty}^{\infty} f(t) e^{-st}\, dt$ $\sigma_1 < \mathrm{Re}(s) < \sigma_2$ exists if $\int_{-\infty}^{\infty} \lvert f(t) \rvert e^{-\sigma t}\, dt < \infty$ $F(\omega) = F(s)$ with** $s = j\omega$	$f(t) = \dfrac{1}{2\pi j} \int_{\sigma - j\infty}^{\sigma + j\infty} F(s) e^{st}\, ds$ $t > 0$ $f(t) = \sum \left(\begin{array}{c} \text{residues of poles} \\ \text{to left of } \sigma \end{array} \right)$ $t < 0$ $f(t) = -\sum \left(\begin{array}{c} \text{residues of poles} \\ \text{to right of } \sigma \end{array} \right)$	$S_{xx}(s) = \mathcal{L}[R_{xx}(\tau)]$ $S_{xy}(s) = \mathcal{L}[R_{xy}(\tau)]$ $S_{yx}(s) = \mathcal{L}[R_{yx}(\tau)]$
One-sided z transform $F(z)$	$F(z) = \sum_{0}^{\infty} f(n) z^{-n} = F_B(z)$ if no initial conditions $\rho > \rho_1$		
One-sided Laplace transform $F(s)$	$F(s) = \int_{0}^{\infty} f(t) e^{-st}\, dt = F_B(s)$ if no initial conditions		

PROPERTIES	POWER TRANSFER	$\dfrac{f(t)+n(t)}{f(k)+n(k)}$ $\quad$ $\begin{array}{c}H(\omega)\\ \text{or } H(z)\end{array}$ $\quad$ $\dfrac{g(t)+m(t)}{g(k)+m(k)}$
$S_{xx}(z)=S_{xx}(z^{-1})$ $\overline{x^2}=\dfrac{1}{2\pi j}\oint_C S_{xx}(z)\,dz$ $=\Sigma\left(\begin{array}{c}\text{residues of poles}\\ \text{inside } C\end{array}\right)$ $S_{xy}(z)=S_{yx}(z^{-1})$	$T(z)=H(z)H(z^{-1})$ Same properties as $S_{xx}(z)$ If a pole at z also a pole at z^{-1}	If $R_{fn}(k)=0$ $S_{yy}(z)=S_{ff}(z)T(z)+S_{nn}(z)T(z)$ If $f(k)$ is deterministic and $R_{fn}(k)=0$ $G(z)=H(z)F(z)$ $S_{mm}(z)=S_{nn}(z)T(z)$
$S_{xx}(\omega)=S_{xx}(-\omega)$ $S_{xx}(\omega)\geq 0$ and even If $S_{xy}(\omega)=R(\omega)+jk(\omega)$ $S_{yx}(\omega)=R(\omega)-jk(\omega)$ $R(\omega)$ even, $k(\omega)$ odd	$T(\omega)=H(\omega)H(-\omega)$ If $T(z)=\dfrac{N(z)}{D(z)}$ then $T(\omega)=\dfrac{N(\cos\omega)}{D(\cos\omega)}$	If $R_{fn}(k)=0$ $S_{yy}(\omega)=S_{ff}(\omega)T(\omega)+S_{nn}(\omega)T(\omega)$ If $f(k)$ is deterministic $G(\omega)=H(\omega)F(\omega)$ $S_{mm}(\omega)=T(\omega)S_{nn}(\omega)$
$S_{xx}(\omega)=S_{xx}(-\omega)$ $S_{xx}(\omega)\geq 0$ and even If $S_{xy}(\omega)=R(\omega)+jk(\omega)$ $S_{yx}(\omega)=R(\omega)-jk(\omega)$ $R(\omega)$ even, $k(\omega)$ odd	$T(\omega)=H(\omega)H(-\omega)$ If $H(s)=\dfrac{N(s)}{D(s)}$ $T(\omega)=\dfrac{N_1(\omega^2)}{D_1(\omega^2)}$	If $R_{fn}(k)=0$ $S_{yy}(\omega)=S_{ff}(\omega)T(\omega)+S_{nn}(\omega)T(\omega)$ If $R_{fn}(\tau)=0$ and $f(t)$ is deterministic $G(\omega)=H(\omega)F(\omega)$ $S_{mm}(\omega)=T(\omega)S_{nn}(\omega)$
$S_{xx}(s)=S_{xx}(-s)$ $\overline{x^2}=\dfrac{1}{2\pi j}\oint_C S_{xx}(s)\,ds$ $\Sigma\begin{bmatrix}\text{residues of poles}\\ \text{of } S_{xx}(s)\text{ to left}\\ \text{of }\sigma\end{bmatrix}$ $=R_{xx}(\tau)$ $S_{xy}(s)=S_{yx}(-s)$	$T(s)=H(s)H(-s)$ Poles and zeros symmetrically located about center-line of strip of convergence $\sigma=0.5(\sigma_1+\sigma_2)$	If $R_{fn}(\tau)=0$ $S_{yy}(s)=S_{ff}(s)T(s)+S_{nn}(s)T(s)$ If $R_{fn}(\tau)=0$ and $f(t)$ deterministic $G(s)=H(s)F(s)$ $S_{mm}(s)=T(s)S_{nn}(s)$

Hence, for large n,

$$(S/N)_{out} = \frac{2.5}{1.08} = 2.3$$

SUMMARY

Chapter 8 dealt with the transform domain or spectral analysis of at least wide-sense stationary random processes, in three stages:

1. The main transforms used in random process theory.
2. The power spectral and cross spectral densities plus their interpretation and properties.
3. The analysis of linear systems with random inputs.

All of these stages are shown summarized in Table 8.8. Each transform, its definition and its inverse are defined in the first three columns. Then the use of a particular transform to denote spectral densities plus their properties is given in the next two columns. Finally, in the last two columns are shown the properties of the power transfer function and the input-output relations for a linear system. The one-sized z and Laplace transforms are indicated as special cases of their bilateral counterparts because for assumed second-order stationary inputs the energy is not considered. The specific initial energy could be a factor in the case of a deterministic signal-plus-noise input.

PROBLEMS

1. (a) Find the two-sided z transform of the following functions:

$$x(n) = n(0.5)^n \qquad n \geq 0$$
$$= 2n \qquad n \leq 0$$
$$x(n) = 0.6^{|n|} \qquad \text{all } n$$

(b) Find and plot the inverse z transform of

$$F(z) = \frac{-z}{(z-0.5)(z-2)} \qquad |z| > 2$$

$$F(z) = \frac{-z}{(z-0.5)(z-2)} \qquad 0.5 < |z| < 2$$

$$F(z) = \frac{-z}{(z-0.5)(z-2)} \qquad |z| < 0.5$$

Do all three parts using the inverse formula and partial fractions.

2. (a) Find the one-sided z transform of the following functions:

$$x(n) = -0.6^n u(n)$$
$$x(n) = -n^2 0.6^n u(n)$$
$$x(n) = -(n-2)^2 0.6^{n-2} u(n-2)$$
$$x(n) = -(n-2)^2 0.6^{n-2} u(n)$$

(b) Find and plot the inverse z transform of

$$F(z) = \frac{z}{z-0.4} \qquad |z| > 0.4$$

$$F(z) = \frac{z^2 + 4z - 3}{z - 0.4} \qquad |z| > 0.4$$

$$F(z) = \frac{z}{(z-0.4)^4} \qquad |z| > 0.4$$

3. (a) If $f(n) = f(n)u(n)$ and $g(n) = g(n)u(n)$, show

$$Z\left[\sum_{p=0}^{n} f(p)g(n-p) \right] = F(z)G(z)$$

$$Z\left[\sum_{\text{all } p} f(p)f(p-n) \right] = F(z)F(z^{-1})$$

(b) Evaluate

$$u(n) * 0.5^n u(n)$$
$$(0.5)^n u(n) \oplus (0.5)^n u(n)$$

4. (a) Solve classically and by the one-sided z transform

   (i) $3y(n) - 0.3y(n-1) = u(n) \qquad n > 0$
   $y(o) = 3$
   (ii) $y(n) + y(n-1) + 0.25y(n-2) = 2^n \qquad n \geq 0$
   $y(o) = 2, \; y(-1) = 1$

(b) Find $h(n)$, the pulse response, when

$$y(n) + 0.5y(n-1) = 2x(n)$$
$$y(n) + 0.5y(n-1) = 3x(n) + x(n-1)$$

5. (a) Find the two-sided Laplace transforms of the following functions:

$$f(t) = e^{3t} u(-t) + te^{-t} u(t)$$
$$f(t) = t^2 e^{3t} u(-t)$$
$$f(t) = t^2 e^{3t} u(t)$$

(b) Find the inverse Laplace transforms of

$$F(s) = \frac{6}{(s+2)^2} \qquad \text{Re}(s) > -2$$

$$F(s) = \frac{6}{(s+2)^2} \qquad \text{Re}(s) < -2$$

$$F(s) = \frac{2s}{(s+1)(s+3)} \qquad -3 < \text{Re}(s) < -1$$

6. Find the one-sided Laplace transforms of the following functions:

$$f(t) = e^{-5t}u(t)$$
$$f(t) = e^{-5(t-2)}u(t-2)$$
$$f(t) = t^2 e^{-5(t-2)}u(t-2)$$

7. (a) Solve, using the one-sided Laplace transform,

$$y''(t) + 3y'(t) + 2y(t) = tu(t) \qquad t > 0$$
$$y(o) = 2, \ y'(o) = 1$$

(b) Find $h(t)$, the impulse response, when

$$y'(t) + 3y(t) = 2x(t)$$
$$y'(t) + 3y(t) = 2x(t) + x'(t)$$

Problems 8 through 13 are review problems involving the Fourier transform.

8. Using only the two basic Fourier transform pairs

$$\sqcap(t) \longleftrightarrow \frac{\sin \pi f}{\pi f}$$

and

$$e^{-a|t|} \longleftrightarrow \frac{2a}{a^2 + 4\pi^2 f^2}$$

and the theorems of Table 8.6 derive the following transforms in the order shown and sketch your results.

(a) $\dfrac{\sin 4t}{t}$

(b) 1

(c) $\cos t$ $\qquad$ using (b) plus theorems

(d) $\dfrac{\cos t}{t - (\pi/2)}$

(e) $\sin \pi t \cos \frac{1}{4}\pi t$

(f) $\displaystyle\int_{-\infty}^{\infty} \frac{\sin \pi P}{\pi P} \cos \frac{\pi}{4}(t - P) dP$

9. Given $e^{-\pi t^2} \longleftrightarrow e^{-\pi f^2}$, use the reciprocal spreading and moment theorems to evaluate the following:

(a) $\int_0^\infty e^{-t^2} \, dt$

(b) $\int_0^\infty t^2 e^{-t^2} \, dt$

10. Use Fourier transform theory to show
    (a) $\delta(t-a) * \delta(t-b) \equiv \delta(t-a-b)$
    (b) $x(t) * [y(t) * z(t)] = [x(t) * y(t)] * z(t)$

11. Two functions of time $x(t)$ and $y(t)$ are band limited to f_x and f_y Hz, respectively. Show that their product $x(t)y(t) = z(t)$ is band limited to $f_z = f_x + f_y$ Hz and that $z(t)$ can be written as

$$z(t) = \sum_{\text{all } n} z\left(\frac{n}{2f_z}\right) \sin C 2\pi f_z\left(t - \frac{n}{2f_z}\right)$$

12. Given that $y(t)$ is band limited to $|f| = \frac{1}{2}$ Hz and has sampled values $y(0) = 1$ and $y(n\tau) = 0$ for all n, where $\tau = 1$, find $y(t)$.

13. An important formula for the Fourier transform of a periodic function is

$$X(f) = \sum_{\text{all } n} \frac{1}{T} X_T(nf_0)\delta(f - nf_0)$$

where $X_T(f)$ is the Fourier transform of one period of $x(t)$, T is the period, and $f_0 = 1/T$. Given that the power spectral density $S_{xx}(f)$ is such that it indicates the distribution of power with frequency, find and sketch $S_{xx}(f)$ for the following periodic functions:
    (a) $\cos t$
    (b) $\cos t \sin 2t$
    (c) $x(t) = \sum_{\substack{-\infty \\ n \text{ an integer}}} g(t - 2n)$ $g(t) = 1 \qquad -\frac{1}{2} < t < \frac{1}{2}$
                                                $= 0 \qquad$ otherwise

14. If $S(f)$ is the power spectral density of a random process show whether or not $d/df[S(f)]$ and $d^2/df^2[S(f)]$ could be power spectral densities.

15. Which of the following functions could be a power spectral density:

    (a) $\dfrac{af}{b + f^2}$ (b) $\dfrac{a}{b + f^2}$ (c) $\dfrac{a + cf}{b + f^2}$

    (d) $\dfrac{a}{f^2 - b}$ (e) $\dfrac{a}{a^2 + f + f^2}$ (f) $\delta(f) + \dfrac{a}{b + f^2}$

16. For each of the processes of Problem 15 in which $S(f)$ qualified as a power spectral density evaluate the mean and the mean square value of the process.

17. Given $x(t) = A\cos(t + \phi)$, where A and ϕ are random variables with

$$f_{A\phi}(\alpha, \beta) = \tfrac{1}{4\pi} \qquad 0 < \alpha < 2, 0 < \phi < 2\pi$$

find

$$S_{xx}(j\omega) \triangleq \overline{S_{x_i x_i}(j\omega)}$$

by taking the ensemble average of the power spectral densities for each member.

18. The input to the system shown is white noise with $S_{xx}(j\omega) = N$. Find $S_{yy}(j\omega)$, $R_{yy}(\tau)$, $\bar{X}$, and $\bar{X^2}$.

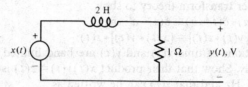

19. Given the following power spectral densities

(a) $S_{xx}(\omega) = \dfrac{1}{4+\omega^2}$

(b) $S_{xx}(\omega) = \dfrac{1+\omega^2}{4+4\omega^2+\omega^4}$

(c) $S_{xx}(\omega) = \dfrac{4\omega^2}{(1+\omega^2)^3}$

use residue theory to find the average power in the process $x(t)$.

20. If the input to the system of Problem 18 is

$$x(t) = 10\cos 5t + n(t)$$

where $n(t)$ is independent white noise with a power spectral density $S_{nn}(j\omega) = 0.07$, find the input and output signal-to-noise ratios and the system noise figure in decibels. If, in addition, the resistor in $H(S)$ generates independent thermal white noise with a power spectral density of 10^{-3}, how does this affect the noise figure?

21. Consider two independent zero-mean random processes $x(t)$ and $y(t)$ with power spectral densities $S_{xx}(j\omega)$ and $S_{yy}(j\omega)$, respectively. Define new random processes

$$z(t) = x(t) + y(t)$$
$$u(t) = x(t) - y(t)$$
$$w(t) = x(t)y(t)$$

Find formulas for $S_{zz}(j\omega)$, $S_{uu}(j\omega)$, and $S_{ww}(j\omega)$. Do you have to assume that these are ergodic processes, or is the condition of wide-sense stationariness sufficient for your results?

22. Find the cross-correlation function

$$R_{xy}(\tau) \triangleq \overline{x(t)y(t-\tau)}$$

and the cross power spectral density

$$S_{xy}(j\omega) \triangleq F[R_{xy}(\tau)]$$

for the input and output of Problem 18. What do they tell us?

23. Find the cross-correlation function $R_{xy}(\tau)$ and the cross power spectral density $S_{xy}(j\omega)$ for the input and output of Problem 20.

24. (a) The input to a system with system function $H(j\omega)$ is a member of an ergodic random process $x(t)$. The error function $\varepsilon(t) = y(t) - x(t)$. Find the power spectrum $S_{\varepsilon\varepsilon}(j\omega)$.

    (b) If the input $x(t)$ is a member of a random binary process

$$x(t) = \sum_{-\infty}^{\infty} A_n g(t - nb - \phi)$$

where the A_n's and ϕ are independent random variables with density functions

$$f_{A_i}(\alpha) = \tfrac{1}{2}\delta(\alpha) + \tfrac{1}{2}\delta(\alpha - 1)$$

and

$$f_\phi(\alpha) = \frac{1}{b} \qquad 0 < \alpha < b$$

and if

$$H(j\omega) = \frac{1}{1 + j\omega RC}$$

find and roughly sketch $\overline{\varepsilon^2(t)}$, $R_{\varepsilon\varepsilon}(\tau)$, and $S_{\varepsilon\varepsilon}(j\omega)$ as the time constant of the filter varies.

25. (a) Prove for a stationary discrete random process that

$$S_{xx}(\omega) = S_{xx}(-\omega)$$
$$S_{xx}(\omega) \geq 0$$
$$S_{xx}(\omega) = S_{xx}(\omega + n\pi)$$

    (b) Prove for jointly stationary discrete random processes that if

$$S_{xy}(\omega) = R(\omega) + jK(\omega)$$

then

$$S_{yx}(\omega) = R(\omega) + jK(-\omega)$$

26. Consider the system function

$$H(z) = \frac{z}{z - 0.5} + \frac{z}{z + 0.25}$$

    (a) Find the power transfer function $T(z) = H(z)H(z^{-1})$ and sketch a pole-zero diagram and indicate the annulus of convergence.

    (b) Find the power transfer function $T(\omega)$ as a rational function of $\cos\omega$, where ω denotes the discrete Fourier transform.

    (c) If the input to a linear system with this power transfer function has $R_{xx}(K) = 2\delta(K)$, find the output mean square fluctuations.

27. Assume that

$$x(n) = \sum x(k)\delta(n-k)$$

where the $x(k)$'s are zero-mean uncorrelated random variables with a variance of 2 is the input to
(a) An ideal low-pass filter with

$$H(\omega) = 2 \qquad 0 < |\omega| < 0.5\pi$$
$$= 0 \qquad 0.5\pi < |\omega| < \pi$$

and $H(\omega) = H(\omega + 2)$.
(b) An ideal band-pass filter with

$$H(\omega) = 2 \qquad 0.25\pi < |\omega| < 0.5\pi$$
$$= 0 \qquad \text{otherwise}$$

and $H(\omega) = H(\omega + 2\pi)$
Evaluate and sketch $R_{xx}(k)$, $S_{xx}(\omega)$, $R_{yy}(k)$, and $S_{yy}(\omega)$, and find $y^2(n)$.

28. For the system shown, $f(k) = 2ku(k)$ and $n(k)$ is a zero-mean uncorrelated essentially white-noise random process with a variance of 2.
(a) Calculate and sketch $g(k)$, $S_{mm}(\omega)$, and $R_{mm}(k)$.
(b) What has the filter done?

$$\underrightarrow{f(k) + n(k)} \boxed{y(k) = x(k) - x(k-1)} \underrightarrow{y(k) = g(k) + m(k)}$$

29. Given

$$y(n) = ay(n-1) + bx(n)$$

where $S_{xx}(\omega) = 1$, find $S_{yy}(\omega)$ and $R_{yy}(k)$.

30. Given that two filters have the same input $x(n)$, with

$$S_{xx}(\omega) = 1 - \frac{|\omega|}{\pi} \qquad -\pi < \omega < \pi$$

and that the outputs are

$$y(n) = 2x(n) + 4x(n-1)$$
$$z(n) = -0.5z(n-1) + 0.5x(n)$$

find
(a) $S_{xy}(\omega)$, $S_{xz}(\omega)$, $R_{xy}(n)$, $R_{xz}(n)$.
(b) $S_{yy}(\omega)$, $S_{zz}(\omega)$, $R_{yy}(k)$, $R_{zz}(k)$.

31. Consider the discrete system shown, with input

$$x(n) = \sum_K x(k)\delta(n-k)$$

Find $\overline{y(n)}$, $\overline{z(n)}$, $\overline{y(n)y(k)}$, $z(n)z(k)$ for all n and k under the following conditions:

(a) The $x(k)$'s are independent, with

$$f_x(\alpha) = 0.5\delta(\alpha) + 0.5\delta(\alpha-1)$$

(b) The $x(k)$'s are dependent with joint mass function

$x(k)$ $\diagdown$ $(k+1)$	0	1
0	$\frac{1}{3}$	$\frac{1}{6}$
1	$\frac{1}{6}$	$\frac{1}{3}$

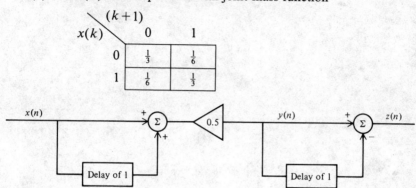

Chapter 9
The Gaussian Random Process and Some Noise Problems

9.1 THE GAUSSIAN RANDOM PROCESS

Gaussian Random Variables

Section 5.5 treated the general case of jointly gaussian random variables, and the main results will be reiterated. That is, n random variables, $\mathbf{X} = (X_1, X_2, \ldots, X_n)$, are jointly gaussian if their joint density function is given by

$$f_{\mathbf{X}}(\boldsymbol{\alpha}) = \frac{1}{(2\pi)^{n/2} |\Lambda_x|^{1/2}} \exp\left\{ \tfrac{1}{2} \left[(\boldsymbol{\alpha} - M_x(\boldsymbol{\alpha})) (\Lambda_x)^{-1} (\boldsymbol{\alpha} - M_x(\boldsymbol{\alpha}))^T \right] \right\}$$

(9.1)

where (Λ_x) is the $n \times n$ covariance matrix with general elements

$$\Lambda_{ij} = \overline{(X_i - \overline{X}_i)(X_j - \overline{X}_j)}$$

$M_x(\boldsymbol{\alpha})$ is a $1 \times n$ matrix that gives the means, $\overline{X}_1, \ldots, \overline{X}_n$; that is,

$$M_x(\boldsymbol{\alpha}) = (\overline{X}_1, \overline{X}_2, \ldots, \overline{X}_n),$$

$|\Lambda_x|$ denotes the determinant of the (Λ_x) matrix and T indicates a transpose. It can be shown that each random variable X_i is gaussian with mean $\bar{X}_i$ and variance Λ_{ii}, and any k random variables are jointly gaussian with a Λ matrix obtained by finding the appropriate $k \times k$ submatrix from (Λ) by deleting rows and columns.

Definition of a Gaussian Random Process

Consider some random process $x(t)$ and define n random variables $X_1, X_2, \ldots, X_n$ that sample it at times $t = t_1, t_2, \ldots, t_n$. By this we mean that when we sample the first member waveform, X_1 is assigned the value $x_1(t_1)$, X_2 is assigned the value $x_1(t_2)$, and similarly for each member waveform of the process. The random process is called gaussian if the joint density function of the n random variables is jointly gaussian or satisfies Eq. 9.1.

If a gaussian random process is stationary, it is completely characterized by its autocorrelation function $R_{xx}(\tau)$ and its mean value $\bar{X}$. The covariance matrix (Λ_x) that characterizes the N random variables $X_1, X_2, \ldots, X_n$ in general is

$$(\Lambda_x) = \begin{pmatrix} \sigma_{x_1}^2 & \cdots & \overline{(X_1 - \bar{X}_1)(X_n - \bar{X}_n)} \\ \vdots & & \vdots \\ \overline{(X_1 - \bar{X}_1)(X_n - \bar{X}_n)} & & \sigma_{x_n}^2 \end{pmatrix} \qquad (9.2)$$

and if $x(t)$ is stationary, this becomes

$$(\Lambda_x) = \begin{pmatrix} R_{xx}(0) - \bar{X}^2 & \cdots & R_{xx}(t_1 - t_n) - \bar{X}^2 \\ \vdots & & \vdots \\ R_{xx}(t_n - t_1) - \bar{X}^2 & \cdots & R_{xx}(0) - \bar{X}^2 \end{pmatrix} \qquad (9.3)$$

In general, for a stationary process,

$$\Lambda_{ii} = R_{xx}(0) - \bar{X}^2$$

and

$$\Lambda_{ij} = R_{xx}(|t_i - t_j|) - \bar{X}^2 = \Lambda_{ji}$$

Corresponding to the theorem in Chapter 5, which relates the joint density function of m random variables $Y_1, \ldots, Y_m$ to the joint density function of n random variables $X_1, \ldots, X_n$, of which they are a linear combination, there is a similar theorem relating the output random process $y(t)$ of a linear system to a gaussian random process $x(t)$, which is the input.

THEOREM 9.1

If the input $x(t)$ to a linear system is a member of a gaussian random process, then the output $y(t)$ is also a member of a gaussian random process.

We will now solve a problem indicating how the joint density function of an output random process $y(t)$ may be found.

Example 9.1

Consider the circuit shown in Figure 9.1a, where a sample waveform of the random process $x(t)$ is applied as input. Assume that $x(t)$ changes amplitude at random instants and that the probability of k amplitude changes in τ s is Poisson with

$$P(k,\tau) = \frac{(\mu\tau)^k}{k!} e^{-\mu\tau}$$

where μ is a known constant for the process. Further assume that the noise is gaussian with the following joint density function:

$$f_{X_1 X_2}(\alpha_1, \alpha_2) = \frac{1}{2\pi\sqrt{1-\rho^2}} \exp\left[-\frac{(\alpha_1^2 - 2\rho\alpha_1\alpha_2 + \alpha_2^2)}{2(1-\rho^2)}\right]$$

where $\rho = e^{-\mu|\tau|}$, $\tau = |t_2 - t_1|$, and the random variables X_1 and X_2 are associated with $x(t)$ at times t_1 and t_2, respectively.

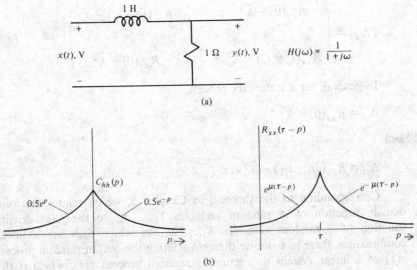

(a)

(b)

Figure 9.1 (a) The system of Example 9.1. (b) Graphical aids for evaluating $C_{hh}(\tau) * R_{xx}(\tau)$.

(a) Find the output autocorrelation function and power spectral density.

(b) Find the joint density function of the three random variables Y_1, Y_2, and Y_3, which sample $y(t)$ at times $t = 1$, 2, and 4, respectively. Assume in this part of the problem that $\mu = 5$.

SOLUTION

(a) From the problem statement we are given

$$R_{xx}(\tau) = e^{-\mu|\tau|}$$

and $\overline{X} = 0$. (Can we evaluate these?)

From Theorem 9.1 we assume that the output $y(t)$ is also gaussian. Considering our linear system, it is easily found that

$$H(j\omega) = \frac{1}{j\omega + 1}$$

and

$$h(t) = e^{-t}u(t)$$

where $u(t)$ is the unit step function. As in previous problems, we can evaluate the impulse response correlated with itself

$$C_{hh}(\tau) \triangleq h(\tau) \oplus h(\tau)$$

$$= \int_{-\infty}^{\infty} h(u)h(u - \tau)\,du$$

as

$$C_{hh}(\tau) = \tfrac{1}{2}e^{-|\tau|}$$

This function characterizes the system for a random input, and the formula for the output autocorrelation function is

$$R_{yy}(\tau) = C_{hh}(\tau) * R_{xx}(\tau)$$

$$= \int_{-\infty}^{\infty} C_{hh}(p)R_{xx}(\tau - p)\,dp$$

where

$$C_{hh}(\tau) = \tfrac{1}{2}e^{-|\tau|} \quad \text{and} \quad R_{xx}(\tau) = e^{-\mu|\tau|}$$

In order to carry out this convolution integral a sketch of $C_{hh}(p)$

and $R_{xx}(\tau - p)$ is shown in Figure 9.1b. The calculation is now carried out.

For $\tau > 0$,

$$R_{yy}(\tau) = \int_{-\infty}^{0} \tfrac{1}{2} e^{p} e^{\mu(\tau - p)} \, dp + \int_{0}^{\tau} \tfrac{1}{2} e^{-p} e^{-\mu(\tau - p)} \, dp$$

$$+ \int_{\tau}^{\infty} \tfrac{1}{2} e^{-p} e^{-\mu(\tau - p)} \, dp$$

After substantial algebraic manipulation the final expression for $R_{yy}(\tau)$ is

$$R_{yy}(\tau) = e^{-\mu\tau} \left[\frac{-2\mu}{2(1-\mu^2)} \right] + e^{-\mu\tau} \left[\frac{2}{2(1-\mu^2)} \right]$$

Obviously $R_{yy}(\tau)$ is an even function of τ since $C_{hh}(\tau)$ and $R_{xx}(\tau)$ are both even functions. Therefore, for all $\tau < 0$,

$$R_{yy}(\tau) = e^{\mu\tau} \left[\frac{-2\mu}{2(1-\mu^2)} \right] + e^{\mu\tau} \left[\frac{2}{2(1-\mu^2)} \right]$$

(b) Since $y(t)$ is a gaussian random process, the joint density function associated with the three random variables Y_1, Y_2, and Y_3 that sample the process at times $t = 1$, $t = 2$, and $t = 4$ will be given by Eq. 9.1 as

$$f_Y(\beta) = \frac{1}{(2\pi)^{3/2} |\Lambda_y|^{3/2}}$$

$$\times \exp\left\{ -\tfrac{1}{2} \left[(\beta - M_Y(\beta))(\Lambda_y)^{-1} (\beta - M_y(\beta))^T \right] \right\}$$

Since $x(t)$ is a zero-mean process, then $M_y(\beta) = 0$. Therefore $f_Y(\beta)$ will be completely characterized by (Λ_y). For our random variables,

$$(\Lambda_y) = \begin{pmatrix} R_{yy}(0) & R_{yy}(1) & R_{yy}(3) \\ R_{yy}(1) & R_{yy}(0) & R_{yy}(2) \\ R_{yy}(3) & R_{yy}(2) & R_{yy}(0) \end{pmatrix}$$

With $\mu = 5$ the values of these elements are

$$R_{yy}(0) = \frac{2(1-\mu)}{2(1-\mu^2)} = 0.17$$

$$R_{yy}(1) = e^{-5}\left[\frac{-10}{2(1-25)}\right] + e^{-5}\left[\frac{2}{2(1-25)}\right] = 0.17e^{-5}$$

$$R_{yy}(2) = 0.17e^{-10}$$

$$R_{yy}(3) = 0.17e^{-15}$$

and

$$(\Lambda_y) = \begin{pmatrix} 0.17 & 0.17e^{-5} & 0.17e^{-15} \\ 0.17e^{-5} & 0.17 & 0.17e^{-10} \\ 0.17e^{-15} & 0.17e^{-10} & 0.17 \end{pmatrix}$$

$$|\Lambda_y|^{1/2} \simeq 0.07$$

The joint density function may be found by substituting these expressions in Eq. 9.1.

9.2 SOME LINEAR NOISE PROBLEMS

In the preceding chapter, Section 8.4 discussed the general problem of a linear system with a random input and focused on the application to the cases where the input was either purely random, a deterministic signal plus noise, or a random signal plus noise. In this section the problem of minimizing noise or signal error will be discussed from two points of view. In Figure 9.2 is shown a schematic diagram of a linear system with a signal-plus-noise input and a means of indicating the error defined as $y(t) - f(t)$ or $y(k) - f(k)$. The error is a measure of the deviation of the output from the signal. If $\varepsilon(t)$ or $\varepsilon(k) = 0$, then the system has completely eliminated the noise. Depending on the application, the expression to minimize noise has different connotations. For example, in radar detection the task may be to choose the system function to maximize the output signal-to-noise ratio at a specified instant of time, whereas if one is interested in the least mean square estimation where $f(t)$ or $f(k)$ is unknown, the task is to choose $H(\omega)$ or $H(z)$ to make $\overline{\varepsilon^2(t)}$ or $\overline{\varepsilon^2(k)}$ a minimum. These two problems of (a) obtaining maximum signal-to-noise ratio at a specified time, and (b) minimizing the least mean square error, lead to the "matched" and "Wiener" filter, respectively, and will now be treated.

Minimizing the Least Mean Square Error: The Continuous Wiener Filter

The basic problem is as follows: "Given an unknown or random stationary signal $f(t)$, whose autocorrelation function and power spectral density $R_{ff}(\tau)$ and $S_{ff}(\omega)$, respectively, are known, and to which is added zero-mean independent stationary noise with autocorrelation function and power spectral density $R_{nn}(\tau)$ and $S_{nn}(\omega)$, design a filter or network to minimize the mean square error $\overline{\varepsilon^2(t)}$ as indicated in Figure 9.2."

Design

The mean square error $\overline{\varepsilon^2(t)}$ is

$$\overline{\varepsilon^2(t)} = R_{\varepsilon\varepsilon}(0) = \int_{-\infty}^{\infty} S_{\varepsilon\varepsilon}(\omega)\,df$$

$$R_{\varepsilon\varepsilon}(\tau) = \{\,\overline{[f(t)-y(t)][f(t+t)-y(t+t)]}\,\}$$
$$= R_{ff}(\tau) + R_{yy}(\tau) - R_{fy}(\tau) - R_{yf}(\tau) \tag{9.4}$$

From Chapter 7, if $x(t) = f(t) + n(t)$ and $R_{fn}(\tau) = 0$, then

$$R_{yy}(\tau) = R_{ff}(\tau) * C_{hh}(\tau) + R_{nn}(\tau) * C_{hh}(\tau)$$
$$R_{fy}(\tau) = R_{ff}(\tau) \oplus h(\tau) \quad \text{or} \quad R_{ff}(\tau) * h(\tau)$$
$$R_{yf}(\tau) = R_{ff}(\tau) \oplus h(-\tau) \quad \text{or} \quad h(-\tau) * R_{ff}(\tau)$$

Substituting these results in Eq. 9.4 yields

$$R_{\varepsilon\varepsilon}(\tau) = R_{ff}(\tau) + [R_{ff}(\tau) + R_{nn}(\tau)] * C_{hn}(\tau)$$
$$- R_{ff}(\tau) \oplus [h(\tau) + h(-\tau)] \tag{9.5}$$

The power spectral density of the error signal, using the derivations and theorems of the bilateral Laplace transform is

$$S_{\varepsilon\varepsilon}(s) = S_{ff}(s) + [S_{ff}(s) + S_{nn}(s)] H(s)H(-s)$$
$$- S_{ff}(s)[H(s) + H(-s)] \tag{9.6}$$

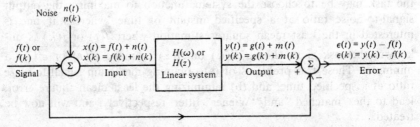

Figure 9.2   A linear continuous or discrete system with signal-plus-noise input.

We must now choose $H(s)$ to minimize

$$\overline{\varepsilon^2(t)} = \frac{1}{2\pi j} \oint_C S_{\varepsilon\varepsilon}(s)\, ds \tag{9.7}$$

where we know that $S_{\varepsilon\varepsilon}(s)$ has symmetrical poles to the left and right of $\sigma = 0$ and C encloses the poles to the left or right. $\varepsilon^2(t)$ equals the residues of the poles to the left of or minus the residues of the poles to the right of σ. Now

$$S_{xx}(s) = S_{nn}(s) + S_{ff}(s) = F_x(s)F_x(-s) \tag{9.8}$$

where $F_x(s)$ is that part of $S_{xx}(s)$ with poles in the left half plane (just like $H(s)$ in $T(s) = H(s)H(-s)$ for the power transfer function). With some ingenuity, Eq. 9.7 may be rewritten as

$$\overline{\varepsilon^2(t)} = \frac{1}{2\pi j} \oint \left\{ \left[F_x(s)H(s) - \frac{S_{ff}(s)}{F_x(-s)} \right]\left[F_x(-s)H(-s) - \frac{S_{ff}(s)}{F_x(s)} \right] \right.$$
$$\left. + \frac{S_{ff}(s)S_{nn}(s)}{S_{xx}(s)} \right\} ds \tag{9.9}$$

Since all the residues from a power spectral function must be positive, Eq. 9.9 is minimized when

$$F_x(s)H(s) - \frac{S_{ff}(s)}{F_x(-s)} = 0$$

or when

$$H(s) = \frac{S_{ff}(s)}{S_{ff}(s) + S_{nn}(s)} \tag{9.10}$$

Since $H(s)$ is the ratio of two power spectral densities, it is not a causal or physically realizable filter, since it contains poles and zeros in both the right and left half planes. A discussion of obtaining the best possible causal, minimum mean square error filter is covered in many texts on communication theory. A very plausible initial treatment is given by Cooper and McGillem.[1]

The Fourier transform version of the Wiener filter result is

$$H(\omega) = \frac{S_{ff}(\omega)}{S_{ff}(\omega) + S_{nn}(\omega)} \tag{9.11}$$

Whenever an abstract result such as Eq. 9.10 or 9.11 is encountered, it is a good idea to develop a feeling for it by applying it to trivial cases for which the answer should be obvious. We will do so with a simple example.

[1] George R. Cooper and Clare McGillem, *Probabilistic Methods of Signal System Analysis* (New York: Holt, Rinehart and Winston, 1971), pp. 212–215.

Example 9.2

Design and comment on your result for the Wiener filter design in the following situations:

(a) $S_{nn}(\omega) = 0$.
(b) $S_{nn}(\omega) > S_{ff}(\omega)$, or when the noise is dominant.

· SOLUTION

(a) The statement $S_{nn}(\omega) = 0$ implies that the noise is negligible. The required system function is

$$H(\omega) = \frac{S_{ff}(\omega)}{S_{ff}(\omega) + 0} = 1$$

which implies a situation in which the signal is passed through an all-pass filter or in which no filter is used since there is no noise to eliminate.

(b) If $S_{nn}(\omega) > S_{ff}(\omega)$, then the required $H(\omega)$ is

$$H(\omega) = \frac{S_{ff}(\omega)}{0 + S_{nn}(\omega)} \qquad \text{(What range of } \omega?)$$

which implies the use of a filter that passes those frequencies at which the signal frequencies predominate or are most significant.

THE DISCRETE WIENER FILTER

Given that $x(k) = f(k) + n(k)$ is a discrete random process, in a completely analogous manner it is possible to derive

$$H(z) = \frac{S_{ff}(z)}{S_{ff}(z) + S_{nn}(z)} \qquad (9.12)$$

to minimize

$$\overline{\varepsilon^2(n)} = \overline{[y(n) - f(n)]^2}$$

Example 9.3

(a) Design a filter $H(z)$ to minimize the mean square error for $x(k) = f(k) + n(k)$, where

$$R_{ff}(k) = \left(\tfrac{1}{2}\right)^{|k|} \quad \text{and} \quad R_{nn}(k) = 2\delta(k)$$

(b) Evaluate the mean square error $\overline{\varepsilon^2(k)}$.

SOLUTION

(a) The power spectral densities $S_{ff}(z)$ and $S_{nn}(z)$ may be found as

$$S_{ff}(z) = \left[1 + \tfrac{1}{2}z^{-1} + \left(\tfrac{1}{2}\right)^2 z^{-2} + \cdots\right]$$

$$+ \left[1 + \tfrac{1}{2}z + \left(\tfrac{1}{2}\right)^2 z^2 + \cdots\right] - 1$$

$$S_{ff}(z) = \frac{1}{1 - \tfrac{1}{2}z^{-1}} + \frac{1}{1 - \tfrac{1}{2}z} - 1$$

$$= \frac{z}{z - \tfrac{1}{2}} + \frac{-2}{(z-2)} - 1 \qquad 0.5 < |z| < 2$$

$$= \frac{z^2 - 2z - 2z + 1}{(z - \tfrac{1}{2})(z-2)} - \frac{z^2 - 2.5z + 1}{(z - \tfrac{1}{2})(z-2)}$$

$$= \frac{-1.5z}{(z - \tfrac{1}{2})(z-2)} \qquad 0.5 < |z| < 2$$

The Wiener filter is

$$H(z) = \frac{\dfrac{-1.5z}{(z - \tfrac{1}{2})(z-2)}}{\dfrac{-1.5z}{(z - \tfrac{1}{2})(z-2)} + 2} \qquad 0.5 < |z| < 2$$

$$= \frac{-1.5z}{-1.5z + 2(z - \tfrac{1}{2})(z-2)}$$

$$= \frac{-1.5z}{2z^2 - 6.5z + 2}$$

$$= \frac{-0.75z}{z^2 - 3.25z + 1}$$

$$= \frac{-0.75z}{(z - 0.34)(z - 2.9)}$$

(b) We now evaluate the mean square error.

$$\overline{\varepsilon^2(k)} = \frac{1}{2\pi j}\oint_C \frac{S_{ff}(z)S_{nn}(z)}{S_{ff}(z)+S_{nn}(z)}\,dz \qquad \text{(Be sure of this)}$$

$$= \frac{1}{2\pi j}\oint_C \frac{-1.5z}{(z-0.34)(z-2.9)}\,dz \qquad 0.34<|z|<2.9$$

$$\therefore \overline{\varepsilon^2(k)} = \text{residue of the pole at } 0.34$$

$$= \frac{-1.5(0.34)}{-2.56}$$

$$= 0.20$$

As in the continuous Wiener filter case, $H(z)$ is not causal since it contains poles outside the unit circle. For a discussion of a causal Wiener filter the reader is referred to Papoulis.[2] If a Wiener filter is given in terms of discrete Fourier transforms, then Eq. 9.11 applies; that is,

$$H(\omega) = \frac{S_{ff}(\omega)}{S_{ff}(\omega)+S_{nn}(\omega)}$$

and this yields the ratio of two polynomials of $\cos\omega$.

Maximizing the Signal-to-Noise Ratio: The Matched Filter

"Given a known deterministic signal $f(t)$ or $f(k)$, with zero-mean independent stationary noise, with autocorrelation function $R_{nn}(\tau)$ or $R_{nn}(k)$ added, design a filter to maximize the signal-to-noise ratio at some specified instant of time." Make use of the Schwarz inequality from Section 5.5, which states:

For real functions $p(t)$ and $q(t)$,

$$\left|\int_{-\infty}^{\infty} p(t)q(t)\,dt\right|^2 \le \left[\int_{-\infty}^{\infty} p^2(t)\,dt\right]\left[\int_{-\infty}^{\infty} q^2(t)\,dt\right]$$

or

$$\left|\sum p(k)q(k)\right|^2 \le \left[\sum p^2(k)\right]\left[\sum q^2(k)\right]$$

and for complex functions $P(\omega)$ and $Q(\omega)$ of frequency

$$\left|\int_{-\infty}^{\infty} P(\omega)Q(\omega)\,df\right|^2 \le \left[\int_{-\infty}^{\infty}|P(\omega)|^2\,df\right]\left[\int_{-\infty}^{\infty}|Q(\omega)|^2\,df\right]$$

and if $P(\omega)$ and $Q(\omega)$ are discrete Fourier transforms with period 2π

[2]A. Papoulis, *Signal Analysis* (New York: McGraw-Hill, 1977), pp. 344–350.

radians per second, then

$$\left| \int_{-\pi}^{\pi} P(\omega)Q(\omega)\,d\omega \right|^2 \leq \left[\int_{-\pi}^{\pi} |P(\omega)|^2\,d\omega \right]\left[\int_{-\pi}^{\pi} |Q(\omega)|^2\,d\omega \right]$$

Comment

As far as Schwarz's inequality is concerned, we note that the equality signs hold for real functions if $p(t) = cq(t)$ or $p(k) = cq(k)$, for any constant c. As well as believing this from the proof involving using the positiveness of

$$\int_{-\infty}^{\infty} [p(t) \pm q(t)]^2\,dt \quad \text{or} \quad \sum_{-\infty}^{\infty} [p(n) \pm q(n)]^2$$

we should intuitively appreciate the result from playing around with numbers. For example, for sequences of only one positive number, say $f(0) = 2$ and $g(0) = 3$, the equality sign holds, since

$$(2 \times 3)^2 = 2^2 \times 3^2$$

For sequences of two numbers, say $f(0) = 2$, $f(1) = 3$, and $g(0) = 3$, $g(1) = 4$,

$$(2 \times 3 + 3 \times 4)^2 \leq (2^2 + 3^2)(3^2 + 4^2)$$

and with thinking we see that the two quantities are equal only for $g(0) = 2c$, $g(1) = 3c$. We may give this result a geometric interpretation or with our background a correlation interpretation.

We will now design the matched filter for the discrete case in Figure 9.2.

DESIGN OF THE DISCRETE MATCHED FILTER

If the input $x(k) = f(k) + n(k)$, where $f(k)$ is a deterministic function and $n(k)$ independent zero-mean stationary noise, then

$$g(k) = \sum_{n=-\infty}^{\infty} f(n)h(k-n)$$

and

$$G(\omega) = F(\omega)H(\omega) \quad \text{or} \quad G(z) = F(z)H(z)$$

yield the deterministic output and its discrete Fourier or bilateral z transform.

The output noise autocorrelation function is

$$R_{mm}(k) = R_{nn}(k) * C_{hh}(k)$$

and the power spectral density is

$$S_{mm}(\omega) = S_{nn}(\omega)H(\omega)H(-\omega)$$

At any time $k = k_0$ the output signal-to-noise ratio is

$$(S/N)_{out} = \frac{g^2(k_0)}{\dfrac{1}{2\pi}\displaystyle\int_{-\pi}^{\pi} S_{nn}(\omega)H(\omega)H(-\omega)\,d\omega}$$

$$= \frac{|\Sigma f(n)h(k_0 - n)|^2}{\dfrac{1}{2\pi}\displaystyle\int_{-\pi}^{\pi} S_{nn}(\omega)H(\omega)H(-\omega)\,d\omega}$$

$$= \frac{1}{4\pi^2}\frac{\left|\displaystyle\int_{-\pi}^{\pi} F(\omega)H(\omega)e^{j\omega k_0}\,d\omega\right|^2}{\dfrac{1}{2\pi}\displaystyle\int_{-\pi}^{\pi} S_{nn}(\omega)H(\omega)H(-\omega)\,d\omega} \tag{9.13}$$

We want to maximize this quantity with respect to $H(\omega)$. Writing

$$\left|\int_{-\pi}^{\pi} F(\omega)H(\omega)e^{j\omega k_0}\,d\omega\right|^2$$

$$= \left|\int_{-\pi}^{\pi} \underbrace{\left(\frac{F(\omega)}{\sqrt{S_{nn}(\omega)}}\right)}_{P(\omega)}\underbrace{\left(\sqrt{S_{nn}(\omega)}\,H(\omega)\right)}_{Q(\omega)} e^{j\omega k_0}\,d\omega\right|^2$$

and applying Schwarz's inequality for the complex functions $P(\omega)$ and $Q(\omega)$ as indicated yields

$$\left|\int_{-\pi}^{\pi} F(\omega)H(\omega)e^{j\omega k_0}\,d\omega\right|^2 \le \left[\int_{-\pi}^{\pi} \frac{|F(\omega)|^2}{S_{nn}(\omega)}\,d\omega\right]\left[\int_{-\pi}^{\pi} |H(\omega)|^2 S_{nn}(\omega)\,d\omega\right]$$

and we notice that the motivation for our choice of $P(\omega)$ and $Q(\omega)$ was for the denominator term to cancel out in Eq. 9.13.

Canceling the denominator term,

$$(S/N)_{out} \le \frac{1}{2\pi}\int_{-\pi}^{\pi} \frac{|F(\omega)|^2}{S_{nn}(\omega)}\,d\omega \tag{9.14}$$

and the equality sign, which is the maximum, holds when

$$H(\omega) = c\,\frac{F^*(\omega)}{S_{nn}(\omega)}e^{-j\omega k_0} \tag{9.15}$$

where $F^*(\omega)$ indicates the complex conjugate and c is any constant. Of course, the maximum signal-to-noise ratio is

$$(S/N)_{max} = \frac{1}{2\pi}\int_{-\pi}^{\pi} \frac{|F(\omega)|^2}{S_{nn}(\omega)}\,d\omega \tag{9.16}$$

Using the bilateral z transform, the matched filter system function is

$$H(z) = c\frac{F(z^{-1})}{S_{nn}(z)}z^{-k_0} \qquad (9.17)$$

and

$$(S/N)_{max} = \frac{1}{2\pi j}\oint_{C_1}\frac{F(z)F(z^{-1})}{S_{nn}(z)}dz \qquad (9.18)$$

which yields the residues of the poles inside (or minus the residues of the poles outside) C_1. For both $H(\omega)$ and $H(z)$ to be applicable, $|z|=1$ must be acceptable for the contour C_1. Again, the symbolic notation for transforms must be commented on.

Given $H(z)$, then $H(\omega) = H(z)$ with $z = e^{j\omega}$, and vice versa. For this reason sometimes the notation $H(e^{j\omega})$ is used for the discrete Fourier transform.

Comments about the Matched Filter

From a trivial point of view, the expression for $H(\omega)$ in Eq. 9.15 is reasonable because it says the filter should pass those frequencies for which the spectrum of the signal is large and that of the noise is small. We also notice from the z transform version that $H(z)$ is not causal since $S_{nn}(z)$ contributes poles outside the unit circle.

THE MATCHED FILTER FOR WHITE NOISE

If the input noise may be assumed white, with power spectral density $S_{nn}(z) = N$, then the matched filter is

$$H(z) = \frac{C}{N}F(z^{-1})z^{-k_0} \qquad (9.19)$$

and the pulse response $h(n)$ is

$$h(n) = \frac{C}{N}f(k_0 - n) \qquad (9.20)$$

Since C/N is just a gain constant and does not contribute to the signal-to-noise ratio, it is often chosen to be unity. This matched filter $h(n) = f(k_0 - n)$ is the input signal $f(n)$ reflected and moved k_0 units to the right. If $f(n)$ is of finite duration, then k_0 may be chosen to make $f(k_0 - n)$ causal.

Example 9.4

(a) Design a filter $H(z)$ to maximize the signal-to-noise ratio at $n = 2$, given

$$f(n) = \delta(n) + 4\delta(n-1) + 2\delta(n-2)$$

and $n(k)$ is white noise with $R_{xx}(n) = \delta(n)$.

(b) Compare your result to using an averaging filter with $h(n) = 0.5[\delta(n) + \delta(n-1)]$.

SOLUTION

(a) The matched filter to maximize the signal-to-noise ratio at $n = 2$ is $h(n) = f(2-n)$, which yields

$$h(n) = 2\delta(n) + 4\delta(n-1) + \delta(n-2)$$

The deterministic output $g(n)$ is found by convolving the sequences $f(n)$ and $h(n)$ to yield

$$\{2,4,1\} * \{1,4,2\} = \{2,12,21,12,2\}$$

$$= \{y(0),\ldots,y(4)\}$$

At $n = 2$ the signal mean square value is maximum with $g^2(2) = 21^2 = 441$. The output noise autocorrelation $R_{mm}(n)$ is

$$R_{mm}(n) = C_{hh}(n) * R_{xx}(n)$$

$$= C_{hh}(n)$$

and

$$\overline{m^2(n)} = C_{hh}(0)$$

$$= 2^2 + 4^2 + 1^2$$

$$= 21$$

$$(S/N)_{max} = \frac{441}{21} = 21$$

This problem was solved in the time domain because $f(n)$ and $h(n)$ were finite sequences and the z transform would just involve multiplying series, which is identical to the time domain approach.

(b) The filter to be used now is $h(n) = \{\frac{1}{2},\frac{1}{2}\}$ and the deterministic output is

$$g(n) = \{1,4,2\} * \{\tfrac{1}{2},\tfrac{1}{2}\}$$

$$= \{0.5, 2.5, 3, 1\}$$

$$\overline{m^2(n)} = C_{hh}(0)$$

$$= 0.5$$

and the signal-to-noise ratio is maximum at $n = 2$; that is,

$$(S/N)_{max} = \frac{9}{0.5} = 18$$

We notice that the matched filter was vastly superior in indicating the presence of the signal. However it might have been fairer to compare the matched filter to averaging twice, and the reader is encouraged to do so.

Figure 9.3 shows a sketch of how $x(n)$ and S/N versus ω might look for the two filters of this problem.

THE CONTINUOUS MATCHED FILTER

If the input $x(t)$, as in Figure 9.2, is $x(t) = f(t) + n(t)$, where $f(t)$ is a deterministic signal and $n(t)$ is independent zero-mean stationary noise, then in a completely analogous manner to the discrete derivation it can be

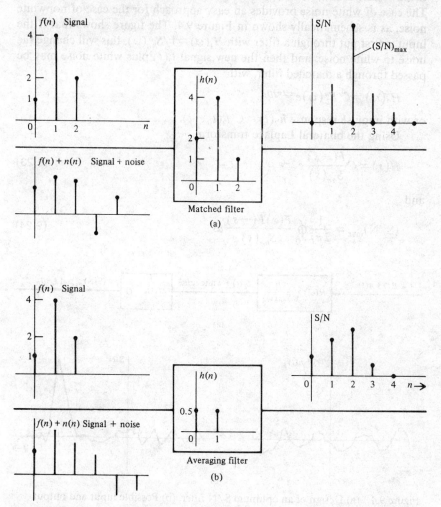

Figure 9.3   (a) The discrete matched filter of Example 9.4. (b) The averaging filter of Example 9.4.

shown that

$$H(\omega) = c\frac{F^*(\omega)}{S_{nn}(\omega)}e^{-j2\pi ft_0} \qquad (9.21)$$

maximizes the signal-to-noise ratio at t_0 and that

$$(S/N)_{max} = \int_{-\infty}^{\infty}\frac{|F(\omega)|^2}{S_{nn}(\omega)}df \qquad (9.22)$$

where the Fourier transform is used.

CASE WHEN THE INPUT NOISE IS NOT WHITE

The case of white noise provides an easy approach for the case of nonwhite noise, as is schematically shown in Figure 9.4. The figure shows that if the input is first put through a filter with $H_1(\omega) = 1/S_{nn}(\omega)$, this will change the noise to white noise, and then the new signal $f_2(t)$ plus white noise may be passed through a matched filter, with

$$H_2(\omega) = C'F_2^*(\omega)e^{-2\pi jft_0}$$

or with impulse response $h_2(t) = C'f_2(t_0 - t)$.

Using the bilateral Laplace transform,

$$H(s) = C\frac{F(-s)}{S_{nn}(s)}e^{-st_0} \qquad (9.23)$$

and

$$(S/N)_{max} = \frac{1}{2\pi j}\oint_C\frac{F(s)F(-s)}{S_{nn}(s)}ds \qquad (9.24)$$

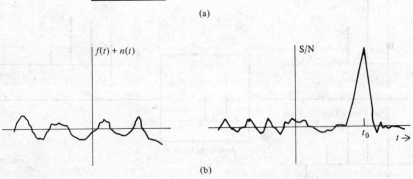

(a)

(b)

Figure 9.4   (a) Design of an optimum S/N filter. (b) Possible input and output signals.

which may be found as the sum of the residues to the left of σ or minus the sum of the residues to the right of σ, and either the Fourier or Laplace transform may be used if $\sigma = 0$ is in the range of convergence.

In the case of white or wide-band noise,

$$h(t) = f(t_0 - t) \tag{9.25}$$

to maximize the signal-to-noise ratio at $t = t_0$ and in the time domain,

$$(S/N)_{max} = \frac{1}{N_0} \int_{-\infty}^{\infty} f^2(t_0 - t) \, dt \tag{9.26}$$

Example 9.4

Design a filter (noncausal) to maximize the signal-to-noise ratio at time $t = 2$ when the input signal $f(t) = e^{-t}u(t)$ is coupled with independent zero-mean noise with power spectral density

$$S_{nn}(s) = \frac{-4}{s^2 - 4}$$

SOLUTION

The noncausal matched filter that maximizes the signal-to-noise ratio is

$$H(s) = \frac{F(-s)}{S_{nn}(s)}$$

where

$$F(s) = \mathcal{L}[e^{-t}u(t)] = \frac{1}{s+1}$$

and

$$F(-s) = \frac{-1}{s-1}$$

$$\therefore H(s) = \frac{-1(-4)}{(s-1)(s^2 - 4)}$$

$$(S/N)_{max} = \frac{1}{2\pi j} \int_{\sigma - j\infty}^{\sigma + j\infty} \frac{F(s)F(-s)}{S_{nn}(s)} \, ds$$

$$= \frac{1}{2\pi j} \int_{\sigma - j\infty}^{\sigma + j\infty} \frac{4}{(s-1)(s+1)(s-2)(s+2)} \, ds \qquad -1 < \sigma < +1$$

$$= \sum \text{residues at } s = -2 \quad \text{and} \quad s = -1$$

$$= \frac{4}{(-3)(-1)(-4)} + \frac{4}{(-2)(-3)1}$$

$$= 0.33$$

9.3 COMMENTS ON SIGNAL PROCESSING AND ESTIMATES

The processing of signals—that is, the transmitting and recovery of them or the interpreting of what can be deduced from operating on waveforms—is common to many fields among which are statistical communications, telecommunications, satellite communications, and biomedical engineering. Each field requires its own special prerequisite language and background, and it is foolhardy to feel that a knowledge of probability theory and random process theory allow for immediate productive entry in any discipline. The preceding remark was intended as a curb on the aggressive tendencies of people with a good system plus statistical background to immediately overanalyze and overprocess data. It is imperative, however, with the availability of extremely cheap and powerful computing facilities to convince nonengineers and even some engineers to rethink their analysis procedure and expand on their interpretation of signals.

Initially we will simplistically list some typical signal processing problems:

1. The recovery of a "known signal" from data.
2. The interpretation of a signal from a pattern of signals.
3. The investigation of fundamental inherent noise.
4. The investigation of correlation between important signal patterns.
5. The approximations introduced by choosing a length of data and a rate at which to consider sampled values of that length.

Some comments will now be made about these typical signal situations. In the case of "type 1," a specifically known signal is often deliberately spread out in frequency and immersed in noise so that the sender may only be in a position to efficiently recover it using specially tapped delay lines or surface acoustic wave devices specifically set to recognize the signal. The field of "Spread Spectrum analysis" is of paramount military importance. In the case of type 2, a typical waveform pattern such as an EEG, EKG, or EMG waveform is analyzed by trying to filter out extraneous noise due to imperfections in sampling and electronic equipment noise. Kalman, Wiener, and low-pass filters are most commonly used. In type 3, based on a knowledge of the field an attempt is made to correlate different interrelated quantities. Some examples are the current and voltage fluctuations in diode devices, the EMG pattern and pressure in a uterine tract, and comparing an actual waveform pattern to an assumed normal or typical pattern. The case of type 4 is more common to research. For example, due to the statistical nature of matter and materials, there is always an inherent noise limitation, particularly at high frequencies, in devices such as diodes and transistors. The state-of-the-art limitation on a device such as "4 dB at 6 GHz" is being continually moved downward to so-called physically realizable minima by improved processing and constantly smaller geometric configurations. More and more sophisticated Monte Carlo calculations based on statistical assumptions are being carried out to shed light on the fundamental nature of inherent noise.

We will now devote some time to the problem of estimates from a finite length of data.

ESTIMATES

Whenever it is assumed that random processes are ergodic, then estimates of various parameters, such as autocorrelation and cross-correlation functions and power and cross power spectral densities, can be made. Assume that we have a section of a member of a continuous random process from $-T < t < +T$ and that we indicate its values every Δ s from $t = -N\Delta$ to $t = +N\Delta$ by $x(-N),\ldots,x(N)$. The "discrete function $x_T(n)$" is either the sampled values of $x(t)$ or a discrete function. We call the section of $x(t)$ from $-N$ to N or $x(n)$ a "window" of $x(t)$ or $x(n)$. Assume also that we define a window $y_T(n)$ for another function $y(t)$ or $y(n)$.

CORRELATION FUNCTIONS

From Chapter 7 we obtain estimates of the correlation functions for any $n \ll N$ as

$$R_{\hat{x}\hat{x}}(n\Delta) = \frac{\Delta}{2N-n+1} \sum_{k=-N}^{N-n} x(k)x(k+n) \tag{9.27}$$

and

$$R_{\hat{x}\hat{y}}(n\Delta) = \frac{\Delta}{2N-n+1} \sum_{k=-N}^{N-n} x(k)y(k+n) \tag{9.28}$$

$R_{\hat{x}\hat{x}}(n\Delta)$ and $R_{\hat{x}\hat{y}}(n\Delta)$ are random variables for each specific n, and Eqs. 9.27 and 9.28 each represent one value from a trial of the phenomenon. We would like in general to know the density function and statistics, such as the mean and variance for $R_{\hat{x}\hat{x}}(n\Delta)$. Such a task is virtually impossible, and much effort is devoted to finding the mean and variance of $P_n = R_{\hat{x}\hat{x}}(n\Delta)$. It can easily be demonstrated that

$$\overline{P_n} = R_{\hat{x}\hat{x}}(n\Delta) \tag{9.29}$$

and that the variance of P_n is inversely proportional to N or the length of the sample window. Also, a smaller n leads to a smaller variance. The reader is directed in particular to the text by Papoulis (*Signal Analysis*, Chapter 10) in the reference list at the end of the chapter. Similar results may be obtained for the cross-correlation function.

SPECTRAL ESTIMATES

If $S_{\hat{x}\hat{x}}(\omega)$ is found as the Fourier transform of $R_{\hat{x}\hat{x}}(n\Delta)$, where $R_{\hat{x}\hat{x}}(n\Delta)$ has been obtained for $-m < n < m$, then serious errors may result. Figures 9.5a

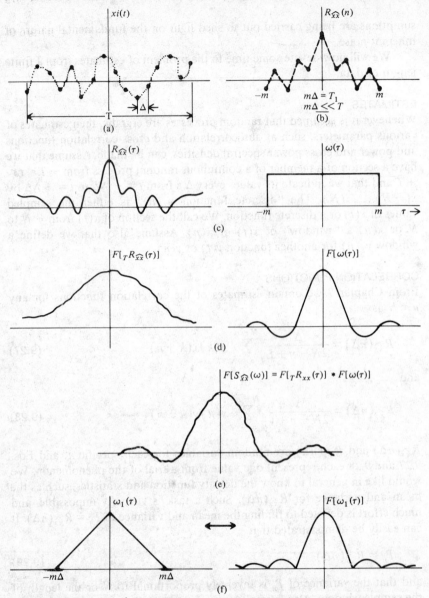

Figure 9.5 (a) A segment of a noise waveform. (b) Estimates of the autocorrelation function. (c) An estimated autocorrelation function and a weighting window. (d) The Fourier transforms. (e) An estimated power spectral density obtained by convolving the transforms of part (d). (f) A "better" weighting window and its Fourier transform.

and b show sketches of $x(t)$ and of $R_{\hat{xx}}(n\Delta)$, where it is assumed that

$$R_{\hat{xx}}(n\Delta) = 0 \qquad |n| > m$$

Let us define $R_{\hat{xx}}(\tau)$ as the continuous function formed by interpolation of $R_{\hat{xx}}(n)$. Now, as shown in Figure 9.5b, $R_{\hat{xx}}(\tau) = 0$, $\tau > m\Delta$, and we say that

$$R_{\hat{xx}}(\tau) = {}_T R_{\hat{xx}}(\tau)\omega(\tau) \tag{9.30}$$

where

$$\omega(\tau) = 1 \qquad |\tau| < m\Delta$$
$$= 0 \qquad \text{otherwise}$$

and $_T R_{\hat{xx}}(\tau)$ denotes an estimate from $-\infty < \tau < \infty$.

Using the convolution theorem for Fourier transforms, we take the Fourier transform of Eq. 9.30 to obtain

$$S_{\hat{xx}}(\omega) = F\left[{}_T R_{\hat{xx}}(\tau)\right] * F\left[\omega(\tau)\right]$$

$$= {}_T S_{\hat{xx}}(\omega) * 2T_1 \frac{\sin 2\pi T_1 f}{2\pi T_1 f}$$

where $T_1 = m\Delta$. This is one estimate of the power spectral density

$$S_{\hat{xx}}(\omega) \triangleq E\left[{}_T S_{\hat{xx}}(\omega)\right]$$

and it is not a good estimate, as we can see from Figure 9.4e. Figure 9.4e shows that $F[\omega(\tau)]$ gives rise to negative values of $S_{\hat{xx}}(\omega)$. No matter how large $T = N\Delta$ the sample length of $x(\tau)$, or $T_1 = m\Delta$ are made, we cannot obtain a good estimate $S_{\hat{xx}}(\omega)$ by taking the Fourier transform of $R_{\hat{xx}}(\tau)$, for values close to m are in error. An improved estimate for $S_{\hat{xx}}(\omega)$ is obtained by placing less weight on the values of $R_{\hat{xx}}(\tau)$ for τ close to $m\Delta$, and this is done by multiplying $_T R_{\hat{xx}}(\tau)$ by a different window function. Figure 9.4f shows one such window $\omega_1(\tau)$, which achieves a better estimate for $S_{\hat{xx}}(\omega)$. If

$$\omega_1(\tau) = 1 - \frac{|\tau|}{T_1} \qquad |\tau| < T_1$$

then

$$S_{\hat{xx}}(\omega) = {}_T S_{\hat{xx}}(\omega) * T_1 \frac{\sin^2 \pi T_1 f}{\pi^2 T_1^2 f}$$

This window function is called the Barlett window, and many other data windows exist, among them the Hamming window.

In practice, however, the autocorrelation function and power spectral density are not generally found by using correlation to obtain

$$R_{xx}(n\Delta) = \frac{\Delta}{2N - n + 1} \sum_{K = -N}^{N} x(k)x(k + n) \tag{9.31}$$

and then determining

$$S_{\hat{x}\hat{x}}(\omega) = F[R_{TC}(\tau)\omega_x(\tau)] \tag{9.32}$$

where $\omega_x(\tau)$ is some window function.

Because of the speed of the fast Fourier transform algorithms, it is common to estimate power spectral density by finding

1. $X_{\hat{F}T}(\omega)$, the fast Fourier transform of $x_T(\tau)$.
2. $S_{\hat{x}x}(\omega) = \dfrac{1}{2T}|X_T(\omega)|^2$.
3. $R_{\hat{x}\hat{x}}(\tau)$, the inverse fast Fourier transform of $\dfrac{1}{2T}|X_T(\omega)|^2$.
4. $S_{\hat{x}\hat{x}}(\omega)$, the fast Fourier transform of $[R_{\hat{x}\hat{x}}(\tau)\omega_x(\tau)]$.

These four steps are faster than the two steps above, where $R_{\hat{x}\hat{x}}(n\Delta)$ is first found by Eq. 9.31 and $S_{\hat{x}\hat{x}}(\omega)$ by Eq. 9.32, since time-consuming convolution has been avoided.

SUMMARY

Chapter 9 treated applications of the system ensemble average manipulations developed in Chapters 7 and 8. It was seen that if the input to a linear system is gaussian, then the output is gaussian and the joint density function for any n random variables sampling the process may be specified.

The problem of designing a filter was discussed next. With our previous experience, both discrete and continuous versions were developed in parallel. The noncausal Wiener and matched filters were derived, and the simplifications due to assumed independent white noise added to the signal were noted. The form of the system function for the Wiener filter may be very difficult to achieve or simulate, and for this reason Kalman filtering is much used.

Finally the chapter concluded with a number of comments on the different considerations and requirements in signal processing. We are now at the stage of applying and utilizing the theory of random processes developed in the text. It is hoped that our probabilistic thinking has been honed and our manipulation of ensemble calculations for systems in the time and frequency domains perfected. The important statistical problem of appreciating that a correlation or spectral value obtained from a data segment is just one trial value of a random variable, with unexpected errors in the spectral variance, was mentioned. A number of important references are listed at the end of the chapter.

PROBLEMS

1. Consider two jointly gaussian random variables defined by

$$f_{XY}(\alpha, \beta) = \frac{1}{2\pi(0.75)}$$
$$\times \exp\left\{-\frac{1}{1.5}\left[(\alpha-1)^2 + (\alpha-1)(\beta-2) + (\beta-2)^2\right]\right\}$$

(a) Find the covariance matrix

$$(\Lambda) = \begin{pmatrix} \sigma_X^2 & \overline{(X-\overline{X})(Y-\overline{Y})} \\ \overline{(X-\overline{X})(Y-\overline{Y})} & \sigma_Y^2 \end{pmatrix}$$

Are X and Y independent?
(b) Find
   (1) $f_X(\alpha)$
   (2) $f_X[\alpha/(Y=2)]$
(c) A new random variable is defined by $Z = 3X - 2Y$.
   (1) Find $(\Lambda)_z$.
   (2) Find $f_Z(\gamma)$.
   (3) $f_{XZ}(\alpha, \beta)$.

2. Given three jointly gaussian random variables with the following matrices

$$\mathbf{M}(\alpha) = (3, -2, 1)$$

and

$$(\Lambda)_x = \begin{pmatrix} 2 & 1 & 0.6 \\ 1 & 4 & 1.4 \\ 0.6 & 1.4 & 2 \end{pmatrix}$$

find $f_{X_1, X_2, X_3}(\alpha, \beta, \gamma)$.

3. (a) Given that gaussian noise $n(t)$ has a power spectral density

$$S_n(f) = \frac{N_0}{2} \qquad |f| < f_c$$

and that $\overline{n(t)} = 0$, find $f_{n(t)}(\alpha)$.
(b) Repeat part (a) for

$$S_n(f) = \frac{N_0}{2}\left(1 - \frac{f}{f_c}\right) \qquad |f| < f_c$$

4. Consider the system shown, which has as its input a member of a gaussian random process with zero mean and

$$R_{xx}(\tau) = 1 - |\tau| \qquad 0 < \tau < 1$$
$$= 0 \qquad \text{otherwise}$$

$$x(t) \quad \boxed{\begin{array}{l} h(t) = 1 \quad 0 < t < 1 \\ = 0 \quad \text{otherwise} \end{array}} \quad y(t)$$

(a) Find $f_x(\alpha)$ for the random variables $x(1)$, $x(1.5)$, and $x(2.5)$.
(b) Find $f_Y(\alpha)$ for the random variables $y(1)$, $y(1.5)$, and $y(2.5)$.
5. Given $x(t)$ and $y(t)$ are real, ergodic, *independent* random processes with the same gaussian density functions

$$f_X(\alpha) = f_Y(\alpha) = \frac{1}{\sqrt{2\pi}} e^{-\alpha^2/2}$$

and that $x(t)$ and $y(t)$ have the same autocorrelation functions.

$$R_{xx}(\tau) = R_{yy}(\tau) = Ae^{-a|\tau|} + B$$

define another random process

$$z(t) = 2x(t) + y(t)$$

(a) Find $R_{zz}(\tau)$ and $S_{zz}(\omega)$ and $f_z(\alpha)$.

(b) If another random process is defined as

$$k(t) = \int_0^t z(r)e^{-a(t-r)}\,dr$$

find $R_{xx}(\tau)$ and $f_k(\alpha)$. Construct a system that produces $k(t)$ as its output.

6. If $x(t)$ is a stationary gaussian random process with

$$S_{xx}(j\omega) = \frac{2}{3 + \omega^2}$$

and X_1, X_2, and X_3 are random variables defined by sampling the process at $t = 0$, $t = 1$, and $t = 4$ s,

(a) Find $f_{X_1 X_2 X_3}(\alpha_1, \alpha_2, \alpha_3)$.

(b) Find $f_{X_1 X_2}[\alpha_1, \alpha_2 / (X_3 = \alpha_3)]$.

(c) Design a filter so that if white noise is applied to it, $x(t)$ is the output.

7. Consider the signal $x(t) = A$; $0 < t < T$, and $x(t) = 0$ otherwise as the input to a low-pass RC filter, as shown. Design τ so that the signal energy divided by the mean square noise is maximum at some time t_0. Evaluate this $(S/N)_{max}$.

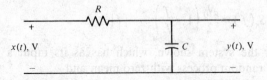

8. The random process

$$x(t) = A_0 \sin \omega_0 t + n(t)$$

which consists of a sinusoidal signal plus white noise is applied to a first-order system

$$H(j\omega) = \frac{1}{\alpha + j\omega}$$

Let the output be

$$y(t) = B_0 \sin(\omega_0 t + \theta) + n_0(t)$$

Find the output signal-to-noise ratio and show that it is maximum for $\alpha = \omega_0$.

9. A coded radar signal $x(t)$ shown is received in the presence of white noise with a power spectral density N_0. Design a filter that maximizes the peak signal-to-noise ratio at some time instant. What is the peak signal-to-noise ratio? Sketch the signal output from the filter.

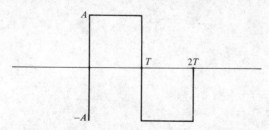

10. (a) Design a filter to maximize the S/N ratio where the signal is

$$f(t) = 1 - \tfrac{1}{2}t \qquad 0 < t < 2$$
$$= 0 \qquad \text{otherwise}$$

and the noise is independent and white with $R_{nn}(\tau) = \tfrac{1}{2}\delta(\tau)$.

   (b) Evaluate the output signal and output mean square noise fluctuations and find $(S/N)_{out}$.

11. Consider a random signal $f(k)$ mixed with independent noise $n(k)$ where the power spectral densities are

$$S_{ff}(z) = \frac{-z}{(z+0.5)(z+2)} \qquad \text{and} \qquad S_{nn}(z) = \frac{-z}{(z+0.2)(z+5)}$$

design a nonphysically realizable filter to minimize the mean square error. Find this least mean square error.

12. (a) Design a filter to maximize the signal-to-noise ratio at $n = 4$, given $f(k) = (-0.5)^k u(k)$ and $n(k)$ is white noise with $R_{xx}(x) = 2\,\delta(k)$. Find this maximum signal-to-noise ratio.

   (b) If you ignore $h(n)$ for $n < 0$ in part (a) to obtain a causal filter, how much is the signal-to-noise ratio affected.

13. Derive the formulas for a discrete matched filter and S/N_{max}.

References

Clarke, A. B., and R. L. Disney. *Probability and Random Processes for Engineers and Scientists*. New York: Wiley, 1970.

Cooper, George R., and Clare D. McGillem. *Probabilistic Methods of Signal and System Analysis*. New York: Holt, Rinehart and Winston, 1971.

Davenport, W. B., Jr., and W. L. Root. *Introduction to Random Signals and Noise*. New York: McGraw-Hill, 1967.

Davenport, Wilbur B. *Probability and Random Processes*. New York: McGraw-Hill, 1970.

Drake, A. W. *Fundamentals of Applied Probability Theory*. New York: McGraw-Hill, 1967.

Larson, Harold J., and O. Shubert Bruno. *Probabilistic Models in Engineering Sciences*, Volume I and Volume II. New York: Wiley, 1979.

Lathai, B. P. *An Introduction to Random Signals and Communication Theory*. Scranton, Pa.: International Textbook Company, 1968.

Papoulis, A. *Probability, Random Variables and Stochastic Processes*. New York: McGraw-Hill, 1965.

Papoulis, A. *Signal Analysis*. New York: McGraw-Hill, 1977.

Papoulis, A. *Circuit and Systems*. New York: Holt, Rinehart and Winston, 1980.

Parzen, E. *Modern Probability Theory and Its Applications*. New York: Wiley, 1960.

Schwarz, M., and L. Shaw. *Signal Processing*. New York: McGraw-Hill, 1980.

Appendix A
Singularity Functions

Definition Delta Functions

The Dirac delta function $\delta(t)$ is defined by

$$\int_{-\infty}^{\infty} f(t)\delta(t)\,dt \triangleq f(0) \qquad\qquad\qquad (A.1)$$

where $f(t)$ is any function that is continuous at $t = 0$. The requirements on $\delta(t)$ are extreme, since if $f(t)$ is changing rapidly at $t = 0$, we must still obtain the answer $f(0)$. We say that in order to satisfy Eq. A.1, $\delta(t)$ must possess the following properties:

$$\delta(0) = \infty$$
$$\delta(t) \to 0 \qquad \text{if } t \neq 0$$

and

$$\int_{-\varepsilon}^{\varepsilon} \delta(t)\,dt = 1 \qquad \text{for any } \varepsilon > 0$$

The "devil spike" notation for $\delta(t)$ is shown in Figure A.1, as are two popular mathematical models $P_1(t)$ and $P_2(t)$. Depending on the rapidity of change of the function $f(t)$ in Eq. A.1, $P_1(t)$ or $P_2(t)$ must be such that the

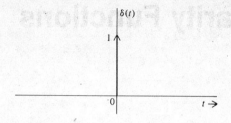

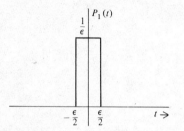 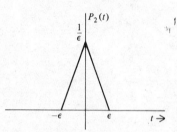

Figure A.1   The notation for a delta function and two practical models.

value of $f(t)$ varies little from $f(0)$ over the range $-\varepsilon < t < \varepsilon$, so that

$$\int_{-\infty}^{\infty} f(t)P_1(t)\,dt = \int_{-\varepsilon/2}^{\varepsilon/2} f(t)\frac{1}{\varepsilon}\,dt$$

$$= f(0)\int_{-\varepsilon/2}^{\varepsilon/2} \frac{1}{\varepsilon}\,dt$$

$$= f(0)$$

and similarly for $P_2(t)$ or for any other models. Indeed if $P_n(t)$ has area A and is located about the origin and $f(t)$ varies very little (subjective) over the duration of $P_n(t)$ from the value $f(0)$, then as far as $f(t)$ is concerned, $P_n(t)$ behaves like a weighted delta function $A\delta(t)$.

PROPERTIES OF $\delta(t)$

Some of the more important properties of delta functions which may easily be demonstrated are:

Property 1 $\displaystyle\int_{-\infty}^{\infty} f(t)\delta(t-a)\,dt = f(a)$

Property 2 $f(t)\delta(t-a) = f(a)\delta(t-a)$

Property 3 $\delta(at+b) = \dfrac{1}{|a|}\delta\left(t+\dfrac{b}{a}\right)$

Property 4 $f(t)*\delta(t-a) = \displaystyle\int_{-\infty}^{\infty} f(u)\delta(t-u-a)\,du$

$$= f(t-a)$$

Property 5 $\delta(t-a)*\delta(t-b) = \delta(t-a-b)$

We will "prove" two of these.

Example A.1

Prove that

(a) $f(t)\delta(t-a) = f(a)\delta(t-a)$.
(b) $\delta(t-a)*\delta(t-b) = \delta(t-a-b)$.

SOLUTION

(a) The identity is true if we can demonstrate that for some $g(t)$, which is continuous at $t = a$,

$$\int_{-\infty}^{\infty} g(t)f(t)\delta(t-a)\,dt = \int_{-\infty}^{\infty} [g(t)]f(a)\delta(t-a)\,dt$$

From Eq. A.1, or Property 1, which is a slight extension of it, the right-hand side of the above equation is equal to $g(a)f(a)$ and the left-hand side is

$$g(t)f(t)|_{t=a}$$

which also equals $g(a)f(a)$, and the identity is established.

(b) By definition, since

$$f(t)*g(t) = \int_{-\infty}^{\infty} f(u)g(t-u)\,du$$

then

$$\delta(t-a)*\delta(t-b) = \int_{-\infty}^{\infty} \delta(u-a)\delta(t-u-b)\,du$$

$$= \delta(t-u-b)|_{u=a} \quad\quad \text{by Property 1}$$

$$= \delta(t-a-b)$$

DELTA FUNCTIONS IN INTEGRATION AND DIFFERENTIATION
By Property 1,

$$\int_{-\infty}^{t} \delta(u-a)\,du = 1 \quad\quad t > a$$

$$= 0 \quad\quad t < a$$

Therefore,

$$\delta(t-a) \longleftrightarrow u(t-a) \tag{A.2}$$

are an integral-derivative pair. This implies that if a function has a piecewise discontinuity at $t = a$, its derivative at $t = a$ is $[f(a^+) - f(a^-)]\delta(t-a)$ or if we integrate across a delta function with weighting factor A_0 at $t = a$, that is, $A_0\delta(t-a)$, we obtain a jump or piecewise discontinuity of A_0 at $t = a$. This is illustrated in Figure A.2.

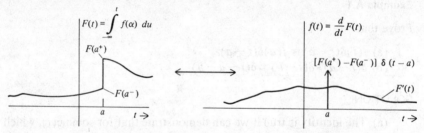

Figure A.2 Differentiating at a piecewise discontinuity or integrating across a weighted delta function.

Density, Cumulative Distribution, and Joint Density Functions

We must be very relaxed in probability theory with the use of delta function notation. If the cumulative distribution function $F_X(\alpha)$ contains a piecewise discontinuity of A_0 at $\alpha = \alpha_0$, which is the case when the probability $p_X(\alpha_0) = A_0$, then its derivative, the density function $f_X(\alpha)$ must be $A_0\delta(\alpha - \alpha_0)$ at $\alpha = \alpha_0$. A delta function in $f_X(\alpha)$ at α_0 indicates a finite probability of the event $(X = \alpha_0)$ occurring.

Example A.3

Interpret the density function $f_X(\alpha)$ shown in Figure A.3a.

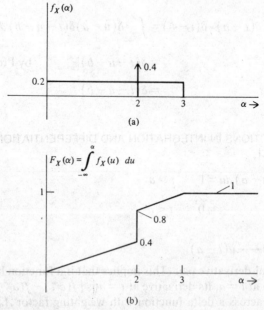

Figure A.3 (a) The density function for Example A.3. (b) The cumulative distribution function.

SOLUTION

Since $f_X(\alpha)$ has a density function with weighting factor 0.4 at $\alpha = 2$, this says the probability of $X = 2$ is 0.4, and since the density function is infinity at $\alpha = 2$, the relative likelihood of the occurrence of $X = 2$ to any other point $X = \alpha$ is infinite. There is only probabilistic meaning to events such as $(\alpha_1 < X < \alpha_2)$ for any range not including the point $\alpha = 2$. The cumulative distribution function $F_X(\alpha)$ is shown plotted in Figure A.3b.

JOINT DENSITY FUNCTIONS

The occurrence of delta functions in a joint density function indicates a constraint on the random variable Y to take on a specific value with a finite probability given the value of X. For example, if

$$f_{XY}(\alpha, \beta) = 0.5\delta(\beta - \alpha) \qquad 0 < \alpha < 2, \ \beta = \alpha$$

this implies that if $X = \alpha$, then Y must also be α, with a probability of 1. We can show this formally by integration:

$$f_X(\alpha) = \int_{-\infty}^{\infty} 0.5\delta(\beta - \alpha)\, d\beta$$

$$= 0.5 \qquad 0 < \alpha < 2$$
$$= 0 \qquad \text{otherwise}$$

$$f_Y[\beta/(X = \alpha)] = \frac{0.5\delta(\beta - \alpha)}{0.5} \qquad \text{if } \beta = \alpha$$

$$= \delta(\beta - \alpha) \qquad \text{if } \beta = \alpha$$
$$= 0 \qquad \text{if } \beta \neq \alpha$$

Appendix B
The Formulas of
Complex Variables

We will now enumerate in logical sequence of development the formulas of complex variables that are used when evaluating inverse Laplace and z transforms.

Definition of Integration

Given a complex function $f(z) = u(x, y) + jv(x, y)$, its integral over a path C from point z_A, where $z_A = x_A + jy_A$, to a point z_B, where $z_B = x_B + jy_B$, is defined as

$$\int_{C^{z_A}}^{z_B} f(z) \, dz \triangleq \lim_{\Delta z_i \to 0} \sum_{i=1}^{N} f(z_i')\Delta z_i \tag{B.1}$$

where z_i' and Δz_i are defined for the curve C in Figure B.1. If C is defined by $y = g(x)$, then Eq. B.1 becomes

$$\int_{C^{z_A}}^{z_B} f(z) \, dz = \left[\int_C u(x, y) \, dx - v(x, y) \, dy \right] + j\left[\int_C v(x, y) \, dx + u(x, y) \, dy \right] \tag{B.2}$$

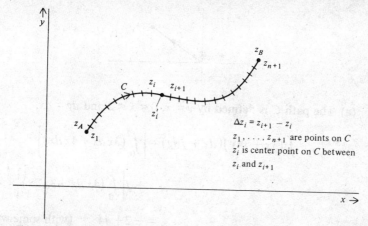

Figure B.1   A curve C used in defining complex integration.

and putting $y = g(x)$ and $dy = g'(x)\,dx$, this becomes

$$\int_{C}^{z_B}{}_{z_A} f(z)\,dz = \int_{x_A}^{x_B} u[x, g(x)]\,dx - \int_{x_A}^{x_B} v[x, g(x)]\,g'(x)\,dx$$
$$+ j\left[\int_{x_A}^{x_B} v(x, g(x))\,dx + \int_{x_A}^{x_B} u(x, g(x))\,g'(x)\,dx \right]$$

Example B.1

Given $f(z) = 3x + j4y$

(a) Evaluate

$$\int_{C}^{2+j2}{}_{0} f(z)\,dz$$

where C is the straight-line path joining the endpoints.

(b) Evaluate

$$\int_{C_1+C_2}^{2+j2}{}_{0} f(z)\,dz$$

where C_1 is given by $y = 0$; $0 < x < 2$, and C_1 is given by $x = 2$; $0 < y < 2$.

SOLUTION

Strictly in terms of z, we could write $f(z) = 3z + j\,\mathrm{Im}(z)$ where "Im" stands for the "imaginary part of."

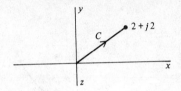

(a) The path C is defined by $y = x$, $0 < x < 2$ and $dy = dx$.

$$\therefore \int_C (3x + j4y)(dx + j\,dy) = \left[\int_0^2 (3x\,dx - 4x\,dx)\right]$$

$$+ j\left[\int_0^2 (4x\,dx + 3x\,dx)\right]$$

$$= -2 + j4 \qquad \text{(with some work)}$$

(b) $\displaystyle\int_{C_1 + C_2} f(z)\,dz = \int_{C_1} f(z)\,dz + \int_{C_2} f(z)\,dz$

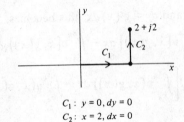

$C_1 : y = 0,\, dy = 0$
$C_2 : x = 2,\, dx = 0$

From our thumbnail sketch, for path C_1,

$$\int_{C_1} (3x + j4y)(dx + j\,dy) = \int_0^2 3x\,dx + j0\,dy$$

$$= 6$$

For path C_2,

$$\int_{C_2} (3x + j4y)(dx + j\,dy) = -\int_0^2 4y\,dy + j\int_0^2 6\,dy$$

$$= -8 + j12$$

Finally,

$$\int_{C_1 + C_2} (3x + j4y)\,dz = -2 + j12$$

We note that the answers from parts (a) and (b) are different, and so the integral is path-dependent.

Complex Integrals That Are Independent of Path

The derivative of a complex function $f(z)$ at $z = z$ is defined as

$$f(z) = \lim_{\Delta z \to 0} \frac{f(z + \Delta z) - f(z)}{\Delta z}$$

and $f'(z)$ exists at $z = z$ if the Cauchy-Riemann conditions for $f(z) = u(x, y) + jv(x, y)$, given by

$$\frac{\partial u}{\partial x} = \frac{\partial v}{\partial y} \quad \text{and} \quad \frac{\partial v}{\partial x} = \frac{\partial u}{\partial y} \tag{B.3}$$

are satisfied, plus the fact all second partials are also satisfied at $z = z$. If $f(z)$ possesses a derivative or is analytic at $z = z$, then

$$f'(z) = \frac{\partial u}{\partial x} + j\frac{\partial v}{\partial x} \quad \text{or} \quad f'(z) = \frac{\partial v}{\partial y} - j\frac{\partial u}{\partial y} \tag{B.4}$$

For polynomial functions of z plus products and quotients of polynomials all the derivative formulas from calculus carry over to complex functions. In addition, the exponential, trigonometric, and hyperbolic functions defined with their derivatives are

$$e^z \triangleq e^x \angle y$$

$$= e^x \cos y + je^x \sin y$$

$$\frac{d}{dz}(e^z) = e^z$$

$$\cos z \triangleq \tfrac{1}{2}(e^{jz} + e^{-jz})$$

$$\sin z \triangleq \frac{-j}{2}(e^{jz} - e^{-jz})$$

$$\frac{d}{dz}\sin z = \cos z \qquad \frac{d}{dz}\cos z = -\sin z$$

$$\cosh z \triangleq \tfrac{1}{2}(e^z + e^{-z})$$

$$\sinh z \triangleq \tfrac{1}{2}(e^z - e^{-z})$$

$$\frac{d}{dz}\cosh z = \sinh z \qquad \frac{d}{dz}\sinh z = \cosh z$$

Cauchy's Theorem

If a function $f(z)$ is analytic on and inside a contour, then

$$\oint_C f(z)\, dz = 0 \tag{B.5a}$$

and

$$\int_{C_1} f(z)\,dz = \int_{C_2} f(z)\,dz \qquad\qquad \text{(B.5b)}$$

where C_1 and C_2 make up a closed contour C. This theorem is illustrated in Figure B.2a.

Example B.2

(a) Evaluate

$$\oint_{C_1+C_2} \frac{1}{z}\,dz$$

(b) Evaluate

$$\int_{C_1} \frac{1}{z}\,dz$$

where C_1 and C_2 are shown sketched in Figure B.2b.

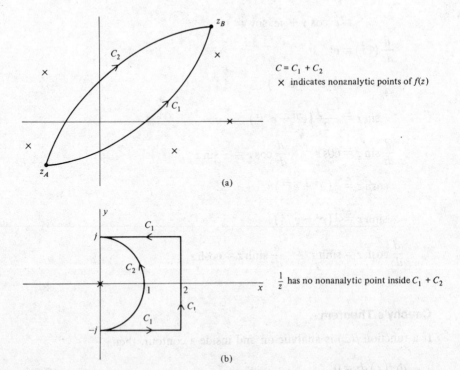

$C = C_1 + C_2$

$\times$ indicates nonanalytic points of $f(z)$

(a)

$\frac{1}{z}$ has no nonanalytic point inside $C_1 + C_2$

(b)

Figure B.2   (a) The contours C, C_1, and C_2 for Cauchy's theorem. (b) The contours C_1 and C_2 for Example B.2.

SOLUTION

(a) The denominator is 0 at $z = 0$ and is not analytic there. Since this point is not included in the closed contour $C_1 + C_2$, then the answer is 0 by Cauchy's theorem.

(b) Since C_1 and C_2 make up a closed contour that does not include a nonanalytic point, then, by Cauchy's theorem,

$$\int_{C_1} \frac{1}{z} dz = \int_{C_2} \frac{1}{z} dz$$

C_2 is a much simpler contour than C_1 and is defined by $z(\theta) = e^{j\theta}$ and $dz = je^{j\theta}$; $-\pi/2 < \theta < \pi/2$.

$$\therefore \int_{C_2} \frac{1}{z} dz = \int_{-\pi/2}^{\pi/2} j \frac{e^{j\theta}}{e^{j\theta}} d\theta$$

$$= j\pi$$

The Fundamental Theorem of Integration

If a complex function $f(z)$ is analytic everywhere, then

$$\int_{z_A}^{z_B} f(z) dz = F(z_B) - F(z_A)$$

via any simple path, where

$$\frac{d}{dz} F(z) = f(z)$$

For example,

$$\int_0^{j4} z \, dz = 0.5z^2 + C \Big|_0^{j4} = -8$$

A simple path is one that does not intersect itself.

The Residue Theorem

If a function $g(z)$ is analytic inside a closed contour except at a finite number of points called poles, where, for example, a pth-order pole at $z = z_0$ is defined as follows:

If $(z - z_0)^n g(z)$ is not analytic, $n = 1, 2, \ldots, (p-1)$ at z_0, but $(z - z_0)^p g(z)$ is analytic at z_0, then

$$\oint_C g(z) \, dz = 2\pi j \sum \left(\begin{array}{c} \text{residues of the poles} \\ \text{of } g(z) \text{ inside } C \end{array} \right) \tag{B.6}$$

where the residue of a pth-order pole of $g(z)$ at $z = z_0$ is

$$\frac{1}{(p-1)!}\left[\frac{d^{p-1}}{dz^{p-1}}(z-z_0)^p g(z)\right]_{z=z_0} \tag{B.7}$$

Example B.3

Evaluate for a fixed t using the residue theorem,

(a) $\dfrac{1}{2\pi j}\displaystyle\oint_C \dfrac{ze^{zt}}{(z-1)(z-2)^2}\,dz$

where C is defined by $1.5e^{j\theta} = z(\theta)$.

(b) $\dfrac{1}{2\pi j}\displaystyle\oint_C \dfrac{e^{zt}}{z(z-1)^2}\,dz$

where C is defined by $3e^{j\theta} = z(\theta)$.

SOLUTION

(a) Since only the pole at $z = 1$ is included inside the contour, the answer is

$$\frac{e^t}{(1-2)^2} = e^t \qquad \text{for any } t$$

(b) Since both the first-order pole at $z = 0$ and the second-order pole at $z = 1$ are included inside C, the answer is

$$\frac{e^{0t}}{(-1)^2} + \frac{d}{dz}\left(\frac{e^{zt}}{z}\right)\bigg|_{\text{at } z=1} = 1 + te^t - e^t$$

for any fixed t.

The Inside-Outside Theorem

If the integrand for a closed contour C is of the form $g(z) = N(z)/D(z)$, where $D(z)$ is of order more than one higher than $N(z)$, then

$$\oint_C g(z)\,dz = 2\pi j \sum \left(\begin{array}{l}\text{residues of the poles}\\ \text{of } g(z) \text{ inside } C\end{array}\right)$$

$$= -2\pi j \sum \left(\begin{array}{l}\text{residues of the poles}\\ \text{of } g(z) \text{ outside } C\end{array}\right)$$

This theorem may be proved by constructing the contour C_1, defined by

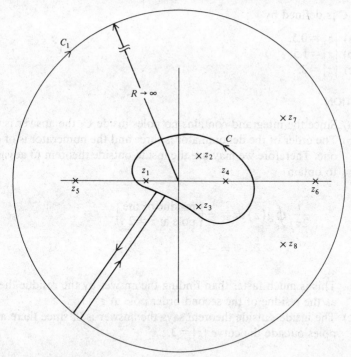

Figure B.3   A function $g(z)$ with four poles inside and outside a contour C.

$z(\theta) = Re^{j\theta}$, where R approaches infinity, and showing that

$$\oint_{C_1} g(z)\, dz \longrightarrow 0$$

Then, by contour subdivision,

$$\oint_C g(z)\, dz = \oint_C g(z)\, dz + \oint_{C_1} g(z)\, dz$$

$$= -2\pi j \sum \left(\begin{array}{c} \text{residues of the} \\ \text{poles outside } C \end{array} \right)$$

These contours are shown in Figure B.3.

___

Example B.4

Use the inside-outside theorem, if possible, to evaluate

$$\frac{1}{2\pi j} \oint_C \frac{2z+1}{(z-1)^2(z-2)}\, dz$$

where C is defined by

  (a) $|z| = 0.5$.
  (b) $|z| = 1.5$.
  (c) $|z| = 3$.

SOLUTION

  (a) Since the integrand contains no poles inside C, the answer is 0.
  (b) The order of the denominator is three and the numerator is of order
      one. Therefore we may use the inside-outside theorem to advantage
      to obtain

$$\frac{1}{2\pi j} \oint_C g(z)\, dz = -\left(\begin{array}{c} \text{residue of the} \\ \text{pole at } z = 2 \end{array} \right)$$

$$= -5$$

      This is much faster than finding the answer by the residue theorem
      as the residue of the second-order pole at $z = 1$.
  (c) The inside-outside theorem says the answer is 0, since there are no
      poles outside the curve $|z| = 3$.

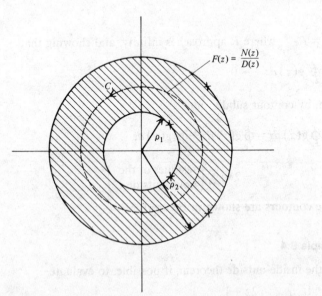

$$F(z) = \frac{N(z)}{D(z)}$$

Figure B.4   Given that $F(z)$ is analytic, $\rho_1 < |z| < \rho_2$, it can be expressed in a Laurent series.

The Laurent Series

If $F(z) = N(z)/D(z)$, then it may be expressed in a Laurent series,

$$F(z) = \sum_{n=-\infty}^{\infty} A_n z_n \tag{B.8}$$

where the Laurent coefficients are

$$A_n = \frac{1}{2\pi j} \oint_C \frac{F(z)}{z^{n+1}} dz \tag{B.9}$$

and the series will converge in an annulus $\rho_1 < |z| < \rho_2$, where there are consecutive poles of $F(z)$ located at $|z| = \rho_1$ and $|z| = \rho_2$, as indicated in Figure B.4. An example will be studied to familiarize ourselves with Eqs. B.8 and B.9, and then some general properties of Laurent series will be discussed.

Example B.5

Express

$$F(z) = \frac{1}{(z-1)(z-2)}$$

in a Laurent series for the following regions:

(a) $|z| < 1$.
(b) $1 < |z| < 2$.
(c) $|z| > 2$.

SOLUTION

(a) For $|z| < 1$

$$A_n = \frac{1}{2\pi j} \oint_C \frac{1}{z^{n+1}(z-1)(z-2)} dz$$

may be interpreted for $n < 0$ and $n \geq 0$ as follows:
For $n \geq 0$, it is advantageous to use the inside-outside theorem and find A_n as

$$A_n = -[\text{residue at } z = 1 + \text{residue at } z = 2]$$

$$= -\left[\frac{1}{1(-1)} + \frac{1}{2^{n+1}} \right]$$

$$= 1 - \left(\tfrac{1}{2}\right)^{n+1}$$

For $n < 0$,

$$A_n = \frac{1}{2\pi j} \oint_C \frac{1}{z^{n+1}(z-1)(z-2)} \, dz$$

$$= \frac{1}{2\pi j} \oint_C \frac{z^{|n|-1}}{(z-1)(z-2)} \, dz$$

$$= 0$$

by the residue theorem, since there are no poles inside C.

We may now write the Laurent series for $0 < |z| < 1$ as

$$\frac{1}{(z-1)(z-2)} = \tfrac{1}{2} + \tfrac{3}{4}z + \tfrac{7}{8}z^2 + \cdots + \left(1 - \left(\tfrac{1}{2}\right)^{n+1}\right)z^n + \cdots$$

This Laurent series and its region of convergence are shown plotted in Figure B.5a.

(b) We now consider evaluating

$$A_n = \frac{1}{2\pi j} \oint_C \frac{1}{z^{n+1}(z-1)(z-2)} \, dz$$

in the region $1 < |z| < 2$ for $n \geq 0$ and $n < 0$.

For $n \geq 0$, using the inside-outside theorem,

$$A_n = -\text{residue of the pole at } z = 2$$

$$= -\left(\tfrac{1}{2}\right)^{n+1}$$

For $n < 0$,

$$A_n = \frac{1}{2\pi j} \oint_C \frac{z^{|n|-1}}{(z-1)(z-2)} \, dz$$

$$= \text{residue of the pole at } z = 1$$

$$= -1$$

We may now write the Laurent series for $1 < |z| < 2$ as

$$\frac{1}{(z-1)(z-2)} = \sum_{n=0}^{\infty} -\left(\tfrac{1}{2}\right)^{n+1} z^n + \sum_{n=-\infty}^{-1} -z^n$$

$$= \left(-\tfrac{1}{2} - \tfrac{1}{4}z - \tfrac{1}{8}z^2 + \cdots\right)$$

$$- \left(z^{-1} + z^{-2} + z^{-3} + \cdots\right)$$

This Laurent series and region of convergence are shown plotted in Figure B.5b.

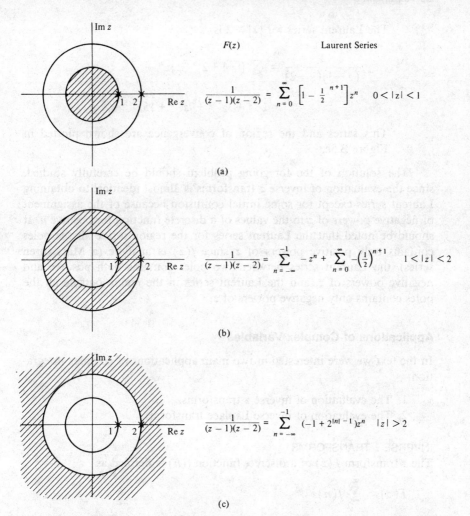

$$F(z) \qquad\qquad \text{Laurent Series}$$

$$\frac{1}{(z-1)(z-2)} = \sum_{n=0}^{\infty}\left[1-\frac{1}{2}^{n+1}\right]z^n \qquad 0<|z|<1$$

(a)

$$\frac{1}{(z-1)(z-2)} = \sum_{n=-\infty}^{-1}-z^n + \sum_{n=0}^{\infty}-\left(\frac{1}{2}\right)^{n+1}z^n \qquad 1<|z|<2$$

(b)

$$\frac{1}{(z-1)(z-2)} = \sum_{n=-\infty}^{-1}(-1+2^{|n|-1})z^n \qquad |z|>2$$

(c)

Figure B.5   The Laurent series for Example B.5. (a) For $|z|<1$. (b) For $1<|z|<2$. (c) For $|z|>2$.

(c) To find the Laurent series for the region $|z|>2$, we proceed as in parts (a) and (b).

For $n \geq 0$,

$$A_n = 0 \qquad \text{(Why ?)}$$

For $n < 0$,

$$A_n = -1 + \frac{1}{2^{n+1}}$$

$$= -1 + 2^{|n|-1}$$

The Laurent series for $|z| > 2$ is

$$\frac{1}{(z-1)(z-2)} = \sum_{-\infty}^{-1} (-1 + 2^{|n|-1}) z^n$$

$$= z^{-2} + 3z^{-3} + 7z^{-4} + 15z^{-5} + \cdots$$

This series and the region of convergence are shown plotted in Figure B.5c.

The solution of the foregoing problem should be carefully studied, since the evaluation of inverse z transforms is almost identical to obtaining Laurent series except for some initial confusion because of the assignment of negative powers of z to the values of a discrete function for positive n. It should be noted that the Laurent series for the region inside all the poles contains only positive powers of z since $f(z)$ is analytic (a MacLauren series), the Laurent series between two poles contains both positive and negative powers of z, and the Laurent series in the region outside all the poles contains only negative powers of z.

Applications of Complex Variables

In the text we were interested in two main applications of complex integration:

1. The evaluation of inverse z transforms.
2. The evaluation of inverse Laplace transforms.

INVERSE z TRANSFORMS

The z transform $F(z)$ of a discrete function $f(n)$ is defined as

$$F(z) = \sum_{-\infty}^{\infty} f(n) z^{-n}$$

$$= \left[f(0) + f(1)z^{-1} + \cdots + f(n)z^{-n} \right]$$

$$+ \left[f(-1)z^{1} + \cdots + f(-n)z^{n} \right] \tag{B.10}$$

and, if $F(z)$ exists, it will do so for some $\rho_1 < |z| < \rho_2$. If $F(z)$ is known in the form $F(z) = N(Z)/D(Z)$; $\rho_1 < |z| < \rho_2$, then by the previous theory we find its Laurent coefficients A_{-n} or the values of $f(n)$.

For $n > 0$,

$$f(n) = A_{-n} = \frac{1}{2\pi j} \oint_C F(z) z^{n-1} dz$$

$$= \sum \begin{bmatrix} \text{residues of the poles of} \\ F(z)z^{n-1} \text{ inside } C \end{bmatrix}$$

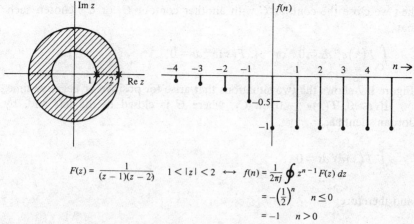

$$F(z) = \frac{1}{(z-1)(z-2)} \quad 1 < |z| < 2 \quad \longleftrightarrow \quad f(n) = \frac{1}{2\pi j} \oint z^{n-1} F(z)\, dz$$
$$= -\left(\tfrac{1}{2}\right)^n \quad n \le 0$$
$$= -1 \quad n > 0$$

Figure B.6   A z transform and its discrete time function.

For $n \le 0$,

$$f(n) = A_{|n|} = \frac{1}{2\pi j} \oint_C \frac{F(z)}{z^{|n|+1}}\, dz$$

$$= -\Sigma \begin{bmatrix} \text{residues of the poles of} \\ F(z)z^{n-1} \text{ outside } C \end{bmatrix}$$

For Example B.5(b), if

$$F(z) = \frac{1}{(z-1)(z-2)} \quad 1 < |z| < 2$$

then

$$f(n) = -\left(\tfrac{1}{2}\right)^n \quad n \le 0$$
$$= -1 \quad n > 0$$

and $F(z)$ and $f(n)$ are shown plotted in Figure B.6.

INVERSE LAPLACE TRANSFORMS

The bilateral Laplace transform of a function $f(t)$ is

$$F(s) = \int_{-\infty}^{\infty} f(t)e^{-st}\, dt$$

and if $F(s)$ exists, it does so for $\sigma_1 < \mathrm{Re}(s) < \sigma_2$. The inverse Laplace transform is

$$f(t) = \frac{1}{2\pi j} \int_{C^{\sigma - j\infty}}^{\sigma + j\infty} F(s)e^{+st}\, ds$$

If $F(s) = N(s)/D(s)$ and the order of $D(s)$ is at least one higher than $D(s)$, then we close the contour C with another contour C_1 or C_2, chosen such that

$$\int_{C_1} F(s)e^{st}\,ds = 0 \quad \text{or} \quad \int_{C_1} F(s)e^{st}\,dt = 0$$

Figure B.7 shows the two situations that arise for positive or negative time.

If $t > 0$, $F(s)e^{st} \to 0$ on C_1, where C is closed to the left and, by Jordan's lemma,

$$\int_{C_1} F(s)e^{st}\,ds = 0$$

and therefore

$$f(t) = \frac{1}{2\pi j}\oint_{C+C_1} F(s)e^{st}\,ds$$

$$= \Sigma\left[\begin{array}{l}\text{residue of the poles of}\\ F(s)e^{st}\text{ to the left of }\sigma\end{array}\right]$$

If $t < 0$, $F(s)e^{st} \to 0$ on C_2, where C is closed to the right and, by Jordan's lemma,

$$\int_{C_2} F(s)e^{st}\,ds = 0$$

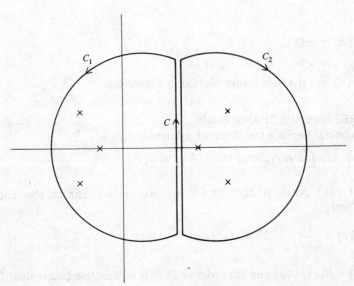

Figure B.7   Closing the contour C for the inverse Laplace transform.

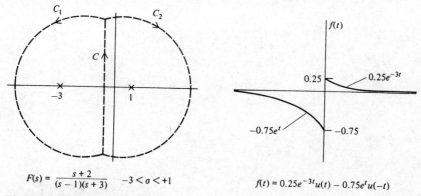

$$F(s) = \frac{s+2}{(s-1)(s+3)} \quad -3 < \sigma < +1$$

$$f(t) = 0.25e^{-3t}u(t) - 0.75e^{t}u(-t)$$

Figure B.8   The Laplace transform and its time function of Example B.6.

and therefore

$$f(t) = \frac{1}{2\pi j} \oint_{C+C_2} F(s)e^{st}\,ds$$

$$= -\sum \begin{bmatrix} \text{residues of the poles of} \\ F(s) \text{ to the right of } \sigma \end{bmatrix}$$

A good discussion of Jordan's lemma is given in many texts.[1]

Example B.6

Find the inverse Laplace transform of

$$F(s) = \frac{s+2}{(s-1)(s+3)} \quad -3 < \sigma < +1$$

SOLUTION

$$f(t) = \frac{1}{2\pi j} \int_{C^{\sigma-j\infty}}^{\sigma+j\infty} \frac{s+2}{(s-1)(s+3)} e^{st}\,ds$$

For $t > 0$,

$$f(t) = \text{residue of the pole at } s = -3$$

$$= \frac{-1}{-4} e^{-3t}$$

$$= 0.25e^{-3t}$$

For $t < 0$,

$$f(t) = -\text{residue of the pole at } s = +1$$

$$= -0.75e^{t}$$

$F(s)$ and $f(t)$ are shown plotted for this example in Figure B.8.

[1] In particular, the reader is referred to S. C. Gupta, *Transform and State Variable Methods in Linear Systems* (New York: Wiley, 1966), pp. 368–371.

Answers to Drill Sets

DRILL SET: SET THEORY, page 12

1. (a) $(4,6,8)$ (b) $(1,3,5,7,9,10)$ (c) $(1,3,4,5,6,7,8,9,10)$
   (d) (9) (e) S (f) $(3,5,7)$
3. (a) True   (b) True
   (c) $\overline{\overline{A} \cap (\overline{A \cap B})} = A \cup (A \cap B) = A = \overline{\overline{A}} \therefore$ True

DRILL SET: THE MATHEMATICS OF COUNTING, page 24

1. (a) $(10)_4 = 5040$ (b) $(9)_3 = 504$ (c) $4(9)_3 = 1512$
   (d) $\dfrac{4!}{1!1!2!}(8)_2 = 572$ (Careful, the answer is not 1144.)
   (e) $4(8)_3 + 4(8)_3 + \dfrac{4!}{1!1!2!}(8)_2 = 3360$
2. (a) $\dbinom{10}{4} = 210$ (b) $\dbinom{8}{2} = 28$ (c) $2\dbinom{8}{3} = 56$
   (d) Multiply all answers by four.

503

DRILL SET: SAMPLE AND EVENT SPACES FOR A RANDOM PHENOMENON, page 31

1. Your tree diagram should indicate 20 results:
   (AAA, AABA $\cdots$ BBB)
2. (a) (A in 4 or less) = (AAA)$\cup$(AABA)$\cup$(ABAA)$\cup$(BAAA)
   (b) (not B in 2d) is the union of ten results
   (c) (series lasts at least 4) is the union of 18 of the 20 possible results.
3. $P(\text{AAA}) = P(\text{BBB}) = \frac{1}{8}$
   $P(\text{any specific 4-game result}) = \frac{1}{16}$, $P(\text{any 5-game result}) = \frac{1}{32}$

DRILL SET: DEFINITION OF PROBABILITY AND THE AXIOMS, page 39

1. (a) $P(\text{3 zeros}) = \dfrac{\dbinom{16}{3}}{2^{16}} = 0.01$

   (b) $P(\text{3 white}) = \dfrac{(7)_3}{(10)_3}$ or $\dfrac{\dbinom{7}{3}}{\dbinom{10}{3}} = 0.29$

   (c) $P(\text{sum is 7}) = \dfrac{6}{36} = 0.17$
2. $P(A\cup B\cup C) = P(A) + P(B) + P(C) - P(A\cap B) - P(A\cap C) - P(B\cap C) + P(A\cap B\cap C)$
3. $P(\text{1st or 2d or 3d is a head}) = \frac{1}{2}+\frac{1}{2}+\frac{1}{2}-3[\frac{1}{4}]+\frac{1}{8} = \frac{7}{8}$

DRILL SET: RANDOM PHENOMENA WITH EQUALLY LIKELY OUTCOMES, page 44

1. (a) $S = \{(H,1),\dots,(H,6),(T,1),\dots,(T,6)\}$, $P(s_i) = \frac{1}{12}$
   (b) $(\text{die} < 3.6) = (H,1)\cup(H,2)\cup(H,3)\cup(T,1)\cup(T,2)\cup(T,3)$
       $P(\text{die} < 3.6) = 6[\frac{1}{12}] = \frac{1}{2}$
   (c) $\frac{3}{12} = \frac{1}{4}$
2. (a) $\dfrac{(26)_3}{(52)_3} = 0.12$

   (b) $1 - \dfrac{(39)_3}{(52)_3} = 0.6$

DRILL SET: CONDITIONAL PROBABILITY; DEPENDENT AND INDEPENDENT EVENTS, page 50

1. (a) $\dfrac{\dbinom{7}{3}}{2^7} = 0.28$

(b) $\dfrac{2\left[\binom{6}{4}+\binom{5}{3}+\binom{4}{1}\right]}{\binom{8}{4}}=0.56$ (Tough)

2. Almost identical

3. $P(5 \text{ black}) = \frac{26}{52} \times \frac{25}{51} \times \frac{24}{50} \times \frac{23}{49} \times \frac{22}{48} = 0.03$

DRILL SET: RELATIONSHIP BETWEEN DIFFERENT EVENT SPACES, page 63

(a)

A_1B_1	A_2B_1	A_3B_1
A_1B_2	A_2B_2	A_3B_2
A_1B_3	A_2B_3	A_3B_3

$P(A_1B_1)=0.35,\ P(A_1B_2)=0.1,\ P(A_1B_3)=0.05$

$P(A_2B_1)=0.18,\ P(A_2B_2)=0.06,\ P(A_2B_3)=0.06$

$P(A_3B_1)=0.16,\ P(A_3B_2)=0.04,\ P(A_3B_3)=0$

$P(B_1)=0.69,\ P(B_2)=0.20,\ P(B_3)=0.11$

(b) $P(A_1/B_1)=\dfrac{0.35}{0.69}=0.51,\ P(A_2/B_1)=\dfrac{0.18}{0.69}=0.26,$

$P(A_3/B_1)=\dfrac{0.16}{0.69}=0.23$

$P(A_1/B_2)=\dfrac{0.1}{0.2}=0.5,\ P(A_2/B_2)=\dfrac{0.06}{0.2}=0.3,$

$P(A_3/B_2)=\dfrac{0.04}{0.2}=0.2$

$P(A_1/B_3)=\dfrac{0.05}{0.11}=0.45,\ P(A_2/B_3)=\dfrac{0.06}{0.11}=0.55,\ P(A_3/B_3)=0$

DRILL SET: MASS, DENSITY, AND DISTRIBUTION FUNCTIONS, page 111

1. $P(1)=0.33,\ P(2)=0.66,\ f_X(\alpha)=0.33\delta(\alpha)+0.66\delta(\alpha-1)$

   $F_X(\alpha)=0,\ \alpha<1,\ F_X(\alpha)=0.33,\ 1<\alpha<2,\ F_X(\alpha)=1,\ \alpha\geq 2$

2. $f_X(\alpha)=0.5\delta(\alpha)+0.25;\ -1<\alpha<1,$ otherwise 0

   $f_Y(\beta)=0.5\delta(\alpha)+0.5;\ 0\leq\alpha<1,$ otherwise 0

   $F_X(\alpha)=0;\ \alpha<-1,\ F_X(\alpha)=0.25(\alpha+1);\ -1<\alpha<0$

   $F_X(\alpha)=0.75+0.25\alpha;\ 0<\alpha<1,\ F_X(\alpha)=1,\ \alpha\geq 1$

   $F_Y(\beta)=0;\ \beta<0,\ F_Y(\beta)=0.5+0.5\beta,\ 0\leq\beta<1,\ F_Y(\beta)=1,\ \beta\geq 1$

3. $y(t)=0;\ t<0,\ y(t)=0.166+0.5t;\ 0\leq t<1$

   $y(t)=0.42+0.5t;\ 1\leq t<2,\ y(t)=1.42;\ t\geq 2$

DRILL SET: PROPERTIES AND USES OF MASS, DENSITY, AND DISTRIBUTION FUNCTIONS, page 117

1. (a) $P[X<2]=F_X(2^-)$ (b) $P[X>3]=1-F_X(3^+)$

   (c) $P[-1<X\le 2]=F_X(2^+)-F_X(-1^+)$

   (d) $P[X>1/X\le 5]=\dfrac{P[1<X\le 5]}{P[X\le 5]}=\dfrac{F_X(5^+)-F_X(1^+)}{F_X(5^+)}$

   $P[(X<-1)\cup(X>2)/(-3<X<1)]$

   $=\dfrac{P[(-3<X<-1)\cup\varnothing]}{P[-3<X<1]}=\dfrac{F_X(-1^-)-F_X(-3^+)}{F_X(1^-)-F_X(-3^+)}$

2. (a) $P[0.1<X<2/X<1]=\dfrac{\int_{0.1}^1\frac16 d\alpha}{\int_0^1\frac16 d\alpha}=0.9$

   (b) $f_X(\alpha/X<1)=0;\ \alpha<0\cup\alpha<1,\ f_X(\alpha/X<1)=1;\ 0<\alpha<1$

   $f_X(\alpha/X>3)=0;\ \alpha<3\cup\alpha>4,\ f_X(\alpha/X>3)=1;\ 3<\alpha<4$

   $f_X(\alpha/X=\frac14)=\delta(\alpha-\frac14)$

   (c) $F_X(\alpha/X<1)=0;\ \alpha<0,\ =1;\ \alpha>1,\ =\alpha;\ 0<\alpha<1$

   $F_X(\alpha/X>3)=0;\ \alpha<3,\ =1;\ \alpha>4,\ =(\alpha-3);\ 3<\alpha<4$

   $F_X(\alpha/X=\frac14)=u(\alpha-0.25)$

DRILL SET: STATISTICS OF A RANDOM VARIABLE, page 124

1. $\Sigma(\alpha_i-\overline X)^2 p_X(\alpha_i)=\overline{X^2}+\overline X^2-2\overline{XX}=\overline{X^2}-\overline X^2$
2. $\ddot X=3.17,\ \overline{X^2}=10.33,\ \sigma_x^2=0.31$
3. $\overline X=6.17,\ \overline{X^2}=38.33,\ \sigma_x^2=0.31$

DRILL SET: THE FUNDAMENTAL THEOREM, page 130

(a) $x(t)=2t;\ -1<t<1,\ x(t)=0,\ 1<|t|<2$

   $y(t)=2|t|;\ -1<t<1,\ y(t)=0,\ 1<|t|<2$

   $z(t)=4t^2;\ -1<t<1,\ z(t)=0,\ 1<|t|<2$

(b) $\overline X=0,\ \overline{X^2}=0.67,\ \sigma_x^2=0.67$

   $\overline Y=0.5,\ \overline{Y^2}=0.67,\ \sigma_y^2=0.42$

   $\overline Z=0.67,\ \overline{Z^2}=1.6,\ \sigma_z^2=1.16$

DRILL SET: THE DENSITY FUNCTION FOR A FUNCTION OF A RANDOM VARIABLE, page 137

1. $f_Y(\beta)=0.33e^{-0.33(\beta-4)}$ $4<\beta<\infty$

   $=0$ otherwise

2. $f_Y(\beta)=0.17\beta^{-0.5}$ $0<\beta<4$

   $=0.08\beta^{-0.5}$ $4<\beta<16$

   $=0$ otherwise

DRILL SET: JOINT DISTRIBUTION, DENSITY, AND MASS FUNCTIONS, page 175

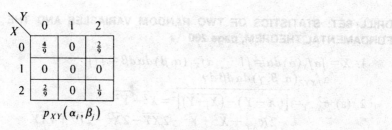

$$p_{XY}(\alpha_i, \beta_j)$$

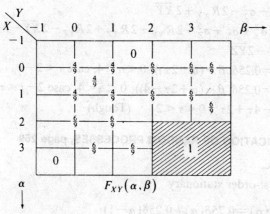

$$F_{XY}(\alpha, \beta)$$

DRILL SET: PROPERTIES OF JOINT DISTRIBUTION, DENSITY, AND MASS FUNCTIONS, page 184

1. (a) $P[X < 4 \cap Y < 5] = F_{XY}(4,5)$

   (b) $P[(X < 3) \cap Y > 2] = F_{XY}(3, \infty) - F_{XY}(3,2)$

   (c) $P\{[(2 < X < 4) \cup (X > 7)]/(3 < Y < 5)\}$
   $= F_{XY}(4,5) + F_{XY}(2,3) - F_{XY}(4,3) - F_{XY}(2,5)$
   $\quad + F_{XY}(\infty,5) + F_{XY}(7,3) - F_{XY}(\infty,3) - F_{XY}(7,5)$

   (d) $P\{[(2 < X < 4) \cap (Y < 1)]/(X < 3)\} = \dfrac{F_{XY}(4,1) - F_{XY}(3,0)}{F_{XY}(3, \infty)}$

   (e) $\dfrac{F_{XY}(4,4) + F_{XY}(1,2) - F_{XY}(4,2) - F_{XY}(2,4)}{F_{XY}(4,7) + F_{XY}(-1,2) - F_{XY}(4,2) - F_{XY}(-1,7)}$

2. (a) $\sum\limits_{j} \sum\limits_{k} p_{XYZ}(\alpha_i, \beta_j, \gamma_k)$

   (b) $\sum\limits_{\alpha_i = 1^+}^{2.6^-} \sum\limits_{\beta_j = 3^+}^{4.1^-} p_{XYZ}(\alpha_i, 4, \gamma_k)$

   (c) $\sum\limits_{\beta_j = 1^+}^{3} p_{XYZ}(\alpha_i, \beta_j, 2) \div \sum\limits_{\text{all } \alpha_i} \sum\limits_{1^+}^{3} p_{XYZ}(\alpha_i, \beta_j, 2)$

   (d) 0

3. (a) 1

(b) 0

DRILL SET: STATISTICS OF TWO RANDOM VARIABLES AND THE FUNDAMENTAL THEOREM, page 200

1. $\overline{X} = \int \alpha f_X(\alpha)\,d\alpha = \int\int \quad \alpha f_{XY}(\alpha,\beta)\,d\alpha\,d\beta = \int\int\int$
$\alpha f_{XYZ}(\alpha,\beta,\gamma)\,d\alpha\,d\beta\,d\gamma$

2. (a) $\sigma^2_{X-Y} = \overline{[(X-Y)-\overline{(X-Y)}]^2} = \overline{X^2} + \overline{Y^2}$
$\qquad -2R_{XY} + \overline{X}^2 + \overline{Y}^2 - 2\overline{XY} - 2\overline{X}^2 - 2\overline{Y}^2 + 4\overline{XY}$
$\qquad = \sigma^2_X + \sigma^2_Y - 2R_{XY} + 2\overline{XY}$

(b) $\sigma^2_{X+Y+Z} = \sigma^2_X + \sigma^2_Y + \sigma^2_Z + 2R_{XY} + 2R_{XZ} + 2R_{YZ} - 2\overline{XY} - 2\overline{XZ}$
$\qquad - 2\overline{YZ}$

3. (a) $f_{XY}(\alpha,\beta) = 0.25\delta(\beta-(\alpha+2\tau)); \ 0 < \alpha < 4, \text{ case } \tau < 2-t$
$\qquad\qquad = 0.25\delta(\beta-(\alpha+2\tau-4)); \ 0 < \alpha < 4, \text{ case } 2-t < \tau < 2$

$R_{XY} = \frac{32}{3} - 4\tau + 2\tau^2; \ 0 < \tau < 2$ (Tough)

DRILL SET: CLASSIFICATION OF RANDOM PROCESSES, page 259

1. $x(t)$ is ergodic

2. $x(t)$ is not first-order stationary

$$f_{x(0^+)}(\alpha) = 0.75\delta(\alpha) + 0.25\delta(\alpha-1)$$

$$f_{x(0.5^+)}(\alpha) = \delta(\alpha)$$

3. $x(t)$ is not first-order stationary

$$f_{x(0^+)}(\alpha) = 0.75\delta(\alpha) + 0.125\delta(\alpha+1) + 0.125\delta(\alpha+1)$$

$$f_{x(0.5^+)}(\alpha) = \delta(\alpha)$$

4. $x(t)$ is ergodic

DRILL SET: SPECIAL DRILL PROBLEM, page 295

(a) $R_{xx}(\tau) = 1 - \dfrac{|\tau|}{b}; \ |\tau| < b$

$\quad = 0 \qquad$ otherwise

(b) $R_{xx}(0) = 1$, $R_{xx}(\pm b) = \frac{1}{3}$, $R_{xx}(\pm 2b) = \frac{1}{9}$ and it is linear in between

$$R_{xx}(\tau) = 1 - \frac{2|\tau|}{3b}; \ |\tau| < b$$

$$R_{xx}(\tau) = \frac{5}{9} - \frac{2|\tau|}{9b}; \ b < |\tau| < 2b \quad \text{etc.}$$

(c) $R_{xx}(0) = 1$, $R_{xx}(\pm b) = -\frac{1}{3}$, $R_{xx}(\pm 2b) = +\frac{1}{9}$ and it is linear in between

$$R_{xx}(\tau) = 1 - \frac{4|\tau|}{3b}; \ |\tau| < b$$

$$R_{xx}(\tau) = -\frac{7}{9} + \frac{4|\tau|}{9b}; \ b < |\tau| < 2b \ \text{ etc.}$$

(d) $R_{xx}(\tau) = 1$

(e) $R_{xx}(0) = 1$, $R_{xx}(\pm b) = -1$, $R_{xx}(\pm 2b) = +1$ and it is linear in between

DRILL SET: CONVOLUTION AND CORRELATION INTEGRALS FOR DETERMINISTIC FUNCTIONS, page 328

(a) In general

$$x(p - t) = p - t - 1; \ t < p < t + 1$$

$$= 0 \qquad \text{otherwise (Plot it)}$$

For $t = -3$

$$x(p - t) = p + 2; \ -3 < p < -2$$

$$= 0 \qquad \text{otherwise}$$

For $t = 1$

$$x(p - t) = p - 2; \ 1 < p < 2$$

$$= 0 \qquad \text{otherwise}$$

For $t = 4$

$$x(p - t) = p - 5, \ 4 < p < 5$$

$$= 0 \qquad \text{otherwise}$$

(b) In general

$$x(t - p) = -p + t - 1; \ t - 1 < p < t$$

$$= 0 \qquad \text{otherwise (Plot it)}$$

For $t = -3$

$$x(t - p) = -p - 4; \ -4 < p < -3$$

$$= 0 \qquad \text{otherwise}$$

For $t = 1$

$$x(t - p) = -p; \quad 0 < p < 1$$
$$= 0 \quad \text{otherwise}$$

For $t = 4$

$$x(t - p) = -p + 3; \quad 3 < p < 4$$
$$= 0 \quad \text{otherwise}$$

(c) $x(p)x(p - t) = 0, \quad |t| > 1$

$$x(p)x(t - p) = 0, \quad t < 0 \cup t > 2$$

(d) $x(t) \oplus x(t) = \frac{1}{6}|t|^3 - \frac{1}{2}|t| + \frac{1}{3}; \quad |t| < 1$
$$= 0 \quad \text{otherwise}$$

$x(t) * x(t) = \frac{1}{6}t^3 - t^2 + t; \quad 0 < t < 1$
$$= -\frac{1}{6}t^3 + t^2 - 2t + \frac{4}{3}; \quad 1 < t < 2$$
$$= 0 \quad \text{otherwise}$$

(e) $C_{yx}(t) = \int_{-\infty}^{\infty} x(p)y(p - t)\,dp$
$$= 0; \quad t < -3 \cup t > 0$$
$$= \frac{1}{2}t^2 + 2t + \frac{3}{2}; \quad -3 < t < -2$$
$$= -\frac{1}{2}; \quad -2 < t < -1$$
$$= \frac{1}{2}t^2; \quad -1 < t < 0$$

$C_{xy}(t) = 0; \quad t < 0 \cup t > 3$
$$= \frac{1}{2}t^2; \quad 0 < t < 1$$
$$= -\frac{1}{2}; \quad 1 < t < 2$$
$$= \frac{1}{2}t^2 - 2t + \frac{3}{2}; \quad 2 < t < 3$$

Answers to Selected Problems

CHAPTER 1

3. Untrue, True

4. (a) No   (b) No   (c) Yes

5. (a) $\binom{m}{p} \times \binom{n}{p}$ (b) 658

7. (a) $\binom{13}{6}\binom{39}{7} = 2.639 \times 10^{10}$

   (b) $\binom{13}{6}\binom{13}{0}\binom{26}{7} = 1.129 \times 10^{9}$

9. (a) $\binom{50}{3}\binom{50}{2} = 2.4 \times 10^{7}$ samples, $P(A) = 0.32$

   (b) $\binom{50}{3}\binom{30}{1}\binom{20}{1} = 1.176 \times 10^{7}$ samples, $P(B) = 0.16$

   (c) $\binom{44}{2}\binom{30}{1}\binom{20}{1} = 7.056 \times 10^{5}$ samples, $P(C) = 0.19$

CHAPTER 2

1. (a) 0.31   (b) 0.50   (c) 0.20   (d) 0.18
3. (a) 0.25   (b) 0.35   (c) 0.77   (d) 0.13
8. (b) 0.648   (c) 0.380
9. (a) $0.75, 0.50, 0.25, 1 - \dfrac{\tau}{T}$ (b) $0.25, 0.50, 0.75, \dfrac{\tau}{T}$

CHAPTER 3

2. (a) 0.267   (b) 0.33
4. (a) 0.86   (b) 0.90
5. 0.81
6. (a) $\dfrac{12!}{3!4!5!}(0.2)^3(0.4)^4(0.4)^5 = 0.06$
7. (a) 0.94   (b) 0.31   (c) $(0.94)^3 + \dbinom{3}{2}(0.94)^2(0.06) = 0.99$
9. (a) 0.17   (b) 0.69   (c) 0.16   (d) 0.50   (e) 0.16
      (f) 0.00024   (g) 0.67
10. (b) $P(0,0) = 0.5$, $P(0,1) = P(1,0) = 0.17$, $P(1,1) = 0.17$

   (c) $0 < \tau < 1$, probabilities are $\dfrac{2}{3} - \dfrac{\tau}{3}, \dfrac{\tau}{3}, \dfrac{\tau}{3}, \dfrac{1}{3} - \dfrac{\tau}{3}$

   $1 < \tau < 2$, probabilities are $\dfrac{1}{3}, \dfrac{1}{3}, \dfrac{1}{3}, 0$

   $2 < \tau < 3$, probabilities are $\dfrac{\tau}{3} - \dfrac{1}{3}, 1 - \dfrac{\tau}{3}, 1 - \dfrac{\tau}{3}, \dfrac{\tau}{3} - \dfrac{2}{3}$

   (d) 0.5

CHAPTER 4

2. $P_X(0) = P_X(3) = \frac{1}{8}$

   $P_X(1) = P_X(2) = \frac{3}{8}$

   $F_X(\alpha) = \frac{1}{8}, \quad 0 < \alpha < 1^-$

   $= \frac{3}{8}, \quad 1 < \alpha < 2^-$

   $= \frac{7}{8}, \quad 2 < \alpha < 3^-$

   $= 1, \quad \alpha > 3$

3. $f_X(\alpha) = 0.5\delta(\alpha - 1) + 0.25[u(\alpha) - u(\alpha - 2)]$

   $F_X(\alpha) = 0.25\alpha, \quad 0 < \alpha < 1^-$

   $F_X(\alpha) = 0.75 + 0.25(\alpha - 1), \quad 1 < \alpha < 2$

   $= 1, \quad \alpha \geq 2$

5. (a) 1   (b) 0.13   (c) 0.34

   (d) 0.98, 0.47, 0.80, 0.56, 0.20, 0.044

6. $P_X(\alpha_i / X \geq 0) = 0, \ \alpha_i < 2$

   $\qquad\qquad = 0.46, \ \alpha_i = 2$

   $\qquad\qquad = 0.30, \ \alpha_i = 3$

   $\qquad\qquad = 0.15, \ \alpha_i = 4$

   $\qquad\qquad = 0.06, \ \alpha_i = 5$

7. $\overline{X} = 0.67, \ \overline{X^2} = 0.78, \ \sigma_x^2 = 0.28$

9. (a) $\overline{X}_1 = 1.25, \ \overline{X_1^2} = 4.17, \ \sigma_{x_1}^2 = 2.6$ (b), (c) Same answers

   (d) First-order statistics give no information about the rapidity of change of a waveform.

11. $\overline{Y} = 0.64, \ \overline{Y^2} = 0.50, \ \sigma_y^2 = 0.10$

12. $f_X(\alpha) = \dfrac{1}{\pi\sqrt{1-\alpha^2}}, \ -1 < \alpha < 1$

   $$F_Y(\beta) = 1 - \frac{2\cos^{-1}\beta}{\pi}, \ \ 0 < \beta < 1$$

   $$F_Y(\beta) = 1, \qquad \alpha > 1$$

   $$f_Y(\beta) = \frac{2}{\pi\sqrt{1-\beta^2}}, \ \ 0 < \beta < 1$$

13. (b) $\overline{X} = 0, \ \overline{X^2} = \sigma_X^2 = 0.22, \ \overline{Y} = 0.22, \ \overline{Y^2} = 0.13, \ \sigma_Y^2 = 0.09$

   $$F_Y(\beta) = 0.33 + 0.67\beta^{0.5}, \ \ 0 < \beta < 1$$

   $$F_Y(\beta) = 1, \ \ \beta > 1$$

   $$f_Y(\beta) = 0.33\delta(\beta) + 0.33\beta^{-0.5}, \ \ 0^- < \beta < 1$$

CHAPTER 5

1. (a) $f_{XY}(\alpha, \beta) = 0.5, \ 0 < \alpha < 1, \ 0 < \beta < 2; \ = 0$ otherwise

   $\qquad F_{XY}(\alpha, \beta) = 0, \ \ \alpha < 0 \cup \beta < 0;$

   $\qquad\qquad\qquad = 0.5\alpha\beta, \ \ 0 < \alpha < 1, \ \ 0 < \beta < 2;$

   $\qquad\qquad\qquad = 0.5\beta, \ \ 1 < \alpha < \infty, \ \ 0 < \beta < 2;$

   $\qquad\qquad\qquad = \alpha, \ \ 0 < \alpha < 1, \ \ 2 < \beta < \infty;$

   $\qquad\qquad\qquad = 1, \ \ \alpha > 1, \ \ \beta > 2$

(b) $f_{XZ}(\alpha, \beta) = 0.5,\ 0 < \alpha < 1,\ \alpha < \beta < \alpha + 2$

$F_{XZ}(\alpha, \beta) = 0, \qquad \alpha < 0;$

$\qquad = 0.5(\alpha\beta - 0.5\alpha^2),\ \ 0 < \alpha < 1,\ \ \alpha < \beta < \alpha + 2;$

$\qquad = 0.5(\beta - 0.5),\ \ 1 < \alpha < \infty,\ \ \alpha < \beta < \alpha + 2;$

$\qquad = \alpha,\ 0 < \alpha < 1,\ \ \alpha + 2 < \beta < \infty;$

$\qquad = 1;\ \ \alpha > 1,\ \ \alpha + 2 < \beta < \infty$

3. (a) $f_X(\alpha) = f_Y(\alpha) = 0.67\delta(\alpha) + 0.33\delta(\alpha - 1)$

   (c) Case $\tau < 1,\ P(0,0) = \dfrac{2-\tau}{3},\ P(0,1) = P(1,0) = \dfrac{\tau}{3},\ P(1,1) = \dfrac{1-\tau}{3}$

4. (b) $P(0,0) = \dfrac{2-\tau}{3},\ P(0, Y \neq 0) = P(X \neq 0,0) = \dfrac{\tau}{3},$

   $P(X \neq 0,\ Y \neq 0) = \dfrac{1-\tau}{3}$

   (c) $f_X(\alpha) = f_Y(\alpha) = 0.67\delta(\alpha) + 0.33,\ 0^- < \alpha < 1;\ = 0 \qquad$ otherwise

   $f_Y(\beta/X = 0) = (1 - 0.5\tau)\delta(\beta) + 0.5,\ \ \beta = 0,\ \ 1 - \tau < \beta < 1$

   $f_Y(\beta/X = \alpha) = \delta[\beta - (\alpha - \tau)],\ \ \alpha > \tau,\ \ 0 < \beta < 1 - \tau$

   $\qquad = \delta(\beta),\ \ \alpha < \tau$

6. (a) $A = 0.5$

   (b) (1) $f_X(\alpha) = 1.33(1 - \alpha^3),\ 0 < \alpha < 1$

   $\qquad f_Y(\beta) = 0.25\beta^3,\ \ 0 < \beta < 2$

   $\qquad f_X(\alpha/Y = \beta) = 2\beta^{-1},\ \ 0 < \alpha < 0.5\beta$

   (2) $P(2X + Y < 1) = 0.018,\ P[(2X + Y) < 1/(X > 0.5)] = 0$

8. $P(2,1) = P(3,1) = P(3,2) = 0.2,\ P(2,2) = 0.3,\ P(2,3) = 1$

   $P(\alpha_i, \beta_j) = 0 \qquad$ otherwise

10. (a) $A = 1.5$ \qquad (b) $f_{XY}(\alpha, \beta) = 1.5(1 - 0.5\beta),\ \ 0 < \alpha < \beta < 2;\ \ 0$
    otherwise;

    $f_{XZ}(\alpha, \gamma) = 1.5(2\gamma - \alpha)\gamma,\ \ 0 < \alpha < 2\gamma < 2;\ 0 \qquad$ otherwise

11. $f_X(\alpha) = 0.5,\ -1 < \alpha < 1;\ 0 \qquad$ otherwise

    $\overline{X} = 0, \qquad \overline{Y} = 0$

    $f_{XY}(\alpha, \beta) = 0.5\delta[\beta - (\alpha + 1)],\ \ -1 < \alpha < 0;$

    $\qquad = 0.5\delta[\beta - (\alpha - 1)],\ \ 0 < \alpha < 1;$

    $R_{XY} = -0.17, \qquad L_{XY} = -0.17$

14. $f_{x+y}(\gamma) = 0.5\gamma,\ 0 < \gamma < 1,\ = 0.5,\ 1 < \gamma < 2,\ = 1.5 - 0.5\gamma,\ 2 < \gamma < 3$

15. $f_{x-y}(\gamma) = 1 + 0.5\gamma,\ -2 < \gamma < -1,\ = 0.5,\ -1 < \gamma < 0,\ = 0.5 - 0.5\gamma,$
$0 < \gamma < 1.$

CHAPTER 6

1. (c) At $t = 0$, $p(0) = 0.5$, $p(1) = 0.5$;

At $t = 1$, $p(1) = 1$;

At $t = 2$, $p(1) = 0.5$, $p(2) = 0.33$, $p(4) = 0.17$;

At $t = 3$, $p(1) = 0.5$, $p(3) = 0.33$, $p(9) = 0.17$

(d) $\overline{x(2)} = 1.83$, $\overline{x^2(1)} = 1$, $\sigma^2_{x(2)} = 1.14$

(e) $p(1, 1) = 0.5$, $p(2, 3) = 0.33$, $p(4, 9) = 0.17$

Otherwise $p(\alpha_i, \beta_j) = 0$, $R_{xx}(2, 3) = 8.5$

4. Nonstationary

6. Nonstationary

7. Ergodic

8. (b) First-order stationary since $p(-1) = p(1) = 0.5$ for any t

(c) Not wide-sense stationary

9. (b) $f_x(\alpha) = 0.5\delta(\alpha) + 0.5\delta(\alpha - 1)$

$f_y(\alpha) = 0.33\delta(\alpha) + 0.33\delta(\alpha - 0.5) + 0.33\delta(\alpha + 0.5)$

$f_z(\alpha) = 0.67\delta(\alpha) + 0.17\delta(\alpha - 0.17) + 0.17\delta(\alpha + 0.17)$

(1) $\overline{X} = 0.5$, $\overline{X^2} = 0.5$, $\sigma_x^2 = 0.25$

(2) $\overline{Y} = 0$, $\overline{Y^2} = 0.17$, $\sigma_y^2 = 0.17$

(3) $\overline{Z} = 0$, $\overline{Z^2} = 0.084$, $\sigma_z^2 = 0.084$

(4) $\overline{y(n)y(n+1)} = \frac{19}{36} = 0.53$

$\overline{z(n)z(n+1)} = \frac{5}{36} = 0.28$

10. (b) $f_X(\alpha) = 0.3\delta(\alpha) + 0.7\delta(\alpha - 1)$

$P(0/0) = \frac{2}{3}$, $P(0/1) = \frac{1}{7}$, $P(1/1) = \frac{6}{7}$, $P(1/0) = \frac{1}{3}$

$$P(X_4 X_5 = 1/X_3 = 0) = \frac{P(0, 1, 1)}{0.3} = \frac{0.3 \times \frac{1}{3} \times \frac{6}{7}}{0.3} = \frac{6}{21} = 0.29$$

(c) Stationary

CHAPTER 7

1. (a) $f_{x(t)}(\alpha) = \dfrac{1-t}{2}\delta(\alpha - 1) + \dfrac{1+t}{2}\delta(\alpha)$, $0 < t < 1$, which depends
on t. Process is not first-order stationary.

(b) $m_x(t) = 0.5(1-t)$, $0 < t < 1$; $= 0.5(t-1)$, $1 < t < 2$; $= 0.5$, $2 < t < 3$ and $m_x(t) = m_x(t \pm 3)$

3. No, the process of problem 1 is a counter example.

4. (b) It is first-order stationary: $\overline{X} = 0$, $\overline{X^2} = 50.7 = \sigma^2$

   (c,d) $R_{xx}(\tau) = 25.3 \cos \omega_0 \tau$

6. (b) $R_{xx}(\tau) = \dfrac{b - \frac{2}{3}\tau}{b}$, $0 < \tau < b$, $= \dfrac{5b - 2\tau}{9b}$, $b < \tau < 2b$

   (c) $R_{xx}(\tau) = \dfrac{b - \tau}{b}$; $0 < \tau < b$; $= 0$, $\tau > b$, for no memory

   $R_{xx}(\tau)$ goes from 1 at $\tau = 0$ to -0.33 at $\tau = b$, etc.

8. $f_z(\alpha) = 0.125\delta(\alpha) + 0.375\delta(\alpha - 1) + 0.375\delta(\alpha - 2) + 0.125\delta(\alpha - 4)$

   (a) $\overline{Z} = 1.5$, $\overline{Z^2} = 3$, $\sigma_z^2 = 0.75$

   (b) Some values for $0 < \tau < 1$ are $P(0,1) = P(1,0) = \frac{3}{64}\tau$

   $P(0,3) = P(3,0) = \frac{1}{64}\tau$, etc.

   $R_{zz}(\tau) = 3R_{xx}(\tau) + 6\overline{X}^2$

15. (c) $p(s)q(t-s)$ is nonzero $5 < t < 8$

   $p(s)p(s-t)$ is nonzero $-2 < t < 2$

   $p(s)q(s-t)$ is nonzero $0 < t < 3$

   (d) $r(t) = 2(t-5)$, $5 < t < 6$; $= 2$, $6 < t < 7$; $= 2(8-t)$, $7 < t < 8$; $= 0$ otherwise;

   $C_{pp}(t) = 4(2 - |t|)$, $|t| < 2$; $= 0$ otherwise;
   $C_{pq}(t) = 2(t+3)$, $-3 < t < -2$; $= 2$, $-2 < t < -1$; $= -2t$, $-1 < t < 0$

   $C_{pq}(t) = C_{qp}(-t)$

20. (a) $R_{xx}(0) = 0.5$

   $R_{xx}(1) = 0.5 \times 0.8 = 0.4$

   $R_{xx}(2) = 0.5(0.8^2 + 0.2^2) = 0.34$

   $R_{xx}(3) = 0.5[\ 0.8(0.8^2 + 0.2^2) + 0.2(2 \times 0.2^2 \times 0.8)] = 0.29$

   $R_{xx}(n) \to 0.25$, for $n > 3$

   $L_{xx}(0) = 0.25$

   $L_{xx}(1) = 0.15$

   $L_{xx}(2) = 0.09$

   $L_{xx}(3) = 0.04$

   $L_{xx}(n) \to 0$, $n > 3$

22. (a) $g(h) = k\,u(k) * [2\delta(k) + \delta(k-1)] = 2k\,u(k) + (k-1)u(k-1)$

For example, $g(0) = 0$, $g(1) = 2$, $g(2) = 5$, etc.

$$R_{mm}(k) = C_{hh}(k) * 2\delta(k)$$
$$= [2\delta(k+1) + 5\delta(k) + 2\delta(k-1)] * 2\delta(k)$$
$$= 4\delta(k+1) + 10\,\delta(k) + 4\,\delta(k-1)$$

(b) $\overline{m^2(k)} = 10$

23. (a) A purely random input

$$R_{xy}(\tau) = 2\,\delta(\tau) * e^{-2\tau}u(\tau)$$
$$= 2e^{-2\tau}u(\tau)$$
$$R_{yx}(\tau) = 2e^{2\tau}u(-\tau)$$
$$R_{yy}(\tau) = 2\delta(\tau) * C_{hn}(\tau)$$
$$= 0.5e^{-2|\tau|}$$

CHAPTER 8

1. (a) $\dfrac{z^4 + 1.5z^3 - 2.75z^2 + z}{(z-0.5)^2(z-1)^2}$, $0.5 < |z| < 1$; $\dfrac{-1.1z}{(z-0.6)(z-1.7)}$, $0.6 < |z| < 1.7$

(b) $|z| > 2$, $f(n) = \dfrac{1}{1.5}[(0.5)^n - (2)^n]u(n)$;

$0.5 < |z| < 2$, $f(n) = \dfrac{1}{1.5}\{(0.5)^n u(n) + (0.5)^{|n|}u[-(n+1)]\}$;

$|z| < 0.5$, $f(n) = \dfrac{1}{1.5}[-2^{|n|} + (0.5)^{|n|}]u(-n)$

5. (a) $\dfrac{-s^2 - s - 4}{(s-3)(s+1)^2}$, $-1 < \sigma < 3$; $\dfrac{-2}{(s-3)^3}$, $\sigma < 3$; $\dfrac{2}{(s-3)^3}$, $\sigma > 3$

(b) $-6t\,e^{-2t}u(t)$, $-6t\,e^{-2t}u(-t)$, $3e^{-3t}u(t) + e^{-t}u(-t)$

8. (a) $\pi\,\square(\dfrac{\pi}{4}f)$ (b) $\delta(f)$ (c) $0.5[\delta(f + \tfrac{1}{2}\pi) + \delta(f - \tfrac{1}{2}\pi)]$

(d) $\pi e^{-j\pi^2 f}\square(\pi f)$

(e) $j0.25[-\delta(f-0.75) - \delta(f-0.25) + \delta(f+0.25) + \delta(f+0.75)]$

(f) $X(f) = 0.5$, $-\tfrac{5}{8} < f < -\tfrac{3}{8}$

$X(f) = 1$, $-\tfrac{3}{8} < f < \tfrac{3}{8}$

$X(f) = 0.5$, $\tfrac{3}{8} < f < \tfrac{5}{8}$

12. $\dfrac{\sin \pi t}{\pi t}$.

15. (a, c, d, e) No   (b, f) Yes

18. $S_{yy}(j\omega) = \dfrac{N}{1+\omega^2}$

$R_{yy}(\tau) = 0.5Ne^{-|\tau|}$

$\overline{X} = 0$

$\overline{X^2} = 0.5N$

22. $\dfrac{0.5N}{j\omega + 0.5}$, $(0.5Ne^{-0.5t})u(t)$

24. (a) $S_{xx}(\omega)(1 + |H(\omega)|^2) - S_{xx}(\omega)[H(\omega) + H(-\omega)]$

(b) $R_{xx}(\tau) = 0.25 + 0.25 \wedge \left(\dfrac{\tau}{b}\right)$

$$S_{xx}(\omega) = 0.25\delta(f) + \frac{\sin^2 \pi bf}{\pi^2 b^2 f^2}$$

$$H(\omega) = \frac{1}{1 + j\omega\tau_0}$$

$$H(-\omega) = \frac{1}{1 - j\omega\tau_0}$$

where $\tau_0 = RC$

28. (a) $g(k) = 2k\,u(k) - 2(k-1)u(k-1) = 2\,u(k-1)$

$S_{mm}(z) = -z + 2 - z = S_{mm}(\omega) = 2 - 2\cos\omega$

$R_{yy}(k) = -\delta(k+1) + 2\delta(k) - \delta(k-1)$

CHAPTER 9

1. (a) $(\Lambda) = \begin{pmatrix} 1 & 1 \\ 0.5 & 0.5 \end{pmatrix}$ No

(b) (1) $f_X(\alpha) = \dfrac{1}{\sqrt{2\pi}} e^{-1(\alpha-1)^2/2}$

    (2) $f_X[\alpha/(Y=2)] = \dfrac{1}{\sqrt{1.5\pi}} e^{-(\alpha-1)^2/1.5}$

(c) (1) $(\Lambda)_z = (\Lambda_{11}) = (7)$

    (2) $f_Z(\alpha) = \dfrac{1}{\sqrt{14\pi}} e^{-(\alpha+1)^2/14}$

    (3) $(\Lambda)_{XZ} = \begin{pmatrix} 1 & 2 \\ 2 & 7 \end{pmatrix}$, $\overline{X} = 1$, $\overline{Z} = 3 - 4 = -1$

2. $f_{X_1 X_2 X_3}(\alpha_1, \alpha_2, \alpha_3)$

$$= \frac{1}{(2\pi)^{1.5} 3.4} e^{-0.5(0.56\alpha_1^2 - 0.23\alpha_1\alpha_2 - 0.18\alpha_1\alpha_3 + 0.22\alpha_2^2 - 0.41\alpha_2\alpha_3 + 0.68\alpha_3^2)}$$

6. (a) $R_{xx}(\tau) = \dfrac{2}{\sqrt{3}} e^{-\sqrt{3}|\tau|}$

$$(\Lambda) = \begin{pmatrix} 1 & 0.18 & 0.001 \\ 0.18 & 1 & 0.006 \\ 0.001 & 0.006 & 1 \end{pmatrix} \frac{2}{\sqrt{3}}$$

$$|\Lambda| = 1.06$$

$$(\Lambda)^{-1} \approx \begin{pmatrix} 1 & -0.18 & 0 \\ -0.18 & 1 & 0 \\ 0 & 0 & 1 \end{pmatrix}$$

10. (a) Causal $h(t) = 0.5t, \ 0 < t < 2$

$$h(t) = 2 - 0.5t, \ 2 < t < 4$$

(b) Maximum output at $t = 2$ and $y(2) = 1.33$

$$(S/N)_{out} = \frac{1.33}{0.5(1.33)} = 2$$

11. $H(z) = \dfrac{0.5(z+0.2)(z+5)}{(z+0.28)(z+3.6)}, \ \overline{\varepsilon^2(t)} = 0.08$

12. $H(z) = \dfrac{2z^{-4}}{z+2}, \ S/N = 0.33$

Index

Random Processes

$x(t), x(n)$ A continuous or discrete random process

$x_i(t), x_i(n)$ The ith member of $x(t)$ or $x(n)$

$x(t_1), x(k)$ The random variable describing sampling the process at $t = t_1$ or $n = k$; $x(t)$ and $x(n)$ are also used

$f_{X_1}(\alpha)$ The density function associated with the random variable $x(t_1)$ or $x(n_1)$; $p_{X_1}(\alpha_i)$ and $p_{X_1}(\alpha_i)$ are used if the processes assume discrete values

$\overline{X}, \overline{X^2}, \sigma_X^2$ Notation for statistics of a first-order stationary process

$M_x(t), M_x(n)$ The mean function $\overline{x(t)}$ or $\overline{x(n)}$ of a nonstationary process

$f_{X_1 X_2}(\alpha, \beta)$ The joint density function associated with $x(t_1)$ and $x(t_2)$ or $x(n_1)$ and $x(n_2)$, $p_{x_1 x_2}(\alpha_i, \beta_j)$ is used for a process assuming discrete values

$\dot{R}_{xx}(t_1, t_2), R_{xx}(n, k)$ The autocorrelation function $\overline{x(t_1)x(t_2)}$ or $\overline{x(n)x(k)}$; if the process is at least second-order stationary, $R_{xx}(\tau)$ and $R_{xx}(k)$ are used

$L_{xx}(t_1, t_2),$ The covariance function, $\overline{\left[x(t_1) - \overline{x(t_1)}\right]\left[x(t_2) - \overline{x(t_2)}\right]}$ or
$R_{xx}(n, k)$ $\overline{\left[x(n) - \overline{x(n)}\right]\left[x(k) - \overline{x(k)}\right]}$. If the process is at least second-order stationary, $L_{xx}(\tau)$ and $L_{xx}(k)$ are used

$R_{xy}(t_1, t_2), R_{xy}(n, k)$ The crosscorrelation function of $x(t)$ with $y(t)$ for two random processes, $\overline{x(t_1)y(t_2)}$ or $\overline{x(n)y(k)}$

$\overline{x(t)x(t+\tau)},$
$\overline{x(n)x(n+k)}$ The autocorrelation function on a time average basis $\overline{x(t)y(t+\tau)}$ and $\overline{x(n)y(n+k)}$ denote the cross correlation on a time average basis

$r_{xy}(t), r_{xy}(k)$ The convolution of two deterministic functions,
$\int_{-\infty}^{\infty} x(p)y(t - p)\,dp$ or
$$\sum_n x(n)y(k - n);$$
also denoted $x(t) * y(t)$ and $x(k) * y(k)$

$C_{xy}(\tau), C_{xy}(k)$ The cross correlation of two deterministic functions x and y,
$\int_{-\infty}^{\infty} x(p - t)y(p)\,dp$ or
$$\sum_n x(n - k)y(n);$$
when $y = x$, we obtain the autocorrelation function, also denoted $\oplus y$

$h(t), h(n)$ The impulse or pulse response of a linear system with zero initial energy, when $\delta(t)$ or $\delta(n)$ is the input

$y(t), y(k)$ The output of a linear system to $x(t)$ or $x(k)$, given as $x(t) * h(t)$ or $x(k) * h(k)$

$R_{yy}(\tau), R_{yy}(k)$ The output autocorrelation of a linear system to an at least second-order stationary input, given as $R_{xx}(\tau) * C_{hh}(\tau)$ or $R_{xx}(k) * C_{hh}(k)$